Insecticides and Pesticides: Agricultural Crop Protection

Insecticides and Pesticides: Agricultural Crop Protection

Editor: Nancy Cahoy

R CALLISTO REFERENCE

www.callistoreference.com

Callisto Reference,
118-35 Queens Blvd., Suite 400,
Forest Hills, NY 11375, USA

Visit us on the World Wide Web at:
www.callistoreference.com

ISBN: 978-1-63239-794-2 (Hardback)

Cataloging-in-publication Data

Insecticides and pesticides : agricultural crop protection / edited by Nancy Cahoy.
        p. cm.
Includes bibliographical references and index.
ISBN 978-1-63239-794-2
    1. Pesticides. 2. Insecticides. 3. Agricultural pests--Control. 4. Crops--Nutrition. 5. Plants, Protection of. I. Cahoy, Nancy.
SB951 .I57 2017
632.95--dc23

Table of Contents

Preface

Agricultural crop protection is the method of managing plant diseases, weeds and other pests that affect the agricultural cultivation and forestry. Insecticides and pesticides are an important component of this field of study. This book elucidates new techniques and their applications in a multidisciplinary approach and also discusses the various techniques that are involved in crop protection like biological pest control, barrier based approaches, biotechnology based approaches, etc. It not only sheds light on the different techniques and strategies of crop production but also focuses on its effects on the plant and human health making it a comprehensive text. This book includes contributions of experts and scientists which will provide innovative insights into this field.

This book aims to highlight the current researches and provides a platform to further the scope of innovations in this area. This book is a product of the combined efforts of many researchers and scientists from different parts of the world. The objective of this book is to provide the readers with the latest information in the field.

I would like to express my sincere thanks to the authors for their dedicated efforts in the completion of this book. I acknowledge the efforts of the publisher for providing constant support. Lastly, I would like to thank my family for their support in all academic endeavors.

Editor

Outbreaks of the Brown Planthopper *Nilaparvata lugens* (Stål) in the Yangtze River Delta: Immigration or Local Reproduction?

Gao Hu[1]*[◉], Fang Lu[1][◉], Bao-Ping Zhai[1], Ming-Hong Lu[2], Wan-Cai Liu[2], Feng Zhu[3], Xiang-Wen Wu[4], Gui-Hua Chen[5], Xiao-Xi Zhang[1]

1 Key Laboratory of Integrated Management of Crop Diseases and Pests (Ministry of Education), College of Plant Protection, Nanjing Agricultural University, Nanjing, China, 2 Division of Pest Forecasting, China National Agro-Tec Extension and Service Center, Beijing, China, 3 Plant Protection Station of Jiangsu Province, Nanjing, China, 4 Plant Protection Station of Shanghai City, Shanghai, China, 5 Plant Protection Station of Jinhua City, Jinhua, China

Abstract

An effective control strategy for migratory pests is difficult to implement because the cause of infestation (i.e., immigration or local reproduction) is often not established. In particular, the outbreak mechanisms of the brown planthopper, *Nilaparvata lugens* (Stål), an insect causing massive losses in rice fields in the Yangtze River Delta in China, are frequently unclear. Field surveys of *N. lugens* were performed in Jiangsu and Zhejiang Provinces in 2008 to 2010 and related historical data from 2003 onwards were collected and analyzed to clarify the cause of these infestations. Results showed that outbreaks of *N. lugens* in the Yangtze River Delta were mostly associated with an extremely high increase in population. Thus, reproduction rather than immigration from distant sources were the cause of the infestations. Although mass migration occurred late in the season (late August and early September), the source areas of *N. lugens* catches in the Yangtze River Delta were mainly located in nearby areas, including the Yangtze River Delta itself, Anhui and northern Jiangxi Provinces. These regions collectively form the lower-middle reaches of the Yangtze River, and the late migration can thus be considered as an internal bioflow within one population.

Editor: Robert B. Srygley, USDA-Agricultural Research Service, United States of America

Funding: Funding provided by the National '973' Program of China (2010CB126200), Shanghai City Sci-Tech Joint Research Project in Yangtze River Delta of Shanghai Municipal Science and Technology Commission (13395810100). The funders had no role in study design, data collection and analysis, decision to publish, or preparation of the manuscript.

Competing Interests: The authors have declared that no competing interests exist.

* E-mail: hugao@njau.edu.cn

◉ These authors contributed equally to this work.

Introduction

Numerous migratory insects, such as *Schistocerca gregaria* (Forskal), *Chortoicetes terminifera* (Walker), *Locusta migratoria manilensis* (Meyen), and *Cnaphalocrocis medinalis* (Guenée), are pests to agricultural crops, and their population outbreaks often cause immense crop damage and serious economic problems [1–5]. The population growth of migratory insects has been attributed to migration and local reproduction. Local pest populations could grow exponentially, at least for a time, because of their high intrinsic rate of increase. In contrast, host plants could be devastated immediately after mass immigration, such as in locusts [2,3]. Therefore, determining the intrinsic cause of pest population outbreaks under the simultaneous effect of immigration and local reproduction is difficult. Here, we were specifically concerned with determining the cause of Nilaparvata lugens (Stål) infestations in the Yangtze River Delta in order to significantly improve pest forecasting and control of this pest.

N. lugens is one of the most serious rice pests in the temperate and tropical regions of East and Southeast Asia. Since 1968, *N. lugens* has greatly damaged the rice industry in China and in other countries in Asia [6,7]. In the past few years, new outbreaks of *N.*

lugens have been continuously recorded. For example, the loss of rice yield in China caused by *N. lugens* was approximately 1,880,000 t in 2005 [8–10]. The Yangtze River Delta, including Shanghai Municipality, southern Jiangsu, and northern Zhejiang Provinces (Fig. 1), is one of the most important rice-producing regions in China. The occurrences of *N. lugens* outbreaks in this region, including those in 2005 and 2006, are more severe and more frequent compared with those in other regions. The problem of *N. lugens* outbreaks in the Yangtze River Delta has elicited more attention and investment compared with other areas; thus, it is the primary focus of this paper.

N. lugens has an annual return migration in East Asia. *N. lugens* cannot overwinter in temperate zones, such as mainland China, Japan, and the Korean Peninsula. Instead, an infestation is initiated by windborne spring/summer migrants from the south [6,7,11–14]. Infestation begins with a northward migration that occurs in late March every year. The distribution of *N. lugens* then further expands northward by progeny of the migrant population and can cover the entire rice-growing region in China, as well as Japan and the Korean Peninsula. From September onwards, the general direction of planthopper migration becomes predominantly southbound [15,16]. At the end of October, most *N. lugens*

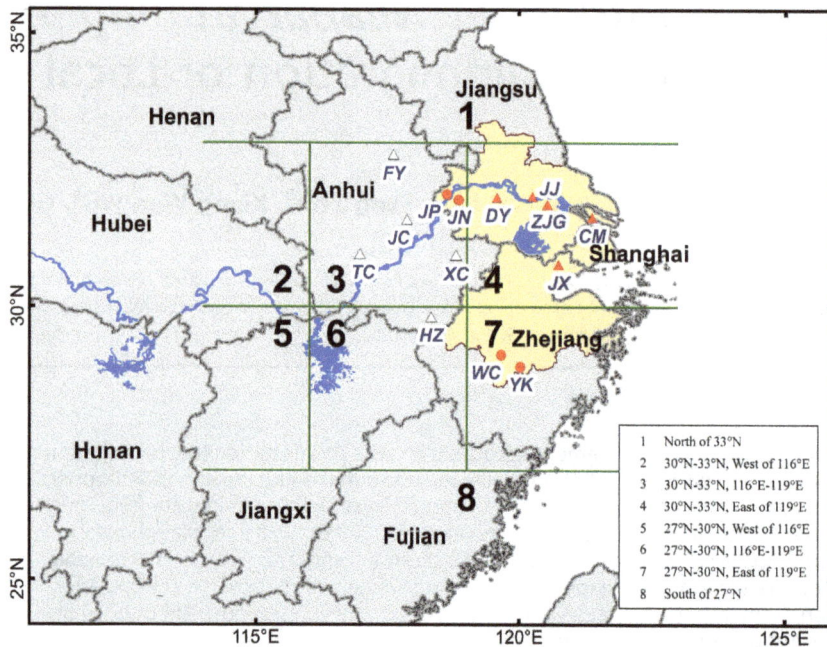

Figure 1. Location of experimental sites and the probable source area of *N. lugens* in the Yangtze River Delta. Notes: The region filled with yellow represents the Yangtze River Delta. Area 1 includes northern Jiangsu and northern Anhui; Area 2 includes southeastern Henan and eastern Hubei; Area 3 includes Southern and central Anhui; Area 4 includes the Yangtze River Delta; Area 5 includes northeastern Hunan and northwestern Jiangxi; Area 6 includes northeastern Jiangxi, northwestern Fujian, and western Zhejiang; Area 7 includes southern and central Zhejiang; Area 8 includes southern Jiangxi and southern and central Fujian. "●" represents the experimental sites in Nanjing and Jinhua City, JP-Jiangpu, JN-Jiangning, WC-Wucheng, and YK-Yongkong. "▲" represents stations in the Yangtze River Delta, CM-Chongming, DY-Danyang, JJ-Jingjiang, ZJG-Zhangjiagang, and JX-Jiaxing; "△" represents stations in the probable source areas of *N. lugens*, FY-Fengyang, JC-Juchao, TC-Tongcheng, XC-Xuancheng, and HZ-Huizhou.

populations are only present in the safe overwintering areas, mainly in the Indo–China Peninsula, within which a proportion of successful return migrants is present [8,11,12,14]. The immigration peaks of *N. lugens* always appear in mid- to late June and mid- to late July in the Yangtze River Delta when the early season rice ripens in the southern and northern areas of South China. *N. lugens* reproduces three generations and causes considerable damage on single-crop late-maturing rice in late September [8,12].

The lower-middle reaches of the Yangtze River are divided into two types of cultivation based on the rice planting system used. One area comprises the Yangtze River Delta, wherein single cropping rice has been dominant since 1997. The other area, including Hubei, southern and central Anhui, northern Hunan, and northern Jiangxi, consists of mixed cultivation wherein both single and double cropping techniques are used. When *N. lugens* in South China begins to emigrate because of the decline in rice availability, the large areas of single-crop rice in mixed planting areas are suitable for their continued survival and breeding. Consequently, the population rapidly grows, which leads to a mass late migration in the Yangtze River Delta where the single-crop rice is at the heading stage during late August to early September [8]. By contrast, migration before mid-August is called early migration. Hu et al. (2011) believe that outbreaks of *N. lugens* in the Yangtze River Delta are frequent because of mass late migration caused by the increased area of the single-crop rice in the mixed planting area in the lower-middle reaches of the Yangtze River [8,10,17–19].

However, *N. lugens* can reproduce two to three generations from colonizing in June and July or until the rice reaches the yellow ripe stage in early to mid-October in the Yangtze River Delta. *N. lugens* is an "r-strategy" organism; thus, the population size of this species

could become very large after two or three generations. Based on this perspective, Cheng and Zhu (2006) concluded that the high growth rate of *N. lugens* is the key to outbreaks in the lower-middle reaches of the Yangtze River, wherein the immigrants provide the initial population and the high number of *N. lugens* catches by light trap in the late season comes from the local population [20]. However, this hypothesis contradicts the fact that emigrating planthoppers (produced by the build-up of local populations) are not thought to be caught by local light traps [21–24].

Field surveys of the population dynamics of *N. lugens* were performed in Jiangsu Province and Zhejiang Province from July to September (2008 to 2010) to resolve the conflicting hypotheses. The data analyzed here were from 2003 to 2010 including systematic field investigation to define the important factors of the outbreak. This study aims to provide insights into ways of forecasting and controlling the population outbreaks of *N. lugens* by studying the outbreak mechanisms of this pest in the Yangtze River Delta.

Materials and Methods

Systematic field investigation

Systematic field investigations were conducted to explore the population dynamics of *N. lugens* in the Yangtze River Delta. In each experimental site, the one or two paddies selected were moderately fertile with routine cultural practices, and, ideally, no pesticides were used to control pests during the rice-growing season. The rice varieties in most areas of the Yangtze River Delta are the local commercial varieties of japonica rice. Systematic field investigation of *N. lugens* populations was performed once every 3 d by the plant-shaking method. A plate (39 cm×29.5 cm×2 cm) was inserted at the base of the rice plants, and *N. lugens* samples were obtained by shaking the plants. The number of *N. lugens* per hill (i.e. all tillers growing from a seedling or a "clump") was counted manually [8].

Systematic field investigations were conducted in Nanjing City, Jiangsu Province from 2008 to 2010. Similar investigations were also carried out in Jinhua City, Zhejiang Province, since 2009 (Fig. 1). The experimental sites were set by the Plant Protection Station of each city; thus, the actual locations of the paddies were different for each year (Table 1). The selected paddies were leased from the local farmers. Farmers always tried to control the pest when no observers were present, and thus pesticides were applied in 2010 (Table 1).

According to the rules of investigation and forecast for *N. lugens* and *Sogatella furcifera* Horváth [25], the degree of occurrence was classified into the following five levels based on the number of *N. lugens* per 100 hills: light (<500), moderate (between 500 and 1000), high (between 1000 and 2000), outbreak (between 2001 and 3000), and severe outbreak (>3000).

Ovary dissection

Macropterous females of *N. lugens* were collected from experimental paddies and dissected every 3 d to estimate the level of ovary development. Using this method, Chen et al. (1979) [26] classified the ovarian development into five levels and related it to the migration status of the population. According to Chen et al.'s criteria, the level of ovary development of the insects gathered from the source areas or during the emigration period was at level I. By contrast, the ovaries of most female insects collected at the landing areas or during the immigration periods were mature with ovary development level III or above because the oocytes underwent continuous and immediate development upon landing of the insect [8,26,27].

Light trap

In each experimental site, a black light trap was placed less than 20 m away from a rice paddy. The trap consisted of a 20 W black light lamp with a top cover, set on top of a pole, 2 m above ground. These lamps were switched on at 1900 h (Beijing Time, same thereafter) and off at 0700 h next morning every day during the systematic field investigation. The catches in the light traps were collected every morning at 0900 h and were identified using a stereomicroscope. *N. lugens* catches were collected and counted every 2 h each night in Nanjing City in September 2010 to observe the nocturnal flight dynamics of the planthoppers.

Other light trap data and field observation data

The daily *N. lugens* light trap data and 5-day field systematic survey data during 2003 to 2010 from plant protection stations of 10 counties, including Danyang (Jiangsu), Zhangjiagang (Jiangsu),

Jingjiang (Jiangsu), Chongming (Shanghai), Jiaxing (Zhejiang), Huizhou (Anhui), Tongcheng (Anhui), Xuancheng (Anhui), Fengyang (Anhui), and Juchao (Anhui), were obtained by the China National Agro-Tec Extension and Service Center (Fig. 1). The five stations in the Yangtze River Delta were studied in mid- to late August and mid- to late September since 2003 to ascertain the occurrence pattern of *N. lugens* in the Yangtze River Delta.

Traditional black light traps (20 W black light lamp) were used to catch planthoppers before 2005. From 2005 onwards, the black light traps were replaced with frequoscillation lamps (20 W black light lamp, Jiaduo Brand, Jiaduo Science, Industry and Trade Co. Ltd., China). The frequoscillation lamps were switched on at 1900 h and off at 0700 h in the next morning every day at the same time from April 1 to November 15.

Population increase rate of *N. lugens* in the Yangtze River Delta

N. lugens migrated from South China to the Yangtze River Delta before mid-August. Rice plants in the Yangtze River Delta were at the mature stage after mid-September. *N. lugens* mainly causes severe damage to rice crops in this region from mid-September. Thus, the population increase rate (R) in the late season was defined as the ratio of the total number of nymphs and adults when the number of early instar nymphs (instar I-III) peaked in mid- and late September to the corresponding peak in mid- and late August. The period between the early instar nymph peak in August and September was approximately equal to one generation duration of N. lugens. A high population increase rate was defined as a ratio of >40 [28,29].

Trajectory analysis

Probable source areas of migratory *N. lugens* were defined by constructing backward/forward trajectories, which were based on the following assumptions: (i) planthoppers are displaced downwind [8,12,15,21,30–32]; (ii) migration normally starts at dusk and partly at dawn [8, 12, 115, 21, 30–32]; (iii) migrants frequently concentrate at the height of approximately 1,000 m [15,32]; and (iv) the planthoppers cannot fly in an atmosphere of <16.5°C [8,15,21,30]. The NOAA ARL HYSPLIT model [8,33,34] was applied to calculate backward trajectories. The backward trajectories for light trap locations were calculated every 2 h during peak periods (1900 h to 0500 h of the next day) and terminated at the take-off time of *N. lugens* (viz. 1900 h on the day and 1900 on the day before that day), with the initial height at 500, 1,000, and 1,500 m aboveground. The duration of trajectories did not exceed 34 h [12,35,36]. An endpoint of a trajectory was considered a probable source if it was located in a rice planting area where the crop was at a later growth stage.

For the five stations (Chongming, Danyang, Jingjiang, Zhangjiagang, and Jiaxing), 90 peak days from August 21 to September

Table 1. Information on experimental paddies from 2008 to 2010.

Locations	Survey periods	Pesticide used
Jiangning, Nanjing	2008/8/3-10/6	No
Jiangning, Nanjing	2009/7/9-9/27	No
Yongkang, Jinhua	2009/7/11-9/27	No
Jiangpu, Nanjing	2010/7/31-9/23	8/3, 8/20: pesticide was applied to control rice planthopper
Wucheng, Jinhua	2010/7/11-9/25	8/13, 9/3: pesticide was applied to control rice planthopper

10 (2003 to 2010) were selected to calculate backward trajectories. The number of trajectory endpoints was counted in each $1° \times 1°$ grid to represent the source area. The whole probable source area was divided into eight zones (Fig. 1), and the probability of valid trajectory endpoints in each zone was then calculated.

Meteorological data

Data on meteorological factors (2003 to 2010), including temperature and wind, were obtained from the China Meteorological Data Service System (http://cdc.cma.gov.cn). The temperature data were expressed as the daily mean value. Wind direction data at the height of 850 hPa were obtained from the Shanghai Municipality at 0800 and 2000 h every day.

Statistical analysis

For a 5% rejection threshold, the association between population increase rate and size of light-trap catch in outbreak and non-outbreak years in five stations from 2003 to 2010 were tested using Chi-square analysis and obtained using R (version 3.0.0, http://www.r-project.org/). When sample sizes were less than 5, Fisher's exact test was then applied.

Ethics statement

No specific permits were required for the described field studies. The brown planthopper $N.$ $lugens$ (Stål) is a major pest of rice in Asia, and huge amounts of manpower and resources are used to control the damage it causes every year.

In this study, we confirmed the following: (i) the location is not privately owned or protected and (ii) the field studies did not involve endangered or protected species.

Results

Population dynamics of N. lugens in the Yangtze River Delta

The $N.$ $lugens$ population in all experimental paddies was at a low density before late July and generally peaked in the second half of September (Fig. 2). In three out of the five paddies, the $N.$ $lugens$ population exceeded 3,000 insects per 100 hills in late September. The three paddies were in Jiangning (the maximum population was 51,140 insects per 100 hills, September 20, 2008), Yongkang (15,673 insects per 100 hills, September 18, 2009), and Wucheng (6,885 insects per 100 hills, September 16, 2010). In the first two paddies, the "hopper burn" (browning of the leaves or withering of the whole plant) areas were approximately 60% and 10% (Fig. 3 and Table 2). The damage in the paddy in Wucheng was minimal. However, excluding the paddy in Wucheng, the increase rates of $N.$ $lugens$ population in the other two paddies with high density were much greater than 40, which was defined by previous studies as a high increase rate (Table 2) [29]. In the paddy in Wucheng, the population increase rate was kept low after August 10. However, pesticide control was performed twice to suppress the increase in $N.$ $lugens$ population at this site (Tables 1 and 2).

During the late season (after mid-August), three main peak periods of $N.$ $lugens$ light trap catches were recorded in the experimental sites. The first two periods were noted in late August and early September, i.e. before the population densities in the paddies reached the maximum value. However, whether these catches were immigrants from distant sources was unclear. Among these cases, the catches of $N.$ $lugens$ in Jiangpu in 2010 were the highest (3,681 specimens caught from August 21 to September 10) (Fig. 2 and Table 2). In the second peak periods (September 7 to September 9) in Jiangpu in 2010, 2,697 individuals were caught in these three nights. Up to 1,397 out of these 2,697 individuals

(48.20%) were caught during 0200 h to 0500 h, and 1,114 (41.31%) were caught around midnight (2100 h to 0200 h). The individuals caught by light trap at midnight and later must have been immigrants.

In each experimental paddy, >100 macropterous females were collected and dissected during the first two peak periods of light trap catches, viz. in late August and early September. Excluding 2010, the individuals with ovaries at levels I and II accounted for >70% of the entire population, whereas those at level III and above accounted for <30% (Table 3). According to the classification criteria established by Chen et al. (1979), these values characterize an emigrating population. Thus, the macropterous adults were from relatively local populations, and the high increase rates were attributed to reproduction rather than immigration. In 2010, the individuals with ovaries at level III and above accounted for a distinctly higher proportion of the population (>30% but <60%) (Table 3). Thus, according to the classification criteria established by Chen et al. (1979) [26,27], the population consisted of local insects that continued to breed in approximately the same area or that mixed with immigrants.

These results showed that the population growth of $N.$ $lugens$ in the late season in the Yangtze River Delta was evidently caused by reproduction rather than immigration in 2008 and 2009. The outbreaks in Jiangning (2008) and Yongkong (2009) resulted from the high increase rate of the local population. In 2010, numerous immigrants maintained the large population, whereas the local population was suppressed by pesticide control.

Relationship between autumn temperature and outbreaks of N. lugens in the late season

The temperature in Jinhua was consistently higher than in Nanjing (Table 4). Accordingly, the population sizes of $N.$ $lugens$ were larger, and the population peaks were earlier in Jinhua (Table 2). Excluding Nanjing in 2009, the temperatures were near or above 26°C before the population densities in the paddies reached the maximum value and were suitable for the population growth of $N.$ $lugens$ (Table 4). In Nanjing in mid-September of 2009, the temperature was only 22.8°C and the $N.$ $lugens$ population may have been suppressed by these cool temperatures. The population of $N.$ $lugens$ in 2010 was controlled by pesticides, so whether a suitable temperature simulated the increase in $N.$ $lugens$ population in this season is unclear.

Occurrence of N. lugens in the Yangtze River Delta from 2003 to 2010

According to the field survey data in 2003 to 2010, most stations in the Yangtze River Delta suffered outbreaks in 2005, 2006, and 2007, whereas occurrences were light in other years. In the five stations (Chongming, Danyang, Jingjiang, Zhangjiagang, and Jiaxing) during these eight years, 14 out of 40 cases (35.0%) reached the outbreak level, viz. the number of $N.$ $lugens$ was >3,000 per 100 hills (Table 5).

The population increase rates of $N.$ $lugens$ in these five stations were calculated for each year (Table 5). Among the 14 outbreak cases, 10 (71.4%) occurred when population increased with a high increase rate ($R>40$), whereas only four cases (28.6%) occurred with a low increase rate (Table 6). The results of Chi-square test ($\chi 2 = 14.70$, $p<0.001$; Fisher's exact test, $p<0.001$) indicated that the outbreak of $N.$ $lugens$ in the Yangtze River Delta was significantly associated with the population increase rate.

In the five stations, the cumulative light-trap catches of $N.$ $lugens$ in early migrations were <1,000. However, the numbers in the cumulative catches during the late season varied over a wide range

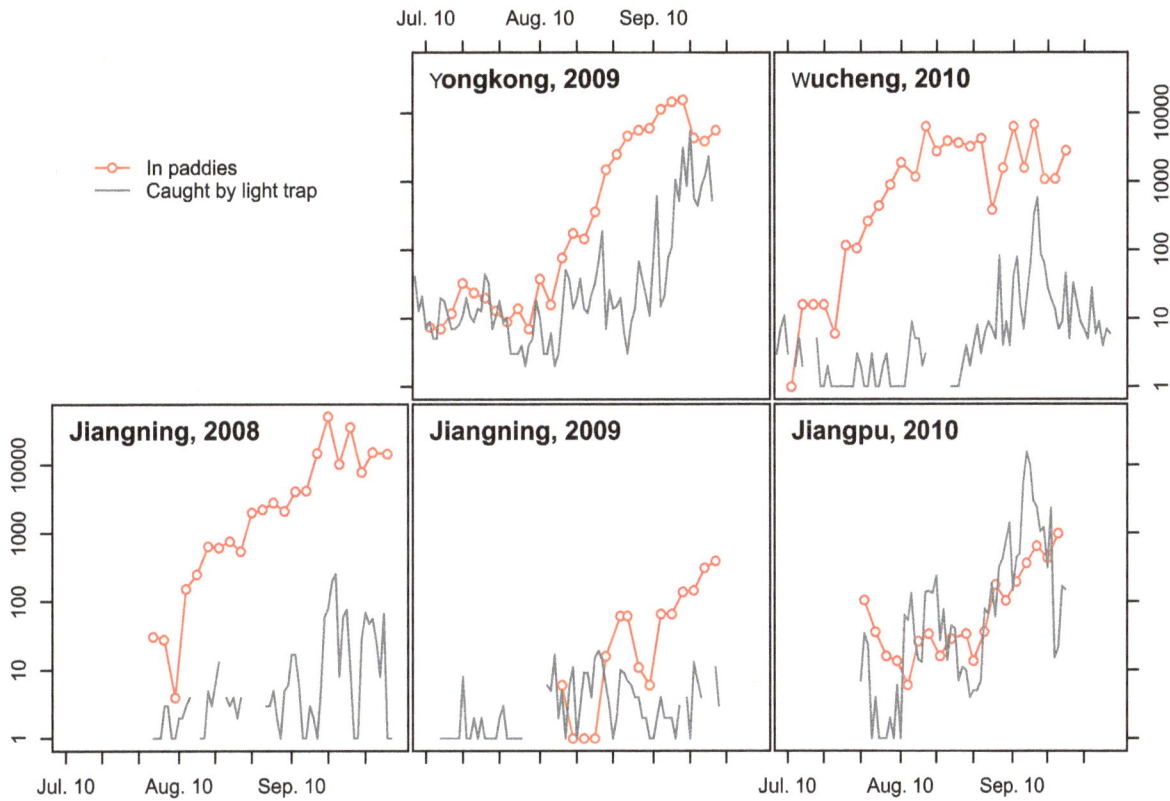

Figure 2. Number of *N. lugens* in paddies and caught by light trap at experimental sites.

(Fig. 4). In the outbreak years (2005, 2006, and 2007), mass migrations occurred during late August to early September. In 2006, the cumulative catches in the late season were the highest ever recorded at several stations, with 10,000 catches per night in peak days. The maximum catch in 2006 was 790,000, which was recorded in Zhangjiagang on August 30. In the light occurrence years, the cumulative catches in the late migration were consistently minimal. In 2003 and 2009, the catches on peak days were tens, while very few rice planthoppers were caught in the late season of 2004, 2008, and 2010 (Fig. 4). Among the 14 outbreak cases, nine cases (64.29%) occurred with numerous catches in the late season, and five cases (35.71%) occurred with a few catches (Table 6). The results of Chi-square test ($\chi2 = 6.26$,

$p = 0.012$; Fisher's exact test, $p = 0.007$) indicated that the outbreak of *N. lugens* in the Yangtze River Delta was significantly associated with light trap catches in the late season.

Among the 14 outbreak cases, five cases occurred with both high population increase rate and numerous catches in the late season, five cases occurred with only a high population increase rate, and the other four cases occurred with only numerous catches (Table 6). These five cases with only the high population increase rate occurred from 2008 to 2009 when the occurrence level of *N. lugens* was light in the whole Yangtze River Delta, and few migrants were noted during that period. Thus, the high increase rate in these five cases was confirmed to be mainly caused by reproduction. Thus, to summarize, high reproductive rates in

Figure 3. "Hopper burn" (A) and macropterous adults swarming on rice leaves (B) in Jiangning District, Nanjing City, in 2008. Note: Pictures were taken on October 2 (yellow ripe stage). Rice was harvested on October 20.

Table 2. Population increase rate of investigation fields from 2008 to 2010.

Year	Station	number of N. lugens per 100 hills		Proliferation multiple	Area of hopper burn (%)	Catches of N. lugens (Aug. 21–Sep. 10)
		Aug.	Sept.			
2008	Jiangning	645 (18)*	51140 (20)	79.29	60	58
2009	Jiangning	15 (28)	395 (27)	26.33	0	110
2009	Yongkang	124 (19)	15070 (18)	121.53	10	688
2010	Jiangpu	28 (24)	980 (23)	35.00	0	3681
2010	Wucheng	6335 (17)	6885 (16)	1.09	0	177

* The numbers in the brackets represent the date when the nymphal population was in early instars (I–III). The periods between the Aug. and Sept. dates were almost one month long and were nearly equal to the generation time of N. lugens.

local or relatively local populations seemed to be the most important factors giving rise to outbreaks.

Source areas of light trap catches in the Yangtze River Delta in the late season

Up to 1620 backward trajectories were calculated for 90 peak days in five stations, and 2,970 endpoints were obtained. After elimination of invalid endpoints (i.e. ones not ending in suitable rice cultivations), the number of available endpoints was 1,802, which accounted for 60.7% of the original 2,970.

The source areas within 2 h to 10 h durations were still in the Yangtze River Delta (83.66% of the total), followed by southern and central Anhui (10.93% of the total). The source areas within 24 h to 34 h durations were mainly in southern and central Anhui (22.71% of the total), the Yangtze River Delta (15.58%), and northern Jiangxi (14.15%). Moreover, only a few endpoints were located in northern Jiangsu, southern Zhejiang, northern Fujian, northeastern Hunan, eastern Hubei, etc. (Table 7, Fig. 5).

Whether the occurrence was light or severe, the source areas within 2 h to 10 h durations were therefore mainly in the surrounding area and partly in southern-central Anhui. The source areas in the Yangtze River Delta and southern-central Anhui accounted for 73.81% and 24.29%, respectively, in the years with light occurrence (2003 and 2009), and 86.42% and 7.19% in the years with outbreaks (2005, 2006, and 2007) (Table 7).

The source areas within 24 h to 34 h durations were different in the years with light occurrence compared to those with outbreaks. The endpoints during the light occurrence years were mostly in southwestern and western regions, including southern and central Anhui, northern Jiangxi, northeastern Hunan, northwestern

Fujian, and western Zhejiang (65.33% of the total) (Table 7). The endpoints during the outbreak years were mostly distributed in southern and central Anhui, the Yangtze River Delta, northern Jiangsu, and northeastern Jiangxi (70.72% in total) (Table 7). Altogether, the source areas of N. lugens within 24 h to 34 h durations across all the years were mainly in Anhui and northern Jiangxi, which were located west of the Yangtze River Delta.

The interannual variation of macropterous adult number in Anhui corresponded with the interannual variation of catches in the Yangtze River Delta. For example, in the five locations (Huizhou, Tongcheng, Xuancheng, Fengyang, and Juchao) in Anhui, the numbers of macropterous adults from August 21 to September 10 were high in 2005 to 2007, when numerous catches were also obtained in the Yangtze River Delta (Fig. 6). Thus, Anhui could be the main source area of N. lugens in the Yangtze River Delta.

Wind directions in the late season in Yangtze River Delta

The summer (southwest) monsoon is replaced by the winter (north-east) monsoon in eastern Asia in late August and early September. In the Yangtze River Delta, the northeasterlies are gradually established at this time. However, the winds change to southwesterly/westerly under the influence of oscillation in subtropical anticyclones. For example, in Shanghai, the wind direction at 850 hPa was mainly southeast (17.6%), southwest (17.6%), and west (18.2%) in late August and was mainly northwest (20%), north (18.8%), and northeast (18.8%) in early September. During late August to early September, the probabilities of all wind directions were similar from 2003 to 2010 (Fig. 7A). However, more west or northwest winds were noted on the 21 peak days of N. lugens immigration from August 21 to September 10 (2003 to 2010), with

Table 3. Ovarian development of N. lugens in the experimental paddies from 2008 to 2010.

Year	Time	Station	Number of N. lugens dissected	Grade of ovarian development		
				Level I	Level II	≥ Level III
2008	8/23–9/11	Jiangning	186	168 (90.32%)	9 (4.84%)	9 (4.84%)
2009	9/9–9/18	Jiangning	111	35 (31.53%)	73 (65.77%)	3 (2.70%)
2009	8/19–9/12	Yongkang	106	68 (64.15%)	8 (7.55%)	30 (28.30%)
2010	9/9–9/16	Jiangpu	154	33 (21.43%)	47 (30.52%)	74 (48.05%)
2010	8/20–9/10	Wucheng	195	120 (61.54%)	4 (2.05%)	71 (36.41%)

Table 4. Average temperature from August 21 to September 30 every 10 d during 2008 to 2010 (°C).

Year	Nanjing				Jinhua			
	8/21–8/31	9/1–9/10	9/11–9/20	9/21–9/30	8/21–8/31	9/1–9/10	9/11–9/20	9/21–9/30
2008	25.5	24.4	25.7	23.0	-	-	-	-
2009	26.8	25.2	22.8	22.3	29.7	28.0	26.5	24.5
2010	26.7	27.1	25.5	20.1	29.3	29.2	27.9	21.4
Mean*	27.0	25.2	24.2	22.3	28.7	26.8	25.7	24.0

* Mean value for 2000–2010

33.3% (westerly) and 28.6% (northwesterly) probabilities (Fig. 7B). The results of Chi-square analysis revealed a significant difference in wind direction between migration days and other periods ($\chi2 = 24.91$, p<0.001) (Table 8). Therefore, the late migration occurred more frequently under winds from the west or northwest. Considering individual seasons, during apparent late migrations in the serious outbreak years of 2005, 2006, and 2007, the frequency of wind directions was mainly west and northwest from August 21 to September 10 (Figs. 7C to 7J).

Discussion

Numerous *N. lugens* were caught by light trap in the Yangtze River Delta in the late season, but the source of these planthopper catches has not exactly been identified in previous studies. Thus, it was controversial whether "late migration" occurred, and the cause of *N. lugens* outbreak (i.e., immigration or local reproduction) has also remained unclear. Similar to some previous studies, the present study confirmed that late migration was an actual

Table 5. Number of *N. lugens* per 100 hills and population increase rate in the Yangtze River Delta from 2003–2010.

Station	Month	Number of *N. lugens*							
		2003	2004	2005	2006	2007	2008	2009	2010
Chongming	Aug.	10	28	85	38	30	12	8	86
		(20) §	(20)	(20)	(25)	(25)	(20)	(20)	(20)
	Sept.	319	101	357	24030	1315	291	49	622
		(20)	(20)	(20)	(20)	(20)	(15)	(15)	(15)
	R*	31.90	3.61	4.20	632.37	43.83	24.25	6.13	7.23
Danyang	Aug.	12	56	72	1797	162	170	264	54
		(20)	(15)	(25)	(20)	(20)	(31)	(31)	(31)
	Sept.	114	56	1230	6505	735	180	254	1404
		(20)	(15)	(30)	(20)	(15)	(25)	(25)	(25)
	R	9.50	1.00	17.08	3.62	4.54	1.06	0.96	26.00
Jingjiang	Aug.	42	10	318	1043	138	147	112	45
		(20)	(20)	(25)	(25)	(25)	(25)	(25)	(15)
	Sept.	30	18	21895	12348	10862	10671	5046	5298
		(20)	(25)	(20)	(25)	(25)	(25)	(20)	(20)
	R	0.71	1.80	68.85	11.84	78.71	72.59	45.05	117.73
Zhangjiagang	Aug.	46	108	210	151	13	87	54	5
		(25)	(25)	(25)	(15)	(15)	(20)	(15)	(25)
	Sept.	640	1106	965	4678	54	73	10	10
		(25)	(25)	(20)	(25)	(15)	(25)	(15)	(25)
	R	13.91	10.24	4.60	30.98	4.15	0.84	0.19	2.00
Jiaxing	Aug.	37	37	1	1891	100	13	7	186
		(15)	(20)	(20)	(20)	(20)	(20)	(15)	(25)
	Sept.	414	293	5098	17633	4377	3812	756	8242
		(20)	(20)	(20)	(20)	(15)	(25)	(15)	(25)
	R	11.19	7.92	5098.00	9.32	43.77	293.23	108.00	44.31

* R- Proliferation multiple
§The numbers in the brackets represent the date when the nymphal population was in early instars (I–III). The periods between the Aug. and Sept. dates were almost one month long and were nearly equal to the generation time of N. lugens.

Table 6. Frequency table comparing population increase rate and catch size for outbreak and non-outbreak situations.

	Outbreak	Non-outbreak	Total
High R and lots of catches in light trap	5	1	6
High R and few catches	5	1	6
Low R and lots of catches	4	4	8
Low R and few catches	0	20	20
Total	14	26	40

Note: R- Population increase rate

immigrants (see also [8,19]). Finally, the size of planthopper catches was related to the wind direction. Catches increased with westerly winds; thus, the major sources of the migrant insects were from the west. However, the backward trajectories in this study showed that the source areas of catches in the Yangtze River Delta in the late season were mainly located in relatively local areas, including the Yangtze River Delta itself, Anhui, and northern Jiangxi. These regions were located at a similar latitude (approximately 30°N), and the migration distances were consistently short by rice planthopper standards (mostly <300 km). When mass late migration occurred, the number of *N. lugens* in the Yangtze River Delta was consistently large, as well as that in Anhui, northern Jiangxi, and other regions at a similar latitude in eastern China. Therefore, these regions could be collectively called the lower-middle reaches of the Yangtze River. The initial population of *N. lugens* in these regions all migrated from South China in June and July. Thus, the late migration can be regarded as merely an internal flow within one population.

In the lower-middle reaches of the Yangtze River, some differences were found between the Delta itself and other areas. In the Yangtze River Delta, single-crop late-maturing rice was mostly planted, which is transplanted in mid- and late June and harvested in late October [8,18,19]. In other regions, including Hubei, southern and central Anhui, northern Hunan, and northern Jiangxi, the single-crop rice was subsequently transplanted between late May and early June and then matured in late

phenomenon. The four main findings of this study are the following. First, many catches in the light trap occurred at or after midnight. The flight of these individuals lasted at least 6 hours after their dusk take-off [8,12,15,21,30–32], and thus they must be immigrants. Second, the population increase rates in the field were, in many cases, near or even greater than the maximum fecundity of *N. lugens* (705 eggs per female [37]) observed in some laboratory cases, such as $R = 5,098$ in Jiaxing in 2005 and $R = 632$ in Chongming in 2006. Third, ovary dissection verified that the macropterous females in Nanhui, Shanghai City in 2007 were

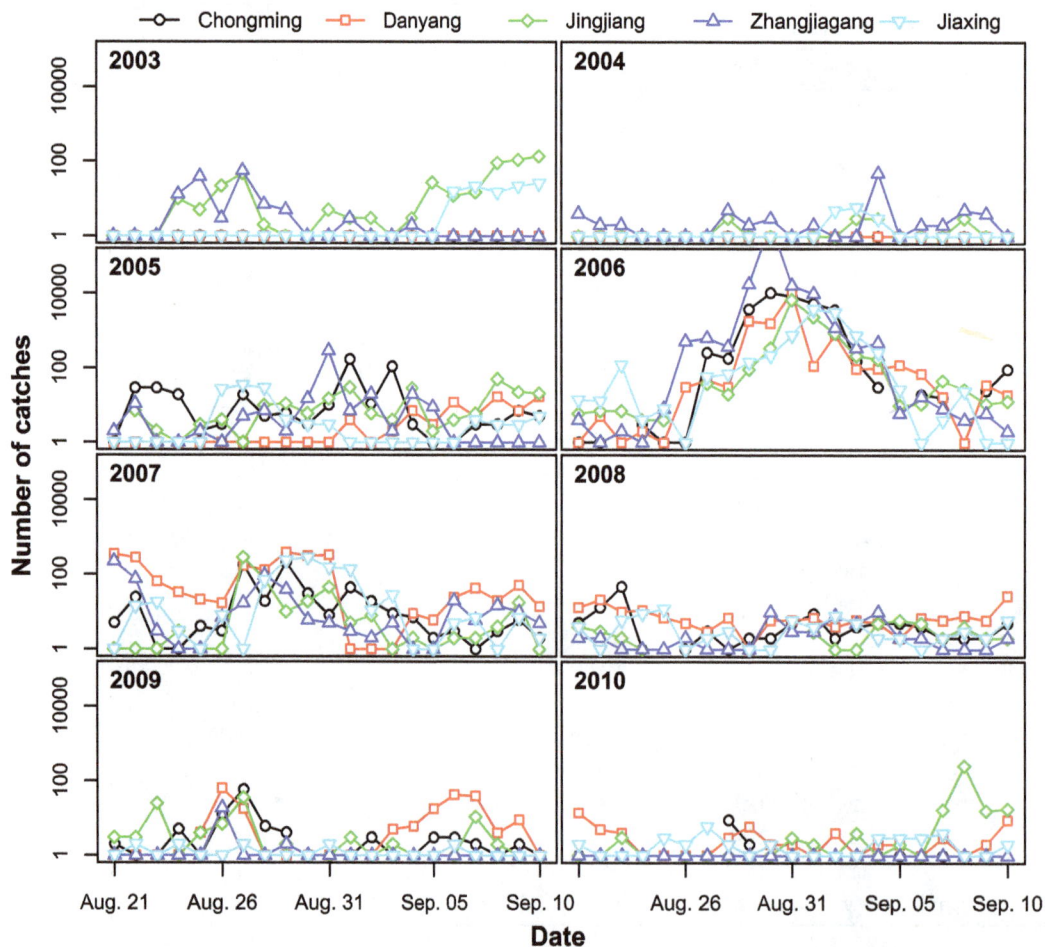

Figure 4. Daily catches of *N. lugens* from August 21 to September 10 from 2003 to 2010.

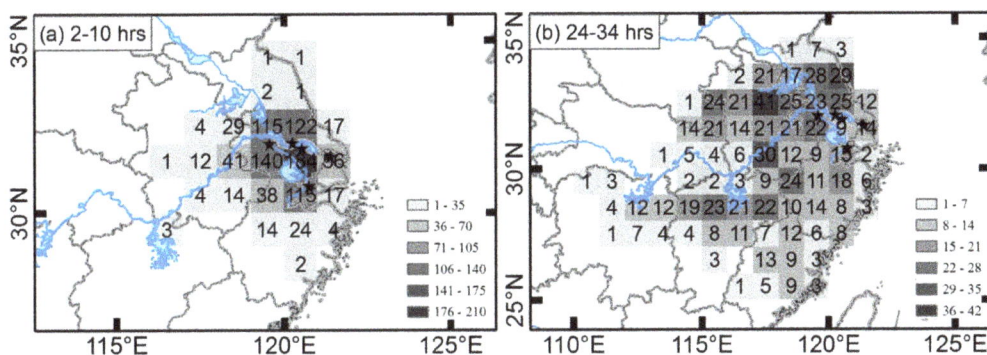

Figure 5. Probability distribution of endpoints of backward trajectories from Chongming, Danyang, Jingjiang, Jiaxing, and Zhangjiagang.

Table 7. Probability distribution of endpoints of backward trajectories from Chongming, Danyang, Jingjiang, Jiaxing, and Zhangjiagang in different available source areas.

Year	Area	2 h–10 h		24 h–34 h	
		Amount	Probability (%)	Amount	Probability (%)
All five years	1	5	0.52	108	12.84
	2	0	0	70	8.32
	3	105	10.93	191	22.71
	4	804	83.66	131	15.58
	5	0	0	102	12.13
	6	3	0.31	119	14.15
	7	44	4.58	74	8.80
	8	0	0	46	5.47
	subtotal	961	100	841	100
Light occurrence (2003 and 2009)	1	0	0	3	1.51
	2	0	0	22	11.06
	3	51	24.29	42	21.11
	4	155	73.81	21	10.55
	5	0	0	59	29.65
	6	3	1.43	29	14.57
	7	1	0.48	22	11.06
	8	0	0	1	0.50
	subtotal	210	100	199	100
Outbreak (2005, 2006, and 2007)	1	5	0.67	105	16.36
	2	0	0	48	7.48
	3	54	7.19	149	23.21
	4	649	86.42	110	17.13
	5	0	0	43	6.70
	6	0	0	90	14.02
	7	43	5.73	52	8.10
	8	0	0	45	7.01
	subtotal	751	100	642	100

Notes: Area 1 includes northern Jiangsu and northern Anhui; Area 2 includes southeastern Henan and eastern Hubei; Area 3 includes Southern and central Anhui; Area 4 includes the Yangtze River Delta; Area 5 includes northeastern Hunan and northwestern Jiangxi; Area 6 includes northeastern Jiangxi, northwestern Fujian, and western Zhejiang; Area 7 includes southern and central Zhejiang; Area 8 includes southern Jiangxi and southern and central Fujian. These probable source area were showed in Fig. 1.

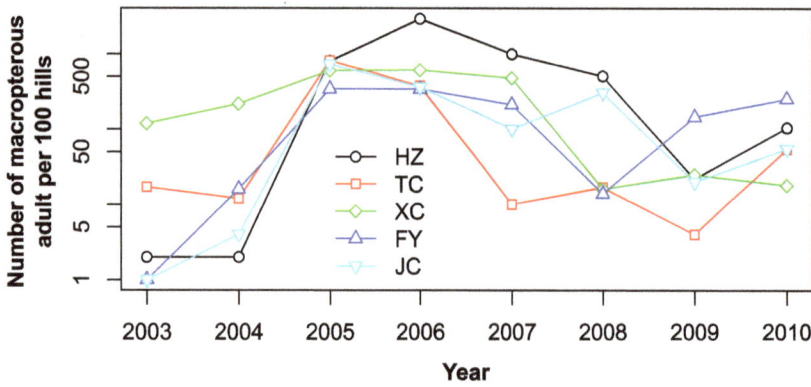

Figure 6. Peak number of *N. lugens* macropterous adults in Anhui Province from August 21 to September 10 from 2003 to 2010. Note: HZ-Huizhou, TC-Tongcheng, XC-Xuancheng, FY-Fengyang, JC-Juchao.

August, with most of the rice harvested in early and mid-September [8,18,19]. These regions are closer to South China than the Yangtze River Delta, and immigrants in June and July were greater because of shorter flight distance for the insects. Moreover, the single-crop rice plants in these regions were at the tillering to booting stage and fit for *N. lugens* whose populations were thus able to grow rapidly, whereas the rice plants in the Yangtze River Delta were still immature. Furthermore, the harvest time of rice in these regions was almost one month earlier than that in the Yangtze River Delta. Thus, large numbers *N. lugens* adults emigrated as the single-crop rice was harvested in these regions, and these insects could invade the Yangtze River Delta where the rice plants were at the heading stage and now suitable for *N. lugens*. Therefore, the late migration would exacerbate the

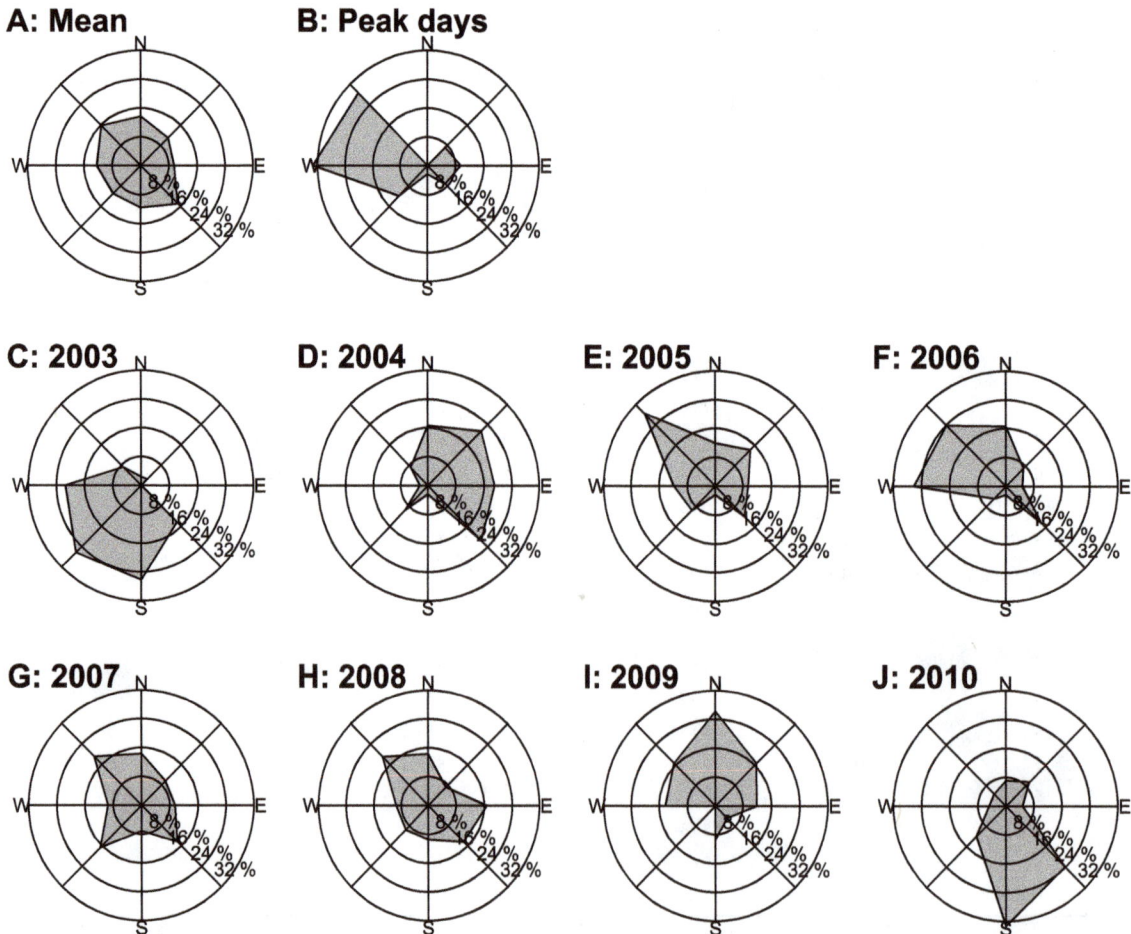

Figure 7. Frequency of wind direction at 850 hPa in Shanghai from 2003 to 2010. Note: (A) Mean value for 2003 to 2010, (B) mean value for peak days of *N. lugens* catches and (C-J) individual value for each year

Table 8. Frequency of wind direction from August 21 to September 10 from 2003 to 2010 in Shanghai.

	On peak days	On nonpeak days	Total
West and northwest wind	26	69	95
Other direction Wind	16	225	241
Total	42	294	336

outbreak of *N. lugens* in the Yangtze River Delta, and vigilance would need to be maintained. In particular, even when the *N. lugens* population in the Yangtze River Delta was suppressed by effective control strategy, mass late migration would negate the results of control efforts.

Our study showed that outbreaks of *N. lugens* mainly occurred along with abnormally high intrinsic rates of increase. Considering 17 outbreak cases (three cases in survey data and 14 cases in historical data), 12 (70.56%) occurred with high increase rates. In other cases, where the rate of increase was low, the number of *N. lugens* was large before late August. Therefore, these outbreaks of *N. lugens* could be divided into two types. One type is called the cardinal type, wherein the number of immigrants in the early season (before late August) is unusually high, thereby causing an early dense population. Thus, this type is also commonly known as the high-immigration type [8,11,38]. In the historical data used for this study, the four cases with low increase rate all occurred in 2006, and the outbreak in this year was previously analyzed as a typical case of high-immigration type [38]. In the other types of *N. lugens* outbreak, called the climate type in previous studies [8,11,38], the population was low in the early season but increased rapidly due, in general, to meteorological factors, especially temperature. In the present study, the high increase rate was mainly attributed to reproduction rather than immigration. Although mass migration was observed in the late season, the source areas were mainly located in relatively nearby areas and the late migration can be regarded as an internal flow within one population. Thus, the other cases in this study could be classified as climate type. However, a complex type could also be present, where a relatively high early immigration and favorable climatic conditions are both contributory factors [8,11,38].

Temperature could control the population of *N. lugens* by direct effects on survival, reproduction, and foraging. These effects are the most crucial factors for *N. lugens* population dynamics [37,39]. The size of the *N. lugens* population in the Yangtze River Delta was affected by temperature, as shown in the analysis of the relationship between the population dynamics of the planthopper and temperature from August to September (2008 to 2010). In the lower-middle reaches of the Yangtze River, a "cooler summer and warmer autumn" is regarded as the favored climatic condition for outbreaks of *N. lugens* [40–42], and this occurred for example in

2005 [8]. The occurrence of *N. lugens* in the Yangtze River Delta in 2005 was the most severe in the past 20 years [9,10]. Hu et al. [8] suggested that steadily warmer autumns have occurred from the 1990s in the lower-middle reaches of the Yangtze River Delta and that such conditions gradually became the norm. In this study, outbreaks of *N. lugens* in the Yangtze River Delta were mainly attributed to the high reproduction rate, thus providing further evidence on why the problems caused by *N. lugens* have become more severe.

N. lugens population can be controlled by pesticide. Limited damage was caused by *N. lugens* in the 1990s due to the use of buprofezin and imidacloprid [11], and the high level of resistance to imidacloprid resulted in chemical control failure and great yield loss in 2005 [43]. Rive cultivars with resistance to *N. lugens* can also suppress this pest significantly [6,44]. But, just as in the case of development of resistance to pesticide, this pest often adapted to resistant cultivars in a short time. The pest adapted to IR26, the first high-yielding cultivar with BPH resistance (with *Bph1* gene), within 2 or 3 years after being released to farmers [6]. High-yielding indica hybrid cultivars without the new resistant gene has promoted the population growth of *N. lugens* in southern China, especially in Hunan Province [45]. Therefore frequent outbreaks of *N. lugens* are not only related to suitable weather conditions, but also with other factors.

In summary, the outbreak mechanisms of *N. lugens* in the Yangtze River Delta were explored by surveying and evaluating historical data, and by determining the sources of catches in light traps in the late season. Although these catches supported the occurrence of migration, the source areas were found to be relatively close at hand and could be collectively considered, for control purposes, with sources in the Yangtze River Delta itself. Therefore, reproduction was the key factor contributing to the high population increase rate of N. lugens, and to the consequent outbreaks in this area of China. These results resolved the conflicting hypotheses on the source of these outbreak populations and will facilitate general control strategies of *N. lugens* in China.

Acknowledgments

We acknowledge the assistance provided by our colleagues Yue Zhao and Xingyun Fu who participated in the field investigations in 2008; Yangyang Hou, Li Wang, Yanbo Xu, He Sui, Yuan He, Yubo Zhu, and Huaijian Liao who participated in the field investigations in 2009; and Yuebo Gao, Yonghui Lei, Haibo Yang, Zhenhua Hao, Haiyan Zhang, Wei Wu, and Yuansong Fang who participated in the field investigations in 2010. We also thank the plant protection stations of Nanjing City and Jinhua City for their assistance in the maintenance of the experiment fields.

Author Contributions

Conceived and designed the experiments: GH BPZ XXZ. Performed the experiments: FL MHL WCL FZ XWW GHC. Analyzed the data: GH FL MHL. Wrote the paper: GH FL XXZ BPZ.

References

1. Pedgley DE (1993) Managing migratory insects pests - a review. Int J Pest Manage 39, 3–12.
2. Day RK, Knight JD (1995) Operational aspects of forecasting migratory pests. pp. 323–334 In Drake VA, Gatehouse AG. (Eds) *Insect migration: Tracking resources through space and time.* Cambridge University Press, New York.
3. Deveson ED, Drake VA, Hunter DM, Walker PW, Wang HK (2005) Evidence from traditional and new technologies for northward migrations of Australian plague locusts (*Chortoicetes terminifera*) (Walker) (Orthoptera: Acrididae) to western Queensland. Austral Ecology 30: 920–935.
4. Ma SC (1958) The population dynamics of the oriental migratory locust (*Locusta migratoria manilensis* Meyen) in China. Acta Entomologica Sinica 8: 1–40.
5. Chang SS, Lo ZC, Keng CG, Li GZ, Chen XL, et al. (1980) Studies on the migration of rice leaf roller *Cnaphalocrocis Medinalis* Guenée. Acta Entomologica Sinica, 23: 130–140.
6. Bottrell DG, Schoenly KG (2012) Resurrecting the ghost of green revolutions past: The brown planthopper as a recurring threat to high-yielding rice production in tropical Asia. Journal of Asia Pacific Entomology 15: 122–140.
7. Zhai BP (2011) Rice planthoppers: A China problem under the international perspectives. Chinese Journal of Applied Entomology 48: 1184–1193.
8. Hu G, Cheng XN, Qi GJ, Wang FY, Lu F, et al. (2011) Rice planting systems, global warming and outbreaks of *Nilaparvata lugens* (Stål). B Entomol Res 101: 187–199.

9. Guo R, Zhao ZH (2006) Crop pests. pp. 376–378 in Ministry of Agriculture of China (Ed.) China Agriculture Yearbook. China Agricultural Press, Beijing.

10. Zhai BP, Cheng JA (2006) The conference summary of workshop on the two primary migratory pests of rice, rice planthopper and rice leaf roller, in 2006. Chinese Bulletin of Entomology 43: 585–588.

11. Cheng XN, Wu JC, Ma F (2003) Brown planthopper: occurrence and control. China Agricultural Press, Beijing.

12. Cheng SN, Chen JC, Xi X, Yang LM, Zhu ZL, et al. (1979) Studies on the migrations of brown planthopper Nilaparvata lugens Stål. Acta Entomologica Sinica 22: 1–21.

13. Otuka A, Matsumura M, Watanabe T, Dinh TV (2008) A migration analysis for rice planthoppers, Sogatella furcifera (Horváth) and Nilaparvata lugens (Stål) (Homoptera: Delphacidae), emigrating from northern Vietnam from April to May. Appl Entomol Zool 43: 527–534.

14. Kisimoto R, Sogawa K (1995) Migration of the Brown Planthopper Nilaparvata lugens and the White-backed Planthopper Sogatella furcifera in East Asia: the role of weather and climate. pp. 93–104 In Drake VA, Gatehouse AG. (Eds) Insect migration: Tracking resources through space and time. Cambridge University Press, New York.

15. Riley JR, Cheng XN, Zhang XX, Reynolds DR, Xu GM, et al. (1991) The long-distance migration of Nilaparvata lugens (Stål) (Delphacidae) in China: radar observations of mass return flight in the autumn. Ecol Entomol 16: 471–489.

16. Riley JR, Reynolds DR, Smith AD, Rosenberg LJ, Cheng XN, et al. (1994) Observations on the autumn migration of Nilaparvata lugens (Homoptera: Delphacidae) and other pests in east central China. Bull. Entomol. Res 84, 389–402.

17. Bao YX, Li JJ, Miao QL, Shen SH, Wang JQ (2008) Simulation of atmospheric dynamical background for a great migration event of the brown planthopper (Nilaparvata lugens Stål): a case study. Chinese Journal of Agrometeorology 39: 347–352.

18. Lu F, Qi GJ, Chen X, Dong XG, Gui YR, et al. (2010) The late immigration of the brown planthopper Nilaparvata lugens (Stål) in Shanghai: Case studies from 2007. Acta Ecologica Sinica 30(12): 3215–3225.

19. Qi GJ, Xiao MK, Wu CL, Jiang C, Zhang XX, et al. (2010) Effect of the change of rice planting system on the formation of outbreak population of brown planthopper, Nilaparvata lugens (Stål). Acta Phytophylacica Sinica, 37:193–200.

20. Cheng JA, Zhu ZR (2006) Analysis on the key factors causing the outbreak of brown planthopper in Yangtze Area, China in 2005. China Plant Protection 32: 1–4.

21. Chen RC, Cheng XN (1980) The take-off behavior of brown planthopper (Nilaparvata lugens Stål) and its synchronous relations to the biological rhythm and environmental factors. Journal of Nanjing Agricultural University 2: 42–49.

22. Ye ZC, He SM, Lu LQ, Zhen HG, Feng BC, et al. (1981) The observation of the take-off behaviour of brown planthopper (Nilaparvata lugens Stål). Chinese Bulletin of Entomology 18: 97–100.

23. Zhong DL (1984) The relationship between the take-off behaviour of brown planthopper (Nilaparvata lugens Stål) and weather condition. Chinese Bulletin of Entomology 22: 97–99.

24. Zhao SF (1992) Behavioral mechanism of phototactic response of natural planthopper population (Sogatella furcifera and Nilaparvata lugens). Acta Agriculturae Jiangxi 4: 74–79.

25. (MAPRC) Ministry of Agriculture of the People's Republic of China. (1995) National standard GB/T 15794–1995: Rules for investigation and forecast of the brown planthopper (Nilaparvata lugens Stål) and the whitebacked planthopper (Sogatella furcifera Horvath). MAPRC, Beijing, China.

26. Chen RC, Cheng XN, Yang LM, Yin XD (1979) The ovarian development of the brown planthopper and it's relation to migration. Acta Entomologica Sinica. 22: 280–288.

27. Zheng DB, Hu G, Yang F, Du XD, Yang HB, et al. (2013) Ovarian development status and population characteristics of Sogatella furcifera(Horváth) and Nilaparvata lugens (Stål): implication for pest forecasting. J. Appl. Entomol.doi: 10.1111/jen.12067

28. Ding ZZ, Chen ML, Li PY (1985) Preliminary study on population reproduction of brown planthopper Nilaparvata lugens (Stål) in single-cropping late-maturing rice. Jiangsu Agricultural Sciences, 13: 1–20.

29. Chen YY, Wang MT (1996) Relationship between population reproduction of Nilaparvata lugens (Stål) and temperature. Jiangsu Agricultural Sciences, 24: 29–31.

30. Ohkubo N, Kisimoto R (1971) Diurnal periodicity of flight behaviour of the brown planthopper, Nilaparvata lugens Stål in the 4th and 5th emergence periods. Jpn. J. App. Entomol. Z. 15: 8–16.

31. Kisimoto R (1984) Meteorological conditions inducing long-distance immigration of the brown planthopper, Nilaparvata lugens (Stål). Chinese Journal of Entomology 4: 39–48.

32. Deng WX (1981) A general survey on seasonal migrations of Nilaparvata lugens (Stål) and Sogatella furcifera (Horváth) (Homoptera: Delphacidae) by means of airplane collections. Acta Phytophylacila Sinica. 8: 73–81.

33. Draxler RR, Hess GD (1998) An overview of the Hysplit_4 modeling system for trajectories, dispersion and deposition. Australian Meteorological Magazine. 47: 295–308.

34. Zhu M, Edward BR, David WR, Ian VM, Mark WS (2005) Low-level jet streams associated with spring aphid migration and current season spread of potato viruses in the U.S. northern Great Plains. Agricultural and Forest Meteorology. 138: 192–202.

35. Chen RC, Wu JR, Zhu SD, Zhang JX (1984) Flight capacity of the brown planthopper Nilaparvata lugens Stål. Acta Entomologica Sinica 27: 121–127.

36. Rosenberg LJ, Magor JI (1987) Prediction wind borne displacements of the brown planthopper Nilaparvata lugens from synoptic weather data. I. Long distance displacements in the northeast monsoon. Journal of Animal Ecology, 56: 39–51.

37. Li RD (1984) Population growth of the brown planthopper, Nilaparvata lugens Stål, as influenced by temperature. Acta Phytophylacica Sinica, 11:101–107.

38. Hu G, Xie MC, Lin ZX, Xin DY, Huang CY, et al. (2010) Are Outbreaks of Nilaparvata lugens (Stål) Associated with Global Warming? Environ. Entomol. 39:1705–1714.

39. Zang W, Hao SG, Wang HK, Cheng XN (1997) A simulation model of brown planthopper population dynamics in Yangtze and Huai River rice area. Journal of Nanjing Agricultural University 20, 32–38.

40. Hou TT, Huo ZG, Wu RF, Ye CL, Wang SY, et al. (2004) Impact of air temperature to the brown planthopper population in late rice crop season in Fuqing Region. Chinese Journal of Agrometeorology, 25, 28–32.

41. Pu MH, Chen JM (1979) The preliminary study of the brown planthopper occurrence degree forecasting by mathematical statistics. Plant Protection (5): 1–9.

42. Cheng JA, Zhang LG, Fan QG, Zhu ZR (1992) Simulation study on effects of temperature on population dynamics of brown planthopper. Chinese Journal of Rice Science 6, 21–26.

43. Wang YH, Gao CF, Zhu YC, Chen J, Li WH, et al. (2008) Imidacloprid susceptibility survey and selection risk assessment in field populations of Nilaparvata lugens (Stål) (Delphacidae). J. Econ. Entomol. 101, 515–522.

44. Brar DS, Virk PS, Jena KK, Khush GS (2009) Breeding for resistance to planthoppers in rice. pp. 401–427. In Heong KL, Hardy B (Eds.), Planthoppers: New Threats to the Sustainability of Intensive Rice Production Systems in Asia. Int. Rice Res. Inst, Los Baños, Philippines.

45. Sogawa K, Liu GJ, Shen JH (2003) A Review on the hyper-susceptibility of Chinese hybrid rice to insect pests. Chinese J. Rice Sci. 17:20–30.

2

Choosing Organic Pesticides over Synthetic Pesticides May Not Effectively Mitigate Environmental Risk in Soybeans

Christine A. Bahlai[1], Yingen Xue[1], Cara M. McCreary[1], Arthur W. Schaafsma[2], Rebecca H. Hallett[1]*

1 School of Environmental Sciences, University of Guelph, Guelph, Ontario, Canada, 2 Department of Plant Agriculture, University of Guelph, Ridgetown, Ontario, Canada

Abstract

Background: Selection of pesticides with small ecological footprints is a key factor in developing sustainable agricultural systems. Policy guiding the selection of pesticides often emphasizes natural products and organic-certified pesticides to increase sustainability, because of the prevailing public opinion that natural products are uniformly safer, and thus more environmentally friendly, than synthetic chemicals.

Methodology/Principal Findings: We report the results of a study examining the environmental impact of several new synthetic and certified organic insecticides under consideration as reduced-risk insecticides for soybean aphid (*Aphis glycines*) control, using established and novel methodologies to directly quantify pesticide impact in terms of biocontrol services. We found that in addition to reduced efficacy against aphids compared to novel synthetic insecticides, organic approved insecticides had a similar or even greater negative impact on several natural enemy species in lab studies, were more detrimental to biological control organisms in field experiments, and had higher Environmental Impact Quotients at field use rates.

Conclusions/Significance: These data bring into caution the widely held assumption that organic pesticides are more environmentally benign than synthetic ones. All pesticides must be evaluated using an empirically-based risk assessment, because generalizations based on chemical origin do not hold true in all cases.

Editor: Stephen J. Johnson, University of Kansas, United States of America

Funding: The authors acknowledge funding from the Natural Sciences and Engineering Council of Canada (http://www.nserc-crsng.gc.ca/), Agriculture and Agri-Food Canada (AAFC) (http://www.agr.gc.ca/index_e.php) and the Ontario Ministry of Agriculture, Food and Rural Affairs - University of Guelph partnership (http://www.uoguelph.ca/research/omafra/). The funders had no role in data collection and analysis, decision to publish, or preparation of the manuscript. AAFC suggested the insecticide list for testing, and this is the only role any funders played in study design.

Competing Interests: In support of other projects unrelated to this study, the authors' research group has received competitive research grants from grower organizations and government bodies and contracts and/or in-kind contributions from manufacturers of both organic and synthetic pesticides. Grant sources for other research projects within the last five years include: Natural Sciences and Engineering Research Council of Canada, Canada Foundation for Innovation, Canada Food Inspection Agency, Agriculture and Agri-Food Canada, Ontario Ministry of Food, Agriculture, and Rural Affairs, United States Department of Agriculture, Instituto Nacional de Investigación Agropecuaria (Uruguay), Grain Farmers of Ontario (formerly Ontario Soybean Growers), Ontario Grape and Wine Research Inc., Ontario Wheat Producers Marketing Board, Ontario Corn Producers Association, Agricultural Adaptation Council of Canada, Romer Labs, Bayer CropScience Canada, Bayer CropScience France, Monsanto Canada, Pioneer Hi-Bred Ltd., Dow AgroSciences, BASF Canada, Syngenta Crop Protection Canada, Syngenta Seeds Canada, DuPont Canada Crop Protection, Natural Insect Control, Woodrill Seeds Ltd.

* E-mail: rhallett@uoguelph.ca

Introduction

A public call for sustainability in agriculture has resulted in numerous government initiatives to develop environmentally friendly agricultural practices [1,2,3,4,5,6]. In 2003, the Canadian government initiated the Pesticide Risk Reduction Program to provide infrastructure for the development and implementation of reduced-risk approaches for managing pests in crops [1]. This program, similar to ones in the UK [4] and USA [3], sought to reduce environmental risk associated with older chemical insecticides by replacing them with low risk alternatives. Though generalizations about the relative safety of natural and synthetic chemicals have been questioned in the past [7], these sustainability programs often continue to emphasize the development of organic and natural insecticides for pest control. These programs make the

assumption that natural insecticides present less risk to the environment than synthetic insecticides, aligning with public opinion [8] and influential scientific papers purporting greater sustainability of organic practice [9].

The sustainability of agricultural practices is a subject of ongoing debate in the literature [10,11,12,13]. Many studies have compared organic, conventional and integrated pest management (IPM) production systems as a whole, but even within a commodity system, the conclusions reached in these studies are widely divergent. A 1999 study [14] of New Zealand apple production suggested an integrated approach was more sustainable, but a 2001 study [9] of the same system in Washington favoured an organic management approach. Differing outcomes may be attributed partially to differing geography, climate and pest complexes at the two locations, but it is likely that differences in assessment methodology and the inconsistencies between specific practices classed as organic or conventional at each location were also influential in obtaining the observed results. Comparing organic, conventional and integrated agriculture is not as simple as it may initially appear [13]: each system is characterized by a suite of practices which are ideologically, rather than empirically defined [12], these systems are not mutually exclusive from each other [9,12], and vary from region to region depending on regulations [14]. Because of these variations, generalizations about the overall sustainability of one system over another are never universal [11]. Pest management practices are often specifically highlighted in the sustainability of organic versus conventional agriculture debate, but much of the debate is fuelled by a fundamental misconception that organic farms do not use pesticides [15]. In fact, organic farms, like conventional farms, have access to a suite of pesticides [15,16]; the primary difference is that organic regulations prohibit all synthetic (i.e.: human-made) chemicals but allow a vast array of mineral and botanical pesticides [17], whereas conventional pesticides can be both naturally and synthetically derived and are regulated individually, on a per active ingredient, per formulation basis [18].

Generalizations about the relative sustainability of one suite of practices over another are dangerous when integrated into policy: government regulations based on faulty assumptions about agricultural systems are expensive and do not effectively reduce the environmental risks they are designed to mitigate [19]. It is thus more productive, and more broadly applicable, to evaluate a given tactic for environmental sustainability on its individual properties and build policy based on results of these individual evaluations [16].

Many national and international initiatives exist to develop environmentally sustainable strategies for managing outbreaks of soybean aphid, including Agriculture and Agri-Food Canada's (AAFC) Pesticide Risk Reduction Program [1]. Soybean aphid is a severe pest of cultivated soybean in North America [20], and approximately 1.2 million hectares of soybean are cultivated each year in Canada alone [21]. Since its introduction to North America 10 years ago [20], numerous studies have examined the role of biological control agents in managing populations of aphids [22,23,24,25,26], but foliar insecticides remain necessary when populations of aphids exceed economic thresholds. The need for reduced risk pesticides in this system is profound: only two foliar insecticides are currently registered for soybean aphid control in Canada [18], one of which is currently under review for re-registration [27]. A broader suite of insecticides with varied mechanisms of action are needed to ensure effective insecticide resistance management can occur [28].

Results

Working with AAFC, we identified four novel products to evaluate as potential reduced risk insecticides to include in integrated pest management programs for soybean aphid (Table 1). Two of these insecticides contained synthetic active ingredients, the other two are natural insecticides permitted for use in certified organic crops in Canada [17]. We included formulations of the two currently registered insecticides in the experiments as conventional controls.

We completed laboratory assays to estimate the direct contact toxicity of these insecticides to several natural enemy species when applied at field rates (Table 2). We used two of the soybean aphid's primary predator species in this study, multicoloured Asian ladybeetle *Harmonia axyridis* and insidious flower bug *Orius insidiosus* [25,26]. There were significant differences in mortality by treatment applied for all insect groups $F_{6,657} = 325.25$, $P<.0001$ for ladybeetle adults; $F_{6,993} = 1069.34$, $P<.0001$ for ladybeetle larvae; $F_{6,277} = 228.11$, $P<.0001$ for flower bug adults), but generally, the two currently registered insecticides were most toxic to natural enemies under laboratory conditions. The other four insecticides were much less toxic to the ladybeetle, though it was found that one of the organic insecticides, *Beauveria bassiana*, was slightly more toxic to adults, and one novel synthetic, flonicamid, was slightly more toxic to larvae than the remaining novel insecticides. The four novel pesticides all caused some mortality to the insidious flower bug, but the two organic insecticides had significantly higher toxicity than the two novel synthetic insecticides.

We conducted a two year, five site study to examine the performance of these insecticides against aphids, and selectivity with respect to natural enemies under field conditions (Fig. 1). In addition to efficacy, it is desirable for an insecticide to have a high selectivity for its target pests in order to minimize environmental impact, and to conserve biological control services provided by other organisms residing in the treated area. All synthetic insecticides had similar efficacy one week after treatment ($F_{6,148} = 7.48$, $P<0.0001$), though dimethoate efficacy was reduced in the second assessment week (Fig. 1a), and yield in plots treated with synthetic insecticides did not differ significantly ($F_{6,90} = 3.51$, $P = 0.0036$) (Fig. 2). The two organic insecticides had lower efficacy than the synthetic insecticides (Fig. 1a) at one week ($F_{1,148} = 25.16$, $P<0.0001$) and two weeks ($F_{1,121} = 17.48$, $P<0.0001$) post-treatment and did not offer significant yield protection over the untreated control (Fig. 2). Field selectivity was highest amongst synthetic insecticides, and lowest amongst organic insecticides included in this experiment ($F_{1,119} = 9.00$, $P = 0.0033$),

Table 1. Insecticides evaluated for use in control of the soybean aphid.

Category	Active ingredient (ai)	Trade name (Supplier)	Mode of action	%ai	Rate per ha	EIQ*	EIQ-FUR**
Conventional (synthetic)	Cyhalothrin-λ	Matador 120E® (Syngenta)	Neurotoxin- sodium channels	13.1	83 mL	47.2	0.4
Conventional (synthetic)	Dimethoate	Lagon 480® (Cheminova)	Neurotoxin- acetylcholine esterase inhibitor	43.55	1,000 mL	33.5	12.5
Novel (synthetic)	Spirotetramat	Movento® (Bayer)	Fatty acid biosynthesis inhibitor	22.4	196 mL	34.2	1.3
Novel (synthetic)	Flonicamid	Beleaf® (FMC)	Neurotoxin- potassium channels	50	196 g	8.7	0.8
Novel (organic) [17]	Mineral oil	Superior 70 oil® (UAP)	Oxygen exchange	99	11,000 mL	30.1	280.2
Novel (organic) [17]	*Beauveria bassiana*	Botanigard® (Laverlam)	Entomopathogenic fungus	22	1,000 g	16.7	3.3

*per unit weight environmental impact quotient (EIQ).
**predicted EIQ-field use rating (EIQ-FUR) for a single application of the insecticide, converted to lbs/ac, as convention dictates.

Table 2. Relative direct contact mortality of natural enemies treated with six insecticides at field rate.

Treatment	Relative H-T adjusted % mortality*		
	Harmonia axyridis **adults**	Harmonia axyridis **larvae**	Orius insidiosus **adults**
Cyhalothrin-λ	34.9b	48.2b	99.1a
Dimethoate	70.7a	99.6a	77.2b
Spirotetramat	2.0de	5.3de	20.7e
Flonicamid	0.8e	10.9c	39.5d
Mineral oil	2.6d	6.3d	60.7c
Beauveria bassiana	10.9c	2.7e	59.5c
Untreated control	−0.1e	−0.1f	−0.2f

*Insecticides were applied at 0.5, 1 and 2× field rate using an airbrush sprayer. Mortality was assessed at 18, 24 and 48 h post treatment for *O. insidiosus*, and every 24 h for *H. axyridis* adults and larvae. Mortality data were Henderson-Tilton adjusted [32] and subjected to a mixed model ANOVA by species and life stage, with relative rate incorporated into the model, and assessment time treated as a repeated measure. Observed mortality within a species and life stage followed by the same letter are not significantly different at $\alpha = 0.05$ (LSD).

and though dimethoate had the numerically lowest selectivity amongst the synthetic insecticides, it was still numerically more selective than the organic insecticides (Fig. 1b).

Net environmental impact of applying each insecticide at given rates was estimated using an Environmental Impact Quotient analysis [29]. The per-unit-EIQ was highest for cyhalothrin-λ, a conventional synthetic insecticide (Table 1), but the EIQ-field use ratings were highest amongst the older synthetic, dimethoate, and the two organic insecticides. The high EIQ-field use rating of dimethoate was due to both a high application rate and a relatively high per-unit EIQ. The EIQ-field use rating for the mineral oil insecticide, though, was more than an order of magnitude higher than that of dimethoate, due to its relatively high per-unit-EIQ and its extremely high application rate. The remaining four insecticides had relatively low EIQ-field use ratings compared with mineral oil and dimethoate.

Discussion

EIQ allows relative impact of various control strategies within a crop to be ranked; it is a standard method for indexing the total environmental impact of an application of a given pesticide. EIQ relies on data which is commonly available on MSDS sheets, incorporates the application rate of a pesticide, and is not site or pest-specific, so it provides a less biased estimation than other pesticide ranking systems used to quantify environmental impact [15,30]. Because EIQ is based on a rating system and does not rely on field obtained data, some authors have criticized its use [12]. However, we found a clear inverse relationship between field selectivity and EIQ for insecticides tested in this study when applied at field rates (Fig. 3), suggesting that EIQ rankings are relevant predictors of at least some in-field parameters for environmental impact, and our results strongly support the continued use of EIQ for ranking pesticide impact. Responses of natural enemy communities are strong indicators of ecological impact of an insecticide, because they are arthropods, like the targets, and are thus likely to be biologically similar to the target of the insecticide, and because they are often found alongside the pest at the time of an insecticide application, heightening their exposure compared to other non-target organisms.

Looking at the issue empirically, our results show that with regards to environmental impact, target selectivity and efficacy, the novel synthetic insecticides we tested have better performance than organic insecticides; suggesting that certain organic manage-

ment practices are not more environmentally sustainable than conventional ones. It has been purported that organic systems are not just better for the environment, but are more economically sustainable because of the price premiums associated with organic food [9]. Consumers are often willing to pay more for products they believe are produced in the most sustainable way possible, but we have shown that the organic methods available are not always the most sustainable choice. Carefully designed integrated pest management systems are likely the best strategy for minimizing environmental impact of agriculture: where certified organic systems may reject the technology with the smallest environmental impact based on ideology [11], IPM maintains the flexibility to incorporate any strategy empirically determined to have the smallest impact. In fact, it has been argued that studies which have concluded that IPM has a greater impact than organic management [9] have simply tested a poorly designed IPM strategy in which the efficacy and impact of individual tactics included in the program were not effectively examined [12], did not accurately reflect IPM practice, or employed biased methods of evaluation [15]. Though IPM practice does not typically come with price premiums associated with the production of organic food, IPM strategies are still commonly used by many conventional farmers [31], and given increased consumer awareness of the benefits of IPM practice, adoption rates are likely to rise.

It is for these reasons that we reject the organic-conventional dichotomy and emphasize that, in order to optimize environmental sustainability, individual tactics must be evaluated for their environmental impact in the context of an integrated approach, and that policy decisions must be based on empirical data and objective risk-benefit analysis, not arbitrary classifications.

Materials and Methods

Selection of insecticides for inclusion in experiments

In May 2008, the Pest Management Centre at AAFC provided us with a list of 14 potential insecticides for inclusion in our experiments. We reviewed each insecticide and eliminated those which had the same mode of action as any other insecticide registered for use against soybean aphid in Canada, and then contacted the suppliers to assess the economic feasibility of using these insecticides in field crops. Two novel synthetic and two organic insecticides were identified to be tested for management of soybean aphid, and the two registered insecticides were included in the experiment as conventional controls. Experimental

A

B

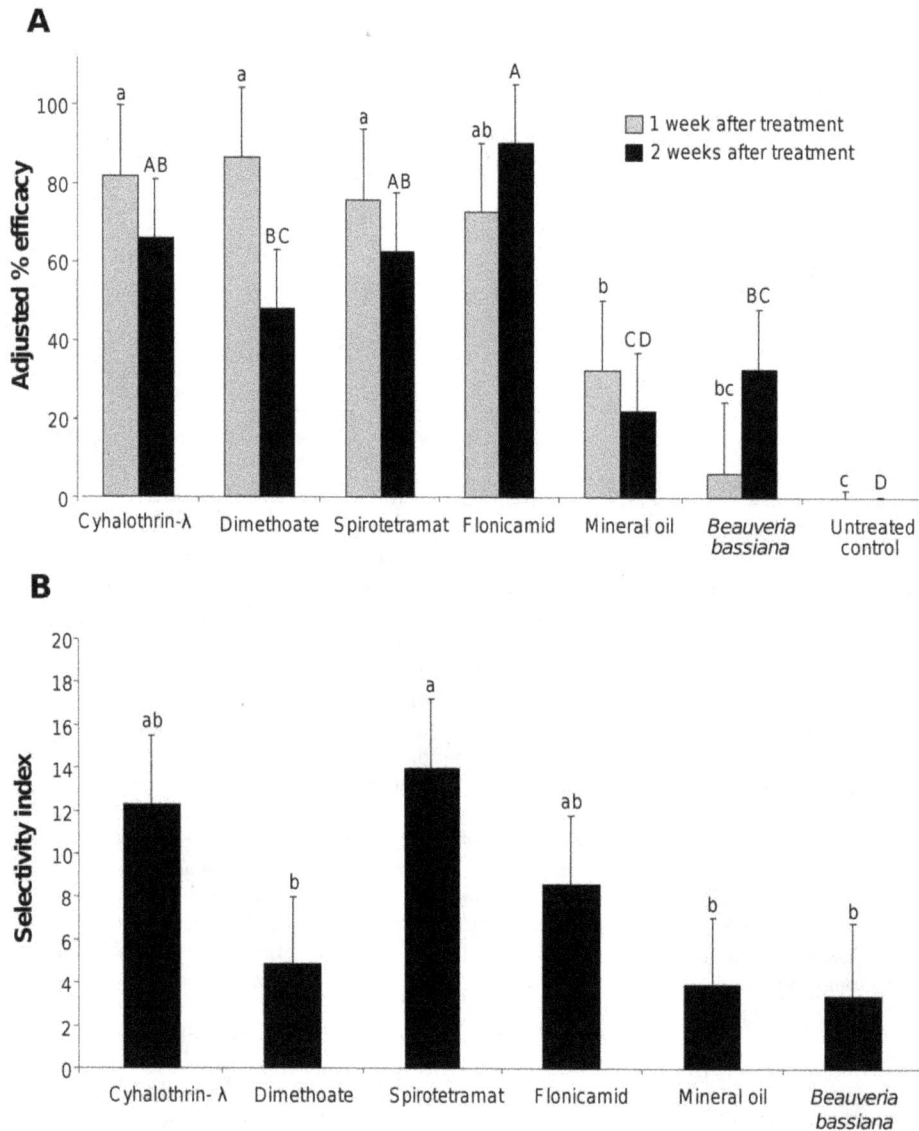

Figure 1. Field efficacy and selectivity observed for six insecticides for aphid control. A) Observed efficacy. Aphid count data were Henderson-Tilton adjusted [32] and subjected to a mixed model ANOVA by post-treatment sampling period with year of experiment, block, pass of tractor, site, and interaction terms between block and pass, block and site, and pass and site incorporated into the model. **b) Observed selectivity.** Field selectivity was determined using the natural enemy-to-aphid ratio in treatment plots, for exact calculation see Materials and Methods. Observed efficacy and selectivity within sampling period marked by the same letter are not significantly different at $\alpha = 0.05$ (LSD).

application rates for novel insecticides were developed in consensus with supplier companies (Table 1). Table S1 provides a complete list of insecticides considered for inclusion in this experiment, and the rationale for products selected.

Determination of direct contact toxicity to natural enemies

Adults and larvae of multicoloured Asian ladybeetle *Harmonia axyridis* and adults of insidious flower bug *Orius insidiosus* were treated with formulated insecticides at the equivalent of 0.5, 1 and 2× field rate using an airbrush spray tower. The untreated control consisted of 1 mL of distilled water. Groups of insects (8–10) were anesthetized using CO_2 then placed in a 50 mm glass Petri plate lined with a piece of 47 mm qualitative filter paper, treated using the spray tower, and then placed in post-treatment containers. Each insecticide-concentration combination was repeated four

times. The spray tower was rinsed with acetone, then distilled water, between each application.

***Orius insidiosus* assays.** *Orius insidiosus* adults were obtained from commercial suppliers (BioBest Biological Systems Canada and MGS Horticultural Inc.). Repetitions of 10 adult *O. insidiosus* were treated, and then placed, post-treatment, in 10 cm plastic Petri plates lined with filter paper moistened with distilled water, and containing 1–2 washed baby spinach leaves, and an excess of frozen *Ephistia* eggs (BioBest Biological Systems Canada) for food. Mortality was recorded at 18, 24 and 48h post treatment.

***Harmonia axyridis* adult assays.** *Harmonia axyridis* were obtained from aggregations on buildings in Guelph, Ontario, Canada, and were reared in laboratory cultures using procedures described by Xue et al [26]. Repetitions of 10 adult *H. axyridis* were treated, and then placed in 10 cm plastic Petri plates lined with filter paper moistened with distilled water, and containing

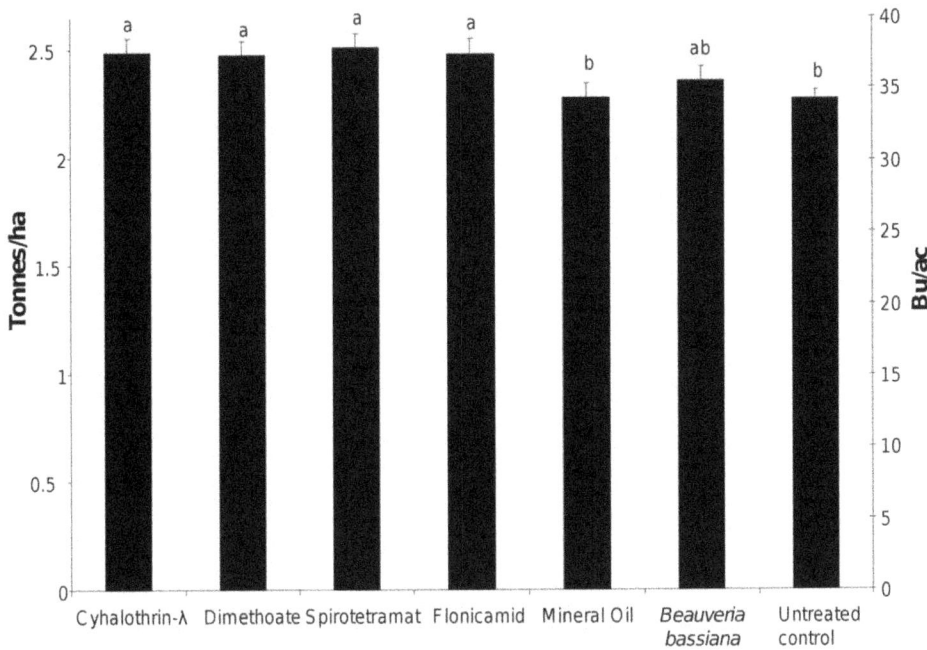

Figure 2. Least-square mean soybean yield in fields treated with six insecticides, 2009. Data were subjected to a mixed model ANOVA with block, site, treatment incorporated into the model. Observed yields marked by the same letter are not significantly different at $\alpha = 0.05$ (LSD). Data from 2008 were excluded from analysis because of low overall aphid populations.

several barley leaves infested with bird-cherry oat aphid (Aphid Banker System; Plant Products, Brampton, Ontario, Canada), and an excess of frozen *Ephistia* eggs (BioBest Biological Systems Canada) for food. Mortality was recorded every 24h for 168 h (7 d).

***Harmonia axyridis* larvae assays.** Second and third instar *H. axyridis* were obtained from the laboratory culture described above. Assays were performed as adult assays above, except repetitions consisted of 8 individuals and instead of being placed

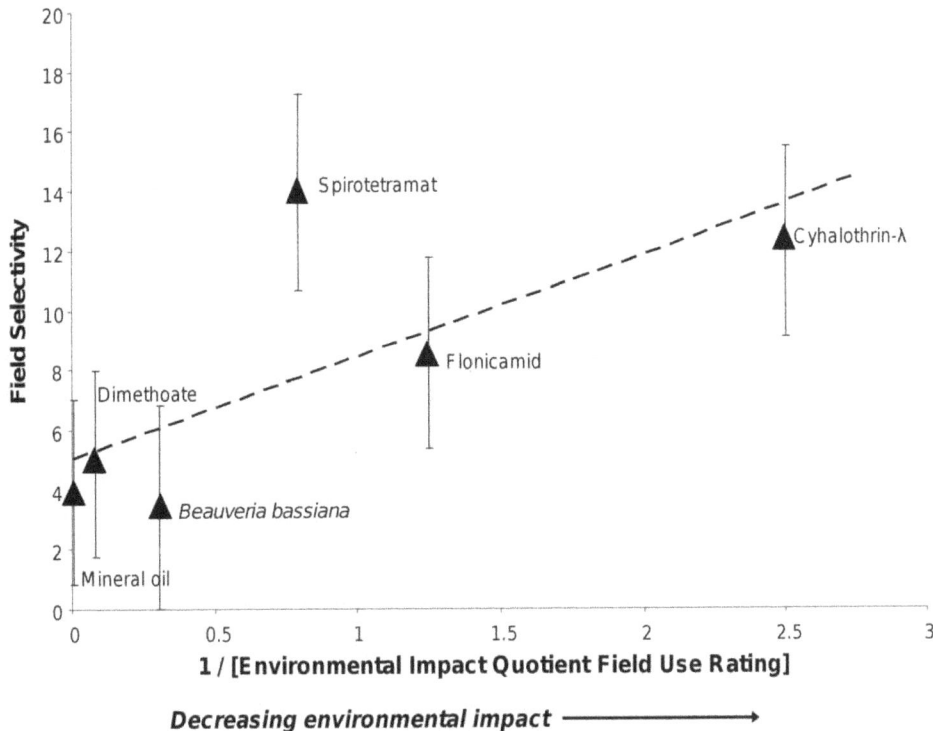

Figure 3. Relationship between observed field selectivity and the inverse of Environmental Impact Quotient at field rates. Field selectivities presented as least square means ($\pm$ SE) of field selectivities observed at four sites in 2009. Equation of regression line is Field selectivity = (3.3 ± 1.7)/EIQ+(0.3 ± 3.1)+site effect, with $F_{93} = 4.23$, $p = 0.0035$.

together in a Petri plate, were placed individually into cells of a rearing tray (BIO-RT-32, C-D International, Inc.) with *Ephistia* eggs and aphid-infested barley to avoid cannibalism.

Statistical analysis of bioassay data. Mortality data was normalized using the Henderson-Tilton adjustment [32], and subjected to a mixed model ANOVA accounting for concentration (relative to field rate), treatment, and assessment time. Assessment time was treated as a repeated measure in the analysis.

Determination of field efficacy and selectivity

In 2009, four soybean fields in southwestern Ontario with aphid populations approaching the action threshold of 250 aphids per plant were identified in collaboration with government extension personnel in July and August, 2009. After obtaining permission from landowners, sites were assessed once weekly until aphid populations exceeded 250 aphids per plant. Upon reaching this threshold, field experiments were initiated. In our initial screening trial in 2008, treatments were applied to a single site with a moderate density of aphids (~120 aphids per plant), due to low aphid populations across our region during that year.

Field experiments employed a RCBD consisting of four blocks of 15 3.7×15.2m beds, with 3 untreated controls per block (one for each tractor pass required), our six insecticides and six other products or formulations not reported in this study. Insecticides were applied using a Teejet Duo nozzle configuration with spray tips #TT11002 at a height of 50cm above the canopy. Spray pressure at the nozzle was 276 kPa and the tractor travelled at a ground speed of 9.7km/h. Fluid delivery rate was maintained at 187 L/ha for all treatments. 2–3 soybean plants were destructively sampled from each bed at each assessment, and assessments were completed 1) immediately before treatment, 2) one week after treatment and 3) two weeks after treatment. Total numbers of aphids, ladybeetles, lacewings, parasitized aphid mummies, syrphid larvae, and flower bugs were assessed on each plant.

Aphid counts were transformed using Henderson-Tilton adjustments to account for population changes in the control between time of treatment and time of assessment, then subjected to a mixed model ANOVA accounting for site, year, tractor pass, replicate, and treatment.

Field Selectivity Calculation

Field selectivity of each insecticide was estimated by calculating the change in the ratio of natural enemies to aphids in each plot, and subjecting these data to a mixed model ANOVA as above. We defined field selectivity as the relative change in the natural-enemy-to-pest population ratio observed after treatment. We standardized the counts of natural enemies of different species by defining a Natural Enemy Unit (NEU), where 1 NEU is the number of predators or parasitoids required to kill 100 pest insects

in 24h. Thus,

$$NEU_{total} = \sum_{i=1}^{N} n_i V_i$$

where N is the total number of natural enemy species, n_i is the total number of individuals of natural enemy species i observed on 10 plants, and V_i is the average voracity of natural enemy species i, that is, the number of pest insects it can kill in 24 h divided by 100. Using functional response data obtained by Xue et al. [26], we defined our soybean aphid ecosystem specific calculation as:

$$NEU_{total} = 1 \times n_{ladybeetles} + 0.08 \times n_{mummies} + 0.15 \times n_{syrphids}$$
$$+ 0.08 \times n_{Orius} + 0.35 \times n_{lacewings}$$

where $n_{ladybeetles}$ is the total number of adult and larvae of ladybeetles of *Harmonia axyridis* or *Coccinella septempunctata*, $n_{mummies}$ is the total number of parasitized aphids, $n_{syrphids}$ is the total number of Syrphidae larvae, n_{Orius} is the total number of *Orius* spp., and $n_{lacewings}$ is the total number of Chrysopidae observed on 10 soybean plants.

Field selectivity was defined as the ratio of NEU/Aphids (NEU/A) after treatment to NEU/A before treatment, normalized by the control, as in the Henderson-Tilton adjustment [32], and took the form:

$$Selectivity = \left[\frac{(NEU/A)_{post-treatment}}{(NEU/A)_{pre-treatment}} \right]_{treated}$$
$$\times \left[\frac{(NEU/A)_{pre-treatment}}{(NEU/A)_{post-treatment}} \right]_{control}$$

This selectivity index results in values <1 if a treatment kills more natural enemies than target pests, and values >1 if a treatment kills more target pests than natural enemies. Larger numbers will indicate a more target-selective pesticide. The selectivity index assumes the applied treatment has at least some efficacy against the target pest.

Environmental Impact Assessment

EIQs were estimated using established methodology [29,33] incorporating data from MSDS sheets provided by the supplier of the insecticides, an EIQ-field use rating was calculated for each insecticide, using the assumption that one application at field rate per season would provide equivalent aphid control. See Table 3

Table 3. Toxicity ratings used to calculate Environmental Impact Quotient for *Beauveria bassiana*, which does not have a published EIQ value.

	Variables from EIQ Equation*										
	DT	D	F	Z	L	R	S	SY	C	P	B
Active ingredient	Dermal toxicity	Bird toxicity	Fish toxicity	Bee toxicity	Leaching potential	Runoff potential	Soil residue half life	Mode of action	Chronic health effects	Plant surface half life	Toxicity to beneficials
Beauveria bassiana	1	1	1	3	1	1	3	1	1	1	5

*Ratings were developed in accordance with methodology presented in Kovach et al. [29].

for values used in the calculation of EIQ for *Beauveria bassiana*, which does not have an existing published EIQ value.

Supporting Information

Table S1 Complete list of insecticides under consideration provided by Agriculture and Agri-Food Canada (AAFC).

Acknowledgments

We sincerely thank L. Des Marteaux, D. Makynen, J. Smith, T. Phibbs, D. Hooker, A. Gradish and T. Baute for technical assistance on this project, R. Norris and B. Stirling for the use of their respective farms, C. Scott-Dupree for use of the airbrush spray tower, M.K. Sears and J.A. Newman for providing comments on this manuscript, and C. Petzoldt and D. Marvin for providing support with the EIQ method. We would also like to thank Syngenta, Bayer, FMC, UAP, and Laverlam for providing insecticides for our experiments, and BioBest Canada and MGS Horticultural for providing insects.

Author Contributions

Conceived and designed the experiments: CB AWS RHH. Performed the experiments: CB YX CMM. Analyzed the data: CB. Contributed reagents/materials/analysis tools: CB YX. Wrote the paper: CB RHH.

References

1. Agriculture and Agri-Food Canada (2003) Pesticide risk reduction and minor use programs: improving ways to manage pests with new technology. Government of Canada.
2. Lynch S, Greene C, Kramer-LeBlanc C (1996) Proceedings of the third national IPM symposium/workshop: Broadening support for 21st century IPM. U.S. Department of Agriculture, Economic Research Service, Natural Resources and Environment Division.
3. Jones E (2004) Grants awarded to develop pesticide risk reduction programs. United States Environmental Protection Agency, Press release 10/14/04.
4. U.K. Department for Environment Food and Rural Affairs and the Forestry Commission (2005) Departmental report 2005. United Kingdom Department for Environment, Food and Rural Affairs.
5. European Commission (1991) Council directive of 15 July 1991 concerning the placing of plant protection products on the market. European Union.
6. Ministry of Science and Technology of the People's Republic of China (2001) National High-tech R&D Program (863 Program).
7. Ames BN, Profet M, Gold LS (1990) Nature's chemicals and synthetic chemicals: comparative toxicology. Proceedings of the National Academy of Sciences, USA 87: 7782–7786.
8. James KH (1990) Risk perceptions and food choice: An exploratory analysis of organic- versus conventional-produce buyers. Risk Analysis 10: 367–374.
9. Reganold JP, Glover JD, Andrews PK, Hinman HR (2001) Sustainability of three apple production systems. Nature 410: 926–930.
10. Lynch D (2009) Environmental impacts of organic agriculture: A Canadian perspective. Canadian Journal of Plant Science 89: 621–628.
11. Trewavas A (2001) Urban myths of organic farming. Nature 410: 409–410.
12. Elliot SL, Mumford JD (2002) Organic, integrated and conventional apple production: why not consider the middle ground? Crop Protection 21: 427–429.
13. Shepherd M, Pearce B, Cormack B, Philipps L, Cuttle S, et al. (2003) An assessment of the environmental impacts of organic farming. United Kingdom Department for Environment, Food and Rural Affairs.
14. Suckling DM, Walker JTS, Wearing CH (1999) Ecological impact of three pest management systems in New Zealand apple orchards. Agriculture, Ecosystems & Environment 73: 129–140.
15. Avery AA (2006) Organic pesticide use: What we know and don't know about use, toxicity, and environmental impacts. Crop Protection Products for Organic Agriculture. Washington, DC: American Chemical Society. pp 58–77.
16. Thompson DG, Kreutzweiser DP (2006) A review of the environmental fate and effects of natural "reduced-risk" pesticides in Canada. Crop Protection Products for Organic Agriculture. Washington, DC: American Chemical Society. pp 245–274.
17. Canadian General Standards Board (2008) Organic productions systems permitted substances list. Government of Canada.
18. Ontario Ministry of Agriculture Food and Rural Affairs (OMAFRA) (2005) Field crop protection guide 2005–2006. Ontario Ministry of Agriculture, Food and Rural Affairs.
19. Kleijn D, Berendse F, Smit R, Gilissen N (2001) Agri-environment schemes do not effectively protect biodiversity in Dutch agricultural landscapes. Nature 413: 723–725.
20. Ragsdale DW, Voegtlin DJ, O'Neil RJ (2004) Soybean aphid biology in North America. Annals of the Entomological Society of America 97: 204–208.
21. Statistics Canada (2009) Cereals and oilseeds review. Ministry of Industry, Government of Canada.
22. Heimpel GE, Ragsdale DW, Venette R, Hopper KR, Neil RJ, et al. (2004) Prospects for importation biological control of the soybean aphid: anticipating potential costs and benefits. Annals of the Entomological Society of America 97: 249–258.
23. Rutledge CE, Neil RJ, Fox TB, Landis DA (2004) Soybean aphid predators and their use in integrated pest management. Annals of the Entomological Society of America 97: 240–248.
24. Costamagna AC, Landis DA, Difonzo CD (2007) Suppression of soybean aphid by generalist predators results in a trophic cascade in soybeans. Ecological Applications 17: 441–451.
25. Desneux N, O'Neil RJ, Yoo HJS (2006) Suppression of population growth of the soybean aphid, *Aphis glycines* Matsumura, by predators: the identification of a key predator and the effects of prey dispersion, predator abundance, and temperature. Environmental Entomology 35: 1342–1349.
26. Xue Y, Bahlai CA, Frewin A, Sears MK, Schaafsma AW, et al. (2009) Predation by *Coccinella septempunctata* and *Harmonia axyridis* (Coleoptera: Coccinellidae) on *Aphis glycines* (Homoptera: Aphididae). Environmental Entomology 38: 708–714.
27. Health Canada Pest Management Agency (2009) PMRA Re-evaluation Workplan (April 2009 to March 2010). Government of Canada.
28. Brattsten LB, Holyoke CWJ, Leeper JR, Raffa KF (1986) Insecticide resistance: Challenge to pest management and basic research. Science 231: 1255–1260.
29. Kovach J, Petzolt C, Degnil J, Tette J (1992) A method to measure the environmental impact of pesticides. New York's Food and Life Sciences Bulletin 139: 1–8.
30. Levitan L, Merwin I, Kovach J (1995) Assessing the relative environmental impacts of agricultural pesticides: the quest for a holistic method. Agriculture, Ecosystems & Environment 55: 153–168.
31. Olson KD, Badibanga T, DiFonzo C (2008) Farmers' awareness and use of IPM for soybean aphid control: report of survey results for the 2004, 2005, 2006, and 2007 crop years. University of Minnesota, Department of Applied Economics.
32. Henderson CF, Tilton EW (1955) Tests with Acaricides against the brown wheat mite. Journal of Economic Entomology 48: 157–161.
33. Kovach J, Petzolt C, Degnil J, Tette J A method to measure the environmental impact of pesticides: Table 2, List of pesticides.

3

A Modeled Comparison of Direct and Food Web-Mediated Impacts of Common Pesticides on Pacific Salmon

Kate H. Macneale*, Julann A. Spromberg, David H. Baldwin, Nathaniel L. Scholz

Northwest Fisheries Science Center, National Marine Fisheries Service, National Oceanic and Atmospheric Administration, Seattle, Washington, United States of America

Abstract

In the western United States, pesticides used in agricultural and urban areas are often detected in streams and rivers that support threatened and endangered Pacific salmon. Although concentrations are rarely high enough to cause direct salmon mortality, they can reach levels sufficient to impair juvenile feeding behavior and limit macroinvertebrate prey abundance. This raises the possibility of direct adverse effects on juvenile salmon health in tandem with indirect effects on salmon growth as a consequence of reduced prey abundance. We modeled the growth of ocean-type Chinook salmon (*Oncorhynchus tshawytscha*) at the individual and population scales, investigating insecticides that differ in how long they impair salmon feeding behavior and in how toxic they are to salmon compared to macroinvertebrates. The relative importance of these direct vs. indirect effects depends both on how quickly salmon can recover and on the relative toxicity of an insecticide to salmon and their prey. Model simulations indicate that when exposed to a long-acting organophosphate insecticide that is highly toxic to salmon and invertebrates (e.g., chlorpyrifos), the long-lasting effect on salmon feeding behavior drives the reduction in salmon population growth with reductions in prey abundance having little additional impact. When exposed to short-acting carbamate insecticides at concentrations that salmon recover from quickly but are lethal to invertebrates (e.g., carbaryl), the impacts on salmon populations are due primarily to reductions in their prey. For pesticides like carbaryl, prey sensitivity and how quickly the prey community can recover are particularly important in determining the magnitude of impact on their predators. In considering both indirect and direct effects, we develop a better understanding of potential impacts of a chemical stressor on an endangered species and identify data gaps (e.g., prey recovery rates) that contribute uncertainty to these assessments.

Editor: Christopher Joseph Salice, Texas Tech University, United States of America

Funding: The National Oceanic and Atmospheric Administration provided funding for the work. The funders had no role in study design, data collection and analysis, decision to publish, or preparation of the manuscript.

Competing Interests: The authors have declared that no competing interests exist.

* E-mail: Kate.Macneale@gmail.com

Introduction

Throughout California and the Pacific Northwest, pesticides are frequently detected in aquatic habitats that support threatened and endangered Pacific salmon (*Oncorhynchus* spp.) [1,2]. Whether these pesticides pose a risk to salmon depends on multiple factors, including the biochemical properties of the pesticides and the effects they have on salmon physiology. For instance, different pesticides that inhibit the same critical enzyme in juvenile salmon can vary in their long-term effects. Such is the case for two classes of insecticides, carbamates and organophosphates, which differ in how long they reduce a salmon's ability to swim and feed normally. Both classes inhibit acetylcholinesterase (AChE), an enzyme required for the proper functioning of cholinergic synapses in vertebrate and invertebrates. Sublethal exposures of AChE-inhibiting insecticides can cause juvenile salmon to feed less [3,4] and swim irregularly [5–7]. These effects may persist for only a few hours after an exposure if the insecticide is a carbamate [8], yet it may take many weeks to months for a fish to resume normal feeding after exposure to an organophosphate [9,10]. Although a difference in feeding recovery time may seem subtle when evaluating toxicological effects, a prolonged reduction in feeding can affect the growth and survival of juveniles and ultimately impact the population [11]. Reduced growth is especially critical for juvenile salmon, because smaller fish have lower first-year survival [12–14].

How quickly salmon resume normal feeding is just one factor to consider when assessing whether pesticides pose a risk to salmon populations. An additional factor is the relative sensitivity of salmon and their prey to various pesticides. For some insecticides, a concentration that kills invertebrates may also cause sublethal effects in fish (i.e., reducing AChE activity); however, for others, concentrations that are lethal for invertebrates may have few if any sublethal effects on fish (e.g., Table 1). Therefore, insecticides found in surface waters may affect a salmon's ability to feed but may also kill much of their prey [1,15,16]. Juvenile salmon feed opportunistically on invertebrates drifting in the water column [12], so reductions in these invertebrates may affect salmon growth and survival as much or more than a reduced ability to feed [17,18]. Because the sensitivities of salmon and their prey differ (e.g., Table 1), each insecticide may have a different potential to affect salmon via reducing their capacity to feed or through the reduction of prey itself.

Table 1. Effects concentrations (μg/L) and slopes for salmon AChE activity, and prey abundance dose-response curves for several organophosphate (OP) and carbamate (CB) insecticides.

Insecticide	Class	Salmon AChE activity		Prey abundance		AChE EC$_{50}$ Prey EC$_{50}$
		EC$_{50}$	slope	EC$_{50}$	slope	
Chlorpyrifos	OP	2.0	1.50	2.30	1.8	0.9
Diazinon	OP	145.0	0.79	1.38	1.8	105.1
Carbaryl	CB	145.8	0.81	4.33	5.5	33.7

The ratio of the AChE EC$_{50}$ to prey EC$_{50}$ illustrates the relative sensitivities of the salmon AChE activity and their prey abundances to the insecticide. Salmon AChE values are from Laetz et al. [29]. Details of how prey abundance values were derived are given in Supporting Information S1.

Considering how insecticides affect the dynamics of aquatic invertebrate communities may also be relevant for assessing impacts on salmon [19]. Insecticides can cause catastrophic invertebrate drift [20], with dead or moribund invertebrates leaving the benthos and flowing downstream in the water column at rates more than 1000 times greater than normal levels (e.g., [21,22]). Juvenile salmon may feed on this temporary "spike", or excess in prey [23], but prey may be depleted for many months following such events [20,21,23,24]. The amount of prey available over time will depend on how vulnerable the invertebrate community is (or how low to some "*prey floor*" it is driven), and how quickly the invertebrate community can rebound (*prey recovery rate*). Because it is unlikely that there is a single prey recovery rate or a single value that reflects the proportion of prey that could persist following an extreme exposure, a range of these values should be evaluated when considering the ways fluctuations in prey affect salmon feeding and growth.

The properties of an insecticide that contribute to its toxicity may also influence the likelihood that individual salmon and their prey are exposed. The environmental persistence of an insecticide, as well as the dynamics of the targeted pest, may influence how frequently it is applied. Repeated applications may be needed for controlling some pests, while a highly persistent insecticide may be applied only once because it remains toxic for several months. Consequently, there may be sustained or repeated exposures depending on how often pesticides are applied throughout a watershed. In addition, the application technique (e.g., applied aerially vs. on the ground), as well as weather (e.g., the frequency and intensity of rain, wind), will influence how likely insecticides contaminate aquatic habitats. Therefore, when evaluating whether an insecticide could harm salmon and their prey, researchers must consider the timing, frequency and duration of a likely exposure.

Although data gaps and uncertainties remain, there is increasing recognition that a more comprehensive approach is needed for evaluating the potential effects of pesticides on non-target communities [16,25–28]. Notably, the National Research Council [28] recently recommended including potential population-level impacts of any sublethal (e.g., impaired behavior) and indirect (e.g., reduced prey) effects of pesticide exposures when assessing impacts on threatened and endangered species, including Pacific salmon. This is challenging, as it requires quantifying the interactions among salmon and their prey as they are altered by the type, timing, frequency, duration and intensity of pesticide applications. When evaluating the population-level impacts of all of these factors, an ecological modeling approach can provide valuable insight.

In evaluations of pesticides and their potential effects on Pacific salmon listed as threatened or endangered under the US Endangered Species Act (ESA) [17,18], the National Marine Fisheries Service (NMFS) has laid the groundwork for modeling sublethal and indirect impacts of these chemicals on salmon populations. Here, using and expanding upon these models, we compare the relative importance of direct and indirect effects of several insecticides on salmon populations, and explore how population-level effects are influenced by the dynamics of salmon prey. To do this, we incorporate prey dynamics into a model that examines the population-level impacts of direct effects of pesticides [11]. Three parameters describing prey community dynamics following an exposure - a one-day spike, the prey floor or lowest level to which prey abundance can be reduced, and the recovery rate - were incorporated in the model. Here we describe how those parameters affect overall salmon population growth rates. In addition, we examine how the frequency, duration and timing of insecticide exposure interact with prey dynamics in their effect on salmon population growth rates.

Methods

We modified a model previously developed by Baldwin et al. [11] to assess the potential effects of AChE-inhibiting pesticides (n-methyl carbamate and organophosphate insecticides) on Pacific salmon at the individual and population scales. At the individual scale, the modified model links chemical exposure to reductions in feeding behavior and prey abundance, food intake and, by extension, the juvenile somatic growth of ocean-type Chinook salmon (*O. tshawytscha*). At the population scale, the model utilizes the relationship between subyearling size at ocean migration and subsequent size-dependent mortality, to evaluate corresponding consequences for population growth rate across multiple generations. We modeled varying pesticide exposures in freshwater habitats for juvenile salmon to estimate how changes in individual somatic growth may influence population-scale abundance, as indicated by a reduction in the intrinsic rate of increase, or λ. Modeled exposure concentrations spanned the known ranges of toxicological sensitivities for salmon and their prey. The model was constructed using MATLAB 7.9.0 (R2009b) (The MathWorks, Inc. Natick, MA).

Individual-based modeling

The organismal portion of the model tracked the somatic growth of individual salmon to assess how a pesticide exposure may act through effects on salmon feeding and on prey abundance. For the direct effects on the salmon feeding we quantified the physiological pathway between AChE activity and the somatic growth of salmon fingerlings based upon a series of empirical relationships between pesticide exposure, AChE inhibition, feeding behavior, food uptake, and somatic growth rate. The relationship between exposure and AChE inhibition includes the

EC_{50} (the concentration that produces 50% AChE inhibition) and the slope of the exposure response curve. For each pesticide, values for the EC_{50} and slope were from Laetz et al. [29]. Further descriptions of the relationships linking the direct effects of AChE inhibition on salmon feeding and growth can be found in Baldwin et al. [11].

For the indirect effects on salmon via their prey, we used empirical data to develop a relationship between pesticide exposure and prey abundance (i.e. the ration of food available for individual juvenile salmon). This relationship includes the EC_{50} (the concentration that would reduce available prey by 50%) and the slope describing the sensitivity of prey to a range of pesticide exposures. Invertebrate toxicity data were obtained from the U.S. Environmental Protection Agency's Ecotox database (http://cfpub.epa.gov/ecotox/) and from replicated mesocosm experiments (e.g., [30]), and were used to generate a single, representative EC_{50} and slope for each pesticide (Table 1). Only toxicity data from studies on taxa known to be salmon prey, or ecological or physiological surrogates, were included in calculating the prey community EC_{50}s. Details can be found in the Supporting Information S1 and Figure S1.

The relationships in the organismal portion of the model utilize steady state sigmoidal dose-response curves to link pesticide exposure with the effects on salmon AChE activity (see [11]) and relative prey abundance. The sigmoidal curves were defined using specific EC_{50}s and slopes (Table 1, Figures 1 and 2). Pesticide exposures in the organismal portion of the model were defined by pulses of various lengths (i.e. number of days) and timing (i.e. day exposure begins) with the pesticide concentration during a single pulse remaining constant (Figure 1A). The relative prey abundance concentration response curve was derived from the prey EC_{50} and slope, and bound between the control abundance and a defined prey floor (Figure 1B). The prey floor is the portion of the prey community that remains regardless of a pesticide exposure. This accounts for a small but constant input of unaffected terrestrial insects into salmon habitats as well as tolerant (pesticide resistant) aquatic invertebrates. For each scenario the exposure concentration was calculated for each time point (Figure 1A) and, using the exposure to relative prey abundance relationship (Figure 1B), the time course for relative prey abundance was determined (Figures 1C and 2B). The time course for relative prey abundance and related available ration also incorporated a one-day spike in prey drift, with the magnitude of the drift depending on both the toxic potency of the pesticide and the sensitivity of the available prey community. The transient spike was followed by a sustained drop in prey abundance and then a gradual recovery (Figure 1C). The size of the prey spike was estimated as a 20-fold increase over the standing prey abundance on the day prior to pesticide exposure, minus the prey floor. Prey recovery was assumed to be constant, reflecting a constant influx of invertebrates from connected habitats. During an exposure any new invertebrates recruited into habitats were subject to the toxicity and the rate of prey recovery was adjusted to capture the additional losses. Prey recovery continued until control (pre-exposure) drift rates were reached or another exposure occurred. The parameter values defining control baseline conditions and exposure scenarios are listed in Tables 2 and 3, respectively.

The direct physiological effect of an exposure will determine a fish's ability to feed (Figure 2A, [11]). The final ration consumed by a fish is dependent on both how much food it is capable of eating (i.e. potential ration) and on how much food is available (i.e. relative prey abundance). The final ration available each day was the product of potential ration and the relative prey abundance (outputs of Figures 2A and 2B). The amount of prey (Figure 2B)

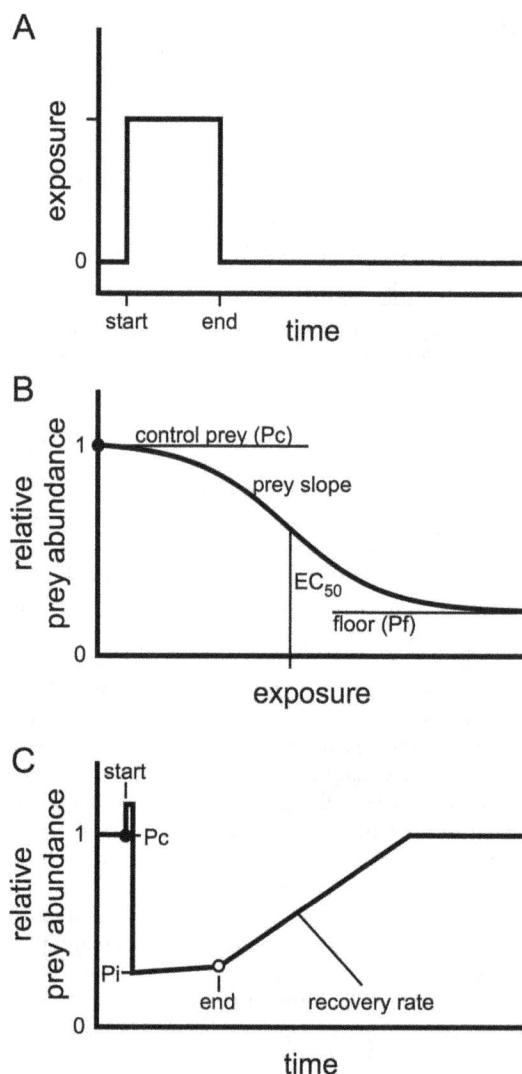

Figure 1. Relationships used to link anticholinesterase exposure to the abundance of prey. A) Step pulse of exposure to an anticholinesterase pesticide. B) Sigmoidal relationship between exposure concentration and relative prey abundance defined by control abundance (Pc), sigmoid slope (prey slope), prey EC_{50}, and a minimum abundance (prey floor, Pf). C) Time course of prey abundance in response to a step exposure. Pc denotes the control prey abundance before the exposure. Pi denotes the reduced prey abundance at the start of the exposure.

that could be consumed during a prey spike was capped at a maximum of 1.5 times the control drift since a fish's maximum feeding capacity limits the amount of excess food it can exploit. The size change for individual fish each day was calculated from the final ration and somatic growth rate (Figure 2C, [31]). An example of the individual somatic growth rate over time is provided in Figure 2D.

The organismal growth model was run for 1000 individual fish, with initial weight selected from a normal distribution with a mean of 1.0 g and standard deviation of 0.1 g. This weight was representative of juvenile Chinook salmon in the early spring, before the seasonal application of insecticides. For each day modeled, the somatic growth rate and fish weight were calculated using parameter values selected from their normal distributions (Table 2). Normal distributions appropriately represented the

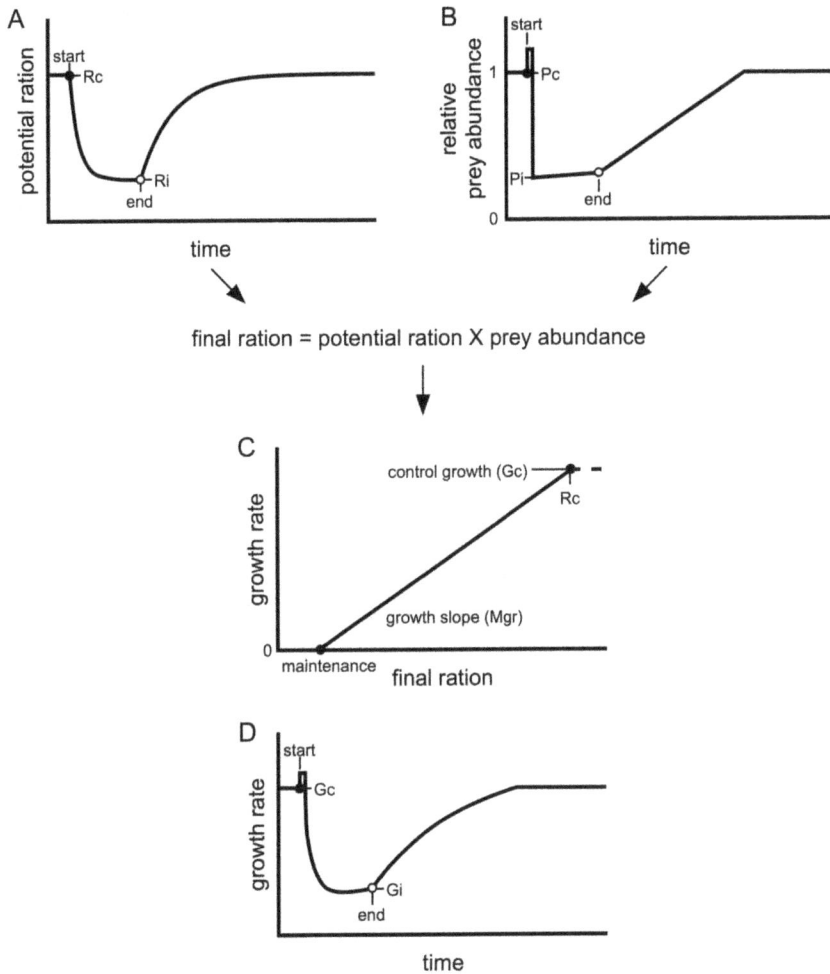

Figure 2. Relationships linking anticholinesterase exposure to individual ration and growth rate. A&B) Relationships describing the time course of the effects of exposure on the organisms ability to capture food (A, potential ration) and the availability of food (B, relative prey abundance). C) Linear model linking final ration (potential ration times relative prey abundance) to growth rate using a line passing through the control condition with a slope denoted by Mgr. D) Time course for effect of exposure on individual growth rate produced by combining A, B, & C. See text for details. Closed circles represent the control condition just prior to exposure, and open circles (e.g. Ai) represent the exposed (inhibited) condition at the end of the exposure.

literature and experimental data and were extended to parameters for which values were estimated. This was repeated each day for the 140 days of the subyearling freshwater growth period across 1000 fish. Further details can be found in Baldwin et al. [11]. The organismal model run produces a mean weight and standard deviation for the subyearling salmon, which is the input for the size-dependent first year survival function of the population model. A sensitivity analysis was run to determine the influence of the organismal model input parameter values on the final somatic size by altering each baseline value by 0.5 to 2 fold (Table 2).

Population-scale modeling

The subyearling weight distribution was used to calculate size-dependent first-year survival for an ocean-type Chinook life-history matrix population model [11]. A brief description is provided here and details can be found in Baldwin et al. [11]. The first-year survival element of the transition matrix incorporates a size-dependent survival rate for a three-month interval, which is the last three months of their first year. This represents the period the subyearling smolts spend in estuarine and nearshore habitats. The weight distributions (based on the calculated mean and

standard deviation) from the organismal model were converted to length distributions by applying condition factor calculations from length and weight relationships collected for subyearling ocean-type Chinook [32].

The relationship between an individual salmon's length and the rate of survival during migration and estuary residence was adapted from Zabel and Achord [14] to match the survival rate for the unexposed Chinook salmon population [33,34]. The relationship is based on the length of a subyearling salmon relative to the mean length of competing subyearling salmon of the same stock, (Δlength), and relates that relative difference to size-dependent survival. A size-dependent survival probability for each fish was generated by randomly selecting length values from the normal distribution calculated from the organismal growth model. This process was replicated for 1000 modeled Chinook salmon for each exposure and a corresponding unexposed cohort to yield mean size-dependent survival rates which were inserted into the first-year survival calculation for the two groups.

To determine population-level responses, an age-structured life history model was constructed for ocean-type Chinook salmon. The model assumed a maximum female age of 5 with reproductive

Table 2. List of values used for control parameters to model organismal growth and the model sensitivity to changes in the parameter.

Parameter	Value[1]	Error[2]	Sensitivity[3]
AChE activity (Ac)	1.0[4,5]	0.06[5]	–0.167
feeding (Fc)	1.0[4,5]	0.05[5]	0.088
ration (Rc)	5% weight/day[6]	0.05[7]	–0.547
feeding vs. activity slope (Mfa)	1.0[5]	0.1[5]	–0.047
ration vs. feeding slope (Mrf)	5 (Rc/Fc)	-	-
growth vs. ration slope (Mgr)	0.35[6]	0.02[6]	–0.547
growth vs. activity slope (Mga)	1.75 (Mfa*Mrf*Mgr)	-	-
initial weight	1 gram[8]	0.1[8]	1.00
control prey drift	1.0[4]	0.05[11]	0.116
AChE impact time-to-effect ($t_{1/2}$)	0.5 day[9]	n/a	0.005
AChE time-to-recovery ($t_{1/2}$)	0.25 days, 30 days[10]	n/a	–0.0001
prey floor	0.05, 0.20, 0.50[11]	n/a	0.178
prey recovery rate	0.005, 0.01, 0.05[12]	n/a	0.323
somatic growth rate (Gc)	1.3[13]	0.06[6]	2.531

[1]mean value of a normal distribution used in the model or constant value when no corresponding error is listed.
[2]standard deviation of the normal distribution used in the model.
[3]mean sensitivity when baseline parameter is changed over range of 0.5 to 2-fold, S = (change in final/baseline weight)/(change in parameter/baseline parameter).
[4]other values relative to control.
[5]derived from [4].
[6]derived from [31].
[7]data from Brett et al. [31] has no variability (ration was the independent variable) so a variability of 1% was selected to introduce some variability.
[8]consistent with field-collected data for juvenile Chinook [64].
[9]estimated from [9].
[10]0.25 days consistent with [8]; 30 days from [65].
[11]range estimated from [21,41,43].
[12]range estimated from [24,43,44].
[13]derived from [31] and adapted for ocean-type Chinook.

maturity at ages 3, 4 or 5 [11]. Transition values were determined from the literature on survival and reproductive characteristics from several ocean-type Chinook populations in the Columbia River system [33,35–39]. The spawner sex ratio was approximately 1:1. Age-based fecundity, number of eggs (standard deviation), of 4511 (65), 5184 (89), and 5812 (102) for 3,4, and 5-year olds was calculated using length data from Howell et al. [33], and the length-fecundity relationships from Healey and Heard [35]. Control survival and reproduction matrix values for the model are listed in Table 4.

Analysis of the transition matrix, A, explored the intrinsic population growth rate as a function of the survival and reproductive rates [40]. The intrinsic population growth rate, i.e. lambda (λ), equals the dominant eigenvalue of A and was calculated using matrix analysis software noted above. Variability was integrated by repeating the calculation of λ 2000 times, selecting the values in the transition matrix from their normal distribution defined by the mean standard deviation for each calculation. Normal distributions appropriately represented the literature data. The population model output consists of the percent change in intrinsic population growth rate (%$\Delta\lambda$; mean and standard deviation) of the pesticide-exposed population from the control population. Change in λ is a population parameter often used by the National Marine Fisheries Service as well as

other federal, state and local agencies in evaluating population productivity, status, and viability. Sensitivity and elasticity analyses were conducted on the control transition matrix as reported previously [11]. The influence of each matrix element, a_{ij}, on λ was assessed by calculating the sensitivity values for A. The sensitivity of matrix element a_{ij} equals the rate of change in λ with respect to a_{ij}, defined by $\delta\lambda/\delta a_{ij}$. The elasticity of matrix element a_{ij} is defined as the proportional change in λ relative to the proportional change in a_{ij}, and equals (a_{ij}/λ) times the sensitivity of a_{ij}. Higher sensitivity and elasticity values indicate greater influence on λ [39].

The populations were assumed to be density independent and closed (i.e. no net migration in or out of the population in the form of straying). Consistent data were lacking to establish any type of density dependent relationship. Since certain forms of density dependence can influence population dynamics through compensation and other demographic processes, using density independence minimizes the likelihood of a Type 2 error arising from inaccurate assumptions of density-dependence. No stochastic impacts are included beyond natural variability in all parameters. This was represented by selecting all parameter values from a normal distribution about their mean for each model step. Nonpesticide influences on population dynamics, including ocean conditions, fishing pressure, and marine food availability were assumed constant and density independent. Each individual fish experienced a pesticide exposure scenario only as a subyearling (during its first spring) and the exposure scenario was assumed to occur annually to all subyearlings in the population.

Scenarios

Using the model, we ran specific scenarios (Table 3) to assess 1) the relative importance of direct vs. food web-mediated pesticide impacts on salmon population growth rates, 2) the influence of prey community sensitivity and response dynamics on salmon population growth rates, and 3) how the frequency, duration and timing of pesticide exposures affect Chinook salmon population growth rates. These scenarios and their parameters are listed in Table 3.

1 The relative importance of direct vs. food web-mediated pesticide impacts on salmon population growth rates. For the first set of scenarios, we considered several insecticides and, for each, compared model runs of either only direct effects or both direct and indirect effects (Table 3). Four pesticides were selected to illustrate how two factors associated with the pesticide – the salmon AChE activity recovery time and the relative toxicity between salmon and their prey - influence the relative importance of the direct and indirect effects of exposure. First, we considered two classes of insecticides, organophosphates and carbamates, which differ in how long it takes salmon AChE activity to recover following exposure (recovery half-lives of 30 days and 6 hours, respectively, Table 3). Second, we used insecticides that differ in their relative toxicity to salmon and their prey. Chlorpyrifos, an organophosphate, has an EC_{50} for brain AChE inhibition in salmon that is similar to the EC_{50} for invertebrate prey mortality (Table 1). By contrast, diazinon (also an organophosphate) is more toxic to invertebrates than to salmon (Table 1). Carbaryl, a carbamate, is similar to diazinon in that it is more toxic to invertebrate prey than salmon (Table 1), but unlike diazinon, salmon can recover rapidly from a sublethal carbaryl exposure (Table 3). In order to make the comparison between the classes of pesticides while controlling for the relative toxicities, we considered a hypothetical carbamate that would be similar to chlorpyrifos in all ways except salmon AChE activity recovery time (Table 3). By comparing this hypothetical carbamate along

Table 3. Model scenarios and the questions they aim to address.

Question	approach	class/compound	Exposure parameters			Salmon AChE parameters			Prey parameters				
			number of pulses	pulse duration, days	pulse start, day(s)	EC₅₀, µg l⁻¹	slope	recovery rate, days	EC₅₀, µg l⁻¹	slope	recovery rate, % day⁻¹	spike	floor
What is the relative importance of direct vs. indirect impacts on salmon populations?	Compare effects of 4 insecticides on salmon alone vs. salmon and their prey	chlorpyrifos	1	4	30	2	1.5	30	2.3	1.8	1	yes	0.2
		diazinon	1	4	30	**145.0**	**0.79**	30	**1.38**	1.8	1	yes	0.2
		carbaryl	1	4	30	**145.8**	**0.95**	**0.25**	**4.33**	**5.5**	1	yes	0.2
		hypothetical carbamate	1	4	30	**2**	**1.5**	**0.25**	**2.3**	**1.8**	1	yes	0.2
How do the dynamics of response of the prey community affect salmon populations?	Vary prey recovery rate	carbaryl	1	4	30	145.8	0.95	0.25	4.33	5.5	**0.5; 1; 5**	yes	0.2
	Vary prey resistance level or floor	carbaryl	1	4	30	145.8	0.95	0.25	4.33	5.5	1	yes	**0.05; 0.2; 0.5**
	Compare with and without prey spike	carbaryl	1	4	30	145.8	0.95	0.25	4.33	5.5	1	**yes & no**	0.2
How do the frequency, duration & timing of exposures affect salmon populations?	Vary frequency, duration and timing of exposures	carbaryl	**1, 2, 4, & 8**	**4, 8, 16, & 32**	**See Fig. 7**	145.8	0.95	0.25	4.33	5.5	**0.5; 1; 5**	yes	**0.05; 0.2; 0.5**

Parameters that were changed in each scenario are in bold.

Table 4. Matrix transition element and sensitivity and elasticity values.

Transition Element	Ocean-type Chinook Salmon		
	Value[1]	Sensitivity	Elasticity
S1	0.0056	57.13	0.292
S2	0.48	0.670	0.292
S3	0.246	0.476	0.106
S4	0.136	0.136	0.0168
R3	313.8	0.0006	0.186
R4	677.1	0.000146	0.0896
R5	1028	1.80E-05	0.0168

[1]Values calculated from data in [32,33,35–39].

with the three actual insecticides, we assessed how salmon AChE activity recovery rates and the relative toxicity of insecticides influence the importance of direct vs. indirect effects.

2. The influence of prey community sensitivity and response dynamics on salmon population growth rates. To address this second point, we developed scenarios using the carbamate carbaryl to examine how specific prey dynamics affect the impact of exposure on salmon population growth rates, since the impacts of this pesticide are primarily due to effects on prey. We ran the model using a range of ecologically relevant values for the prey community's recovery and resistance (Table 3). The rate at which invertebrate communities recover following pesticide exposures varies widely. In some systems, invertebrate densities and biomass return to background or reference levels within weeks of exposure, while in others it can take months to years to recover [24,41–45]. Invertebrate community recovery rates of around 1% per day have been reported frequently [41,43], and therefore, we used that as the prey recovery rate for most of the scenarios. We also used slower and faster rates in additional scenarios to explore how different prey recovery rates affect salmon population growth rates (Table 3; [24,46]).

To explore how the resistance or tolerance of the prey community to insecticide toxicity translates to salmon population growth rates, we set the prey floor (the lowest possible ration available following exposures) at 0.05, 0.20 or 0.50. These values indicate under extreme exposures (i.e., those exceeding the invertebrate EC_{50}), the available ration could be reduced by as much as, but not more than, 95%, 80%, or 50%, respectively. While no studies specify prey floors per se, field studies quantifying the impacts of highly toxic pesticide exposures on invertebrates indicate this range is realistic [21,41,43,47–49]. Because some of the most carefully executed experiments report <10% to 25% of the community may persist after exposures to pesticides [21,41,50], we set the prey floor at 0.20 for most scenarios (Table 3). In addition, to assess the importance of a one-day spike in prey to salmon population growth, we ran scenarios with and without the spike (Table 3).

3. The influence of the frequency, duration and timing of pesticide exposures on salmon population growth rates. To address this, we examined the output of scenarios that varied in the duration, frequency and timing of carbaryl exposures (Table 3). To examine the effect of changing the exposure duration, the model was run using a constant exposure lasting 4, 8, 16, or 32 days. To assess the effect of exposure frequency, a 4-d exposure was repeated either 1, 2, 4, or 8 times. Finally, to assess the impact of exposure timing, different intervals between multiple exposures were used. All combinations were run and those representing the full range of output are presented in the results.

Results

Sensitivity Analyses

A sensitivity analysis conducted on the organismal model revealed that changes in the control somatic growth rate had the greatest influence on the final weights of juvenile salmon (Table 2). While the somatic growth value was experimentally derived for sockeye salmon [31], this value was adapted for ocean-type Chinook salmon and is within the variability reported in the literature for other salmon (reviewed in [51]). Other parameters related to the daily growth rate calculation, including the growth to ration slope (Mgr) and control ration, produced strong sensitivity values. Initial weight, prey recovery rate and prey floor also strongly influenced final weight values (Table 2). Large changes (0.5 to 2 times) in other key parameters produced proportional changes in final weight. The sensitivity analysis of the control population matrix predicted the greatest changes in population growth rate result from changes in first-year survival (Table 4). The standard deviation for the control population matrix was on average 10.2 and for the exposed population matrices ranged from 6.8 to 11.6. The elasticity values for the transition matrix also corresponded with the driving influence of first-year survival.

The relative importance of direct vs. food web-mediated pesticide impacts on salmon population growth rates

The relative importance of direct effects of pesticide exposure on salmon feeding behavior versus indirect effects via reductions in their prey depends not only on how quickly salmon recover from an exposure but also on how sensitive salmon are compared to their prey. If salmon are slow to recover following exposure, and their AChE EC_{50} is similar to the EC_{50} of their prey (e.g., chlorpyrifos, Figure 3A), almost all of the impact on salmon population growth rates (λ) results from the direct effect on fish. Reductions in prey have little additional effect on the percent change in salmon population growth rates (%$\Delta\lambda$), regardless of the exposure concentration (Figure 3A). When the relative toxicity of the organophosphate is much greater for prey than for salmon (e.g., diazinon, Table 1 and Figure 3B), the direct effect on fish is still important, however, reductions in prey are also important. For diazinon, reductions in prey contribute much more to the overall %$\Delta\lambda$, particularly at lower concentrations (e.g. just above the prey EC_{50} but well below the salmon EC_{50}). Therefore, for insecticides like organophosphates that can have sublethal but long-acting effects on individual salmon, the difference in the relative toxicity of the insecticide to the salmon compared to their prey determines how important direct effects on salmon are compared to the reductions in their prey.

In contrast, when salmon are able to recover quickly (e.g., the insecticide is a carbamate instead of an organophosphate), reductions in salmon population growth rates are due almost entirely to reductions in prey and not to sublethal toxicity to the fish (Figure 3C and 3D). The example of the hypothetical carbamate (Figure 3D) illustrates that this is the case regardless of the relative toxicity of the carbamate on salmon versus their prey. For instance, even if we assume the AChE EC_{50} and prey EC_{50} are similar for this hypothetical carbamate, the effect of reduced prey produces a greater %$\Delta\lambda$ (Figure 3D). Thus for insecticides

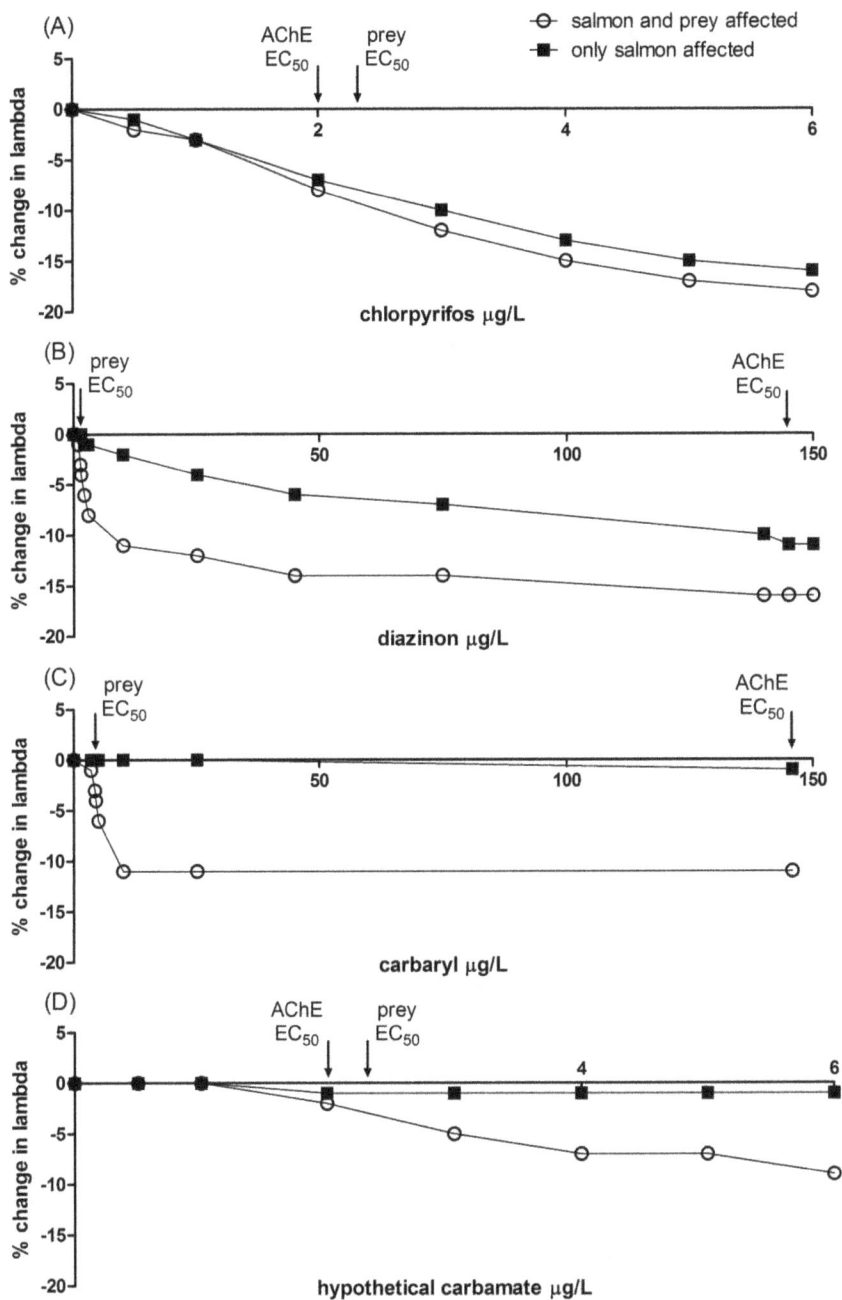

Figure 3. Change in salmon population growth rates due to direct and indirect effects of pesticides. Mean percent change in population growth rates (%Δλ) between unexposed ocean-type Chinook salmon and those exposed to a single, 4-day exposure per generation of chlorpyrifos (A), diazinon (B), carbaryl (C), and a hypothetical carbamate (D) as generated by the model. Scenarios for each insecticide include one with only the direct effects on salmon and another with effects on both salmon and their prey.

that can have sublethal but brief effects on salmon, reductions in prey are more important in reducing salmon population growth rates than direct effects on feeding behavior, regardless of how sensitive salmon are compared to the prey.

The influence of prey community sensitivity and response dynamics on salmon population growth rates

For single, 4-day pulses of an insecticide like carbaryl (in which effects on prey dominate), the magnitude of the %Δλ depends on both the prey recovery and prey floor (Figure 4). For the scenario with the intermediate values for prey recovery rate and the prey

floor, the mean reduction in salmon population growth rates was −11% at the highest concentration considered (145.8 µg/L carbaryl, which is the EC_{50} of the salmon AChE activity and 33.7 times the EC_{50} of the prey). At this same concentration, a scenario with a fast recovery and a high prey floor resulted in %Δλ of only −1%. At the other end of the spectrum, a scenario assuming slow prey recovery and a low prey floor resulted in %Δλ of −21% at the highest concentration considered. The magnitude was lower at more environmentally realistic concentrations, but the range in %Δλs indicated the prey recovery rate and the prey floor remain important parameters in understanding potential

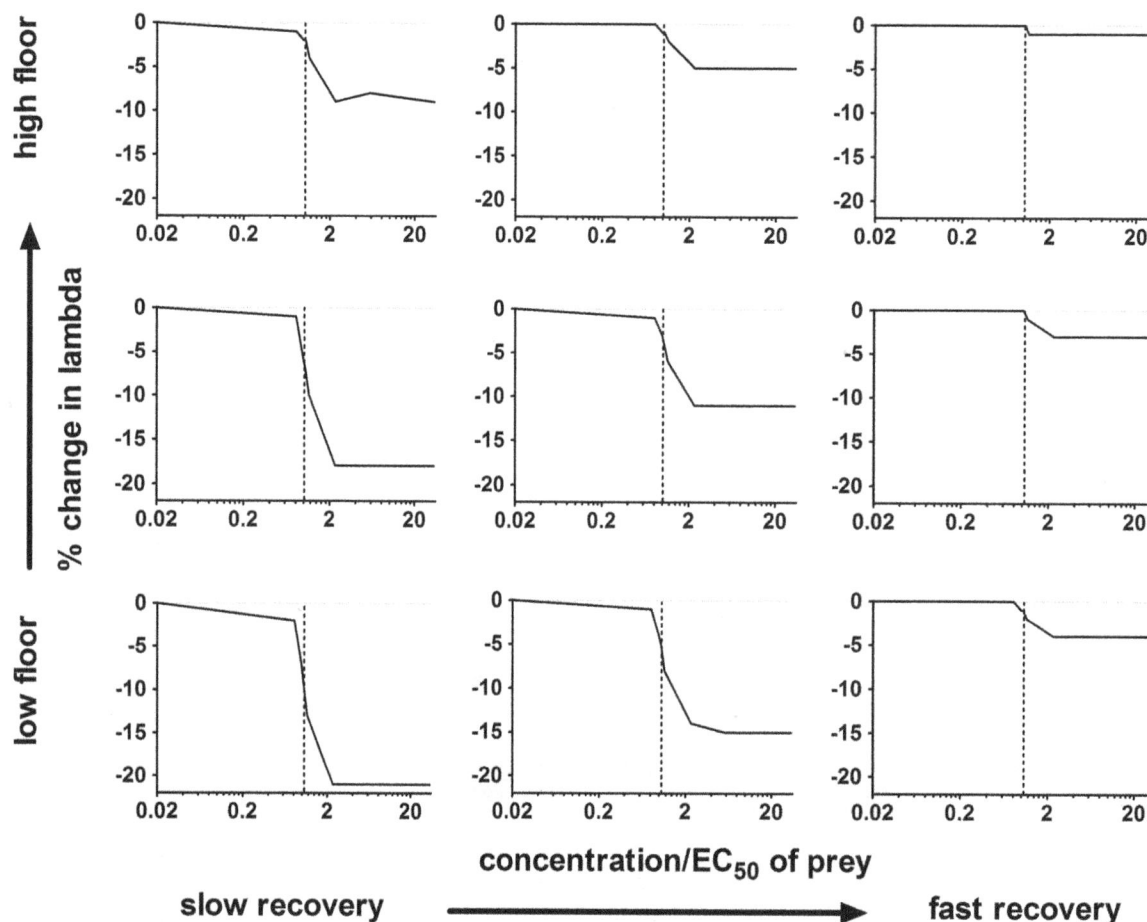

Figure 4. Effect of prey recovery rate and prey floor on salmon population growth rates. Mean percent change in lambda (on the y-axes) between modeled populations of unexposed ocean-type Chinook salmon and those exposed to a single, 4-day exposure per generation of a carbamate insecticide (e.g., carbaryl). The nine scenarios vary in the prey recovery and prey floor parameters (Table 3). Prey recovery rates were 0.5%, 1% and 5% per day, or slow, intermediate and fast, respectively. Prey floors were 0.05, 0.20 and 0.50, or low, intermediate and high, respectively. The vertical dotted line in each panel marks where the exposure concentration equals the EC_{50} of the prey (4.33 µg/L for carbaryl).

effects on salmon population growth rates. For instance after a single, 4-day pulse at a concentration equivalent of 1.15 times the prey EC_{50} (e.g. 5 µg/L carbaryl), model outputs ranged in mean %$\Delta\lambda$ from −1% to −13% (Figure 4) depending on the prey recovery rate and the prey floor.

Although the range of responses across these scenarios indicates both the prey recovery rate and the prey floor affect the %$\Delta\lambda$s, the prey recovery rate may be particularly important. The range of %$\Delta\lambda$s was especially wide across the prey recovery rates considered (0.5, 1 and 5%/day; Figure 4). The slow prey recovery rate resulted in declines in salmon population growth rates (%$\Delta\lambda$s of −9 when the exposure was 2 times the prey EC_{50}) even when as much as 50% of the prey were unaffected (i.e. prey floor was 0.5, Figure 4). In contrast, when the prey floor was at the lowest level considered (0.05) such that 95% of the available prey could be eliminated following an exposure, salmon population growth rates were only moderately reduced if the prey recovery rate was fast at 5%/day (%$\Delta\lambda$ of −4 when the exposure was 2 times the prey EC_{50}).

The inclusion of the prey spike caused interesting fluctuations in prey abundance immediately following exposure(s) (Figure 5), but this short-term effect had almost no lasting effect on salmon population growth rates. Two scenarios were run with and without spikes (a single 4-d pulse and 4, 4-d pulses, each with prey

recovery = 1% and prey floor = 0.20) at seven concentrations (ranging from 0.8 to 33.7 times the prey EC_{50}, or 3.5 to 145.8 µg/L carbaryl). For the single pulse exposure, the spike ameliorated the impact but only minimally and only at the highest concentration (e.g., %$\Delta\lambda$s = −11 and −12 at 145.8 µg/L carbaryl with and without spike, respectively). The inclusion of prey spikes following each of the 4, 4-day pulses in the second scenario also had a minimal but more frequent effect; the inclusion of spikes lessened the impact on λ slightly across all but one of the concentrations (difference between %$\Delta\lambda$s = 1). The apparent effect of the spike parameter was minimal compared to the prey recovery rate and prey floor parameters, largely because the spike only influences ration for 1 day per exposure pulse, out of the subyearling salmon's 140-day growth period.

The influence of the frequency, duration and timing of pesticide exposures on salmon population growth rates

Varying the duration, frequency and timing of the pesticide exposures affected the modeled impacts on salmon population growth rates (Figures 6 and 7). Longer or more frequent exposures resulted in greater reductions in prey abundance and subsequent reductions in salmon population growth rates. For example, increasing the duration of a single pulse increased the effect on

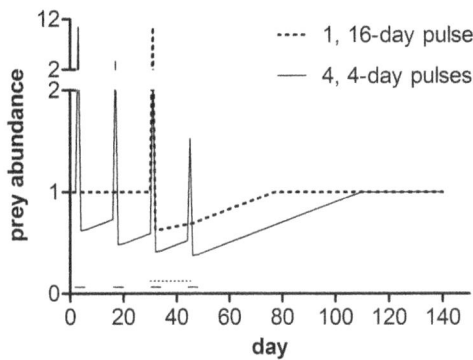

Figure 5. Effect of sustained or pulsed exposure on prey abundance. Prey available per day (relative to control) for juvenile ocean-type Chinook salmon in scenarios in which fish and prey were exposed to carbaryl at the prey EC_{50} (4.33 μg/L) for either 1, 16-day pulse starting at day 30 (dates noted by dotted line), or 4, 4-day pulses starting on days 2, 30, 58 and 86 (noted by solid lines). Prey floor and prey recovery rates were intermediate (0.2 and 1% per day, respectively) for both scenarios. Model output indicated %Δλs for 1, 16-day and 4, 4-day exposure scenarios were –3 (9.8) and –12 (8.9), respectively.

populations; exposing populations for 32 days instead of 4 days resulted in %Δλ –9 compared to %Δλ –6, respectively (at 5 μg/L carbaryl, with prey recovery = 1% and prey floor = 0.20, starting at day 30, Figure 6). Figures 6 and 7 illustrate how short frequent exposures (e.g., 4, 4-d pulses; 16 total days of exposure) had a greater effect on salmon population growth rates than long continuous exposures (e.g., 1, 32-d pulse; 32 total days of exposure), even when the actual duration of the exposure was longer in the latter. This occurs due to the additional time the prey community takes to recover from multiple exposures (Figure 6). For example, prey abundance was reduced to a greater extent for a longer portion of the growing period after four, 4-day pulses than after a single, 16-day pulse (Figure 5). The importance of prey recovery rate is especially clear when comparing salmon population-level impacts from single, long versus multiple, short exposure pulses (Figure 7).

We found the number of repeated exposures influenced salmon population growth rates more than the timing of the exposures. Figure 6 illustrates how the timing of exposures affected the length of time prey were reduced throughout the season. Although not shown, the model output also indicated prey abundance on any given day varied depending on the timing of exposures. While this may have implications for designing monitoring studies to assess prey abundance, for most scenarios the shift in timing of prey abundance had little overall effect on salmon population growth rates (Figure 6).

Discussion

Our analyses show current use pesticides at modeled exposure concentrations may affect salmon population growth rates both directly, though changing their feeding behavior, and indirectly, by reducing their prey. For some insecticides, reductions in prey alone are sufficient to reduce salmon population growth rates. The potential impact depends on the frequency and duration of the exposure concentration, as well as the dynamics of the prey community. Prey communities that are relatively tolerant and can recover quickly may be able to sustain juvenile salmon, while less resistant invertebrate communities that cannot rebound quickly

may not be productive enough to support robust salmon populations.

The class of insecticide and its relative toxicity to salmon vs. prey are key factors affecting the relative importance of direct vs. indirect effects on salmon population growth rates

The differences in salmon population growth rates following modeled exposures to various insecticides illustrate the critical need for understanding the diversity of possible toxicological effects that pesticides may have on a species and its habitat. These examples demonstrate how different classes of insecticides may have very different effects on various components of a food web, and these effects are not generally identified using standard toxicology tests (e.g., physiological responses and food-web mediated responses). The need for considering potential indirect effects as well sublethal direct effects was emphasized in the National Research Council review [28], as it recognized that not doing so may underestimate the true impacts of pesticides on ESA-listed species. While the importance of sublethal indirect effects on populations via trophic interactions has been well documented in field studies and mesocosm experiments [19,52–54], formally incorporating them into risk assessments is an important step forward in ecotoxicology [16,26,28].

Prey community resistance and recovery are key factors affecting susceptibility of salmon populations

While many researchers credit invertebrate communities with a capacity for "rapid" recovery, they often measure that in weeks to months. Our analyses indicate that even reductions in prey abundance for that "short" time can affect individual salmon and ultimately salmon populations, as they must feed throughout their freshwater residence. Resistance of the prey community is also important, but the greatest potential declines in λ occur when the prey community is slow to recover. This suggests indirect effects of pesticides on salmon are influenced by the prey community they are consuming. Prey recovery would be most rapid in communities with high connectivity, where colonizers can easily move in [55–57]. Typically these would be in less disturbed watersheds. However, even in less disturbed areas recovery may take weeks to months if habitats are relatively isolated, such as small tributaries and off-channel habitats, or if the community is dominated by univoltine taxa. Aquatic communities in small tributaries within an agricultural or urban watershed, for example, may have few if any sources of colonists and prey recovery may therefore be delayed because of a watershed's position within a landscape.

The frequency, duration & timing of exposures are important in determining effects on salmon population growth rates

Repeated exposures, even short in duration, produced particularly large reductions in salmon population growth rates. In addition, exposures that occur throughout the salmon growth period can have an equal or greater impact than isolated exposures early in the season. It may be possible to assess how frequently a habitat is exposed, but the long-term effects of multiple exposures on salmon also depend on the dynamics of the prey community (e.g., Figure 7). Salmon feeding in habitats with prey communities that are especially sensitive and slow to recover will be disproportionately impacted by repeated exposures.

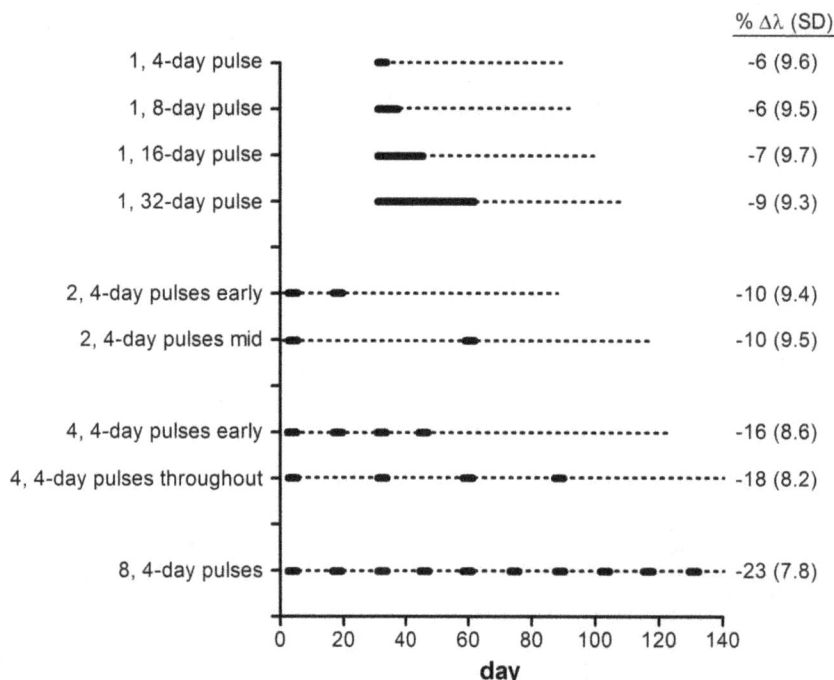

Figure 6. Effect of exposure timing and duration on salmon population growth rates and prey abundance. Timing and duration of exposures and results from 9 scenarios run with 1.15 times prey EC_{50} (5.0 µg/L) of carbaryl. Dark solid lines indicate the timing and duration of exposure(s). Dotted lines indicate period after each exposure in which prey abundance was reduced (ration <1). Values listed after scenarios are the %$\Delta\lambda$ (SD) for ocean-type Chinook salmon, for exposures of 5.0 µg/L. Prey recovery rate was 1% per day and the prey floor was 0.20.

Data gaps and limitations of the model

As with all models, our analyses are limited by the data available. Parameters that are included, such as the prey recovery rates and the floor, were based on relevant literature values (e.g., [21,41,43]) but were not necessarily derived from experiments with pesticides in salmon habitats. For instance, we assume prey communities have a consistent sensitivity and recovery rate, but it is unlikely that after recurrent pesticide exposure, invertebrate communities would be as resilient as they were when first disturbed [58,59]. Cuffney et al. [41] found that repeated seasonal pesticide pulses resulted in additional losses overall (effectively a lower prey floor with each pulse). Even though we assume all prey may be susceptible to repeated pulses, the model may underestimate effects on salmon population growth rates by not accounting for declines in resilience with repeated exposures. Likewise, macroinvertebrate toxicity values were primarily from 48 – 96-hour laboratory tests, and are not necessarily representative of field conditions. These represent data gaps that could be

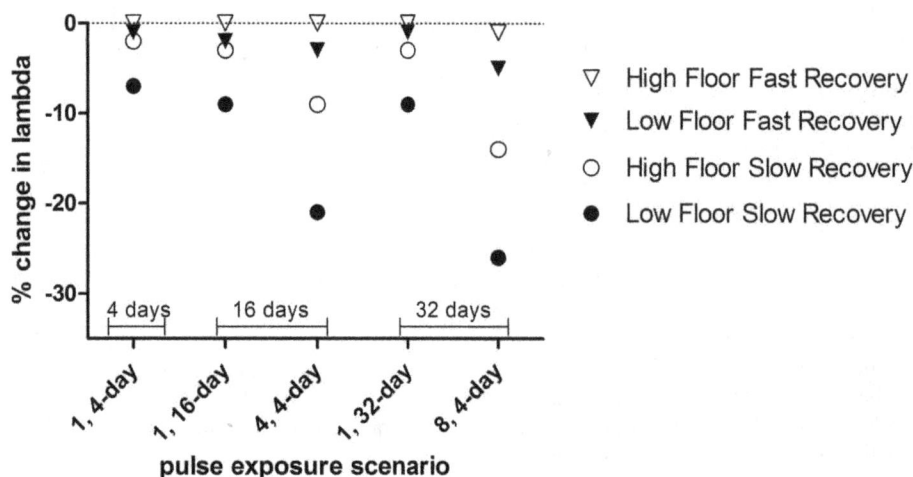

Figure 7. Effect of pulses, prey recovery and prey floor on salmon population growth rates. Mean percent change in lambda of ocean-type Chinook salmon following various exposures to carbaryl. All exposures were at the prey EC_{50} (4.33 µg/L), but the duration and frequency of the exposures varied as well as the floor and recovery rates for the prey. Low and high prey floors were 0.05 and 0.5; slow and fast prey recovery rates were 0.5% or 5% per day. The total days of exposure are noted with brackets.

addressed with targeted research on the sensitivity and dynamics of aquatic invertebrate communities following chemical exposures.

Lack of data also limited which relationships were included in the model. For instance, the model incorporates time explicitly, but how organisms respond in space is not defined. While we assume 100% of a salmon population and its critical habitat are exposed to a particular concentration, monitoring data suggest it is more likely that exposures are patchy across a landscape [1,60], with some fish and invertebrates affected more than others as they develop and move among habitats. Given the frequency and extent at which pesticides are detected, there may be few if any refuges in developed watersheds. Environmental data indicate that not only are pesticides detected at concentrations that exceed individual aquatic–life benchmarks [1], they are often detected multiple times throughout the year and in mixtures with other pesticides. For instance, Werner et al. [2] demonstrated how exposures can be recurrent yet patchy in their survey of 24 sites in the Sacramento and San Joaquin River Delta. They found nearly 10% of 400 water samples scattered across these sites over two years were toxic to the zooplankton *Ceriodaphnia dubia* [2], indicating a range of aquatic habitats may be exposed at some point in any one year. By not including spatial dynamics in exposure scenarios, the model may overestimate potential effects. Alternatively, assuming prey populations would be exposed to discrete exposures and then be able to recover may be unrealistic and cause the model to underestimate long-term effects.

Potential effects of pesticides on additional trophic levels and in combination with other stressors were also not included in the model due to lack of data. For instance, other than discrete pesticide exposures, the model does not include additional stressors (e.g., thermal stress, loss of shelter, additional pesticides, metals) that may additionally or synergistically impact fish and their prey [10,29,61,62]. Furthermore, the model does not include effects on other trophic levels (e.g., primary producers) or ecosystem processes (e.g., organic matter processing, [41]).

Future applications

Integrating potential direct and indirect effects of pesticide exposures in this model has improved our ability to distinguish and compare their relative importance under different environmental conditions. As additional data are available and relationships are better defined, this and similar population models evaluating effects of pesticides on ESA-listed species will be improved [25]. Model analyses are an important component of risk assessments (e.g., NOAA's Biological Opinions [17,18]), providing a transparent framework linking exposure and biological effects. Targeted pesticide monitoring and concurrent toxicity testing are critical for assessing risk of exposure and the toxicity of water and sediments [2,15,48,63], but we also need to better understand how exposures affect prey densities and salmon feeding behavior in the field. By formally including the indirect effects on salmon via their prey and varying the prey community's resistance and rate of recovery [26], the model helps determine which parameters influence salmon populations most and identifies empirical information needed to expand our understanding of community dynamics acting in these habitats.

Acknowledgments

We thank Scott Hecht and Tony Hawkes for critical discussions about model development and Paul van den Brink for generously sharing data regarding impacts and invertebrate community recovery rates. We thank Lyndal Johnson and Walt Dickhoff for constructive comments on the manuscript.

Author Contributions

Conceived and designed the experiments: KHM JAS DHB NLS. Performed the experiments: KHM JAS DHB. Analyzed the data: KHM JAS DHB. Contributed reagents/materials/analysis tools: KHM JAS DHB. Wrote the paper: KHM JAS DHB NLS.

References

1. Gilliom RJ (2007) Pesticides in U.S. streams and groundwater. Environmental Science and Technology 41: 3407–3413.
2. Werner I, Deanovic LA, Connor V, de Vlaming V, Bailey HC, et al. (2000) Insecticide-caused toxicity to *Ceriodaphnia dubia* (Cladocera) in the Sacramento-San Joaquin River Delta, California, USA. Environmental Toxicology and Chemistry 19: 215–227.
3. Morgan MJ, Kiceniuk JW (1990) Effect of fenitrothion on the foraging behavior of juvenile Atlantic salmon. Environmental Toxicology and Chemistry 9: 489–495.
4. Sandahl JF, Baldwin DH, Jenkins JJ, Scholz NL (2005) Comparative thresholds for acetylcholinesterase inhibition and behavioral impairment in coho salmon exposed to chlorpyrifos. Environmental Toxicology and Chemistry 24: 136–145.
5. Beauvais SL, Jones SB, Brewer SK, Little EE (2000) Physiological measures of neurotoxicity of diazinon and malathion to larval rainbow trout (*Oncorhynchus mykiss*) and their correlation with behavioral measures. Environmental Toxicology and Chemistry 19: 1875–1880.
6. Moore A, Waring CP (1996) Sublethal effects of the pesticide diazinon on olfactory function in mature male Atlantic salmon parr. Journal of Fish Biology 48: 758–775.
7. Scholz NL, Truelove NK, French BL, Berejikian BA, Quinn TP, et al. (2000) Diazinon disrupts antipredator and homing behaviors in chinook salmon (*Oncorhynchus tshawytscha*). Canadian Journal of Fisheries and Aquatic Sciences 57: 1911–1918.
8. Labenia JS, Baldwin DH, French BL, Davis JW, Scholz NL (2007) Behavioral impairment and increased predation mortality in cutthroat trout exposed to carbaryl. Marine Ecology Progress Series 329: 1–11.
9. Ferrari A, Venturino A, de D'Angelo AMP (2004) Time course of brain cholinesterase inhibition and recovery following acute and subacute azinphos-methyl, parathion and carbaryl exposure in the goldfish (*Carassius auratus*). Ecotoxicology and Environmental Safety 57: 420–425.
10. Laetz CA, Baldwin DH, Hebert V, Stark JD, Scholz NL (2013) Interactive neurobehavioral toxicity of diazinon, malathion, and ethoprop to juvenile coho salmon. Environmental Science and Technology 47: 2925–2931.
11. Baldwin DH, Spromberg JA, Collier TK, Scholz NL (2009) A fish of many scales: extrapolating sublethal pesticide exposures to the productivity of wild salmon populations. Ecological Applications 19: 2004–2015.
12. Higgs DA, Macdonald JS, Levings CD, Dosanjh BS (1995) Nutrition and feeding habits in relation to life history stage. In: C Groot, L Margolis and W. C Clarke, editors. Physiological Ecology of Pacific Salmon. Vancouver, British Columbia, Canada: University of British Columbia Press. pp. 159–315.
13. Beamish RJ, Mahnken C (2001) A critical size and period hypothesis to explain natural regulation of salmon abundance and the linkage to climate and climate change. Progress in Oceanography 49: 423–437.
14. Zabel RW, Achord S (2004) Relating size of juveniles to survival within and among populations of chinook salmon. Ecology 85: 795–806.
15. Anderson BS, Phillips BM, Hunt JW, Connor V, Richard N, et al. (2006) Identifying primary stressors impacting macroinvertebrates in the Salinas River (California, USA): Relative effects of pesticides and suspended particles. Environmental Pollution 141: 402–408.
16. Macneale KH, Kiffney PM, Scholz NL (2010) Pesticides, aquatic food webs, and the conservation of Pacific salmon. Frontiers in Ecology and the Environment 8: 475–482.
17. NMFS (2009) Endangered Species Act section 7 consultation: Environmental Protection Agency registration of pesticides containing carbaryl, carbofuran and methomyl. Biological opinion. Silver Spring, MD USA.
18. NMFS (2010) Endangered Species Act section 7 consultation biological opinion: Environmental Protection Agency registration of pesticides containing azinphos

methyl, bensulide, dimethoate, disulfoton, ethoprop, fenamiphos, naled, methamidophos, methidathion, methyl parathion, phorate and phosmet. Biological opinion. Silver Spring, MD USA.

19. Fleeger JW, Carman KR, Nisbet RM (2003) Indirect effects of contaminants in aquatic ecosystems. Science of the Total Environment 317: 207–233.

20. Schulz R (2004) Field studies on exposure, effects, and risk mitigation of aquatic nonpoint-source insecticide pollution: A review. Journal of Environmental Quality 33: 419–448.

21. Wallace JB, Lugthart GJ, Cuffney TF, Schurr GA (1989) The impact of repeated insecticidal treatments on drift and benthos of a headwater stream. Hydrobiologia 179: 135–147.

22. Kreutzweiser DP, Sibley PK (1991) Invertebrate drift in a headwater stream treated with permethrin. Archives of Environmental Contamination and Toxicology 20: 330–336.

23. Davies PE, Cook LSJ (1993) Catastrophic macroinvertebrate drift and sublethal effects on brown trout, *Salmo trutta*, caused by cypermethrin spraying on a Tasmanian stream. Aquatic Toxicology 27: 201–224.

24. Colville A, Jones P, Pablo F, Krassoi F, Hose G, et al. (2008) Effects of chlorpyrifos on macroinvertebrate communities in coastal stream mesocosms. Ecotoxicology 17: 173–180.

25. Kohler HR, Triebskorn R (2013) Wildlife ecotoxicology of pesticides: can we track effects to the population level and beyond? Science 341: 759–765.

26. Rohr JR, Kerby JL, Sih A (2006) Community ecology as a framework for predicting contaminant effects. Trends in Ecology and Evolution 21: 606–613.

27. Naiman RJ, Alldredge JR, Beauchamp DA, Bisson PA, Congleton J, et al. (2012) Developing a broader scientific foundation for river restoration: Columbia River food webs. Proceedings of the National Academy of Sciences of the United States of America 109: 21201–21207.

28. NRC (2013) Assessing Risks to Endangered and Threatened Species from Pesticides. Washington, D.C.: The National Academies Press. pp. 142.

29. Laetz CA, Baldwin DH, Collier TK, Hebert V, Stark JD, et al. (2009) The synergistic toxicity of pesticide mixtures: implications for risk assessment and the conservation of endangered Pacific salmon. Environmental Health Perspectives 117: 348–353.

30. Van den Brink PJ, Blake N, Brock TCM, Maltby L (2006) Predictive value of species sensitivity distributions for effects of herbicides in freshwater ecosystems. Human and Ecological Risk Assessment 12: 645–674.

31. Brett JR, Shelbourn JE, Shoop CT (1969) Growth rate and body composition of fingerling sockeye salmon, *Oncorhynchus nerka*, in relation to temperature and ration size. Journal of the Fisheries Research Board of Canada 26: 2363–2394.

32. Johnson LL, Ylitalo GM, Sloan CA, Anulacion BF, Kagley AN, et al. (2007) Persistent organic pollutants in outmigrant juvenile chinook salmon from the Lower Columbia Estuary, USA. Science of the Total Environment 374: 342–366.

33. Howell P, Jones K, Scarnecchia D, LaVoy L, Kendra W, et al. (1985) Stock assessment of Columbia River anadromous salmonids. Final Report to Bonneville Power Administration, Portland, Oregon.: Oregon Department of Fish and Wildlife.

34. Kostow K (1995) Biennial report on the status of wild fish in Oregon. Portland, Oregon, USA: Oregon Department of Fish and Wildlife.

35. Healey MC, Heard WR (1984) Inter- and intrapopulation variation in the fecundity of chinook salmon (*Oncorhynchus tshawytscha*) and its relevance to life history theory. Canadian Journal of Fisheries and Aquatic Sciences 41: 476–483.

36. Roni P, Quinn TP (1995) Geographic variation in size and age of North American chinook salmon. North American Journal of Fisheries Management 15: 325–345.

37. Ratner S, Lande R, Roper BB (1997) Population viability analysis of spring chinook salmon in the South Umpqua River, Oregon. Conservation Biology 11: 879–889.

38. PSCCTC (2002) Pacific Salmon Commission Joint Chinook Technical Committee report: annual exploitation rate analysis and model calibration. Vancouver, British Columbia, Canada.

39. Greene CM, Beechie TJ (2004) Consequences of potential density-dependent mechanisms on recovery of ocean-type chinook salmon (*Oncorhynchus tshawytscha*). Canadian Journal of Fisheries and Aquatic Sciences 61: 590–602.

40. Caswell H (2001) Matrix population models: construction, analysis, and interpretation. Sunderland, MA: Sinauer Associates. 722 p.

41. Cuffney TF, Wallace JB, Webster JR (1984) Pesticide manipulation of a headwater stream: invertebrate responses and their significance for ecosystem processes. Freshwater Invertebrate Biology 3: 153–171.

42. Liess M, Schulz R (1999) Linking insecticide contamination and population response in an agricultural stream. Environmental Toxicology and Chemistry 18: 1948–1955.

43. Van den Brink PJ, Van Wijngaarden RPA, Lucassen WGH, Brock TCM, Leeuwangh P (1996) Effects of the insecticide Dursban(R) 4E (active ingredient chlorpyrifos) in outdoor experimental ditches: II. Invertebrate community responses and recovery. Environmental Toxicology and Chemistry 15: 1143–1153.

44. Ward S, Arthington AH, Pusey BJ (1995) The effects of a chronic application of chlorpyrifos on the macroinvertabrate fauna in an outdoor artificial stream system - species responses. Ecotoxicology and Environmental Safety 30: 2–23.

45. Pusey BJ, Arthington AH, McLean J (1994) Effects of a Pulsed Application of Chlorpyrifos on Macroinvertebrate Communities in an Outdoor Artificial Stream System. Ecotoxicology and Environmental Safety 27: 221–250.

46. Heckmann LH, Friberg N (2005) Macroinvertebrate community response to pulse exposure with the insecticide lambda-cyhalothrin using in-stream mesocosms. Environmental Toxicology and Chemistry 24: 582–590.

47. Kreutzweiser DP (1990) Response of a brook trout (*Salvenius fontinalis*) population to a reduction in stream benthos following an insecticide treatment. Canadian Journal of Fisheries and Aquatic Sciences 47: 1387–1401.

48. Anderson BS, Phillips BA, Hunt JW, Worcester K, Adams M, et al. (2006) Evidence of pesticide impacts in the Santa Maria River watershed, California, USA. Environmental Toxicology and Chemistry 25: 1160–1170.

49. Lugthart GJ, Wallace JB, Huryn AD (1990) Secondary production of chironomid communities in insecticide-treated and untreated headwater streams. Freshwater Biology 24: 417–427.

50. Schulz R, Thiere G, Dabrowski JM (2002) A combined microcosm and field approach to evaluate the aquatic toxicity of azinphosmethyl to stream communities. Environmental Toxicology and Chemistry 21: 2172–2178.

51. Weatherley AH, H. S Gill. (1995) Growth. In: L. M. C Groot, and W. C Clarke, editor editors. Physiological ecology of Pacific salmon. Vancouver, British Columbia, Canada: University of British Columbia Press. pp. 103–158.

52. Brazner JC, Kline ER (1990) Effects of chlorpyrifos on the diet and growth of larval fathead minnows, *Pimephales promelas*, in littoral enclosures. Canadian Journal of Fisheries and Aquatic Sciences 47: 1157–1165.

53. Relyea RA, Diecks N (2008) An unforeseen chain of events: Lethal effects of pesticides on frogs at sublethal concentrations. Ecological Applications 18: 1728–1742.

54. Groner ML, Relyea RA (2011) A tale of two pesticides: how common insecticides affect aquatic communities. Freshwater Biology 56: 2391–2404.

55. Power M (1999) Recovery in aquatic ecosystems: an overview of knowledge and needs. Journal of Aquatic Ecosystem Stress and Recovery 6: 253–257.

56. Hutchens JJ, Chung K, Wallace JB (1998) Temporal variability of stream macroinvertebrate abundance and biomass following pesticide disturbance. Journal of the North American Benthological Society 17: 518–534.

57. Raven PJ, George JJ (1989) Recovery by riffle macroinvertebrates in a river after a major accidental spillage of chlorpyrifos. Environmental Pollution 59: 55–70.

58. Ashauer R, Boxall ABA, Brown CD (2007) Modeling combined effects of pulsed exposure to carbaryl and chlorpyrifos on *Gammarus pulex*. Environmental Science and Technology 41: 5535–5541.

59. Mohr S, Berghahn R, Schmiediche R, Hübner V, Loth S, et al. (2012) Macroinvertebrate community response to repeated short-term pulses of the insecticide imidacloprid. Aquatic Toxicology 110–111: 25–36.

60. Johnson HM, Domagalski JL, Saleh DK (2011) Trends in pesticide concentrations in streams of the western United States, 1993–2005. Journal of the American Water Resources Association 47: 265–286.

61. Chen XD, Stark JD (2010) Individual- and population-level toxicity of the insecticide, spirotetramat and the agricultural adjuvant, Destiny to the Cladoceran, *Ceriodaphnia dubia*. Ecotoxicology 19: 1124–1129.

62. Lydy MJ, Austin KR (2004) Toxicity assessment of pesticide mixtures typical of the Sacramento-San Joaquin Delta using *Chironomus tentans*. Archives of Environmental Contamination and Toxicology 48: 49–55.

63. Weston DP, Lydy MJ (2012) Stormwater input of pyrethroid insecticides to an urban river. Environmental Toxicology and Chemistry 31: 1579–1586.

64. Nelson TS, Ruggerone G, Kim H, Schaefer R, Boles M (2004) Juvenile Chinook migration, growth and habitat use in the Lower Green River, Duwamish River and nearshore of Elliott Bay 2001–2003, Draft Report. Seattle, WA.

65. Chambers JE, Boone JS, Carr RL, Chambers HW, Straus DL (2002) Biomarkers as predictors in health and ecological risk assessment. Human and Ecological Risk Assessment 8: 165–176.

4

Evaluation of the Distribution and Impacts of Parasites, Pathogens, and Pesticides on Honey Bee (*Apis mellifera*) Populations in East Africa

Elliud Muli[1,2,9], **Harland Patch**[3*,9], **Maryann Frazier**[3,9], **James Frazier**[3], **Baldwyn Torto**[1], **Tracey Baumgarten**[3], **Joseph Kilonzo**[1], **James Ng'ang'a Kimani**[1], **Fiona Mumoki**[1], **Daniel Masiga**[1], **James Tumlinson**[3], **Christina Grozinger**[3]

1 The International Centre of Insect Physiology and Ecology (icipe), Nairobi, Kenya, 2 Department of Biological Sciences, South Eastern Kenya University (SEKU), Kitui, Kenya, 3 Department of Entomology, Center for Pollinator Research, Pennsylvania State University, University Park, Pennsylvania, United States of America

Abstract

In East Africa, honey bees (*Apis mellifera*) provide critical pollination services and income for small-holder farmers and rural families. While honey bee populations in North America and Europe are in decline, little is known about the status of honey bee populations in Africa. We initiated a nationwide survey encompassing 24 locations across Kenya in 2010 to evaluate the numbers and sizes of honey bee colonies, assess the presence of parasites (*Varroa* mites and *Nosema* microsporidia) and viruses, identify and quantify pesticide contaminants in hives, and assay for levels of hygienic behavior. *Varroa* mites were present throughout Kenya, except in the remote north. Levels of *Varroa* were positively correlated with elevation, suggesting that environmental factors may play a role in honey bee host-parasite interactions. Levels of *Varroa* were negatively correlated with levels of hygienic behavior: however, while *Varroa* infestation dramatically reduces honey bee colony survival in the US and Europe, in Kenya *Varroa* presence alone does not appear to impact colony size. *Nosema apis* was found at three sites along the coast and one interior site. Only a small number of pesticides at low concentrations were found. Of the seven common US/European honey bee viruses, only three were identified but, like *Varroa*, were absent from northern Kenya. The number of viruses present was positively correlated with *Varroa* levels, but was not correlated with colony size or hygienic behavior. Our results suggest that *Varroa*, the three viruses, and *Nosema* have been relatively recently introduced into Kenya, but these factors do not yet appear to be impacting Kenyan bee populations. Thus chemical control for *Varroa* and *Nosema* are not necessary for Kenyan bees at this time. This study provides baseline data for future analyses of the possible mechanisms underlying resistance to and the long-term impacts of these factors on African bee populations.

Editor: Glenn Francis Browning, The University of Melbourne, Australia

Funding: This study was funded by an NSF-BREAD grant (0965441)(http://www.nsf.gov/bio/bread/index.jsp) to J. Tumlinson, M. Frazier, J. Frazier, C. Grozinger, D. Masiga, E. Muli, H. Patch. The funders had no role in study design, data collection and analysis, decision to publish, or preparation of the manuscript.

Competing Interests: The authors have declared that no competing interests exist.

* E-mail: hmpatch@psu.edu

9 These authors contributed equally to this work.

Introduction

Pollinators are essential contributors to global nutrition and food security. An estimated three-quarters of major global food crops benefit from pollinators [1]. Fruits, vegetables, and nuts, which provide key vitamins, minerals, fats and other micronutrients are particularly dependent on pollinators [2], and thus pollinators form a crucial line of defense against micronutrient deficiencies in developing countries. Furthermore, the productivity of many high value crops grown in the developing world, such as cacao, coffee, and cashews, is strongly tied to pollination services [3–6]. Indeed, the amount of animal pollinated crops grown globally has increased significantly in the last fifty years [7], making both developed- and developing world countries increasingly dependent on pollinator populations for food security and production of economically important crops.

Globally, pollination services amount to $212 billion, corresponding to ~9.5% of the total value of world agriculture production for human consumption in 2005 [3]. Honey bees (*Apis mellifera*) are one of the most important pollinators worldwide, contributing $14.6 billion in pollination services to the US in 2000 [8] and $3.2 billion to the South African economy in 1998 [9]. However, honey bee populations have been in decline in North America and Europe over the last ~30 years, with beekeepers routinely losing 30% of their managed colonies every winter during the last 7 years [10]. Several factors have been shown to negatively impact the longevity of honey bee colonies, including parasites (primarily *Varroa* mites [11] and *Nosema* microsporidia [12]), pathogens (22 different viruses have been identified [13,14], along with several bacterial and fungal brood pathogens [15,16]), pesticide exposure [17], poor nutrition [18], reduced genetic diversity [19], and management practices [20]. Large-scale surveys of managed honey bee populations in the US and Europe have failed to identify a single factor that is consistently strongly correlated with colony losses, leading researchers to believe a combination of factors acts synergistically to reduce survival [21–25].

In East Africa, honey bees provide critical pollination services, nutrition, and income for small-holder farmers and rural families. There is considerable genetic diversity in *Apis mellifera* populations in this region: indeed, five distinct *Apis mellifera* subspecies, each adapted to a specific ecological niche, have been identified in Kenya and in the surrounding region [26–29]. These bee populations are unmanaged: typically beekeepers set out empty receptacles (traditionally, hollowed-out logs), and bee swarms will occupy them as they migrate into the area [30,31]. In western Kenya, pollinators provide USD $3.2 million in ecosystems services to 8 crops (beans, cowpeas, butternuts, sunflower, monkeynut, tomatoes, capsicum and passion fruit, [32]). Furthermore, the honey collected from these colonies serves as an important source of nutrition and income for families. Currently, Kenya is a net importer of honey (over 10 metric tons in 2005, [33]), and thus honey and potentially beeswax production could be improved upon as a viable source of income for many rural communities.

Discussions with beekeepers in 2010, data from the Kenyan National Beekeeping Station, and the personal experience and observations of the Kenyan authors indicate that over the past five to seven years there has been a significant decline in the number of hives that are being colonized, reduction in the size of migratory swarms, and decrease in honey production [34]. Beekeepers noted that in the past, empty hives were colonized in a matter of weeks during the swarming season, whereas now it could take months and many hives remain uncolonized. Recently beekeepers complained about fist-size swarms that they chase out of hives knowing these will not result in productive colonies. According to nation-wide data collected by the National Beekeeping Station in 2005, hives in Kenya numbered 1,356,534 and average honey production was 20.28 kg/hive. In 2006 hives numbered 1,241,604 but honey production dropped to 15 kg/hive and in 2007, hive numbers increased to 1,575,978 but honey production dropped to 9.3 kg/colony [34]. This reduced performance may be disease related because in 2009, we identified *Varroa* mites in honey bee colonies in East Africa for the first time [35]. A previous survey conducted between 1996 and 1998 ([36]; Shi Wei, personal communication, 2011) did not detect *Varroa* and it thus appears to be a relatively recent introduction. The introduction of *Varroa* mites to South Africa in 1997 was associated with large losses of managed honey bee colonies [9]. Furthermore, *Varroa* mites have been shown to vector several honey bee viruses [37,38], which can also negatively impact honey bee health [13].

In 2010, we initiated a nationwide survey to obtain comprehensive information about the distribution of parasites, pathogens and pesticides in honey bee populations throughout Kenya, and determine if these are correlated with honey bee health, as measured in terms of colony size. In 24 locations across the country we assessed the numbers and sizes of honey bee colonies, the presence of *Varroa*, *Nosema*, and viruses, identified and quantified pesticide contaminants in hives, and assayed for levels of hygienic behavior (removal of dead pupae by adult bees, which is a measure of resistance to brood diseases and parasites such as *Varroa*) in colonies (see Figure 1 for apiary locations). We further analyzed this data set to determine if there were associations between *Varroa* loads, viral diversity, hygienic behavior, subspecies, location (which is correlated with elevation), and colony size. Given the relatively recent introduction of *Varroa* into this region, the large degree of genetic and ecological diversity, and the lack of

confounding factors introduced by intensive management practices, this survey provides an unprecedented opportunity to examine the factors affecting honey bee health, lays the groundwork for long-term monitoring of bee populations in this region, and provides important information for the conservation of populations of this key pollinator species in East Africa.

Materials and Methods

General survey information

We surveyed and collected samples from 24 apiaries, comprising 81 colonies total, across Kenya (Figure 1). Compiled detailed information and results related to each apiary and colony are available in Table S1 and summaries of the types of data obtained from each apiary are provided in Table S2. All samples were collected at maintained apiaries and consent was given by the collaborating authority. In all cases the owner, in the case of private land, or relevant authority, in the case of public land, gave permission for collections. See Form S1 for a list of the apiaries and owners. Apiaries 1–15 were surveyed in June 2010, and apiaries 16–24 were surveyed in July - September 2010. For each colony, foragers returning to the hive entrance with pollen were collected in RNAlater (Qiagen, Valencia, CA) or 95% ethanol. Abdomens of the RNAlater-stored bees were pierced with sterile scalpel blades to expose soft tissues to the RNAlater preservative. These samples were used for viral detection. Bees collected in EtOH were used for *Nosema* detection and for subspecies identification. Bees were collected into individual 2 ml cryogenic vials (VWR, Radnor, PA). Samples were collected on ice, stored at −20°C during field collections, and then shipped to Penn State University (University Park, PA) within a month. At Penn State, RNAlater samples were stored at −80°C and all EtOH samples were stored at 4°C. When brood nests were accessible (for example, the interiors of colonies housed in traditional log hives could not be sampled without destroying the colony) wax and stored pollen samples were collected into sterile Whirl-Pak bags (Nasco, Fort Atkinson, WI) for pesticide analysis. When possible, colonies and apiaries were surveyed for population health (percentage of occupied hives in an apiary), colony size (number of frames of adult and immature bees, hygienic behavior (removal of dead pupae), and numbers of *Varroa* mites. Since it was not possible to obtain information on all parameters from each colony/apiary in the survey, it was necessary to use different subsets of colonies/apiaries to examine and statistically analyze correlations among these parameters. Details of the specific assays, collections, and associated statistical analyses are below.

Survey of *Varroa* mites

The presence and quantity of *Varroa* mites were assessed using a standard sugar roll assay described in [39], using a half-cup measuring cup to collect approximately 350 bees. This assay was performed on colonies from 19 apiaries; apiaries 2, 16, 19, 22 and 24 were not assessed.

Colonies at twelve apiaries (sites 1, 3, 4, 10, 11, 12, 13, 14, 15, 20, 21, 23) were used to examine the correlation between *Varroa* presence and elevation. These apiaries were selected because *Varroa* had been sampled and was present (note that *Varroa* was absent from apiaries 17 and 18, but this was likely because these regions are geographically quite distant and isolated and *Varroa* may not yet have been introduced), and the majority had five sampled colonies/apiary (sites 10, 14 and 15 had 3, 2, and 4 colonies, respectively). In order to obtain a normal distribution of the data, *Varroa* counts were converted to logarithmic scale using the following equation: log number of *Varroa* = $\log_{10}$ (*Varroa* count

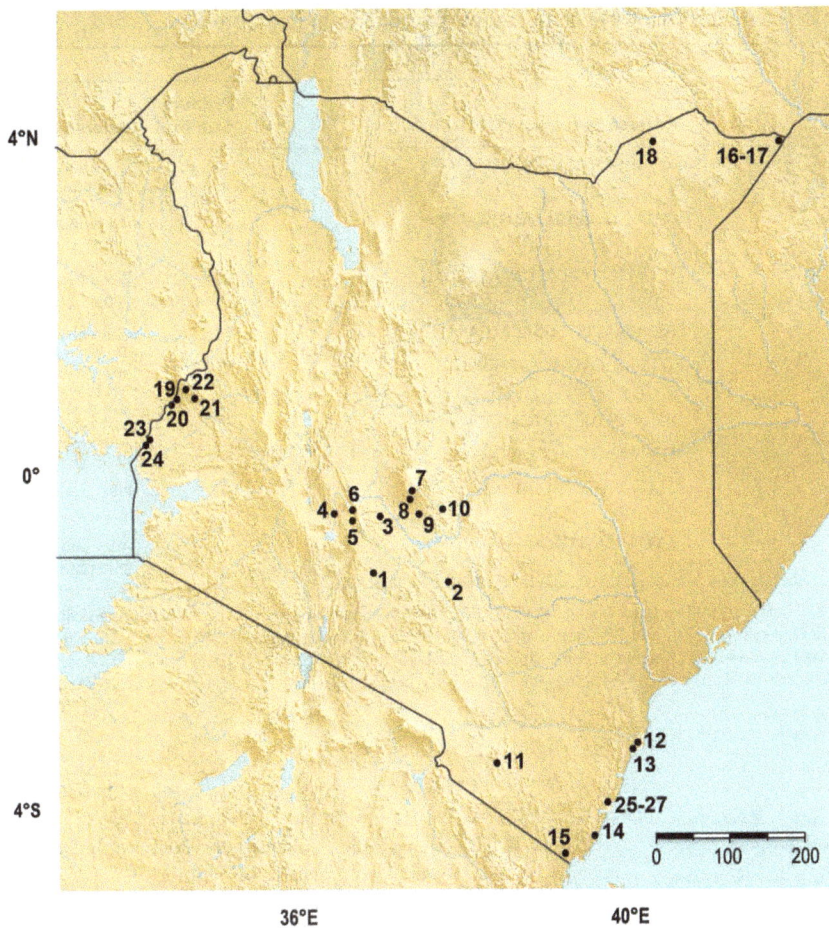

Figure 1. Geographic location of surveyed apiaries. Twenty-four apiaries were surveyed throughout Kenya with an additional three apiaries (25–27), see supplemenatry material, surveyed for ecological effects on colony health. The location and numerical designation of the apiaries is indicated on the map.

+1). All subsequent statistical analyses for *Varroa* counts were performed using logarithmic scale. The log number of *Varroa* in each colony was correlated with elevation using a correlation analysis in JMP 9.0.2 (SAS, Cary, NC).

Eleven apiaries (sites 1, 3, 4, 10, 11, 12, 13, 14, 15, 20, 21) were used to examine the correlation between *Varroa* presence and colony size. Colony size was measured as the number of frames of bees. The log number of *Varroa*/colony was correlated with the number of frames of bees/colony using a correlation analysis in JMP 9.0.2 (SAS, Cary, NC).

Survey of *Nosema*

Sixteen apiaries (sites 1, 3, 4, 10, 11, 12, 13, 14, 15, 16, 17, 18, 20, 21, 22, 23) were screened for the presence of *Nosema* microsporidia. DNA was extracted from pools of 5 foragers/colony (collected in 95% ethanol) using a CTAB buffer (100 mM Tris HCl, pH 8.0; 20 mM EDTA, pH 8.0; 1.4 M NaCl; 2% (w/v) cetyltrimethylammonium bromide; 0.2% (v/v) 2-mercaptoethanol) plus proteinase K overnight incubation at 55°C followed by a phenol/chloroform/isoamyl alcohol (25:24:1) extraction. The species of *Nosema* present in the samples was confirmed using a PCR-RFLP of partial small subunit (SSU) rRNA gene as in [40], see Table 1 for primer sequences. The SSU fragment was amplified in a 25 µl PCR reaction containing $1 \times$ PCR Buffer, 2.5 mM $MgCL_2$, 200 µM of each dNTP, 0.5 µM of forward and

reverse primer, 2.5 units of platinum Taq polymerase (Invitrogen, Grand Island, NY), and 1000 ng of template DNA. The PCR conditions were as follows: 95°C for 4 minutes, followed by 45 cycles of 95°C (60 s), 48°C (60 s), and 72°C (60 s), and a final extension step at 72°C for 4 minutes. The PCR products were separated on a 1% agarose gel and visualized with ethidium bromide. The 400 bp PCR amplicon was subjected to two double digest RFLP reactions with the restriction enzymes MspI and either NdeI or PacI (New England Biolabs, Ipswich, MA) at 37°C for 3 hours. The resulting fragments were separated on a 2% agarose gel and visualized with ethidium bromide to determine their sizes. Furthermore, the 400 bp fragments were also gel purified and extracted with the Qiaquick gel extraction kit (Qiagen, Valencia, CA) and sequenced at the Penn State Genomics Core Facility (University Park, PA). In order to increase our ability to detect *Nosema* infections, we monitored infections in foragers (levels and prevalence of *Nosema* are highest in foragers) and used a molecular approach, which is not only more sensitive than screening for spores using light microscopy, but can also detect the vegetative forms of *Nosema* [41,65]. While several studies have successfully used molecular detection of *Nosema* in pools of 5 bees/colony for large-scale colony screening of *Nosema* infections [40,42], larger pools (25–30 bees) or repeated measurements may have increased the sensitivity of this screen at the individual colony level, though likely not at the apiary level.

Table 1. Primers used for molecular analysis for identification of bee populations, pathogens and parasites.

Primer	Forward Sequence (5'-3')	Reverse Sequence (5'-3')	Product Size (bp)	Reference
ABPV	TTATGTGTCCAGAGACTGTATCCA	GCTCCTATTGCTCGGTTTTTCGGT	900	Benjeddou et al. 2001
BQCV	TGGTCAGCTCCCACTACCTTAAAC	GCAACAAGAAGAAACGTAAACCAC	700	Benjeddou et al. 2001
CBPV	AGTTGTCATGGTTAACAGGATACGAG	TCTAATCTTAGCACGAAAGCCGAG	455	Ribiere et al. 2002
DWV	ATCAGCGCTTAGTGGAGGAA	TCGACAATTTTCGGACATCA	701	Chen et. al. 2005
IAPV	GCGGAGAATATAAGGCTCAG	CTTGCAAGATAAGAAAGGGGG	586	Di Prisco et. al. 2011
KBV	GATGAACGTCGACCTATTGA	TGTGGGTTGGCTATGAGTCA	417	Stoltz et al 1995
SBV	GCTGAGGTAGGATCTTTGCGT	TCATCATCTTCACCATCCGA	824	Chen et. al. 2005
Nosema SSU	GCCTGACGTAGACGCTATTC	GTATTACCGCGGCTGCTGG	400	Klee, Besana et. al. 2007
Mitochon. Markers	TGATAAAAGAAATATTTTGA	GAATCTAATTAATAAAAAA	688	Arias and Sheppard 1996

Abbreviations: Israeli acute paralysis virus (IAPV), acute bee paralysis virus (ABPV), black queen cell virus, (BQCV), chronic bee paralysis virus (CBPV), deformed wing virus (DWV), kashmir bee virus (KBV), and sacbrood virus (SBV). References: Arias MC and WS Sheppard WS (1996) *Molecular Phylogenetics and Evolution* 5: 557–566; Benjeddou et al. (2001) *Applied and Environmental Microbiology* 67:2384–2387; Chen et al. (2005) *Applied and Environmental Microbiology* 71(1):436–441; Di Prisco et. al. (2011) *Journal of General Virology* 92: 151–15; Klee et al. (2007). Journal of Invertabrate Pathology 96: 1–10. Ribiere et al. (2002) *Apidologie* 33: 339–351; Stoltz et al. (1995) *Journal of Apicultural Research* 34: 153–160.

Survey of Viruses

Pools of 5 foragers/colony were collected into RNAlater (Qiagen, Valencia, CA); all 81 colonies in the survey were assayed. RNA was extracted from these pooled samples using Tri-Reagent (Sigma-Aldrich, St. Louis, MO). cDNA was synthesized using 150 ng of RNA for each pooled samples. The presence of seven common honey bee viruses was identified using PCR with previously published primers specific for each virus [37,43–46]. See Table 1 for a listing of the viruses, primers used, and references for the primers. PCR conditions were as follows: 2 minutes at 95°C, followed by forty cycles of 95°C (30 s), 55°C (60 s), and 68°C (120 s) and a final extension at 68°C for 7 minutes. PCR products were separated on a 1% agarose gel and visualized with ethidium bromide. Bees collected from apiaries at Penn State were used as positive controls. A previous large-scale survey of viral infection dynamics in colonies found that molecular analysis of pools of five foragers/colony provided equivalent detection sensitivities as individual analyses of 10–15 bees per colony [14]; however, as in the case of the survey for *Nosema* above, larger pools (25–30 bees) or repeated measurements may have increased the sensitivity of this screen at the individual colony level, though likely not at the apiary level.

Associations between viral diversity (the number of viruses present in a colony) and colony size (measured by the number of frames of adult bees) were assessed in all colonies where colony size measurements were available. Fifteen apiaries (sites 1, 2, 3, 4, 5, 6, 9, 10, 11, 12, 13, 14, 15, 20, and 21) with 58 colonies included in the analysis. A Kruskal-Wallis test was used to determine if there were significant differences in colony size among colonies with 0, 1 or 2 viruses using JMP 9.0.2 (SAS, Cary, NC).

Associations between viral diversity and *Varroa* levels were measured in the 19 apiaries (see above) in which *Varroa* measurements were available; 66 colonies were used. A Kruskal-Wallis test was used to determine if there were significant differences in *Varroa* loads among colonies with 0, 1, or 2 viruses,

followed by a pairwise comparisons using nonparametric Wilcoxon pairwise tests, using JMP 9.0.2 (SAS, Cary, NC).

Because it was not possible to determine which variable (viral diversity versus number of frames or number of *Varroa*) is dependent and which is independent, we also performed a correlation analysis in JMP 9.0.2 (SAS, Cary, NC). To obtain a normal distribution of the data, viral diversity were converted to logarithmic scale using the following equation: log number of viruses $= \log_{10}$ (number of viruses $+1$).

Survey of Hygienic Behavior

When possible, a colony's hygienic behavior (the removal of freeze-killed pupae) was assessed as in [47]. A 3-inch (7.62 cm) diameter PVC cylinder was pressed into a frame of capped brood containing purple-eyed pupae; this area corresponds to approximately 207 cells of naturally drawn African honey bee comb. The percent hygienic behavior for a colony was calculated by taking the final number of fully and partially removed pupae/(207 − number originally uncapped or empty cells) *100 (Table S1).

Hygienic behavior was assessed for 10 apiaries, at sites 1, 3, 4, 11, 12, 13, 14, 15, 20, and 21. In total, 36 colonies were assayed. Associations between hygienic behavior and elevation, colony size, and log *Varroa* counts were determined using a correlation analysis with JMP 9.0.2 (SAS, Cary, NC). A Kruskal-Wallis test was used to determine if there were differences in levels of hygienic behavior among colonies with 0, 1 or 2 viruses, followed by nonparametric Wilcoxon pairwise tests. A correlation analysis was also performed between the log number of viruses and hygienic behavior.

Subspecies Identification

Heads of foragers were dissected and homogenized with a Fastprep instrument (Thermo Fisher, Waltham, MA) for three cycles at maximum time and speed. DNA was extracted using the DNeasy Blood and Tissue Kit (Qiagen, Valencia, CA) according to manufacturer's instructions. Mitochondrial DNA including the tRNA ILE and part of the ND2 gene were amplified as described in [48], see Table 1 for primer sequences. PCR was performed on a Mastercycler Pro (Eppendorf, Hauppauge, NY) using 25 µl reactions consisting of 2.5 units of platinum Taq DNA Polymerase, PCR buffer minus magnesium at a concentration of 1X, 0.2 mM dNTP mix, 1.25 mM $MgCl_2$, 5% DMSO, 0.2 µM primers and 10 ng of extracted DNA; reagents were purchased from Invitrogen (Carlsbad, CA). The PCR was carried out using the thermal profile of 1 minute at 94°C, followed by 40 cycles of 94°C (40 s), 42°C (80 s) and 62°C (120 s) and a final extension at 72°C for 4 minutes. No-template controls were performed with each PCR run. Products were visualized on 1.0% agarose gels, excised and extracted with the QIAquick PCR Purification Kit (Qiagen, Valencia, CA) and submitted for sequencing. Sequencing was preformed at the Genomics Core Facility at Pennsylvania State University. Sequences were aligned using ClustalW [49] package in BioEdit 7.6 [47]. There were 24 unique haplotypes from the 109 individuals sequenced (see Table S3 for the haplotype designation for each individual). These 24 unique haplotypes were used to construct a neighbor-joining tree [50] along with haplotypes corresponding to the subspecies described from [48](see Figure S1 for the tree). The ND2 region variable sites are described in Table S4.

Survey of Pesticides

When possible, brood nest wax and bee bread (stored, fermented pollen mixed with nectar) were collected for pesticide analysis. Approximately 3–10 grams of each matrix was collected into individual sterile, 50 ml centrifuge tubes for each colony. Samples were shipped to Penn State University within three weeks of collection and stored at −80°C. Pesticide analysis was performed on wax (45 colonies) and pollen (25 colonies) samples obtained from 15 (sites 1–15) and 13 (sites 1, 2, 4, 5, 6, 8, 9, 10, 11, 12, 13, 14, 15) of the apiaries respectively. One to five wax or pollen samples were pooled by site to provide a single wax and single pollen sample per site for analysis, averaging 9.45 and 7.1 grams/pooled sample respectively. Samples were shipped to the USDA-AMS-NSL lab in Gastonia, NC for extraction and were screened for the presence of 171 pesticides and toxic metabolites by LC-MS-MS and GC-MS according to methods described in [51].

Results

Varroa

Of the 19 apiaries assayed for *Varroa* mites, 17 (89.5%) had mites present (Table S1). The two assayed apiaries (sites 17 and 18) that did not have *Varroa* were in far northeastern corner of Kenya near the border with Somalia and Ethiopia. In total, 66 colonies were assayed for *Varroa*, and *Varroa* was found in 55 (83%) of them. The levels of *Varroa* were highly variable across colonies and apiaries. Twenty-four (corresponding to 36%) colonies had 5 or fewer mites in samples of ~350 bees, 16 colonies had 6–10 mites, 11 colonies had 11–20 mites, 6 colonies had 21–30 mites, while 9 colonies had more than 30 mites. In the US, it is recommended that colonies are treated for mites when 5–20 mites are found in samples of ~300 bees in the fall [39,52]. *Varroa* levels were positively correlated with elevation ($r(53) = 0.44$, $p = 0.001$; Figure 2A) and with colony size ($r(48) = 0.35$, $p = 0.013$, Figure 2B). There was no correlation between colony size and elevation ($r(48) = −0.02$, $p = 0.87$, data not shown)

Since *Varroa* levels can change seasonally, we repeated this analysis using only colonies sampled in June 2010 (from apiaries 1, 3, 4, 10, 11, 12, 13, 14, 15,). Again, colonies at higher elevations had significantly higher *Varroa* loads ($r(38) = 0.39$, $p = 0.014$; data

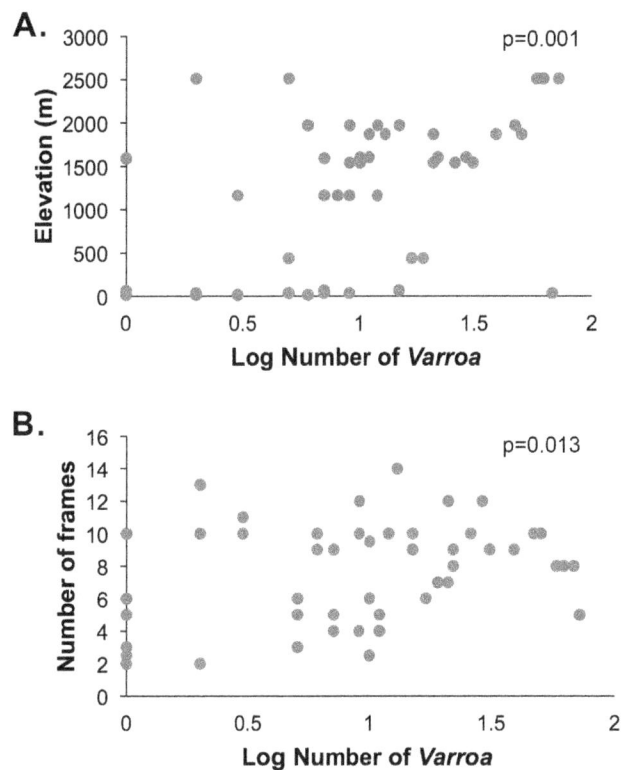

Figure 2. Association of *Varroa* infestation with elevation and colony size. A. Levels of *Varroa* mites were positively correlated with elevation, with colonies at higher elevations having significantly higher average numbers of *Varroa* ($r(53) = 0.44$, $p = 0.001$). **B.** Levels were also positively correlated with colony size (($48) = 0.35$, $p = 0.013$). *Varroa* counts were converted to logarithmic scale.

not shown), and there was a trend for a positive correlation between *Varroa* levels and colony size ($r(38) = 0.31$, $p = 0.054$, data not shown).

Nosema

Sixteen apiaries were assessed for the presence of *Nosema ceranae* and *Nosema apis*. *Nosema* was identified in 4/5 colonies at site 12, in 5/5 colonies at site 13, in 4/5 colonies at site 15, and in 2/3 colonies at site 22 (Table S1). An RFLP approach was used to determine the presence and subspecies of *Nosema* [40]. Interestingly, the fragmentation pattern was not consistent with either species. The 400 bp 16S rRNA gene amplicon from these samples showed an alternate cleavage pattern when double digested with restriction enzymes *Msp*I and *Nde*I. A digestion pattern of 3 fragments (one at 225 bp, one at 100 bp and one at 75 bp) was observed. The predicted digestion pattern for *Nosema apis* of three fragments (at 175, 136, and 91 bp) was observed for the *N. apis* control. The predicted digestion pattern for *Nosema ceranae* of 2 fragments (175 and 225 bp) was observed for the *N. ceranae* control. Sequencing of the 400 bp *Nosema* 16S rRNA gene region used for the RFLP analysis revealed two recombination events. An inversion of TAC from CAT at position 151 removed the predicted *Nde*I cleavage site from the reference and the *N. apis* control sequences. An insertion of a thymine into the sequence CATAG produced an alternate *Nde*I cleavage site of CATATG at position 341. The *Msp*I enzyme cleaved the sequence CCGG as expected at position 242. The amplicons were not cleaved by the enzyme *PacI* which exploits a unique digestion site for *Nosema ceranae*.

Viruses

All colonies in all 24 apiaries were assessed for the presence of seven viruses commonly found in honey bees in North America and Europe: Israeli acute paralysis virus (IAPV), acute bee paralysis virus (ABPV), black queen cell virus, (BQCV), chronic bee paralysis virus (CBPV), Deformed wing virus (DWV), Kashmir bee virus (KBV), and sacbrood virus (SBV). Only DWV, BQCV, and ABPV were detected in Kenyan bee populations. Viruses were found in 20 apiaries; no viruses were detected in apiaries at sites 15, 16, 17, and 18 (Table S1). Sites 16, 17, and 18 are in northeastern Kenya (which were also free of *Varroa*), while site 15 is near the southeastern coast. DWV was found at 12 sites in 36% (29 out of 81) of the colonies, BQCV was found at 18 sites in 60% (49 out of 81) of the colonies, and ABPV was found in one colony at Site 13 (this colony also had DWV and *Nosema* infections).

Average colony size (based on the number of frames of bees) was not associated with viral diversity (the number of viruses in a colony; Kruskal-Wallis: $H(2) = 2.74$, $p = 0.254$, Figure 3A). A correlational analysis between colony size and viral diversity was also not significant ($r(57) = 0.20$, $p = 0.128$). However, there was a significant correlation between viral diversity and *Varroa* loads (Kruskal-Wallis: $H(2) = 13.10$; $p = 0.001$); colonies with 1 or 2 viruses had significantly higher *Varroa* loads than colonies that had no viruses ($p < 0.05$, Wilcoxon pairwise tests, Figure 3B). Similarly, a correlational analysis between *Varroa* loads and viral diversity was significant ($r(65) = 0.49$, $p < 0.001$). Eight out of 66 colonies with *Varroa* infestations had no viruses, while only 2 colonies had one virus and no detectable *Varroa*.

Since levels of viruses can change over time [14], we repeated this analysis using colonies sampled only in June 2010 and obtained similar results. There was no correlation between viral diversity and colony size ($r(47) = 0.11$, $p = 0.456$, data not shown), but there was a significant positive correlation between *Varroa* levels and viral diversity ($r(47) = 0.33$, $p = 0.023$, data not shown).

Hygienic behavior

Hygienic behavior was assessed in 36 colonies from 10 sites (Table S1). There were no significant correlations between hygienic behavior and elevation or colony size ($r(35) = -0.18$, $p = 0.289$, Figure 4A and $r(35) = -0.22$, $p = 0.196$, Figure 4B, respectively). There was, however, a significant negative correlation between hygienic behavior and the numbers of *Varroa* in these colonies ($r(35) = -0.42$, $p = 0.011$, Figure 4C). Hygienic behavior was also significantly associated with viral diversity (Kruskal-Wallis: $H(2, 34) = 6.43$, $p = 0.040$); levels were lower in colonies with 1 type of virus versus colonies with 0 or 2 viruses ($p < 0.05$, Wilcoxon pairwise tests, Figure 4D). Note the correlational analysis of viral diversity and hygienic behavior was not significant in this case ($r(35) = -0.0277$, $p = 0.90$). Similar results were obtained if only colonies sampled in June 2010 were used (elevation: $r(25) = -0.31$, $p = 0.122$; number of *Varroa*: $r(25) = -0.63$, $p = 0.0005$; viral diversity: $H(2) = 5.55$, $p = 0.076$, $r(25) = -0.21$, $p = 0.292$; data not shown), though in this case the negative correlation between hygienic behavior and colony size was significant ($r(25) = -0.46$, $p = 0.019$).

Pesticides

Pesticide analysis was performed on pool wax samples from 15 sites (1 sample per site) and pooled bee bread samples from 13 sites. Samples were screened for the presence of 171 pesticides and toxic metabolites. Only 4 pesticides, 1-naphthol, chlorothalonil,

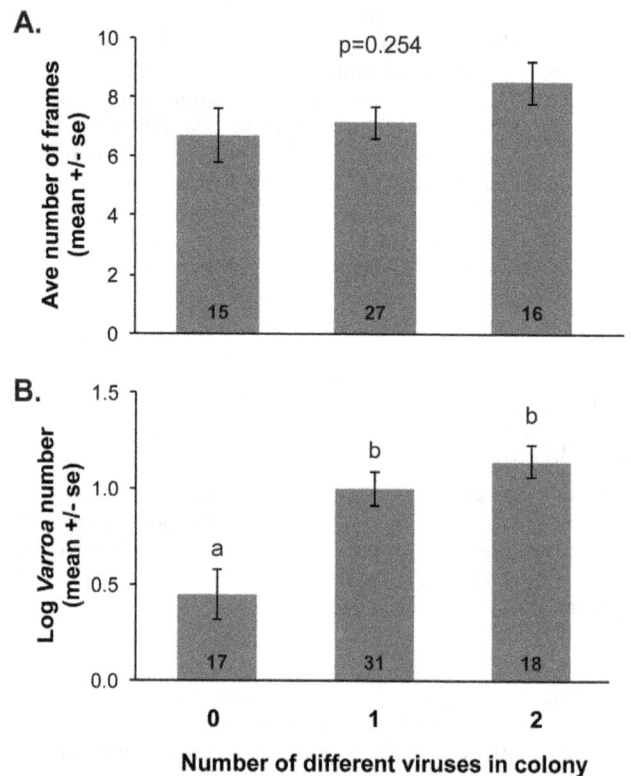

Figure 3. Association of viral diversity with colony size and *Varroa*. A. Colony size (the number of frames of bees) was not affected by viral diversity (the number of viruses in a colony), $H(2) = 2.74$, $p = 0.254$. **B.** However, colonies with different number of viruses had significantly different numbers of *Varroa* ($H(2) = 13.10$; $p = 0.0014$). Colonies with 1 or 2 viruses had significantly higher *Varroa* loads than colonies that had no viruses ($p < 0.05$, Wilcoxon pairwise tests, different letters denote significant differences). The number of colonies in each group is indicated at the bottom of each bar. *Varroa* counts were converted to logarithmic scale.

chlorpyrifos and fluvalinate, were identified and were present at very low levels (below 50 ppb) with the exception of 1-napthol which was found at only one location (site 9, Nji-ini Forest) at 116 ppb (Table S1). However one or more of these pesticides were detected at 14 of the 15 sites sampled (data not shown).

Subspecies

Our attempt to identify subspecies of *A. mellifera* in Kenya was based on the work of Arias and Sheppard (1996) (see Table S1 for haplotype information and Figure S1 for the neighbor-joining tree. Accession Numbers KJ628964-KJ628987). Representative haplotype 3.4.11 is identical in sequence to the previously described *A. m. monticola* (Figure S1). Our analysis suggests *A. m. monticola* is present at high altitude on Mount Elgon as previously suggested by Ruttner [27,53] with some mixing lower on the mountain at Chepkui, but ND2 *monticola* is also present in the Aberdares at Marchorwe. However it is important to note that other bees from all three sites exhibit *A. m. scutellata* haplotypes. The *monticola* clade forms a sister clade to a *scutellata* group (Group A) that includes 15 apiaries from across the Kenya, excluding the northern region and the southeast. This ND2 region is identical to that from honey bees from South Africa (SCUTE2) and *A. m. adansonii* (ADANS1) from Nigeria, and similar to Africanized bees from Brazil (BRASI1 and BRASI2). A distinct clade of *scutellata* honey bees (Group B) with

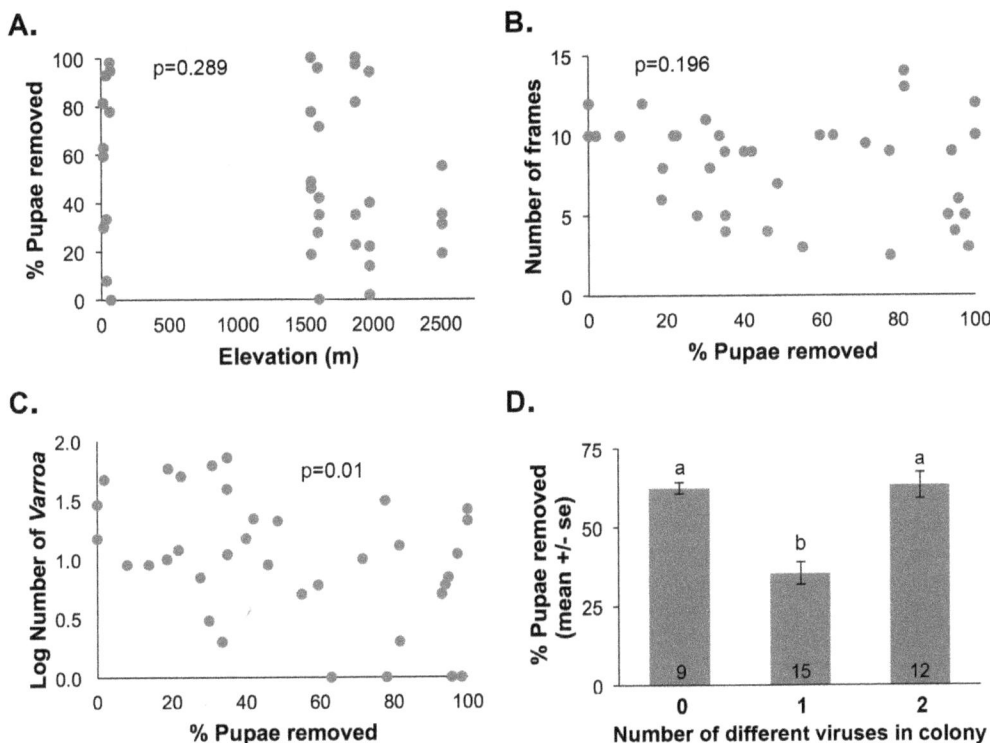

Figure 4. Association of hygienic behavior with colony location, size, parasite and pathogen loads. A. There were no significant correlations between hygienic behavior and elevation (r(35) = −0.18, p = 0.289). **B.** Hygienic behavior was not correlated with colony size (r(35) = −0.22, p = 0.196). **C.** There was, however, a significant negative correlation between hygienic behavior and the numbers of *Varroa* in these colonies (r(35) = −0.42, p = 0.011; *Varroa* counts were converted to logarithmic scale). D. Hygienic behavior was also significantly associated with viral diversity (H(2, 34) = 6.43, p = 0.040); levels were lower in colonies with 1 type of virus versus colonies with 0 or 2 viruses (p<0.05, Wilcoxon pairwise tests, different letter denote significant differences). The number of colonies in each group is indicated at the bottom of each bar.

the dominant haplotype of 2.4.11 (identical to SCUTE1 from Kenya) is found predominantly in the central and eastern portion of the country. The individuals in the haplotype group identified by 12.1.12 are from the coast and the far north of Kenya (identical ND2 regions are found at sites 12, 13, 16, 17). Apiary 12 is in the Arabuko Sokoke Forest Reserve, a remnant of the coastal tropical forest that supports a number of unique endemic species. Although no molecular comparison is available from other studies, we speculate that this could be *A. m. litorea* given the close relationship with *A. m. lamarckii* (LAMARC) and the honey bees in the northern apiaries at Mandera [27]. Individuals from one colony at apiary 18, called Mandera West, were distinct from the other Kenyan bees, although another colony at the same apiary groups with *scutellata* A. These could represent either *A. m. yemenitica* or the recently described *A. m. simensis* [27,28], neither of which have been described with molecular makers. There were 14 unique individual haplotypes whereas many apiaries had multiple ND2 haplotypes (Table S3). We found no correlation between genotype, as measured the ND2 region, and honey bee health although more data will be require to clearly distinguish populations or subspecies.

Discussion

Our survey suggests that several new parasites and pathogens (*Varroa*, *Nosema*, DWV, BQCV, and ABPV) have recently invaded honey bee populations in East Africa (see below for further discussion). However, none of these factors are correlated with colony size, implying that several factors thought to critically

undermine bee populations in the US and Europe (*Varroa*, *Nosema*, and pesticide use) are not yet directly impacting Kenyan bee populations in terms of this metric. However, since there may be a time-lag before newly introduced parasites and pathogens cause substantial negative effects, continuous monitoring of these populations should be conducted to evaluate the long-term dynamics of these host-pathogen interactions. Our data also suggests that chemical control methods for *Varroa* and *Nosema*, which are heavily used by US beekeepers, may be unnecessary at this time for Kenyan honey bees, and indeed, these bees may possess novel resistance mechanisms. Furthermore, we found an intriguing correlation between elevation and *Varroa* levels, suggesting that environmental factors (climate, landscape ecology) may play a key role in mediating this host-parasite interaction, and perhaps honey bee health in general, though the effect of these environmental factors needs to be explored in greater detail in larger scale studies. We did not find any association with subspecies genotype and honey bee health. However, based on the molecular marker that we used (the ND2 region of the mitochondria) there was considerable gene flow between populations, and thus more sensitive markers may be needed to fully characterize subspecies and population differences.

Varroa alone does not appear to strongly impact honey bee colonies in Kenya. Based on recommendations for US beekeepers, approximately 2/3 of the surveyed colonies had *Varroa* levels high enough to warrant treatment to control mite populations [39,52]. However, *Varroa* levels in Kenyan colonies are not correlated with decreases in colony strength; in fact, there is a positive correlation between *Varroa* numbers and colony size. Colonies with higher levels of hygienic behavior did have lower levels of *Varroa*, as expected [54], but hygienic behavior was quite variable across the surveyed colonies. *Varroa* levels were strongly positively correlated with elevation. Based on a limited preliminary study of the impacts of geographic location on colony weight and *Varroa* levels (see Figure S2) this is possibly due to differences in climate or floral resources rather than due to differences in honey bee genetic background, though the association between nutrition, climate and pathogen loads in honey bee populations needs to be more thoroughly assessed. *Varroa* was recently found in honey bee colonies in Nigeria as well, at comparable levels (approximately 80% of colonies infested, with 2–55 mites found on 100 bees), but again no negative impacts on *Varroa* infestation on colony health and productivity were reported [55]. It is unclear what factors contribute to reduce the impacts of *Varroa* on African bees relative to European bees. Previous studies have indicated that Africanized bees in South America have higher levels of hygienic behavior, higher levels of grooming mites off of adult bees, lower levels of mite reproduction on pupae, and are less attractive to *Varroa* mites than European bees [56]. African bee subspecies also tend to abscond (abandoning hives and sometimes migrating) and swarm (where a large fraction of the colony leaves with the queen to form a new colony) more readily than European bees [27], thereby causing breaks in brood rearing that may reduce *Varroa* loads [57]. Indeed, breaking the brood cycle is recommended as a method to reduce *Varroa* loads in European honey bees [58], and removal of drone brood significantly reduces *Varroa* levels [59]. However, all of these factors presumably would lead to lower overall *Varroa* levels in colonies, rather than a reduced impact on worker mortality and colony size with equivalent levels. Thus, other parameters, perhaps physiological or behavioral, may contribute to the higher levels of tolerance of African bees to *Varroa* infestation.

Kenyan bee populations also displayed infections with DWV, BQCV, and ABPV, and *Varroa* numbers are strongly positively correlated with viral diversity (number of viruses present). Previous studies have demonstrated that *Varroa* can vector DWV and IAPV [37,38], and the introduction of *Varroa* to a naïve population of honey bees in Hawaii was correlated with reduction in sequence diversity of DWV [60], suggesting *Varroa* mites vector a specific viral strain. Thus, *Varroa* may have introduced these three viruses to the Kenyan honey bee populations, or is altering the population structure of these viruses. Alternatively, the presence of these viruses may be weakening bees' defenses to *Varroa*; this possibility has not yet been examined. Interestingly, a 2010 study of honey bee colonies in Uganda (which borders western Kenya and was thought to be *Varroa*-free at the time of the study) found BQCV in 30–40% of the colonies, but DWV and ABPV were not detected [61], suggesting that these two viruses may have been recently introduced to Kenya, perhaps by *Varroa*. Indeed, the viruses were not present in the geographically distant apiaries in far northeastern Kenya, suggesting that they, and *Varroa*, have not yet spread to this region. It should be noted that while the number of individual bees assayed for viruses in each colonies was small (5 bees), the number of colonies assayed throughout Kenya (81 colonies) and those assessed for correlation between viral diversity and *Varroa* levels (66 colonies) was fairly substantial. Thus, while the results of the individual colonies should be interpreted with caution (see Pirk et al [62] for more information on pathogen sampling in bee populations), the overall data set suggests that only three common European viruses are circulating, at this point, in the Kenyan populations at detectable levels, and that these are associated with *Varroa* parasitization levels.

As is the case with *Varroa*, DWV and BQCV do not appear to negatively impact Kenyan bee colonies, since viral infection is not correlated with colony size. Kenyan colonies are overall relatively small (7 frames of honey bees on average, compared to US colonies which can easily reach 20+ frames) and thus may not be at their limits of productivity. Better measures of the impacts of viruses, *Varroa*, and *Nosema* on bee health would include the longevity of infected and uninfected colonies, their ability to reproduce and successfully establish new colonies. Indeed, other studies have indicated that colonies may be particularly sensitive to the impacts of viruses during specific stressful periods (such as colony founding or migration) which could result in reduced populations. In European honey bees, high DWV levels in the fall are associated with reduced overwintering survival [63,64]. While there was an effect of viral diversity on hygienic behavior, this was not consistent – colonies with no viruses were as hygienic as colonies with two viruses. Previous studies have suggested that bees more readily remove larvae that are parasitized by *Varroa* with high viral titers [65], but our results suggest that hygienic behavior is not likely functioning as a major mechanism to reduce viral loads in Kenyan colonies.

The gut microsporidia *Nosema ceranae* was linked to colony losses in Spain [66], and high levels of *Nosema bombi* have been associated with declining bumble bee populations in North America [67]. However, other studies have not found a strong correlation with *Nosema* presence and colony declines [14,24], suggesting that *Nosema* may act in concert with other factors, such as pesticides [68], to reduce bee health. While *Nosema* was previously found in Zimbabwe and South Africa [69,70], it was not detected in a previous survey in 1996–1998 in Kenya ([36]; Shi Wei, personal communication, 2011). In our 2010 survey, *Nosema* was only identified in two apiaries at the coast [site 12 and 13], at site 15 near the Tanzania border and, surprisingly, at high elevation interior site on Mt. Elgon (site 22). Using PCR-based detection methods we did not detect *Nosema* in other locations. As in the case of the viral analyses, though the sample sizes for individual colonies was limited (5 bees/colony), our analysis at the apiary level should be quite robust in detecting *Nosema* if it is present. Notably, colonies in coastal *Nosema*-infected apiaries were similar in size to those in non-infected apiaries, and *Varroa* levels were substantially lower than other regions in the country. Thus, *Nosema* does not appear to be affecting honey bee populations in Kenya.

In the US, >90% of honey bee colonies contain pesticide residues [51]. Over 129 different pesticide-related chemicals have been found in US bee colonies, with an average of 6 chemicals per colony. Pesticide exposure has been linked to honey bee population declines [10,71] reduced survival and impaired development of brood [72], impaired cognitive function, [73–75], altered expression of immune genes [76], and increased *Nosema* loads [68,77]. In the Kenyan colonies surveyed in this study we found only four pesticides, and most were at very low levels compared to pesticide levels in North America. The most commonly found pesticide was the fungicide chlorothalonil (12 of 15 sites), while chlorpyrifos, an organophosphate insecticide, was found in five of the apiaries tested. Interestingly, fluvalinate was found in one apiary: fluvalinate is commonly used to control *Varroa* mites in the North America and Europe; however this broad-spectrum pyrethroid is also used to control mosquitos and horticultural pests such as aphids, whiteflies and thrips. The low levels of pesticides in hives from across Kenya, particularly when compared to levels in developed countries, suggests pesticide residues play only a limited role in honey bee health in Kenya at this time.

Conclusions

Honey bees provide critical pollination services to agriculture and natural landscapes, and the honey and wax produced by honey bees represent a potential source of income for families in East Africa and across the world. Our survey suggests that several new parasites and pathogens (*Varroa*, *Nosema*, DWV, BQCV, and ABPV) have recently invaded honey bee populations in East Africa. Our results indicate that these parasites and pathogens are not yet impacting honey bee health in Kenya, at least in terms of colony size. Interestingly, levels of *Varroa* are strongly impacted by elevation/geographic region, suggesting that environmental factors modulate *Varroa* infestation rates. Finally, our phylogenetic analyses suggest there is considerable mixing of honey bee populations in Kenya, and thus newly introduced parasites and pathogens can likely move easily throughout the region. Our studies suggest that honey bee populations in East Africa appear to be largely resistant or tolerant of the parasites and pathogens that threatened honey bee populations in other parts of the world, and are not yet significantly impacted by other stressors, such as exposure to environmental toxins. However, since there may be a time lag before these newly introduced pathogens and parasites significantly impact honey bee populations, additional long-term monitoring is necessary. Finally, our results also highlight the importance of environmental factors in buffering honey bee populations from these stressors, and with increasing habitat fragmentation and destruction and environmental extremes brought on by global climate change, populations of this keystone species, in Africa and throughout the world, will be under increasing pressures.

Supporting Information

Figure S1 Neighbor-joining tree (Saitou and Nei, 1987) comparing representative Kenyan honeybee haplotypes (see Table S2) with subspecies described in Arias and Sheppard 1996 (in capitals). The European subspecies *Apis mellifera mellifera* (MELLI1) is used as the outgroup. The percentage of replicate trees in which the associated taxa clustered together in the bootstrap test (2000 replicates) are shown next to the branches for values greater than 30% (1). Branch length indicates number of SNP differences. The analysis involved 39 nucleotide sequences from the ND2 mitochondrial region. Twenty-four of the 39 are representative unique haplotypes. For a list of all individuals represented by these haplotypes see Table S2. There were a total of 579 nucleotide positions in the final dataset. Phylogenetic analysis was conducted in MEGA5 (2). This analysis suggests as many as 7 clades of *A. mellifera* within Kenya. Whether this is evidence of more subspecies than previously described will require more sampling. Moreover, the weak statistical support suggests a need to expand beyond the ND2 region to describe East African subspecies of *A. mellifera*. Accession Numbers KJ628964-KJ628987.

Figure S2 Effect of location on colony weight and *Varroa* numbers. Colonies of from upland site 1 (icipe) were assayed in June 2010, immediately prior and after moving upland colonies from site 1 to site 25 on the coast. Coastal colonies were assayed at nearby apiaries (sites 26 and 27). There were no significant differences in colony weight (**A**, $F(3,26) = 1.43$, $p = 0.256$), but there were significantly fewer *Varroa* in colonies moved to site 25 (**B**, $F(3,25) = 6.63$, $p = 0.0019$). In August 2010, upland colonies at sites 1 and 25 and coastal colonies at site 25 were assayed. There were again no significant differences in weight (**C**, $F(2, 16) = 1.29$,

$p = 0.301$), but both upland and coastal colonies had significantly fewer *Varroa* at site 25 (**D**, $F(2,16) = 6.96$, $p = 0.0067$). The number of colonies in each group at each timepoint is indicated at the bottom of each bar in B and D. Letters represent groups that were significantly different with a Tukey HSD post-hoc pairwise comparison, at $p<0.05$. While the graphs show the actual average numbers of *Varroa*, counts were converted to logarithmic scale for statistical analysis.

Table S1 Apiary colony list indicating location, *Varroa* load, hygienic behavior, presence of virus, *Nosema*, pesticide and subspecies identification. Each apiary was given a site number and apiary name. Each colony has a unique identification number with the apiary number in the first position and the colony number in the second. Colony size is indicated by numbers of frames with bees. Total *Varroa* counts are based on standard sugar roll assay described in Ellis and Macedo, 2001 (1). The percent hygienic behavior (2) was calculated by taking the final number of fully and partially removed pupae/(207 − number originally uncapped or empty cells) *100. Boxes with UD are undetermined indicating measurements were not taken for these colonies. Positive for virus or *Nosema apis* detection per colony is indicated by an "X". The presence of pesticide detected in wax or bee bread is indicated in parts per billion (ppb). Only four pesticides were detected: CP (chloropyrios), CT (chlorothalonil), N (1-naphtol), F (fluvalinate). ND2 subspecies identification is based on the analysis presented in Figure S1. Individuals from most of the colonies grouped with the scutellata A or B clades. Some were identical in sequence to *A. mellifera monticola* described by Arias and Sheppard, 1996 (3). Some individuals from colonies at sites 12, 13, 16 and 17 were most closely related to *A. mellifera lamarckii* and are thus called "lamarckii-like". Colonies at site 18 were unique and perhaps represent a distinct subspecies (*A. m. simensis* or *yemenitica*). A few individuals did not group with any previously described subspecies and are thus indicated as UD. Interestingly three colonies (2.1, 2.2, and 16.1) had individuals from multiple mitochondrial lineages.

Table S2 Type of data collected at each apiary. An "X" indicates the data type at the top of the column was collected in at least one colony in the corresponding apiary. See Materials and Methods for a full description of the type of data collected and the analyses that were performed.

Table S3 A list of ND2 haplotypes used for comparison. Individuals were sequenced for the ND2 region (as described in 1) from 24 apiaries across Kenya. There were 24 unique haplotypes from 109 individuals sequenced. Each column is headed by the representative haplotype (in bold) used for analysis. Haplotypes identical to the representative type follow in each column. The columns are the far right is a list of unique single haplotypes. Numeric designations for each individual follow the following scheme: apiary.colony.individual.

Table S4 ND2 region variable sites from Kenya honeybee haplotypes compared to the subspecies *A. m. adansonii* (ADANS2) (1). The nucleotide positions starting from the ND2 ATC (isoleucine) are indicated in the top row with corresponding position numbers from the complete honeybee mitochondrial genome (2). The codon position for each SNP is indicated at the bottom of the figure. Over the 579 bp of the ND2 coding region 85% of SNPs were in the third codon position.

Three SNPs were in the second codon position. The first is a transversion at position 53 (T↔A) that results in an amino acid change of isoleucine ↔ asparagine. A second codon position change is a transition at nucleotide position 161 (C↔T) resulting in an amino acid change of threonine ↔ isoleucine. The third second codon change (position 458) is also a transition (C↔T) resulting in threonine ↔ isoleucine. The Kenya honeybee population also shows a first codon transition (position 412; G↔A) that results in an amino acid difference (valine ↔ isoleucine) when compared to the reference sequence.

Acknowledgments

We would like to thank Diana Sammatro for help in early specimen collections, Justin Malloy for assistance with molecular work associated with *Nosema* survey and Daisy Salifu and Elina Lastro Niño for providing advice for the statistical analyses.

Author Contributions

Conceived and designed the experiments: EM HP MF JF JT CG. Performed the experiments: EM HP MF BT TB JK JNK FM DM. Analyzed the data: HP MF TB CG. Wrote the paper: EM HP MF JF BT TB DM JT CG.

References

1. Klein AM, Vaissiere BE, Cane JH, Steffan-Dewenter I, Cunningham SA, et al. (2007) Importance of pollinators in changing landscapes for world crops. Proceedings of the Royal Society B-Biological Sciences 274: 303–313.
2. Eilers EJ, Kremen C, Greenleaf SS, Garber AK, Klein AM (2011) Contribution of Pollinator-Mediated Crops to Nutrients in the Human Food Supply. Plos One 6.
3. Gallai N, Salles JM, Settele J, Vaissiere BE (2009) Economic valuation of the vulnerability of world agriculture confronted with pollinator decline. Ecological Economics 68: 810–821.
4. Klein AM, Steffan-Dewenter I, Tscharntke T (2003) Fruit set of highland coffee increases with the diversity of pollinating bees. Proceedings of the Royal Society of London Series B-Biological Sciences 270: 955–961.
5. Bhattacharya A (2004) Flower visitor and fruitset of Anacardium occidentole. Annales Botanici Fennici 41: 385–392.
6. Roubik DW (2002) The value of bees to the coffee harvest. Nature 417: 708.
7. Aizen MA, Harder LD (2009) The Global Stock of Domesticated Honey Bees Is Growing Slower Than Agricultural Demand for Pollination. Current Biology 19: 915–918.
8. Morse RA, Calderone NW (2000) The value of honey bees as pollinators of U.S. crops in 2000. Bee Culture 128: 1–15.
9. Allsopp M (2004) Cape honeybee (Apis mellifera capensis Eshscholtz) and varroa mite (Varroa destructor Anderson & Trueman) threats to honeybees and beekeeping in Africa. International Journal of Tropical Insect Science 24: 87–94.
10. vanEngelsdorp D, Meixner MD (2010) A historical review of managed honey bee populations in Europe and the United States and the factors that may affect them. Journal of Invertebrate Pathology 103: S80–S95.
11. Sammataro D, Gerson U, Needham G (2000) Parasitic mites of honey bees: life history, implications, and impact. Annu Rev Entomol 45: 519–548.
12. Chen YP, Huang ZY (2010) Nosema ceranae, a newly identified pathogen of Apis mellifera in the USA and Asia. Apidologie 41: 364–374.
13. Chen YP, Siede R (2007) Honey Bee Viruses. In: Maramorosh K, Shatkin AJ, Murphy FA, editors. Advances in Virus Research: Elsevier Academic Press. pp. 34–80.
14. Runckel C, Flenniken ML, Engel JC, Ruby JG, Ganem D, et al. (2011) Temporal Analysis of the Honey Bee Microbiome Reveals Four Novel Viruses and Seasonal Prevalence of Known Viruses, Nosema, and Crithidia. Plos One 6.
15. Aronstein KA, Murray KD (2010) Chalkbrood disease in honey bees. J Invertebr Pathol 103 Suppl 1: S20–29.
16. Genersch E (2008) Paenibacillus larvae and American Foulbrood - long since known and still surprising. Journal fur Verbrauchershutz und Lebensmittelsicherheit 3: 429–434.
17. Desneux N, Decourtye A, Delpuech JM (2007) The sublethal effects of pesticides on beneficial arthropods. Annu Rev Entomol 52: 81–106.
18. Brodschneider R, Crailsheim K (2010) Nutrition and health in honey bees. Apidologie 41: 278–294.
19. Mattila HR, Seeley TD (2007) Genetic diversity in honey bee colonies enhances productivity and fitness. Science 317: 362–364.
20. VanEngelsdorp D, Caron D, Hayes J, Underwood R, Henson M, et al. (2012) A national survey of managed honey bee 2010–11 winter colony losses in the USA: results from the Bee Informed Partnership. Journal of Apicultural Research 51: 115–124.
21. Chauzat MP, Carpentier P, Madec F, Bougeard S, Cougoule N, et al. (2010) The role of infectious agents and parasites in the health of honey bee colonies in France. Journal of Apicultural Research 49: 31–39.
22. Cox-Foster DL, Conlan S, Holmes EC, Palacios G, Evans JD, et al. (2007) A metagenomic survey of microbes in honey bee colony collapse disorder. Science 318: 283–287.
23. Nguyen BK, Ribiere M, vanEngelsdorp D, Snoeck C, Saegerman C, et al. (2011) Effects of honey bee virus prevalence, Varroa destructor load and queen condition on honey bee colony survival over the winter in Belgium. Journal of Apicultural Research 50: 195–202.
24. vanEngelsdorp D, Evans JD, Saegerman C, Mullin C, Haubruge E, et al. (2009) Colony collapse disorder: a descriptive study. Plos One 4: e6481.

25. vanEngelsdorp D, Speybroeck N, Evans JD, Nguyen BK, Mullin C, et al. (2010) Weighing risk factors associated with bee colony collapse disorder by classification and regression tree analysis. J Econ Entomol 103: 1517–1523.
26. Hepburn RH, Radloff SE (1998) Honeybees of Africa. New York: Springer. 386 p.
27. Ruttner F (1987) Biogeography and Taxonomy of Honeybees: Springer. 284 p.
28. Meixner MD, Leta MA, Koeniger N, Fuchs S (2011) The honey bees of Ethiopia represent a new subspecies of Apis mellifera-Apis mellifera simensis n. ssp. Apidologie 42: 425–437.
29. Whitfield CW, Behura SK, Berlocher SH, Clark AG, Johnston JS, et al. (2006) Thrice out of Africa: ancient and recent expansions of the honey bee, Apis mellifera. Science 314: 642–645.
30. Crane E (1999) The World History of Bee Keeping and Honey Hunting New York: Routledge. 720 p.
31. Mbae RM (1999) Overview of beekeeping development in Kenya. In: Raina SK, Kioka EN, Mwanycky SW, editors. The Conservation and Utilization of Commercial Insects. Nairobi: ICIPE Science Press. pp. 103–105.
32. Kasina M, Kraemer M, Martius C, Wittmann D (2009) Diversity and Activity Density of Bees Visiting Crop Flowers in Kakamega, Western Kenya. Journal of Apicultural Research 48: 134–139.
33. UNCTAD/WTO ITC (2004) Kenya: Supply Suvery on Apicultural and Horticultural Products. Geneva, Switzerland: International Trade Centre.
34. National Beekeeping Station (2007) Hive Population and Production in Kenya (2005, 2006, and 2007) Provincial Summaries. Nairobi, Kenya: Ministry of Livestock.
35. Frazier M, Muli E, Conklin T, Schmehl D, Torto B, et al. (2010) A scientific note on Varroa destructor found in East Africa; threat or opportunity? Apidologie 41: 463–465.
36. Wei S (2001) Genetic Variation and Colony Development of Honey Bees Apis mellifera in Kenya. Uppsala: Swedish University of Agricultrual Sciences. 22 p.
37. Di Prisco G, Pennacchio F, Caprio E, Boncristiani HF, Jr., Evans JD, et al. (2011) Varroa destructor is an effective vector of Israeli acute paralysis virus in the honeybee, Apis mellifera. J Gen Virol 92: 151–155.
38. Gisder S, Aumeier P, Genersch E (2009) Deformed wing virus: replication and viral load in mites (Varroa destructor). J Gen Virol 90: 463–467.
39. Ellis MD, Macedo PA (2001) G01-1430 Using the Sugar Roll Technique to Detect Varroa Mites in Honey Bee Colonies. Lincoln, NE: University of Nebraska. 4 p.
40. Klee J, Besana AM, Genersch E, Gisder S, Nanetti A, et al. (2007) Widespread dispersal of the microsporidian Nosema ceranae, an emergent pathogen of the western honey bee, Apis mellifera. J Invertebr Pathol 96: 1–10.
41. Fries I, Chauzat MP, Chen YP, Doublet V, Genersch E, Gisder S, Higes M, McMahon DP, Martin-Hernandez R, Natsopoulou M, Paxton RJ, Tanner G, Webster TC, Williams GR (2013) Standard methods for Nosema research. Journal of Apicultural Research 51(5): http://dx.doi.org/10.3896/IBRA.1.52.1.14.
42. Traver BE, Fell RD (2011) "Prevalence and infection intensity of Nosema in honey bee (Apis mellifera L.) colonies in Virginia." Journal of Invertebrate Pathology 107(1): 43–49.
43. Benjeddou M, Leat N, Allsopp M, Davison S (2001) Detection of acute bee paralysis virus and black queen cell virus from honeybees by reverse transcriptase pcr. Appl Environ Microbiol 67: 2384–2387.
44. Chen YP, Higgins JA, Feldlaufer MF (2005) Quantitative real-time reverse transcription-PCR analysis of deformed wing virus infection in the honeybee (Apis mellifera L.). Appl Environ Microbiol 71: 436–441.
45. Ribiere M, Triboulot C, Mathieu L, Aurieres C, Faucon JP, et al. (2002) Molecular diagnosis of chronic bee paralysis virus infection. Apidologie 33: 339–351.
46. Stoltz D, Shen XR, Boggis C, Sisson G (1995) Molecular diagnosis of Kashmir bee virus infection. Journal of Apicultural Research 34: 153–160.
47. Spivak M, Reuter GS (1998) Honey bee hygienic behavior. American Bee Journal 138: 283–286.

48. Arias MC, Sheppard WS (1996) Molecular phylogenetics of honey bee subspecies (Apis mellifera L.) inferred from mitochondrial DNA sequence. Mol Phylogenet Evol 5: 557–566.

49. Thompson JD, Higgins DG, Gibson TJ (1994) CLUSTAL W: improving the sensitivity of progressive multiple sequence alignment through sequence weighting, position-specific gap penalties and weight matrix choice. Nucleic Acids Res 22: 4673–4680.

50. Saitou N, Nei M (1987) The neighbor-joining method: a new method for reconstructing phylogenetic trees. Mol Biol Evol 4: 406–425.

51. Mullin CA, Frazier M, Frazier JL, Ashcraft S, Simonds R, et al. (2010) High levels of miticides and agrochemicals in North American apiaries: implications for honey bee health. Plos One 5: e9754.

52. Delaplane KS, Hood WM (1997) Effects of delayed acaricide treatment in honey bee colonies parasitized by Varroa jacobsoni and a late season treatment threshold for the southeastern USA. Journal of Apicultural Research 36: 125–132.

53. Ruttner F (1992) Naturgeschichte der Honigbienen: Kosmos Verlags-GmbH.

54. Spivak M, Reuter GS (2001) Varroa destructor infestation in untreated honey bee (Hymenoptera: Apidae) colonies selected for hygienic behavior. Journal of Economic Entomology 94: 326–331.

55. Akinwade KL, Badejo MA, Ogbogu SS (2012) Incidence of the Korea haplotype of Varroa destructor in southwest Nigeria. Journal of Apicultural Research 15: 369–370.

56. Guzman-Novoa E, Vandame Rm, Arechavaleta M, E. (1999) Susceptibility of European and Africanized honey bees (Apis mellifera L.) to Varroa jacobsoni Oud. in Mexico. Apidologie 30: 173–182.

57. Fries I, Hansenb H, Imdorfc A, Rosenkranz P (2003) Swarming In Honey Bees (Apis Mellifera) And Varroa Destructor Population Development In Sweden Apidologie 34: 389–397.

58. National Bee Unit (2010) FAQ 17: Queen trapping. York: The Food and Environmental Research Agency. 1–2 p.

59. Calderone NW (2005) Evaluation of drone brood removal for management of Varroa destructor (Acari: Varroidae) in colonies of Apis mellifera (Hymenoptera: Apidae) in the northeastern United States. J Econ Entomol 98: 645–650.

60. Martin SJ, Highfield AC, Brettell L, Villalobos EM, Budge GE, et al. (2012) Global honey bee viral landscape altered by a parasitic mite. Science 336: 1304–1306.

61. Kajobe R, Marris G, Budge G, Laurenson L, Cordoni G, et al. (2010) First molecular detection of a viral pathogen in Ugandan honey bees. J Invertebr Pathol 104: 153–156.

62. Pirk CWW, de Miranda JR, Kramer M, Murray TE, Nazzi F, et al (2013) "Statistical guidelines for Apis mellifera research." Journal of Apicultural Research 52(4).

63. Highfield AC, El Nagar A, Mackinder LC, Noel LM, Hall MJ, et al. (2009) Deformed wing virus implicated in overwintering honeybee colony losses. Appl Environ Microbiol 75: 7212–7220.

64. Dainat B, Evans JD, Chen YP, Gauthier L, Neumann P (2012) Predictive markers of honey bee colony collapse. Plos One 7: e32151.

65. Schoning C, Gisder S, Geiselhardt S, Kretschmann I, Bienefeld K, et al. (2012) Evidence for damage-dependent hygienic behaviour towards Varroa destructor-parasitised brood in the western honey bee, Apis mellifera. J Exp Biol 215: 264–271.

66. Higes M, Martin-Hernandez R, Botias C, Bailon EG, Gonzalez-Porto AV, et al. (2008) How natural infection by Nosema ceranae causes honeybee colony collapse. Environ Microbiol 10: 2659–2669.

67. Cameron SA, Lozier JD, Strange JP, Koch JB, Cordes N, et al. (2012) Patterns of widespread decline in North American bumble bees. Proc Natl Acad Sci U S A 108: 662–667.

68. Pettis JS, vanEngelsdorp D, Johnson J, Dively G (2012) Pesticide exposure in honey bees results in increased levels of the gut pathogen Nosema. Naturwissenschaften 99: 153–158.

69. Fries I, Slamenda SB, Da Silva A, Pieniazek NJ (2003) African honey bees (Apis mellifera scutellata) and Nosema (Nosema apis) infections. Journal of Apicultural Research 42: 13–15.

70. Swart DJ (2003) The occurence of Nosema apis (Zander), Acarapis woodi (Rennie), and the Cape problem bee in the summer rainfall region of South Africa Rhodes University. 43 p.

71. Faucon J-P, Mathieu L, Ribiere M, Martel A-C, Drajnudel P, et al. (2002) Honey bee winter mortality in France in 1999 and 2000. Bee World 83 14–23.

72. Zhu W, Schmehl DR, Mullin CA, Frazier JL (submitted) Chronic oral toxicity of four commone pesticides and their mixtures to honey bee larvae.

73. Bortolotti L, Montanari R, Marcelino J, Medrzycki P, Maini S, et al. (2003) Effects of sub-lethal imidacloprid doses on the homing rate and foraging activity of honey bees. Bulletin of Insectology 56: 63–67.

74. Ciarlo TJ, Mullin CA, Frazier JL, Schmehl DR (2012) Learning impairment in honey bees caused by agricultural spray adjuvants. Plos One 7: e40848.

75. Decourtye A, Devillers J, Cluzeau S, Charreton M, Pham-Delegue MH (2004) Effects of imidacloprid and deltamethrin on associative learning in honeybees under semi-field and laboratory conditions. Ecotoxicol Environ Saf 57: 410–419.

76. Gregorc A, Evans JD, Scharf M, Ellis JD (2012) Gene expression in honey bee (Apis mellifera) larvae exposed to pesticides and Varroa mites (Varroa destructor). J Insect Physiol 58: 1042–1049.

77. Wu JY, Smart MD, Anelli CM, Sheppard WS (2012) Honey bees (Apis mellifera) reared in brood combs containing high levels of pesticide residues exhibit increased susceptibility to Nosema (Microsporidia) infection. J Invertebr Pathol 109: 326–329.

5

Point Mutations Associated with Organophosphate and Carbamate Resistance in Chinese Strains of *Culex pipiens quinquefasciatus* (Diptera: Culicidae)

Minghui Zhao[1,2], Yande Dong[1], Xin Ran[1,2], Zhiming Wu[1], Xiaoxia Guo[1], Yingmei Zhang[1], Dan Xing[1], Ting Yan[1], Gang Wang[1], Xiaojuan Zhu[1], Hengduan Zhang[1], Chunxiao Li[1]*, Tongyan Zhao[1]*

1 Beijing Institute of Microbiology and Epidemiology, State Key Laboratory of Pathogens and Biosecurity, Department of Vector Biology and Control, Beijing, China, **2** Anhui Medical University, Hefei, China

Abstract

Acetylcholinesterase resistance has been well documented in many insects, including several mosquito species. We tested the resistance of five wild, Chinese strains of the mosquito *Culex pipiens quinquefasciatus* to two kinds of pesticides, dichlorvos and propoxur. An acetylcholinesterase gene (ace1) was cloned and sequenced from a pooled sample of mosquitoes from these five strains and the amino acids of five positions were found to vary (V185M, G247S, A328S, A391T, and T682A). Analysis of the correlation between mutation frequencies and resistance levels (LC_{50}) suggests that two point mutations, G247S ($r^2 = 0.732$, $P = 0.065$) and A328S ($r^2 = 0.891$, $P = 0.016$), are associated with resistance to propoxur but not to dichlorvos. Although the V185M mutation was not associated with either dichlorvos or propoxur resistance, its RS genotype frequency was correlated with propoxur resistance ($r^2 = 0.815$, $P = 0.036$). And the HWE test showed the A328S mutation is linked with V185M, also with G247S mutation. This suggested that these three mutations may contribute synergistically to propoxur resistance. The T682A mutation was negatively correlated with propoxur ($r^2 = 0.788$, $P = 0.045$) resistance. Knowledge of these mutations may help design strategies for managing pesticide resistance in wild mosquito populations.

Editor: Israel Silman, Weizmann Institute of Science, Israel

Funding: This work was funded by the foundation from the Infective Diseases Prevention and Cure Project of National Ministry of Public Health of China (No.2008ZX10004 and No.2012ZX10004219). The funders had no role in study design, data collection and analysis, decision to publish, or preparation of the manuscript.

Competing Interests: The authors have declared that no competing interests exist.

* E-mail: aedes@126.com (CXL); tongyanzhao@126.com (TYZ)

Introduction

Acetylcholinesterase (AChE, EC 3.1.1.7) is a key enzyme in the nervous system of both vertebrates and invertebrates that terminates nerve impulses by catalyzing the hydrolysis of the neurotransmitter acetylcholine (ACh) released from the presynaptic membrane [1]. The inhibition of AChE by organophosphate and carbamate insecticides leads to the desensitization of the ACh receptor, thereby blocking nerve signal transmission. Organophosphates and carbamates have structures analogous to ACh and inhibit AChE competitively at the active site. Hydrolysis of these pesticide compounds retards the reactivation of the enzyme or inactivates it [2]. The extensive use of organophosphate and carbamate insecticides has resulted in the development of high levels of resistance to them among insects [3,4,5,6].

Ace1 is the key AChE gene in insects. Several studies have found evidence that a point mutation in the ace1 gene is associated with resistance to organophosphate and carbamate pesticides. This point mutation changes the structure of AChE making it insensitive to these insecticides. The first report of this mutation conferring insecticide resistance was in the two-spotted spider mite in 1964 [7]. Subsequent studies have demonstrated that many insect species have developed resistance to organophosphate and

carbamate pesticides through decreased sensitivity of AChE [8], including many mosquito species, such as *Anopheles gambiae* [9], *Cx. pipiens* [10,11], *Cx. pipiens quinquefasciatus* [12], *Cx. tritaeniorhynchus* and *Cx. vishnui* [13]. However, so far, only three ace1 mutations, G119S, F331W and F290V (*T. californica* numbering) [13,14,15,16], have been confirmed to be involved in such resistance in mosquito species. Determining the mutations that confer resistance to specific pesticides is important to designing effective strategies for managing pesticide resistance. *Cx. pipiens quinquefasciatus* is the main mosquito species in urban environments in southern China and one of the most studied in terms of insecticide resistance. We here report the results of an investigation of mutations in the ace1 gene in five wild Chinese populations of *Cx. pipiens quinquefasciatus*. Knowledge of these mutations may have practical benefits for reducing pesticide resistance in this species.

Results

Resistance of the Five Mosquito Populations to Dichlorvos and Propoxur

LC_{50} values of the five different populations ranged from 0.266 to 1.67 ppm for dichlorvos, and from 0.279 to 1.27 ppm for propoxur (Table 1). The HC strain had the lowest LC_{50} and was

Table 1. Levels of dichlorvos and propoxur resistance in five populations of Cx. pipiens quinquefasciatus.

Population[1]	Insecticide	LC$_{50}$ and LC$_{90}$ (ppm) (95% CL)[2]	Regression Equation	Slope	Standard Deviation	χ^2	P	RR[3]
LA	Dichlorvos	0.095[4]						1
	Propoxur	0.115[5]						1
GN	Dichlorvos	1.189 (0.923, 1.521) 3.376 (2.475, 5.505)	Y = -0.212+2.827x	2.827	0.200	58.08	<0.01	12.52
	Propoxur	1.266 (1.073, 1.595) 3.672 (2.576, 7.043)	Y = -0.284+2.772x	2.772	0.282	41.21	0.002	11.01
HP	Dichlorvos	0.750 (0.661, 0.853) 2.499 (2.067, 3.164)	Y = 0.306+2.453x	2.453	0.171	8.452	0.934	7.895
	Propoxur	0.531 (0.500, 0.564) 0.894 (0.820, 0.997)	Y = 1.557+5.668x	5.668	0.429	9.720	0.881	4.617
HC	Dichlorvos	0.266 (0.224, 0.309) 1.032 (0.835, 1.366)	Y = 1.252+2.175x	2.175	0.197	7.039	0.900	2.800
	Propoxur	0.279 (0.238, 0.320) 0.947 (0.755, 1.329)	Y = 1.338+2.413x	2.413	0.208	29.80	0.054	2.426
QB	Dichlorvos	1.240 (1.051, 1.464) 6.047 (4.609, 8.661)	Y = -0.174+1.862x	1.862	0.118	38.81	0.038	13.05
	Propoxur	0.598 (0.559, 0.639) 0.895 (0.813, 1.033)	Y = 1.635+7.319x	7.319	0.592	23.91	0.032	5.200
SF	Dichlorvos	1.672 (1.520, 1.822) 4.365 (3.905, 4.999)	Y = -0.687+3.076x	3.076	0.208	17.46	0.737	17.60
	Propoxur	0.785 (0.738, 0.837) 1.423 (1.278, 1.639)	Y = 0.522+4.959x	4.959	0.400	19.02	0.213	6.826

[1]LA = Lab strain; GN = Guangzhou Nansha; HP = Haikou Poxiang; HC = Haikou Changliu; QB = Qionghai Boao; SF = Sanya Fenghuang.
[2]CL = confidence limits.
[3]RR = Resistance Ratio.
[4]and [5]are coming from Li Chunxiao' dissertation [39].

the most susceptible to both dichlorvos and propoxur. The SF strain had an LC$_{50}$ to dichlorvos of 1.67 ppm and was 17.6 times more resistant to dichlorvos than the laboratory strain (LC$_{50}$ 0.095 ppm). The GN strain had an LC$_{50}$ to propoxur of 1.27 ppm and was 11.0 times more resistant to propoxur than the laboratory strain (LC$_{50}$ 0.115 ppm). The HP strain was 7.89 times more resistant to dichlorvos, and 4.62 times more resistant to propoxur, than the laboratory strain. The QB strain was 13.1 times more resistant to dichlorvos, and 5.20 times more resistant to propoxur than the laboratory strain.

Identification of Ace1 Mutations

To identify mutations in the ace1 gene, the cDNA of a pooled sample of mosquitoes from each of the five populations was cloned and sequenced. Five mutations (V185M, G247S, A328S, A391T, and T682A) in the pooled ace1 gene were identified (Figure 1), and the sequence was deposited in GenBank under the accession number KF680946. Note that this identification of 5 mutations does not imply all occur in the same ace1 gene. The V185M mutation was GTG to ATG, the G247S mutation was GGC to AGC, the A328S mutation was GCC to TCC, the A391T mutation was GCC to ACC, and the T682A mutation was ACA to GCA.

Polymorphism of the Ace1 Gene in Natural Population

1. Determination of the allele frequencies. The allele frequencies of each mutation were determined by specific PCR amplification using the primers Cx-ace2-F, Cx-ace2-R and Cx-ace3-F, Cx-ace3-R on the cDNA obtained from individual mosquitoes. Genotypes of each mosquito in each population was determined by sequencing, and mutation frequencies (R%) computed (Table 2). We can see from Table 2 that the V185M, A328S and T682A mutations were present at different frequencies in all five strains. However, the A391T mutation was only found in the HP and QB strains, and the G247S mutation was found in all but the HC strain.

2. Hardy–Weinberg Equilibrium (HWE) test and genetic linkage analysis of the mutations. The results of GENE-POP software analysis of HWE and genetic linkage of the acetylcholinesterase gene mutations are shown in Tables 2 and 3. The HWE test indicates the QB and GN populations have a heterozygote deficit with respect to the T682A mutation (P< 0.05), and the HP population a heterozygote excess with respect to the A391T mutation (P<0.05). Mutations in all other populations did not deviate from the HWE and none of the five mutations deviated from the HWE across all populations (P> 0.05).

Results of linkage disequilibrium analysis of the five mutations are shown in Table 3. Evidence of linkage disequilibrium was found for V185M with respect to the A328S and A391T mutations (P<0.05), The G247S and A328S mutations' linkage disequilibrium P-value was 0.0821, only slightly above 0.05. This suggests that these two mutations might exist in the same gene. Our sequencing data indicated that that these two mutations do indeed occur in the same ace1 gene in some mosquitoes. But the conclusion had to be confirmed by more data. All other gene polymorphism was randomly distributed.

Correlation of Resistance with Mutation Frequencies

The correlation between resistance to dichlorvos and propoxur and the frequencies of four mutations (V185M, G247S, A328S, T682A) are shown in Figure 2 and Table 4. The four mutations' frequencies were all not significantly correlated with dichlorvos resistance. Although the frequency of the V185M mutation was uncorrelated with propoxur resistance (Figure 2 A), its RS genotype frequency was ($r^2 = 0.815$, $P = 0.036$) (Figure 2 B). The correlation between the frequency of the G247S mutation and propoxur resistance was close to significance ($r^2 = 0.732$, $P = 0.065$), and there was a significant linear relationship between the frequency of the A328S mutation and propoxur resistance ($r^2 = 0.891$, $P = 0.016$) (Figure 2 C, D). The frequency of the T682A mutation was negatively correlated with propoxur ($r^2 = 0.788$, $P = 0.045$) resistance (Figure 2 E).

3D Models of Mutations and Structural changes at the Catalytic Site

A 3D model was made of the *Cx. pipiens quinquefasciatus* ace1 gene sequence allowing the location and structure of four mutations to be visualized (Figure 3). The V185M and A391T mutations are distant from the active site of the enzyme-catalytic triad (S327, H567, E453; S200, H440, E327 in *T. californica*) (Figure 3A, B). The other two mutations, G247S and A328S, are close to the catalytic site (Figure 3C, D) and could therefore potentially affect the binding between AChE and its substrates (Ach: ZINC3079336 and propoxur: ZINC1590885). Figure 3E-H illustrates the change in amino acids and H-bonds associated with the G247S and A328S mutations. These two substitutions change the amino acids present at catalytic sites removing the two H-bonds (S327(8) O_γ-O3, S327(8)O_γ-O4) between AChE and Ach (Figure 3E, F) and reducing the three H-bonds between AChE and propoxur (G247(4)-O13, S327(8)O_γ-O11, H567(14)-NH27) to one (S327(10)O_γ-NH27) (Figure 3G, H). Hence, these two mutations could have a major effect on the catalytic activity of the AChE enzyme.

Discussion

The indiscriminate use of insecticides over more than half a century has resulted in high levels of insecticide resistance in many mosquito species [13,17,18]. We tested the resistance of five Chinese *Cx. pipiens quinquefasciatus* populations to dichlorvos and propoxur. Our results show that, compared to a laboratory strain, these five populations displayed a 2.80- to 17.6-fold resistance to dichlorvos and 2.43- to 11.0-fold resistance to propoxur. The frequent use of these insecticides has created an intense selection pressure for traits that confer resistance to them, such as changes in behavior, epidermal structure, metabolic enzymes and target site mutations. Resistance may be conferred by any one, or more than one of these mechanisms. Osta et al (2012) found that the dramatic reduction in the frequency of the G119S (*T. californica* numbering) mutation in *Culex pipiens* mosquitoes was probably due to the increased use of pyrethroids over organophosphate insecticides [19]. Therefore, alternating between different kinds of insecticides is one way of minimizing the development of resistance to any one kind.

We used cloning and sequencing to identify five point mutations in the ace1 gene of Chinese *Cx. pipiens quinquefasciatus*. HWE tests suggest that these five mutations do not deviate from the HWE across all populations. However, the tests also indicated that the QB and GN populations were deficient in heterozygotes with respect to the T682A mutation and that HP population had an excess of heterozygotes with respect to the A391T mutation (P< 0.05). Further work will be required to determine the reasons for these departures from the HWE. Linkage disequilibrium analysis indicated significant linkage between the V185M mutation and the A328S and A391T mutations. Although linkage between other mutations was statistically insignificant, that between the G247S and A328S mutations was nearly so (P = 0.0821). Our sequencing

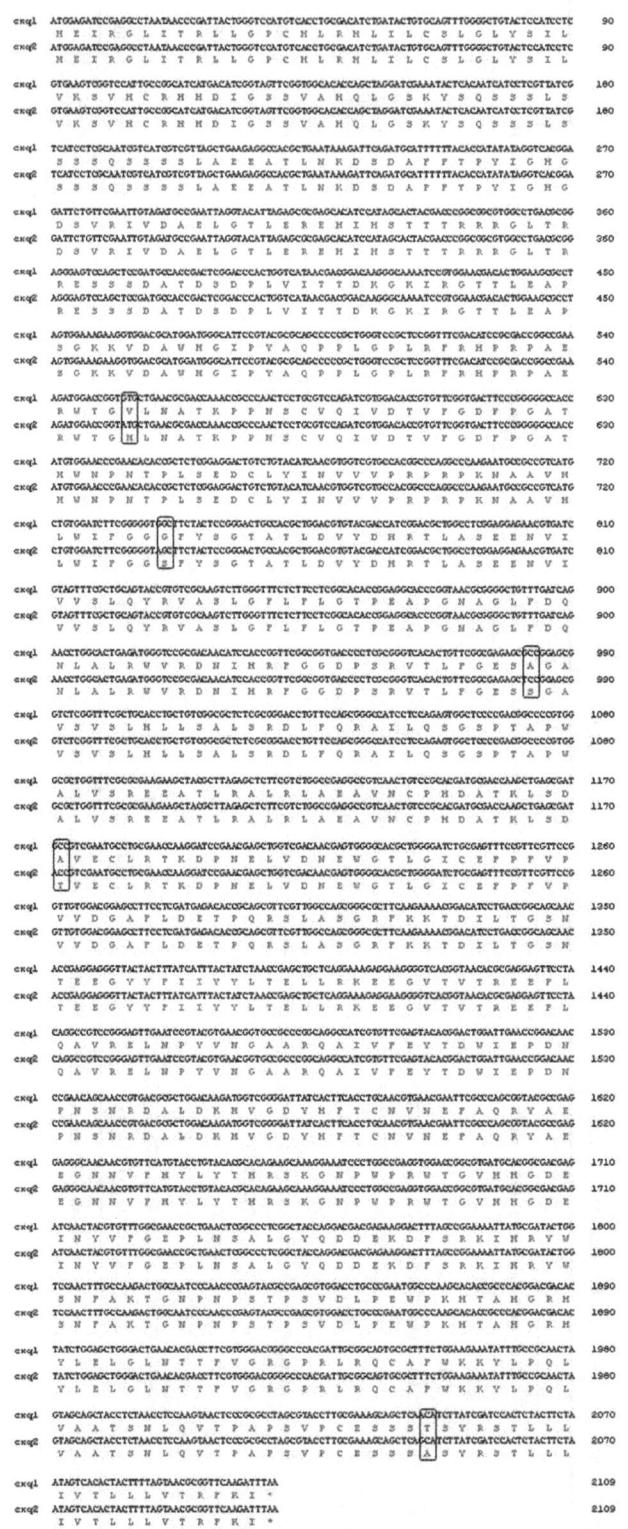

Figure 1. Alignment of nucleotide and amino acid sequences of *Cx. pipiens quinquefasciatus*. Cxq1 is the template nucleotide sequence (no amino acid mutation) and Cxq2 the mutant nucleotide sequence. Nucleotides are numbered on the first line, amino acids on the second. The five mutations are shown in the black frames.

Table 2. Mutation frequencies of five ace1 gene mutations and HWE test in five populations of *Cx. pipiens quinquefasciatus*.

Mutations	Strains	Numbers	Mutation frequency (R %)	P-value of HWE		HWE across strains	
				deficit	excess	χ^2	P
V185M	GN	36	25.0	1.00	0.06	9.84	0.45
	HP	33	6.10	1.00	0.91		
	HC	30	16.7	0.15	0.99		
	QB	31	11.3	1.00	0.68		
	SF	30	23.3	0.50	0.84		
G247S	GN	36	18.1	0.73	0.70	0.00	1.00
	HP	33	1.50	No[1]	No		
	HC	30	0.00	No	No		
	QB	30	11.7	1.00	0.67		
	SF	30	5.00	1.00	0.95		
A328S	GN	36	47.2	0.83	0.39	1.77	1.00
	HP	33	19.7	0.77	0.66		
	HC	34	2.90	1.00	0.98		
	QB	30	11.7	1.00	0.67		
	SF	30	16.7	1.00	0.41		
A391T	GN	15	0.00	No	No	7.79	0.10
	HP	22	47.7	1.00	0.02		
	HC	13	0.00	No	No		
	QB	23	54.3	0.84	0.45		
	SF	22	0.00	No	No		
T682A	GN	35	18.6	0.01	1.00	18.0	0.06
	HP	32	51.6	0.73	0.53		
	HC	36	48.6	0.90	0.28		
	QB	33	39.4	0.03	1.00		
	SF	31	24.2	0.89	0.44		

[1]No is no information, the reasons are because the site is homozygous for one mutation in this sample or because there is a single heterozygote.

Table 3. P-value for linkage disequilibrium of each pair of loci across all populations (Fisher's method).

Locus pair	χ^2	df	P-Value
V185M & G247S	11.237	8	0.1887
V185M & A328S	23.804	10	0.0081
G247S & A328S	13.988	8	0.0821
V185M & A391T	7.5840	2	0.0226
G247S & A391T	3.4992	2	0.1738
A328S & A391T	5.8691	4	0.2091
V185M & T682A	4.8208	10	0.9028
G247S & T682A	2.4273	6	0.8765
A328S & T682A	14.160	10	0.1658
A391T & T682A	1.1946	4	0.8790

results suggest that these two mutations occur within the same ace1 gene in some mosquitoes but further work is required to confirm this hypothesis.

These results are the first report of the V185M mutation in *Cx. pipiens quinquefasciatus*. Although there was no apparent correlation between the frequency of this mutation and resistance to dichlorvos and propoxur, the frequency of its RS genotype was significantly correlated with propoxur resistance ($r^2 = 0.815$, $P = 0.036$). Although the 3D model indicates that V185M is located far from the active site, the positive linear relationship between its RS genotype frequency and propoxur resistance, and its apparent linkage with the A328S mutation suggest that it may be involved in propoxur resistance. Of course, we cannot rule out the possibility that insecticide resistance involves multiple dupli-

cation of the ace1 gene. Further research needs be required to determine how this might affect the catalytic center.

Our results (Figure 2, Table 4) suggest that the G247S mutation is not associated with propoxur resistance and that the A328S mutation is. The G247S mutation corresponds to G119S in *T. californica* which has been associated with insecticide resistance in mosquitoes by several authors [20,21]. The G119 position is part of the oxyanion hole (G118, G119, and A201 in *T. californica*), close to the catalytic Serine (S200) where a G to S substitution would reduce accessibility to inhibitors and substrate by steric hindrance. S119 is close enough to the catalytic residues to alter the presentation of inhibitors and substrates. This could be the reason this mutation confers resistance to some insecticides [22,23]. Although the correlation between the frequency of the G247S

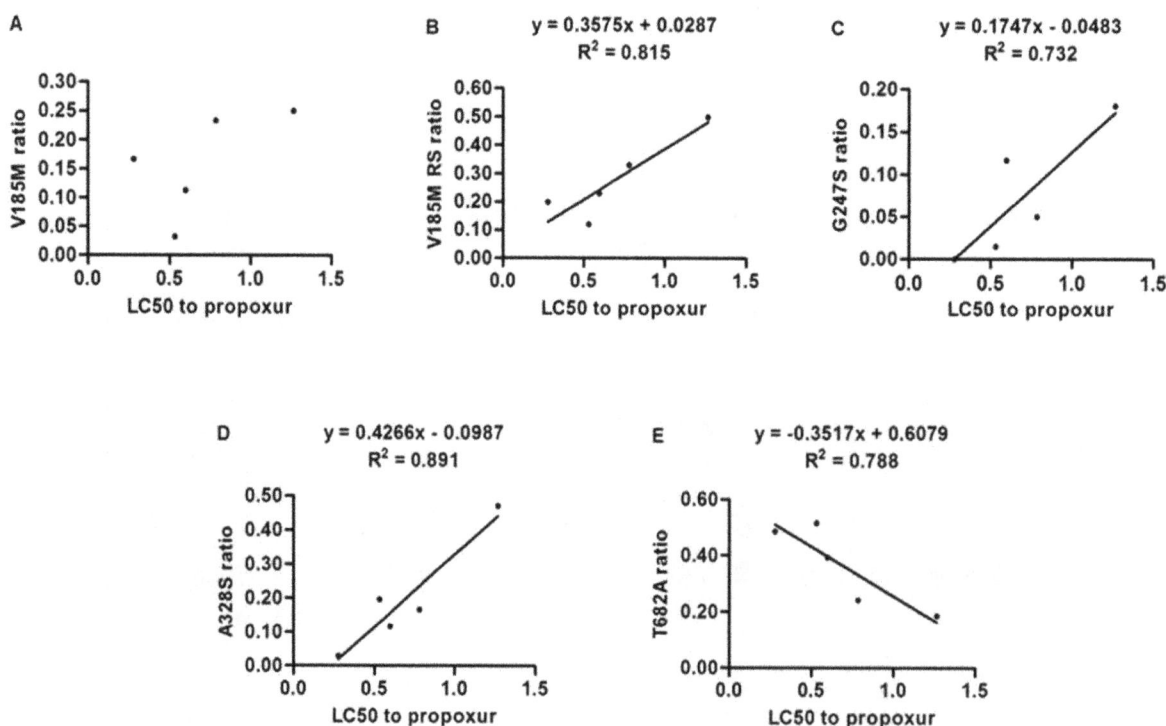

Figure 2. Linear regression of the relationship between resistance levels (LC50) and mutation ratios. Resistance levels to propoxur are plotted against the ratios of V185M (A), the RS ratio of V185M (B), G247S (C), A328S (D), and T682A (E).

Table 4. The analysis results of correlation between propoxur LC50 and mutation frequencies.

Mutations	Insecticide	R (95% CL[1])	R[2]	P	Significance[2]
V185M	Propoxur	0.647(−0.549,0.974)	0.419	0.238	No
V185M (RS%)	Propoxur	0.903(0.101,0.994)	0.815	0.036	Yes
G247S	Propoxur	0.855(−0.110,0.990)	0.732	0.065	No
A328S	Propoxur	0.944(0.366,0.996)	0.891	0.016	Yes
T682A	Propoxur	−0.887(−0.993, −0.023)	0.788	0.045	Yes

[1]CL = confidence limits.
[2]$\alpha = 0.05$.

mutation and propoxur resistance was not statistically significant (P = 0.065), numerous prior publications have reported such an association [9,10,12] and noted that this mutation is often combined with other mutations in resistant strains. Therefore, we suspect that G247S probably is involved in propoxur resistance. We may have failed to detect a significant correlation between the frequency of this mutation and resistance because of its low frequency in our sample, which could be because most mosquitoes carrying it were heterozygotes. Furthermore, the resistance conferred by this mutation may be nearly recessive under certain bioassay conditions [24].

The A328 position corresponds to the A201 position in *T. californica*, which is located within the active gorge of the enzyme, close to the catalytic site,and is a part of the oxyanion hole. Li et al (2009) also found the A328S mutation in *Cx. pipiens pallens* and made a three-dimensional model of AChE to visualize this mutation. However, they did not demonstrate a relationship between the A328S mutation and resistance [25]. Khajehali et al (2010) found the corresponding A201S mutation in *Tetranychus urticae* Koch, and demonstrated that this was possibly involved in resistance to organophosphorus and carbamate insecticides [26]. Our results suggest that this mutation is involved in propoxur resistance ($r^2 = 0.891$, P = 0.016). The linkage disequilibrium and sequencing results indicate that A328S and G247S mutations exist in a same ace1 gene in some mosquitoes, which suggests that they may work synergistically. The G119 and A201 positions (*T. californica* numbering) are both part of the oxyanion hole, and could therefore both contribute an amide nitrogen to form bonds that could stabilize the enzyme–substrate complex. The substitution of serine for glycine and alanine may change the conjunctions, conferring resistance to some insecticides [23,27]. We can see from Figure 3 that these substitutions could decrease the numbers of H-bonds between enzyme and substrate. H-bonds are the strongest force between molecules so a reduction in these could reduce enzyme-substrate stability and interfere with the catalytic reaction.

The G119S mutation was the first mutation found in mosquito vectors [10]. Previous studies indicated that this mutation would incur a high fitness cost [28], however, although the cost of resistance is often high at the beginning of selection when resistance is unstable, the cost reduces and resistance stabilizes with increasing duration of exposure to insecticides [29]. Other mutations can play an important role in this process. Mutero et al (1994) found that high levels of resistance were obtained by the combination of several point mutations [1] and Menozzi et al (2004) demonstrated that combining mutations could increase insecticide resistance in *Drosophila melanogaster* [30]. Our results show that the A328S mutation (A201S in *T. californica*) may work synergistically with the G119S mutation in the oxyanion hole. It's possible that the A328S mutation compensates for some of the

fitness costs incurred by the G119S mutation. This is a fascinating question but further *in vitro* assays are required to confirm this hypothesis.

The A391T mutation was found only in the HP and QB populations, which had moderate LC_{50} values and in which it had a frequency of around 0.500. The genetic linkage analysis indicates a linkage between this mutation and V185M, however, in view of the small sample size further work is required to confirm this. The three-dimensional model revealed that the A391 mutation is distant from the active site. This indicates that this mutation is unlikely to affect catalytic activity and is probably not involved in dichlorvos and propoxur resistance. How this mutation developed and its function, if any, in pesticide resistance requires further investigation.

Our results provide the first evidence of the T682A mutation in *Cx. pipiens quinquefasciatus*. The frequency of this mutation was negatively correlated with propoxur resistance ($r^2 = 0.788$, P = 0.045). Fournier et al (1988) found that AChE in *Drosophila melanogaster* was composed of two, non-covalently associated, polypeptides of 55 and 16 kDa. AChE is an amphiphilic protein linked to the membrane of neuronal cholinergic synapses via a glycolipid anchor at the C-terminal end of the 55 kDa polypeptide [31]. Nabeshima et al (2004) found an I697M replacement near the C-terminus (Ile701) in *Culex tritaeniorhynchus*, but considered that this was unlikely to be the cause of AChE insensitivity [32]. Our results also indicate that the T682A mutation is near the C-terminus of AChE, and that the frequency of this mutation is negatively correlated with propoxur resistance. Despite its negative correlation with resistance, it's possible that this mutation may change the C-terminus structure of AChE thereby reducing its attachment to the membrane and the stability of enzyme. We don't know whether this mutation works in combination with the other four mutations or not, or if its apparent negative relationship with resistance is related to fitness costs.

In conclusion, we found five ace1 gene mutations in *Cx. pipiens quinquefasciatus* that are correlated with propoxur, but not dichlorvos resistance. The V185M mutation was first confirmed in *Cx. pipiens quinquefasciatus* and may be involed in propoxur resistance. The allele frequencies of the G247S and A328S mutations were positively correlated with resistance. So the G247S and A328S mutations are also likely to confer propoxur resistance. The A391T mutation appears unrelated to dichlorvos and propoxur resistance and the T682A mutation appears negatively correlated with resistance to propoxur. Identifying the mutations that confer resistance to specific insecticides can inform the choice of insecticides for a given insect population, thereby reducing the development of resistance and improving the efficacy of control.

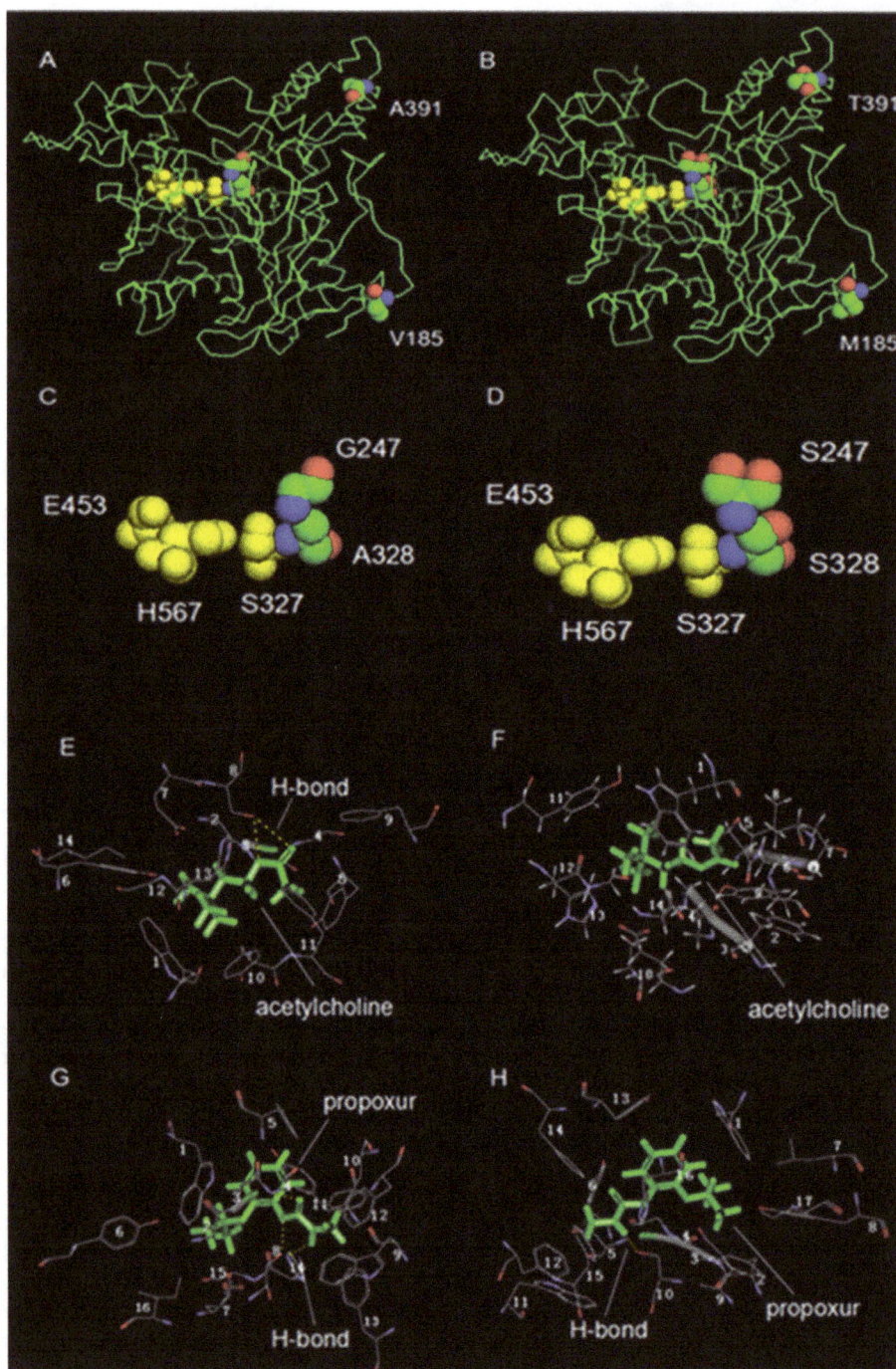

Figure 3. Three dimensional model of the AChE of *Cx. pipiens quinquefasciatus* **based on the structure of** *T. californica* **(PDB: 3ZV7).** The four mutations are shown as red, green and blue van der Waals spheres and the catalytic triad (S327, H567, E453; S200, H440, E327 in *T. californica*) is shown in yellow. A–D illustrates the four mutations. A shows the V185 and A391 positions and B the M185 and T391 mutations. C and D show the G247S and A328S mutations, and the catalytic triad. E–H shows changes in the enzyme–substrate complex; Ach (E, F) and propoxur (G, H) are shown in green and the H-bond as yellow dotted lines. Amino acids are marked with numbers. E1–14 (wild-type enzyme) are W212, G245, G246, G247, Y249, Y258, E326, S327, F416, Y456, F457, H567, G568, I571 respectively; The two H-bonds were composed of S327 Ogamma and O3, S327 Ogamma and O4. F1–14 (G247S/A328S mutant) are W212, F244, G245, G246, S250, G251, T252, L255, Y258, S327, Y456, H567, G568, I571 respectively; G1–16 (wild-type enzyme) are W212, G245, G246, G247, Y249, Y258, E326, S327, W360, F416, Y456, F457, F527, H567, G568, I571 respectively; The three H-bonds were composed of G247(4) NH and O13, S327(8) Ogamma and O11, H567(14) and NH27. H1–17 (G247S/A328S mutant) are W212, F244, G245, G246, S247, Y249, L255, Y258, E326, S327, W360, F416, Y456, F457, F527, H567, I571 respectively. The only H-bond was composed of S327(10) Ogamma and NH27.

2109bp

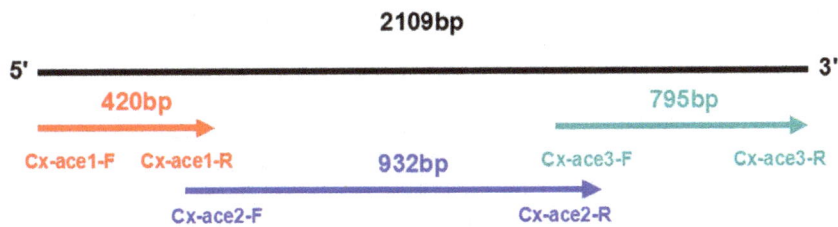

Figure 4. Schematic diagram of the amplification of the ace1 gene. The complete sequence was 2109 bp (black), and the three sections are indicated by red, blue and green arrows.

Materials and Methods

Statement of Ethical Approval

No ethical approval was required as no regulated animals were used in this study. Pre-permission (May–September 2012) was granted for observation, collection and field research on mosquitoes in Guangdong and Hainan Provinces, which was conducted as part of the Infective Diseases Prevention and Cure Project of the China National Ministry of Public Health (No.2008ZX10004 and No.2012ZX10004219). All field studies on *Cx. pipiens quinquefasciatus* were authorized by Guangdong and Hainan Provincial CDC Committees for Animal Welfare and Animal Ethics (address: 176 Xingang West Road, Guangzhou, Guangdong province, and 44 Haifu Road, Haikou, Hainan province, P. R. China).

Mosquito Strains

Specimens of *Cx. pipiens quinquefasciatus* were collected from five different field sites; Guangzhou Nansha (E113°29′29.35″, N22°48′4.13″) and Haikou Poxiang (E110°19′33.79″, N19°59′55.07″) in May 2012, and Haikou Changliu (E110°11′50.36″, N20°0′50.25″), Qionghai Boao (E110°34′57.13″, N19°09′42.07″) and Sanya Fenghuang (E109°26′54.38″, N18°18′2.91″) in September 2012. The susceptible strain had been reared in an insectarium for more than 10 years without exposure to any insecticides.

Cx. pipiens quinquefasciatus larvae were collected at each field site and reared to adulthood. Some wild caught female adults were frozen in liquid nitrogen for subsequent testing.

Bioassay

Bioassays were conducted by putting thirty late 3rd or early 4th instar larvae into pans containing 200 ml water. Measured quantities of insecticides were added to each pan using an automatic pipette according to the methods specified by the WHO [33]. Larval mortality was recorded 24 h after each treatment. No food was offered to larvae during bioassays. Larvae were maintained in the laboratory under a 14L:10D photoperiod, 75% relative humidity and temperature of 26±1°C during bioassays. Bioassays of each insecticide were repeated three times. Statistical analyses were performed using SPSS software version 13.

Extraction of RNA and cDNA Synthesis

Total RNA was extracted from specimens from each population with Trizol reagent (GBT) following the manufacturer's protocol and cDNA synthesized from the total RNA with cDNA synthesis kit (TaKaRa). The cDNA was stored at −20°C.

PCR Amplification

Gene specific primers based on the published insecticide resistant sequence of the *Cx. pipiens quinquefasciatus* ace1 gene (GenBank Accession No.:CQ753634.1, this includes a G119S mutation related to propoxur resistance) were designed in NCBI-Primer-BLAST and used to amplify the ace1 gene of each population. The ace1 gene is 2109 bp and is divided into three sections (Figure 4). The primers used are shown in Table 5.

Cloning and Sequencing of PCR Products

PCR products were purified using a universal DNA purification kit (TIANGEN) and the purified products were ligated into the pEASY-T1 vector (TRANSGEN). The recombinant plasmids were cloned into Trans1-T1 competent cells (TRANSGEN). The microbials were spread on LB solid medium (including ampicillin, X-gal, IPTG) and cultured overnight. White clones were selected, placed in LB liquid medium and cultured to turbidity. Positive clones were identified by PCR using M13 forward and reverse primers and sequenced by Sangon Biotech [25]. Based on the discovery of clones, the genotype of individual mosquitoes was determined for each amino acid position by specific PCR amplification and sequencing. In this procedure, a single mosquito's RNA was extracted and reversed transcribed to cDNA,

Table 5. The primers used to amplify the *Cx. pipiens quinquefasciatus* ace1 gene.

Primers	5′→3′Sequence	length (bp)	PCR parameters
Cx-ace1-F	ATGGAGATCCGAGGCCTAAT	420	94°C,5 min; 94°C,30 s; 62°C,30 s; 72°C,1 min,35 cycles; 72°C,7 min.
Cx-ace1-R	GCCCTTGTCCGTCGTTATG		
Cx-ace2-F	CGGACCCACTGGTCATAACG	932	94°C,5 min; 94°C,30 s; 65°C,30 s; 72°C,1 min,35 cycles; 72°C,7 min.
Cx-ace2-R	ACCCTCCTCGGTGTTGCTG		
Cx-ace3-F	CGCTTCAAGAAAACGGA	795	94°C,5 min; 94°C,30 s; 55°C,30 s; 72°C,1 min,35 cycles; 72°C,7 min.
Cx-ace3-R	TTAAATCTTGAACCGCGT		

then amplified by specific PCR before sequencing. Calculated mutation frequencies were based on the sequencing results.

Hardy–Weinberg Equilibrium (HWE) Test and Genetic Linkage Analysis of the Mutations

The Hardy–Weinberg equilibrium (HWE) describes the theoretical frequency of two alleles of a single locus in the absence of mutation and selection after one generation of random mating in an indefinitely large population with discrete generations [34]. We used GENEPOP software to analyze the HWE and genetic linkage of mutations.

Correlation of Pesticide Resistance with the Allele Frequency of Different Mutations in the Five Mosquito Populations

The resistance (LC_{50}) of the five populations to propoxur and dichlorvos was determined by bioassay and the allele frequencies of the various mutations were determined by gene specific amplification and sequencing as described above. The LC_{50} of a laboratory strain that had not been exposed to either pesticide was also determined to serve as a control. Correlations between resistance and mutation frequency were analyzed using Graphpad Prism 5.

Three-dimensional (3D) Modeling

The ace1 gene sequence of *Cx. pipiens quinquefasciatus* was translated into an amino acid sequence of AChE1. The protein was then modeled against the 3D structure of *T. californica* AChE (PDB accession no. 3ZV7) using the SWISS-MODEL homology modeling server (http://swissmodel.expasy.org/) [35,36,37] and molecular docking using the LibDock utility in Discovery Studio 2.5 [38].

Author Contributions

Conceived and designed the experiments: MHZ CXL YD XXG YMZ DX TYZ. Performed the experiments: MHZ XR ZMW TY XJZ. Analyzed the data: MHZ CXL GW HDZ. Contributed reagents/materials/analysis tools: MHZ CXL. Wrote the paper: MHZ.

References

1. Mutero A, Pralavorio M, Bride JM, Fournier D (1994) Resistance-associated point mutations in insecticide-insensitive acetylcholinesterase. Proceedings of the national academy of sciences 91: 5922–5926.
2. Kono Y, Tomita T (2006) Amino acid substitutions conferring insecticide insensitivity in Ace-paralogous acetylcholinesterase. Pesticide biochemistry and physiology 85: 123–132.
3. Hemingway J, Georghiou GP (1983) Studies on the acetylcholinesterase of Anopheles albimanus resistant and susceptible to organophosphate and carbamate insecticides. Pesticide biochemistry and physiology 19: 167–171.
4. Kwon DH, Im JS, Ahn JJ, Lee JH, Marshall Clark J, et al. (2010) Acetylcholinesterase point mutations putatively associated with monocrotophos resistance in the two-spotted spider mite. Pesticide biochemistry and physiology 96: 36–42.
5. Hemingway J (1982) Genetics of organophosphate and carbamate resistance in Anopheles atroparvus (Diptera: Culicidae). Journal of economic entomology 75: 1055–1058.
6. Chandre F, Darriet F, Doannio JMC, Riviere F, Pasteur N, et al. (1997) Distribution of organophosphate and carbamate resistance in Culex pipiens quinquefasciatus (Diptera: Culicidae) in West Africa. Journal of medical entomology 34: 664–671.
7. Smissaert HR (1964) Cholinesterase inhibition in spider mites susceptible and resistant to organophosphate. Science 143: 129–131.
8. Fournier D (2005) Mutations of acetylcholinesterase which confer insecticide resistance in insect populations. Chemico-Biological Interactions 157: 257–261.
9. Weill M, Fort P, Berthomieu A, Dubois MP, Pasteur N, et al. (2002) A novel acetylcholinesterase gene in mosquitoes codes for the insecticide target and is non–homologous to the ace gene Drosophila. Proceedings of the Royal Society of London Series B: Biological Sciences 269: 2007–2016.
10. Weill M, Lutfalla G, Mogensen K, Chandre F, Berthomieu A, et al. (2003) Comparative genomics: Insecticide resistance in mosquito vectors. Nature 423: 136–137.
11. Alout H, Berthomieu A, Hadjivassilis A, Weill M (2007) A new amino-acid substitution in acetylcholinesterase 1 confers insecticide resistance to Culex pipiens mosquitoes from Cyprus. Insect biochemistry and molecular biology 37: 41–47.
12. Djogbénou L, Akogbéto M, Chandre F (2008) Presence of insensitive acetylcholinesterase in wild populations of Cx.quinquefasciatus from Benin. Acta Tropica 107: 272–274.
13. Alout H, Berthomieu A, Cui F, Tan Y, Berticat C, et al. (2007) Different amino-acid substitutions confer insecticide resistance through acetylcholinesterase 1 insensitivity in Culex vishnui and Culex tritaeniorhynchus (Diptera: Culicidae) from China. Journal of medical entomology 44: 463–469.
14. Ben Cheikh R, Berticat C, Berthomieu A, Pasteur N, Ben Cheikh H, et al. (2009) Genes conferring resistance to organophosphorus insecticides in Culex pipiens (Diptera: Culicidae) from Tunisia. Journal of medical entomology 46: 523–530.
15. Weill M, Malcolm C, Chandre F, Mogensen K, Berthomieu A, et al. (2004) The unique mutation in ace-1 giving high insecticide resistance is easily detectable in mosquito vectors. Insect molecular biology 13: 1–7.
16. Alout H, Labbé P, Berthomieu A, Pasteur N, Weill M (2009) Multiple duplications of the rare ace-1 mutation F290V in Culex pipiens natural populations. Insect biochemistry and molecular biology 39: 884–891.
17. Suman DS, Tikar SN, Parashar BD, Prakash S (2010) Development of insecticide resistance in Culex quinquefasciatus mosquito from different locations in India. Journal of Pesticide Science 35: 27–32.
18. Tantely ML, Tortosa P, Alout H, Berticat C, Berthomieu A, et al. (2010) Insecticide resistance in Cx.quinquefasciatus and Aedes albopictus mosquitoes from La Réunion Island. Insect Biochemistry and Molecular Biology 40: 317–324.
19. Osta MA, Rizk ZJ, Labbé P, Weill M, Knio K (2012) Insecticide resistance to organophosphates in Culex pipiens complex from Lebanon. Parasites & vectors 5: 1–6.
20. Cui F, Raymond M, Berthomieu A, Alout H, Weill M, et al. (2006) Recent emergence of insensitive acetylcholinesterase in Chinese populations of the mosquito Culex pipiens (Diptera: Culicidae). Journal of medical entomology 43: 878–883.
21. Wong DM, Li J, Chen QH, Han Q, Mutunga JM, et al. (2012) Select small core structure carbamates exhibit high contact toxicity to "carbamate-resistant" strain malaria mosquitoes, Anopheles gambiae (Akron). PloS one 7: e46712.
22. Alout H, Weill M (2008) Amino-acid substitutions in acetylcholinesterase 1 involved in insecticide resistance in mosquitoes. Chemico-biological interactions 175: 138–141.
23. Silman I, Millard CB, Ordentlich A, Greenblatt HM, Harel M, et al. (1999) A preliminary comparison of structural models for catalytic intermediates of acetylcholinesterase. Chemico-Biological Interactions 119–120: 43–52.
24. Bourguet D, Prout M, Raymond M (1996) Dominance of insecticide resistance presents a plastic response. Genetics 143: 407–416.
25. Li CX, Dong YD, Song FL, Zhang XL, Zhao TY (2009) An amino acid substitution on the acetylcholinesterase in the field strains of house mosquito, Culex pipiens pallens (Diptera: Culicidae) in China. Entomological News 120: 464–475.
26. Khajehali J, Van Leeuwen T, Grispou M, Morou E, Alout H, et al. (2010) Acetylcholinesterase point mutations in European strains of Tetranychus urticae (Acari: Tetranychidae) resistant to organophosphates. Pest management science 66: 220–228.
27. Sussman JL, Harel M, Frolow F, Oefner C, Goldman A, et al. (1991) Atomic structure of acetylcholinesterase from Torpedo californica: a prototypic acetylcholine-binding protein. Science 253: 872–879.
28. Djogbénou L, Noel V, Agnew P (2010) Costs of insensitive acetylcholinesterase insecticide resistance for the malaria vector Anopheles gambiae homozygous for the G119S mutation. Malar J 9: 12.
29. Shi MA, Lougarre A, Alies C, Frémaux I, Tang ZH, et al. (2004) Acetylcholinesterase alterations reveal the fitness cost of mutations conferring insecticide resistance. BMC Evolutionary Biology 4: 5.
30. Menozzi P, Shi MA, Lougarre A, Tang ZH, Fournier D (2004) Mutations of acetylcholinesterase which confer insecticide resistance in Drosophila melanogaster populations. BMC Evolutionary Biology 4: 4.
31. Fournier D, Bride JM, Karch F, Bergé JB (1988) Acetylcholinesterase from Drosophila melanogaster Identification of two subunits encoded by the same gene. FEBS letters 238: 333–337.
32. Nabeshima T, Mori A, Kozaki T, Iwata Y, Hidoh O, et al. (2004) An amino acid substitution attributable to insecticide-insensitivity of acetylcholinesterase in a Japanese encephalitis vector mosquito, Culex tritaeniorhynchus. Biochemical and biophysical research communications 313: 794–801.
33. Cetin H, Yanikoglu A, Cilek JE (2005) Evaluation of the naturally-derived insecticide spinosad against Culex pipiens L.(Diptera: Culicidae) larvae in septic tank water in Antalya, Turkey. Journal of vector ecology 30: 151.
34. Mayo O (2008) A century of Hardy–Weinberg equilibrium. Twin Research and Human Genetics 11: 249–256.

35. Arnold K, Bordoli L, Kopp J, Schwede T (2005) The SWISS-MODEL workspace: a web-based environment for protein structure homology modelling. Bioinformatics 22: 195–201.

36. Kiefer F, Arnold K, Kunzli M, Bordoli L, Schwede T (2009) The SWISS-MODEL Repository and associated resources. Nucleic Acids Research 37: D387–D392.

37. Peitsch MC (1995) Protein modeling by E-mail Bio/Technology 13: 658–660.

38. Accelrys Inc (2009) Discovery Studio 2.5 Guide. San Diego http://www.accelrys.com.

39. Li CX (2005) Study on the genes responsible for organophosphate and carbamate resistance in Culex pipiens pallen. Beijing Institute of Microbiology and Epidemiology.

6

Cultivation-Independent Screening Revealed Hot Spots of IncP-1, IncP-7 and IncP-9 Plasmid Occurrence in Different Environmental Habitats

Simone Dealtry[1], Guo-Chun Ding[1], Viola Weichelt[1], Vincent Dunon[2], Andreas Schlüter[3], María Carla Martini[4], María Florencia Del Papa[4], Antonio Lagares[4], Gregory Charles Auton Amos[5], Elizabeth Margaret Helen Wellington[5], William Hugo Gaze[5], Detmer Sipkema[6], Sara Sjöling[7], Dirk Springael[2], Holger Heuer[1], Jan Dirk van Elsas[8], Christopher Thomas[9], Kornelia Smalla[1]*

1 Julius Kühn-Institut – Federal Research Centre for Cultivated Plants (JKI), Institute for Epidemiology and Pathogen Diagnostics, Braunschweig, Germany, 2 Division of Soil and Water Management, KU Leuven, Heverlee, Belgium, 3 Center for Biotechnology (CeBiTec), Institute for Genome Research and Systems Biology, Bielefeld University, Bielefeld, Germany, 4 IBBM (Instituto de Biotecnología y Biología Molecular), CCT-CONICET-La Plata, Departamento de Ciencias Biológicas, Facultad de Ciencias Exactas, Universidad Nacional de La Plata, La Plata, Argentina, 5 School of Life Sciences, University of Warwick, Warwick, United Kingdom, 6 Laboratory of Microbiology, Wageningen University, Wageningen, The Netherlands, 7 Södertörns högskola (Sodertorn University), Inst. för Naturvetenskap, Miljö och medieteknik (School of Natural Sciences, Environmental Studies and media tech), Huddinge, Sweden, 8 University of Groningen, Groningen, The Netherlands, 9 School of Biosciences, University of Birmingham, Edgbaston, Birmingham, Warwick, United Kingdom

Abstract

IncP-1, IncP-7 and IncP-9 plasmids often carry genes encoding enzymes involved in the degradation of man-made and natural contaminants, thus contributing to bacterial survival in polluted environments. However, the lack of suitable molecular tools often limits the detection of these plasmids in the environment. In this study, PCR followed by Southern blot hybridization detected the presence of plasmid-specific sequences in total community (TC-) DNA or fosmid DNA from samples originating from different environments and geographic regions. A novel primer system targeting IncP-9 plasmids was developed and applied along with established primers for IncP-1 and IncP-7. Screening TC-DNA from biopurification systems (BPS) which are used on farms for the purification of pesticide-contaminated water revealed high abundances of IncP-1 plasmids belonging to different subgroups as well as IncP-7 and IncP-9. The novel IncP-9 primer-system targeting the *rep* gene of nine IncP-9 subgroups allowed the detection of a high diversity of IncP-9 plasmid specific sequences in environments with different sources of pollution. Thus polluted sites are "hot spots" of plasmids potentially carrying catabolic genes.

Editor: Axel Cloeckaert, Institut National de la Recherche Agronomique, France

Funding: This study was funded by the EU 7th Framework Programme (MetaExplore 222625) and the Inter-University Attraction Pole (IUAP) "µ-manager" of the Belgian Science Policy (BELSPO, P7/25). The funders had no role in study design, data collection and analysis, decision to publish, or preparation of the manuscript.

Competing Interests: The authors have declared that no competing interests exist.

* E-mail: kornelia.smalla@jki.bund.de

Introduction

The search for novel enzymes able to degrade recalcitrant natural contaminants such as chitins and lignins and man-made pollutants such as halogenated aliphatic and aromatic compounds motivated metagenomic explorations of various environments. It has been observed that the microbial metagenomes of open ecosystems, including soils and aquatic habitats, clearly represent rich reservoirs of genes that determine the desired enzymatic reactions in which chitinases, ligninases and dehalogenases are involved [1,2]. By anthropogenic activities, recalcitrant compounds have also been released as environmental pollutants. Typical metagenomic approaches employ genetic or activity screens of cloned large DNA fragments from various environments [3]. However, the idea of capturing complete mobile genetic elements (MGE) into suitable recipients might be an alternative and complementary approach to access the genes coding for novel enzymes or even complete degradative pathways. Mobile genetic elements such as plasmids are often found to play an important role in the adaptation of bacterial communities to changing and, due to pollutants, often challenging environmental conditions. For example, partial or complete degradative pathways were previously reported to be localized on plasmids belonging to the IncP-1, IncP-7 or IncP-9 group [1]. The present study aimed to monitor various environments for the abundance of these plasmids by using a cultivation-independent total community (TC-) DNA based approach to select the most promising habitats for mining plasmids potentially carrying genes coding for novel enzymes. We hypothesized that the frequency of occurrence of genes encoding the desired enzymatic activities is increased in the MGE gene pool. In particular, plasmids belonging to the incompatibility groups (Inc) P-1, P-7 and P-9 often carry genes responsible for the degradation of xenobiotic (man-made) and natural organic pollutants, being essential players in the adaptation of bacterial communities to new toxic compounds released in the environment [1]. Therefore, selected natural or treated environments were analyzed for the prevalence of plasmids belonging to the IncP-1, IncP-7 and IncP-9 groups by a cultivation-independent approach. Some of these environments were enriched for the desired degradation function by adding the relevant substrates, i.e. chitin, lignin and/or organohalogens. The habitats sampled included a variety of soils (one soil sample amended with chitin, peat bogs), biopurification systems (BPS) for pesticide removal from contaminated water, biogas production plants, wastewater, as well as aquatic (river bank sediments, sponges) environments from a wide range of geographic regions. Total community DNA was analyzed for the presence of IncP-1, IncP-7 and IncP-9 plasmids by means of PCR and subsequent Southern blot hybridization. A novel primer system for the specific amplification of IncP-9 plasmids was developed and tested in the present study. Southern blot hybridization using probes derived from reference plasmids belonging to different subgroups of IncP-1 plasmids provided new insights into their environmental dissemination. Our results showed a particularly widespread dissemination of IncP-1 plasmid-specific sequences. Different hot spots of plasmid occurrence were identified.

Materials and Methods

Ethics Statement of Provided Samples

None of the samples used in the present work involved any endangered or protected species. The marine sponges were obtained under legal permits from competent authorities: *Halichondria panicea* was obtained under a permit from authorities given to Wageningen University, while *Corticium candelabrum* and *Petrosia ficiformis* marine sponges were sampled under a Spanish permit to CEAB-CSIC. The sediments and soil originated from UK were taken from a river bed accessed from a public right of way and therefore no permissions were needed. Landsort Deep was sampled from a national environmental monitoring site (BY31) in conjunction with the Baltic Sea monitoring programme. The sampling permission was provided by the Stockholm University marine research Center. The Askö samples were sampled at the Stockholm Marine Research Center (now the Stockholm University Marine Research Center. The sediment and soils from Argentina were obtained from public locations as part of fundamental studies performed through a collaborative project with the agreement of the Facultad de Ciencias Exactas, Universidad Nacional de La Plata, and did not require any specific permission. The biopurification systems (BPS) samples were obtained from private land with permission from the local farmers in Kortrijk, Leefdaal, Lierde and Koksijde, located in Belgium.

Extraction of Total Community DNA (TC-DNA) and Metagenomic DNA (Pooled Fosmid Library) from Different Environmental Samples

The TC-DNA and/or metagenomic DNA (metagenomic DNA represented by the metagenomic pooled fosmid library from Baltic Sea) from different environmental samples originating from various geographic regions were extracted using different methods. The protocols used for TC-DNA extraction of each sample type are given in Table 1.

16S rRNA Gene PCR Amplification and Quantification

16S rRNA gene PCR amplification reaction was done as previously described by Heuer *et al.* (2009) [5] (product size of 1506 bp). The quality of the PCR product was determined by electrophoresis in 1% agarose gel and visualized with ethidium bromide staining and under UV light by comparison with the 1-kb gene-rulerTM DNA ladder (Fermentas, St Leon-Rot, Germany). Quantitative PCR (qPCR) targeting the 16S rRNA gene was performed with the TaqMan system as described by Suzuki *et al.* (2000) [6]. The 16S rRNA gene qPCR standard was made from cloned 16S rRNA gene amplicons (1467 bp) of *E. coli.* and 10^9, 10^8, 10^7 16S rRNA gene copy numbers were used.

Southern Blot-PCR Based Detection of IncP-1 Plasmids

IncP-1 plasmids belonging to the α, β, γ, δ and ε subgroups were detected based on the amplification of the *trfA* region (product size of 281 bp) from TC-DNA and metagenomic DNA using the primers described by Bahl *et al.* (2009) [2]. Digoxigenin-labeled probes targeting different IncP-1 plasmids subgroups were generated from reference plasmids belonging to the IncP-1α, β, γ, δ and ε plasmids (Table 2). The IncP-1 mixed probe was prepared by mixing probes generated for the different subgroups. The random primed digoxigenin labeling of PCR amplicons excised from preparative agarose gels was done according to the Roche manufacturer's protocol (Roche Diagnostics Deutschland GmbH, Mannheim, Germany).

Southern Blot-PCR Based Detection of IncP-7

PCR amplification of the *rep* region of IncP-7 plasmids (product size of 524 bp) from TC-DNA was performed as previously described by Izmalkova *et al.* (2005) [4]. Southern blotted PCR amplicons were hybridized at medium stringency with the dig-labeled IncP-7 probe generated from the reference plasmid pCAR1 isolated from *Pseudomonas resinovorans* according to the manufacturer's instructions (QIAGEN® Plasmid Mini Kit) (Table 2). The randomly primed digoxigenin labeling of PCR amplicons was done as described above.

Analyzing the Diversity and Abundance of IncP-9 Plasmids by a Novel PCR System Targeting the *oriV-rep* Region

To study the abundance and diversity of IncP-9 plasmids, a novel PCR system targeting the *oriV-rep* regions was developed and applied to detect IncP-9 plasmids in TC- and metagenomic DNA from all samples analyzed (Table 1).

Multiple alignments of 28 sequences of *oriV* (EU499619-EU499641, AF078924, AB237655, AJ344068, AB257759 and AF491307) and *rep* (EU499644-EU499666, AF078924, AB237655, AJ344068, AB257760 and AF491307) were performed with Molecular Evolutionary Genetics Analysis (MEGA 4). Conserved regions of sequences belonging to nine IncP-9 subgroups [7] were used for the primer design. The selected primer system consists of 21-mer degenerate forward primer (5-GAG GGT TTG GAG ATC ATW AGA-3) and reverse primer (5-GGT CTG TAT CCA GTT RTG CTT-3). *In silico* analysis showed no mismatch for at least 12 bp at the 3′ end of each primer and 1–4 mismatches for each sequence type at the 5′ end (Fig. S1). The expected amplicon size is 610–637 bp. The primers were further tested with plasmid DNA from the reference plasmids summarized in Fig. S1. None of the plasmids belonging to other incompatibility groups was amplified while the reference plasmids were amplified. The reaction mixture (25 µl) contained 1 µl template DNA (1–5 ng), 1× Stoffel buffer (Applied Biosystems, Foster, CA), 0,2 mM dNTPs, 2,5 mM MgCl$_2$, 2 µg/µl bovine serum albumin, 0.2 µM of each primer, and 2.5 U TrueStartTaq DNA polymerase (Stoffel fragment, Applied Biosystems). Denaturation was carried out at 94°C for 5 min, followed by 35 cycles

Table 1. Description of environmental samples analyzed and TC-DNA extraction applied.

Samples	Description of samples	TC-DNA extraction method
A	Biogas production plant fermentation sample from Bielefeld, Germany	[24]
B.1	*Biopurification system (BPS) from Leefdaal, Belgium	[17]
B.2	BPS from Leefdaal, Belgium	[17]
B.3	BPS from Leefdaal, Belgium	[17]
C.1	*BPS from Belgium (Pcfruit)	[17]
C.2	BPS from Belgium (Pcfruit)	[17]
C.3	BPS from Belgium (Pcfruit)	[17]
C.4	BPS from Belgium (Pcfruit)	[17]
C.5	BPS from Belgium (Pcfruit)	[17]
C.6	BPS from Belgium (Pcfruit)	[17]
D.1	*BPS from Lierde, Belgium	[17]
D.2	BPS from Lierde, Belgium	[17]
D.3	BPS from Lierde, Belgium	[17]
E.1	*BPS from Kortrijk, Belgium	[17]
E.2	BPS from Kortrijk, Belgium	[17]
E.3	BPS from Kortrijk, Belgium	[17]
F.1	*BPS from Koksijde, Belgium	[17]
F.2	BPS from Koksijde, Belgium	[17]
F.3	BPS from Koksijde, Belgium	[17]
G.1	Soil from La Plata, Argentina polluted with industrial residues and petrol	[18]
G.2	Soil from La Plata, Argentina polluted with industrial residues and petrol	[18]
G.3	Soil from La Plata, Argentina polluted with industrial residues and petrol	[18]
H.1	Sediments from La Plata, Argentina polluted with pesticides and petrol	[18]
H.2	Bordering soil from a water channel in La Plata, Argentina polluted with pesticides, residues from paper industry	[18]
H.3	Bordering soil from a water channel in La Plata, Argentina polluted with pesticides, residues from paper industry	[18]
J	Marginal river forest soil from La Plata, Argentina polluted with industrial residues	[18]
L.1	Bordering soil from a water channel in Buenos Aires, Argentina polluted with industrial residues	[18]
L.2	Bordering soil from a water channel in Buenos Aires, Argentina polluted with industrial residues	[18]
L.3	Bordering soil from a water channel in Buenos Aires, Argentina polluted with industrial residues	[18]
M	*Halichondria panicea* (marine sponge) from Oosterschelde, Netherlands	[25]
N	*Corticium candelabrum* (marine sponge) from Punta Santa Anna (Blanes), Spain	[25]
O	*Petrosia ficiformis* (marine sponge) from Punta Santa Anna (Blanes), Spain	[25]
P.1	Askö sediment from Baltic Sea Sweden (bottom fraction - anoxic)	[26]
P.2	Askö sediment from Baltic Sea Sweden (middle fraction - mixed anoxic/oxic)	[26]
P.3	Askö sediment from Baltic Sea Sweden (top fraction - oxic)	[26]
Q	Pooled fosmid library, Askö sediment, Baltic Sea	[3]
R	Landsort in Sweden	[26]
S.1	Sediment from a river in Warwickshire, UK	[27]
S.2	Sediment from a river in Warwickshire, UK	[28]
T	Soil from UK amended with chitin (Test site 1)	[28]

*BPS samples received water contaminated with different types of pesticides from spillage and residue water collected when cleaning the spraying equipment such as ethofumesate, fenpropimorf, fluroxypyr, glyphosate, linuron, metamitron and S-metalochlor (information provided by the farmers).

of 1 min at 94°C, 1 min at 53°C (primer annealing) and 2 min at 72°C and a final extension of 10 min at 72°C.

PCR amplicons of *oriV-rep* regions of nine IncP-9 subgroups IncP-9 plasmids (Table 2) were gel-purified and digoxigenin-labeled as described above. Southern blot hybridization of *oriV-rep* amplicons from different environmental samples listed above was performed with a mixture of these probes under medium stringency following the manufacturer's instructions (Roche Diagnostics Deutschland GmbH, Mannheim, Germany). Clone libraries were generated for these three BPS to confirm primer specificity. *oriV-rep* amplicons were gel-purified, ligated into pGEM vectors, and transformed into *E. coli* JM109 competent cells according to the instructions of the manufacturer. Clones containing the correct inserts were selected for sequencing.

Table 2. Generation of probes for Southern blot hybridization.

Probe	Reference plasmid	Plasmids host strain	Primers
IncP-1α	RP4	*E. coli*	[2]
IncP-1β	R751	*E. coli CM544*	[2]
IncP-1γ	pQKH54	*E. coli DH10B*	[2]
IncP-1δ	pEST4011	*Alcaligenes xylosoxidans EST4002*	[2]
IncP-1ε	p3-408	*E. coli cv601-GFP*	[2]
IncP-7	pCAR1,	*Pseudomonas resinovorans CA10*	[4]
IncP-9 α	pM3	*Pseudomonas putida*	This study
IncP-9 β	pBS2	*Pseudomonas putida BS268*	This study
IncP-9 γ	pSN11	*Pseudomonas putida BS349*	This study
IncP-9 δ	pSN11	*Pseudomonas putida SN11*	This study
IncP-9 ε	pMG18	*Pseudomonas putida AC34*	This study
IncP-9 ζ	pNL60	*Pseudomonas spp. 18d/1*	This study
IncP-9 η	pNL15	*E. coli C600*	This study
IncP-9 θ	pSVS15	*Pseudomonas fluorescens SVS15*	This study
IncP-9 ι	pNL22	*Pseudomonas spp. 41a/2*	This study

BLAST-N analysis was used to identify *oriV-rep* sequences of IncP-9. All sequences analyzed share high similarity with IncP-9 *oriV* or *rep* sequences in NCBI. The sequences and those of known *oriV-rep* sequences in the data base were aligned and phylogenetic tree was calculated according to the neighbor-joining method and bootstrapping analysis using MEGA 4.

Nucleotide Sequence Accession Numbers of Cloned IncP-9 *oriV-rep* Gene Amplicons

Amplicon sequences have been submitted to NCBI SRA with IncP-9 *oriV-rep* gene amplicons under accession numbers KF706553 - KF706633.

Results

Determination of Bacterial 16S rRNA Gene Copies by qPCR

To estimate the bacterial density of the different environmental samples analyzed, 16S rRNA gene copies were determined by quantitative real-time PCR from the TC-DNA. Most of the samples (Table 3) showed a high abundance of bacterial populations ranging from 10^8 to 10^9 16S rRNA gene copy numbers per gram of material. For a few samples significantly lower 16S rRNA gene copy numbers per gram of material (Tukey's test p>0.05) were detected (Table 3).

Distribution of IncP-1 Plasmids in Different Environments

To investigate the presence of IncP-1 plasmids in different habitats a detection system based on Southern blot-PCR was applied. Using the IncP-1 mixed probe from PCR products hybridization signals of the expected size (251 bp) were detected in a very wide range of different habitats (Table 3), indicating that IncP-1 plasmids of different subgroups are widely distributed. By using probes specific for the five different IncP-1 different subgroups (α, β, γ, δ and ε), differences in the composition of IncP-1 plasmids according to the geographic area and sample type were observed. Strong hybridization signals of IncP-1α plasmids

were only observed in one TC-DNA from Askö sediment (Sweden), in TC-DNA from a biogas production plant (Germany) and fosmid DNA from Baltic Sea sediments. Strong hybridization signals were observed using the IncP-1β specific probe in the TC-DNAs of all biopurification system (BPS) samples from Belgium (Table 3, Fig. 1) and most of the sediment samples from Argentina, indicating that in these environments bacterial populations carrying IncP-1β plasmids were highly abundant. The highest IncP-1γ hybridization signal was observed in the TC-DNA of the BPS located in Kortrijk. Less intense IncP-1γ hybridization signals were detected in the TC-DNAs of other BPS from Belgium and in TC-DNA of sediments from Argentina. In all TC-DNAs of BPS from Belgium, strong IncP-1δ hybridization signals were observed and a weaker hybridization signal, compared to BPS TC-DNA, was detected in TC-DNA from sediments in Argentina. Very strong IncP-1ε hybridization signals were again detected in all BPS TC-DNAs from Belgium (Table 3, Fig. 2) and most of the sediments from Argentina. Using IncP-1 mixed-probe, strong hybridization signals were detected in soils from Argentina and soil treated with chitin from the UK, indicating a high abundance of IncP-1 plasmids.

Distribution of IncP-7 Plasmids in Different Environments

To investigate the occurrence of IncP-7 plasmids in different environments, a PCR-based detection approach was applied in combination with Southern blot hybridization. Strong hybridization signals were observed in all TC-DNAs from BPS analyzed (Fig. 3), indicating a high abundance of bacterial populations carrying in BPS IncP-7 plasmid. Less intense hybridization signals were observed in the TC-DNAs of seven sediment river samples from Argentina and in the TC-DNA from soil amended with chitin from the UK. Hybridization signals using the amplicon probe specific for IncP-7 plasmids were not detected in any of the other environmental samples analyzed (Table 3).

IncP-9 Plasmid Occurrence and Diversity in Different Environmental Samples

In order to verify the occurrence and diversity of IncP-9 plasmids in different habitats the new IncP-9 primer system developed in the present work was applied. Very strong hybridization signals were detected in all TC-DNAs of BPS samples, indicating that BPS are reservoirs of bacteria carrying IncP-9 plasmids. Less intense hybridization signals were observed in the TC-DNA of sediment samples from Argentina. A weaker hybridization signal was detected in the soil amended with chitin from the UK (Table 3).

To verify primer specificity and to gain insights into the IncP-9 plasmid diversity from BPS samples (indicated as a "hot spot" of IncP-9 plasmids), a clone library was generated with amplicons from PCR using primers targeting the IncP-9 *oriV-rep* region in TC-DNA of three different BPS. Sequencing revealed the presence of different IncP-9 subgroups while phylogenetic analysis (Fig. 4) showed IncP-9 plasmid types similar to *oriV-rep* sequences of pWWO and pM3 as well as several sequences that could not be affiliated to previously known IncP-9 plasmid groups indicating an undiscovered diversity of this plasmid group.

Discussion

In the present study a PCR-based screening combined with Southern blot hybridization allowed the detection of IncP-1, IncP-7 and IncP-9 plasmids in a wide range of different geographic areas and sample types. The results indicated a high abundance of these plasmids in environments with different sources of pollution.

Table 3. Bacterial densities and PCR-Southern blot hybridization detection of plasmid replicon-specific sequences belonging to the five IncP-1 subgroups, IncP-7 and IncP-9.

Sample	Description of samples	P-1	α	β	ε	γ	δ	P-7	P-9	16S log10/g
A	Biogas production plant from Bielefeld, Germany	+++	++	++	++	-	++	-	-	9,34
B.1	Biopurification system (BPS) from Leefdaal, Belgium	+++	-	++	+++	++	+++	+++	+++	9,32
B.2	BPS from Leefdaal, Belgium	+++	-	++	+++	++	+++	+++	+++	9,25
B.3	BPS from Leefdaal, Belgium	+++	-	++	+++	++	++	+++	+++	8,43
C.1	BPS from Belgium (Pcfruit)	+++	+	+++	+++	++	++	+++	+++	9,32
C.2	BPS from Belgium (Pcfruit)	+++	+	+++	++	++	+++	+++	+++	8,28
C.3	BPS from Belgium (Pcfruit)	++	-	++	++	-	+++	++	+	8,36
C.4	BPS from Belgium (Pcfruit)	+	-	++	+	+	+++	+++	+++	8,54
C.5	BPS from Belgium (Pcfruit)	+++	(+)	+++	+++	++	+++	+++	+++	8,66
C.6	BPS from Belgium (Pcfruit)	+	-	++	++	+	++	++	-	8,15
D.1	BPS from Lierde, Belgium	+++	-	++	-	-	+++	++	++	8,61
D.2	BPS from Lierde, Belgium	+++	-	++	++	-	+++	++	++	8,59
D.3	BPS from Lierde, Belgium	+++	-	++	+++	++	+++	++	++	8,31
E.1	BPS from Kortrijk, Belgium	+++	-	+++	+++	+++	+	+++	++	9,2
E.2	BPS from Kortrijk, Belgium	+++	-	+++	+++	+++	++	+++	+++	9,03
E.3	BPS from Kortrijk, Belgium	+++	-	++	++	+++	-	+++	+++	9,11
F.1	BPS from Koksijde, Belgium	++	(+)	++	+	-	+++	(+)	+++	9,01
F.2	BPS from Koksijde, Belgium	++	+++	++	-	-	++	-	+++	8,9
F.3	BPS from Koksijde, Belgium	++	(+)	++	+	-	+++	-	+++	8,95
G.1	Soil from La Plata, Argentina	+++	(+)	+	+	-	+++	+++	+++	8,55
G.2	Soil from La Plata, Argentina	+++	-	(+)	+	++	+	+++	+++	8,53
G.3	Soil from La Plata, Argentina	+++	-	++	-	++	+	-	(+)	8,22
H.1	Sediments from La Plata, Argentina	+++	++	+++	++	++	+++	+++	+++	8,96
H.2	Bordering soil from a water channel in La Plata, Argentina	+++	++	++	++	++	+++	+	+++	8,49
Sample	Description of samples	P-1	α	β	ε	γ	δ	P-7	P-9	16S log10/g
H.3	Bordering soil from a water channel in La Plata, Argentina	+++	+	++	++	++	+++	-	+++	8,7
I	Sweet-water soil from a river in La Plata, Argentina	+++	+	++	++	++	++	+++	+++	7,91
J	Marginal river forest soil from La Plata, Argentina	++	-	-	-	-	-	-	-	8,32
L.1	Bordering soil from a water channel in Buenos Aires, Argentina	+++	-	++	+	-	++	-	-	8,29
L.2	Bordering soil from a water channel in Buenos Aires, Argentina	+++	+	+++	++	++	+++	-	(+)	8,6
L.3	Bordering soil from a water channel in Buenos Aires, Argentina	+++	-	++	++	++	+++	++	+++	7,66
M	Halichondria panicea (marine sponge) from Oosterschelde, Netherlands	++	-	++	-	-	+++	-	-	7,32
N	Corticium candelabrum (marine sponge) from Punta Santa Anna (Blanes), Spain	++	-	+	-	-	+	-	-	8,18
O	Petrosia ficiformis (marine sponge) from Punta Santa Anna (Blanes), Spain	++	-	++	-	-	+	-	-	8,4
P.1	Askö sediment from Baltic Sea Sweden (bottom fraction - anoxic)	++	-	-	-	-	+	-	-	8,34
P.2	Askö sediment from Baltic Sea Sweden (middle fraction mixed anoxic/oxic)	++	++	+++	++	-	+	-	-	8,43
P.3	Askö sediment from Baltic Sea Sweden (top fraction - oxic)	+++	-	++	+++	-	+	-	-	8,09
Q	Pooled fosmid library, Askö sediment, Baltic Sea	+++	++	+	+	-	+	-	-	5,01
R	Landsort in Sweden	+++	++	-	-	-	-	-	-	8,16
S.1	Sediment from a river in Warwick, UK	+++	/	/	/	/	/	-	(+)	5,78
S.2	Sediment from a river in Warwick, UK	+++	/	/	/	/	/	++	-	6,26
T	Soil from Cuba amended with chitin (Test site 1)	+++	/	/	/	/	/	+++	++	6,95
negative control		-	-	-	-	-	-	-	-	

Table 3. Cont.

Sample	Description of samples	P-1	α	β	ε	γ	δ	P-7	P-9	16S log10/g
RP4 (IncP-1α)		+++	+++							
R751 (IncP-1β)		+++		+++						
pKJK5 (IncP-1ε)		++			++					
pQKH54 (IncP-1γ)		+++				+++				
pEST4011 (IncP-1δ)		+++					+++			
pCAR1 (IncP-7)								+++		
pNF 142 (IncP-9)									+++	

Hybridization signal: (+++) very strong, with exposure time up to five minutes; (++) strong, with exposure time up to one hour; (+) weak, with exposure time up to three hours; (−) none, with exposure time of more than three hours; (/) not analyzed.

It is tempting to speculate that degradative genes localized on the plasmid of these groups might contribute to the bacterial degradation of a variety of pollutants such as pesticides, due to the "metabolic complementation" resulting from the combination of different genes brought together by different plasmids. While IncP-1 plasmids typically host genes associated with the degradation of man-made pollutants (xenobiotics) [8], IncP-7 and IncP-9 plasmids often carry genes responsible for degradation of natural contaminants, such as polyaromatic hydrocarbons [9]. Screening TC-DNA revealed that IncP-1, IncP-7 and IncP-9 specific sequences vary according to sample type and degree of pollution. IncP-1 plasmid specific sequences were detected in a wide range of environments: marine sponges, soils and sediments, Baltic Sea sediment fosmid library, biogas production plant, river sediments, chitin-treated soils and BPS contaminated with pesticides. Very strong hybridization signals for all different IncP-1 subgroups tested except for IncP-1α plasmids were especially observed in the BPS samples heavily contaminated with pesticides, indicating an

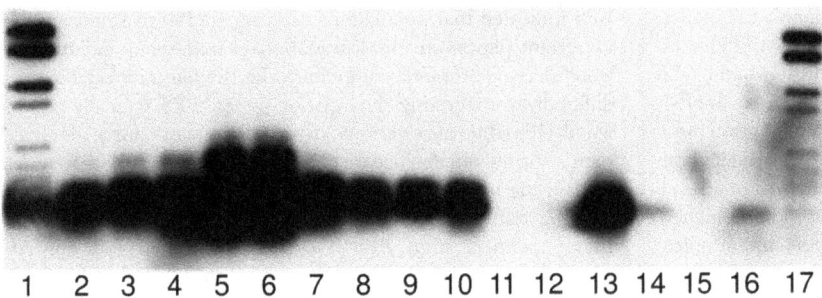

Figure 1. Biopurification systems (BPS). Hybridization of Southern-blotted PCR products obtained with *trfA* primer system from TC-DNA of BPS (IncP-1β specific group). Lanes: 1 and 17, dig ladder; lanes 2 to 4, BPS from Lierde, Belgium; lanes 5 to 7, BPS from Kortrijk, Belgium; lanes 8–10, BPS from Koksijde, Belgium; lane 11, negative control; lanes 12–16, IncP-1 positive controls RP4 (α), R751 (β), pKJK5 (ε), pQKH54 (γ) and pEST4011 (δ). Exposure time of 5 min.

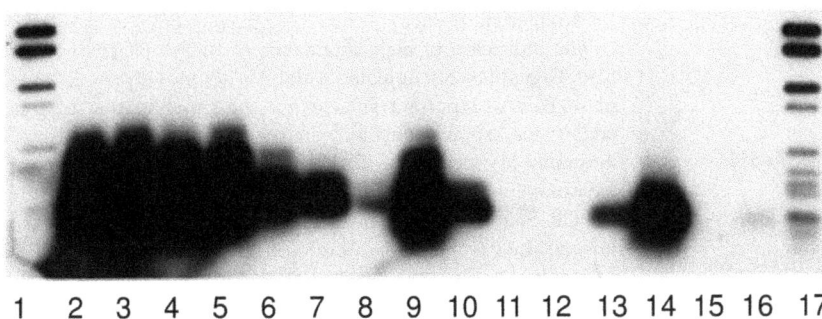

Figure 2. Biopurification systems (BPS). Hybridization of Southern-blotted PCR products obtained with *trfA* primer system from TC-DNA of BPS with the IncP-1ε specific probe. Lanes: 1 and 17, dig ladder; lanes 2 to 4 BPS from Lierde, Belgium; lanes 5 to 7, BPS from Kortrijk, Belgium; lanes 8 to 10, BPS from Koksijde, Belgium; lane 11, negative control; lanes 12 to 15, IncP-1 positive controls RP4 (α), R751 (β), pKJK5 (ε), pQKH54 (γ) and pEST4011 (δ). Exposure time of 5 min.

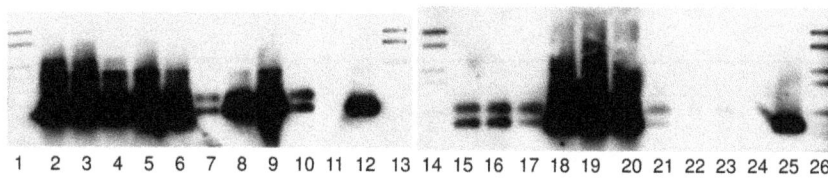

Figure 3. Biopurification systems (BPS). Hybridization of Southern-blotted PCR products obtained with *rep* primer system from TC-DNA of BPS with the IncP-7 probe generated from pCAR1. Lanes: 1, 13 and 26, dig ladder; lanes 2 to 4, BPS from Leefdaal, Belgium; lanes 5 to 10, BPS from Belgium (Pcfruit); lanes 15 to 17, BPS from Lierde, Belgium; lanes 18 to 20, BPS from Kortrijk, Belgium; lanes 21 to 23, BPS from Koksijde, Belgium; lanes 11 and 24, negative control; lanes 12 and 25 IncP-7 positive control pCAR-1. Exposure time of 5 min.

unusual high abundance of bacterial populations carrying IncP-1 plasmids. Indeed, the use of BPS, defined as a pollution control technique employing microorganisms to degrade pesticides through biodegradation processes [10], in on-farm treatment of water contaminated with pesticides has substantially increased and enhanced the degradation rates [11]. Strong hybridization signals of IncP-1β and IncP-1ε plasmids observed in all BPS samples and in some sediments from Argentina contaminated with oil, suggested that IncP-1β and IncP-1ε plasmids might be important in the local adaptation of bacteria to changing environmental conditions [12,13]. Strong IncP-1 plasmid hybridization signals observed in sediments from different regions: Warwick (UK), La Plata (Argentina) and sediments from Sweden indicated that IncP-1 plasmids might also have an important ecological role in the adaptation and biodegradation processes in sediments as previously reported already for mercury-contaminated sediments in Kazakhstan [14]. The apparently high abundance of IncP-1 plasmids in soils from different regions contaminated with different pollutants, such as soils from Argentina polluted with oil and soils from the UK enriched with chitin, also suggested that IncP-1 plasmids might substantially contribute to the adaptation and survival of the soil bacterial communities in response to wide range of environmental pollutants [8,15–17]. The results from several studies suggested a correlation between IncP-1 plasmid abundance and pollution as hypothesized by Smalla *et al.* (2006) and confirm

previously published quantitative data on the abundance of IncP-1 plasmids in BPS samples from one BPS site by means of a qPCR targeting the *korB* gene. Obviously, the relative abundance of IncP-1 plasmids can only be precisely quantified by quantitative real-time PCR. However, the recently developed *korB* quantitative PCR system [18] cannot indicate the relative abundance of the different IncP-1 subgroups which was achieved with specific probes for different IncP-1 groups used in the present study in a semi-quantitative manner.

The study by Sevastsyanovich *et al.* (2008) already showed that IncP-9 plasmid diversity is much broader than previously imagined. In view of this huge plasmid diversity, a novel IncP-9 primer system was developed and established in the present work. Typically, IncP-9 plasmids are related to the degradation of natural pollutants as polyaromatic hydrocarbons [19]. However, the detection of very strong IncP-9 hybridization signals mainly in BPS indicated that populations carrying IncP-9 plasmids are also important players in the degradation of man-made pollutants or wood-derived aromatic compounds. IncP-9 plasmids often possess different aromatic-ring degradation genes. BPS typically contain wood chips but also various aromatic ring-containing pesticides such as bentazon, epoxiconazol and diflufencan [20], which could explain the high abundance of IncP-9 plasmids observed in BPS. Cloning and sequencing of amplicons obtained with the novel IncP-9 primers from BPS TC-DNA confirmed not only the specificity of the primers but also showed the presence of plasmids with high similarity to pWWO, that were previously reported to carry degradative genes (Fig. 4) [21,22]. The presence of several sequences with high similarity to the *oriV-rep* sequence of pM3, an antibiotic resistance plasmid belonging to the IncP-9α subgroup, in BPS 2 might be caused by manure addition in the beginning of every year (on March) by the farmers as a C-source in BPS material. Therefore, the addition of manure in BPS as nutrient source for the microorganisms might be reconsidered and replaced for an alternative one.

The indication of high abundance of IncP-9 plasmids in soils from Argentina contaminated with oil is not too surprising. IncP-9 plasmids are important vehicles for the dissemination of genes coding for enzymes involved in the degradation of polycyclic aromatic hydrocarbons (PAH) and are very often found in environments polluted with oil [23] (Flocco *et al.*, unpublished).

PCR-Southern blot hybridization results showed that bacteria hosting IncP-7 plasmids were also highly abundant in BPS, indicating a role of these plasmids in the degradation of man-made pollutants such as pesticides. It can be concluded that PCR-Southern blot hybridization detection of plasmid-specific sequences from TC-DNA is a suitable and specific but semi-quantitative approach to investigate the occurrence of plasmid-specific sequences in different environments and in a large number of samples. The detection of plasmids was possible independently of

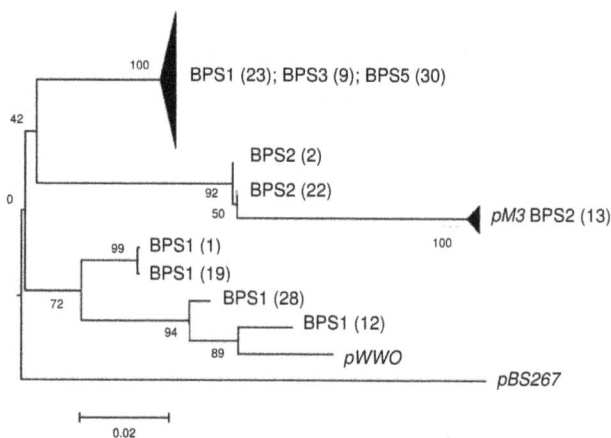

Figure 4. Neighbor-Joining phylogenetic tree based on the multiple alignment of cloned amplicon sequences of the *oriV-rep* IncP-9 gene. Sequences from known IncP-9 plasmids have been included as references. Value at each node is percent bootstrap support of 1,000 replicates. BPS1; BPS2 and BPS5 correspond to three different biopurification systems (BPS), located in Belgium. Numbers in brackets correspond to number of clones and numbers without brackets correspond to the clone designation.

the cultivation of their original hosts [1] and indicated "hot spots" of IncP-1, IncP-7 and IncP-9 plasmids, such as BPS.

Author Contributions

Performed the experiments: SD VW. Analyzed the data: SD GCD HH KS. Contributed reagents/materials/analysis tools: HH CT. Wrote the paper: SD. Contributed sending samples: VD AS MCM MFP AL GCA EMHW WHG D Sipkema SS D Springael JDE. Corrected the writing: KS.

References

1. Heuer H, Smalla K (2012) Plasmids foster diversification and adaptation of bacterial populations in soil. FEMS Microbiology Reviews 36: 1083–1104.
2. Bahl MI, Burmølle M, Meisner A, Hansen LH, Sørensen SJ (2009) All IncP-1 plasmid subgroups, including the novel ε subgroup, are prevalent in the influent of a Danish wastewater treatment plant. Plasmid 62: 134–139.
3. Hardeman F, Sjoling S (2007) Metagenomic approach for the isolation of a novel low-temperature-active lipase from uncultured bacteria of marine sediment. FEMS Microbiology Ecology 59: 524–534.
4. Izmalkova TY, Sazonova OI, Sokolov SL, Kosheleva IA, Boronin AM (2005) The P-7 incompatibility group plasmids responsible for biodegradation of naphthalene and salicylate in fluorescent pseudomonads. Microbiology 74: 290–295.
5. Schauss K, Focks A, Heuer H, Kotzerke A, Schmitt H, et al. (2009) Analysis, fate and effects of the antibiotic sulfadiazine in soil ecosystems. Trac-Trends in Analytical Chemistry 28: 612–618.
6. Suzuki MT, Taylor LT, DeLong EF (2000) Quantitative analysis of small-subunit rRNA genes in mixed microbial populations via 5′-nuclease assays. Applied and Environmental Microbiology 66: 4605–4614.
7. Sevastsyanovich YR, Krasowiak R, Bingle LEH, Haines AS, Sokolov SL, et al. (2008) Diversity of IncP-9 plasmids of Pseudomonas. Microbiology-Sgm 154: 2929–2941.
8. Krol JE, Penrod JT, McCaslin H, Rogers LM, Yano H, et al. (2012) Role of IncP-1 beta plasmids pWDL7::rfp and pNB8c in chloroaniline catabolism as determined by genomic and functional analyses. Applied and Environmental Microbiology 78: 828–838.
9. Jutkina JHE, Vedler E, Juhanson J, Heinaru A (2011) Occurrence of plasmids in the aromatic degrading Bacterioplankton of the Baltic Sea. Genes 2: 853–868.
10. Castillo MdP, Torstensson L, Stenström J (2008) Biobeds for environmental protection from pesticide use - a review. Journal of Agricultural and Food Chemistry 56: 6206–6219.
11. Omirou M, Dalias P, Costa C, Papastefanou C, Dados A, et al. (2012) Exploring the potential of biobeds for the depuration of pesticide-contaminated wastewaters from the citrus production chain: Laboratory, column and field studies. Environmental Pollution 166: 31–39.
12. Trefault N, De la Iglesia R, Molina AM, Manzano M, Ledger T, et al. (2004) Genetic organization of the catabolic plasmid pJP4 from Ralstonia eutropha JMP134 (pJP4) reveals mechanisms of adaptation to chloroaromatic pollutants and evolution of specialized chloroaromatic degradation pathways. Environmental Microbiology 6: 655–668.
13. Oliveira CS, Lazaro B, Azevedo JSN, Henriques I, Almeida A, et al. (2012) New molecular variants of epsilon and beta IncP-1 plasmids are present in estuarine waters. Plasmid 67: 252–258.
14. Smalla K, Haines AS, Jones K, Krögerrecklenfort E, Heuer H, et al. (2006) Increased abundance of IncP-1 beta plasmids and mercury resistance genes in mercury-polluted river sediments: First discovery of IncP-1 beta plasmids with a complex mer transposon as the sole accessory element. Applied and Environmental Microbiology 72: 7253–7259.
15. Top EM, Springael D (2003) The role of mobile genetic elements in bacterial adaptation to xenobiotic organic compounds. Current Opinion in Biotechnology 14: 262–269.
16. Sen DY, Brown CJ, Top EM, Sullivan J (2013) Inferring the Evolutionary History of IncP-1 Plasmids Despite Incongruence among Backbone Gene Trees. Molecular Biology and Evolution 30: 154–166.
17. Dunon VSK, Bers K, Lavigne R, Smalla K, Springael D (2013) High prevalence of IncP-1 plasmids and IS1071 insertion sequences in on-farm biopurification systems and other pesticide polluted environments. FEMS Microbiology Ecology.
18. Jechalke S, Dealtry S, Smalla K, Heuer H (2013) Quantification of IncP-1 plasmid prevalence in environmental samples. Applied and Environmental Microbiology 79: 1410–1413.
19. Gomes NCM, Flocco CG, Costa R, Junca H, Vilchez R, et al. (2010) Mangrove microniches determine the structural and functional diversity of enriched petroleum hydrocarbon-degrading consortia. FEMS Microbiology Ecology 74: 276–290.
20. Fetzner S, Lingens F (1994) Bacterial Dehalogenases: Biochemistry, Genetics, and Biotechnological Applications. Microbiological Reviews 58: 641–685.
21. Greated A, Lambertsen L, Williams PA, Thomas CM (2002) Complete sequence of the IncP-9 TOL plasmid pWW0 from Pseudomonas putida. Environmental Microbiology 4: 856–871.
22. Sota M, Yano H, Ono A, Miyazaki R, Ishii H, et al. (2006) Genomic and functional analysis of the IncP-9 naphthalene-catabolic plasmid NAH7 and its transposon Tn4655 suggests catabolic gene spread by a tyrosine recombinase. Journal of Bacteriology 188: 4057–4067.
23. Izmalkova TY, Mavrodi DV, Sokolov SL, Kosheleva IA, Smalla K, et al. (2006) Molecular classification of IncP-9 naphthalene degradation plasmids. Plasmid 56: 1–10.
24. Zhou Y (1996) Two-phase anaerobic digestion of water hyacinth pretreated with dilute sulphuric acid. Huanjing Kexue 17: 13–92.
25. Sipkema D, Blanch HW (2010) Spatial distribution of bacteria associated with the marine sponge Tethya californiana. Marine Biology 157: 627–638.
26. Edlund A, Hardeman F, Jansson JK, Sjoling S (2008) Active bacterial community structure along vertical redox gradients in Baltic Sea sediment. Environmental Microbiology 10: 2051–2063.
27. Gaze WH, Zhang LH, Abdouslam NA, Hawkey PM, Calvo-Bado L, et al. (2011) Impacts of anthropogenic activity on the ecology of class 1 integrons and integron-associated genes in the environment. ISME Journal 5: 1253–1261.
28. Byrne-Bailey KG, Gaze WH, Kay P, Boxall ABA, Hawkey PM, et al. (2009) Prevalence of sulfonamide resistance genes in bacterial isolates from manured agricultural soils and pig slurry in the United Kingdom. Antimicrobial Agents and Chemotherapy 53: 696–702.

7

Negative Cross Resistance Mediated by Co-Treated Bed Nets: A Potential Means of Restoring Pyrethroid-Susceptibility to Malaria Vectors

Michael T. White[1], Dickson Lwetoijera[2], John Marshall[1], Geoffrey Caron-Lormier[3], David A. Bohan[4], Ian Denholm[5], Gregor J. Devine[6]*

1 MRC Centre for Outbreak Analysis and Modelling, Imperial College, London, United Kingdom, 2 Ifakara Health Institute, Ifakara, Tanzania, 3 University of Nottingham, Sutton Bonington, Leicestershire, United Kingdom, 4 INRA, UMR 1347 Agroécologie, Pôle ECOLDUR, Dijon, France, 5 University of Hertfordshire, Hatfield, Hertfordshire, United Kingdom, 6 QIMR Berghofer Medical Research Institute, Brisbane, Australia

Abstract

Insecticide-treated nets and indoor residual spray programs for malaria control are entirely dependent on pyrethroid insecticides. The ubiquitous exposure of *Anopheles* mosquitoes to this chemistry has selected for resistance in a number of populations. This threatens the sustainability of our most effective interventions but no operationally practicable way of resolving the problem currently exists. One innovative solution involves the co-application of a powerful chemosterilant (pyriproxyfen or PPF) to bed nets that are usually treated only with pyrethroids. Resistant mosquitoes that are unaffected by the pyrethroid component of a PPF/pyrethroid co-treatment remain vulnerable to PPF. There is a differential impact of PPF on pyrethroid-resistant and susceptible mosquitoes that is modulated by the mosquito's behavioural response at co-treated surfaces. This imposes a specific fitness cost on pyrethroid-resistant phenotypes and can reverse selection. The concept is demonstrated using a mathematical model.

Editor: Rick E. Paul, Institut Pasteur, France

Funding: This work was partly funded by the Bill and Melinda Gates Foundation (grant ID OPP52644). The funders had no role in study design, data collection and analysis, decision to publish, or preparation of the manuscript. No additional external funding was received for this study.

Competing Interests: The authors have declared that no competing interests exist.

* E-mail: greg.devine@qimrberghofer.edu.au

Introduction

A recent surge in effort and funding has led to the expansion of insecticide treated bed net (ITN) and indoor residual spray (IRS) programs in many parts of Africa and dramatic decreases in malaria transmission. Although four insecticide classes (carbamates, organophosphates, pyrethroids and the organochlorine DDT) are currently approved for IRS, the vast majority of spraying programs utilise synthetic pyrethroids. This is also the only insecticide class approved for use on ITNs [1]. The ubiquitous presence of pyrethroids in public health and the agricultural sector has resulted in strong selection pressure for mutations that confer resistance to pyrethroids in insect vectors of disease. In the absence of remedial measures, the impacts of this on malaria transmission can be severe [2,3].

Pyrethroid resistance is widely reported in African malaria vectors [4] but there is little that can be done in response. There are few novel insecticidal products nearing commercialisation and the reassessment of old and previously resisted chemistries in new guises is now commonplace. A novel, resistance-beating combination of safe compounds with World Health Organisation (WHO) approval is therefore a timely and exciting proposition. We propose a mechanism to delay or reverse selection for pyrethroid resistance through a phenomenon called negative cross resistance (NCR) in which organisms resistant to one compound of a binary mixture are hyper-susceptible to the other. This imposes a

fitness cost on the resistant genotype that can decrease the frequency of resistant alleles. This is distinct from the conventional use of binary mixtures and rotations where there is no hyper-sensitivity and whose role in resistance management is severely limited if the target pest has already developed resistance to either compound [5].

NCR has long been discussed by agricultural [6,7] and public health entomologists [8] but it has largely eluded attempts at practical implementation. It remains an intriguing alternative to the "treadmill" approach of resistance management (the sequential replacement of one chemical class by another, as insects evolve a succession of protective mechanisms).

In our model, we exploit a potent chemosterilant (pyriproxyfen or PPF) and the differential behaviour of pyrethroid-resistant and susceptible mosquitoes at pyrethroid-treated surfaces. The model draws on the impacts of pyrethroids on susceptible and resistant insects and on recent proofs that PPF exposure dramatically reduces egg viability in *Anopheles gambiae* [9,10].

Assumptions

Our thesis requires unequivocal differences in the mortality and behaviour of pyrethroid-resistant and susceptible *Anopheles* mosquitoes when exposed to binary treatments of PPF and pyrethroids. Host-seeking or resting mosquitoes are more likely to be irritated, repelled or killed by co-treated surfaces if they are pyrethroid-susceptible. Conversely, pyrethroid-resistant insects are more likely to spend time resting or trying to feed at those surfaces. By surviving pyrethroid exposure they will pick up sterilising doses of PPF. This imposes a fitness cost on the pyrethroid-resistant phenotype. We call this phenomenon "behaviourally-mediated NCR", since genotype selection results from a behavioural response rather than from any direct interaction between insecticides and physiological resistance mechanisms.

"Knock-down resistance" (kdr) is the most ubiquitous of the pyrethroid resistance mechanisms described for *An. gambiae* s.l. and other mosquito genera. It involves a modification of the pyrethroid target site and is often found in tandem with other detoxification mechanisms. It remains the best diagnostic for predicting pyrethroid-resistance [11]. The frequency of the allele in resistant field populations commonly ranges from 50–95% [12–15] and, unsurprisingly, resistant homozygotes can account for a large proportion of individuals [13,16]. The mutation is incompletely recessive [17] and, in response to pyrethroids, heterozygotes (SR) suffer intermediate mortality to homozygous resistant (RR) and susceptible (SS) forms [13,14]. Behavioural studies in the laboratory show that individuals carrying kdr alleles maintain contact with pyrethroid-treated surfaces for longer periods than susceptible insects, are less repelled and are more likely to blood-feed (i.e. through a treated net) than their susceptible counterparts. Heterozygotes tend to display intermediate behaviours [12,13,18,19]. These impacts, in the presence of ITNs, have been widely demonstrated under field conditions and are most commonly recorded as differential blood-feeding success. Generally, SS insects are 2–5 fold less likely to feed than their SR and RR counterparts [20–24]. We exploit these behavioural differences to impose a PPF-mediated fitness-cost on pyrethroid-resistant mosquitoes exposed to PPF/pyrethroid co-treatments.

PPF is a juvenile hormone analogue with low toxicity to mammals. It inhibits metamorphosis and embryogenesis in several insects [25] and it is currently under evaluation by the World Health Organisation Pesticide Evaluation Scheme (WHOPES) as a component of a pyrethroid-treated bed net. It is approved as a mosquito larvicide and it may be suitable for autodissemination by mosquitoes for that purpose [26]. It is also a powerful chemosterilant. Exposure to PPF reduces the fecundity of adult female *An. gambiae* s.l. mosquitoes by reducing the number and viability of oviposited eggs. Ohashi *et al* [9] noted that the effects were dose-dependent and also reduced longevity. Harris *et al* [10] observed that *An. arabiensis* were completely sterilised for at least one gonotrophic cycle. Ohba et al [27] showed that both the fecundity and fertility of *Aedes albopictus* were affected when insects were exposed to PPF through a net while feeding on mice. These papers note that the sterilising impacts of PPF depend on the mosquito being exposed close to the time of feeding (the assumption being that PPF interferes with subsequent oogenesis and egg maturation) and suggest that co-treated bed nets may be an effective tool for exposing pyrethroid resistant mosquitoes to sterilising doses of PPF.

Mathematical Models

We compare the reproductive fitness of *Anopheles gambiae* s.s. kdr susceptible (SS) and kdr homozygous resistant (RR) mosquitoes in the presence of PPF/pyrethroid co-treated surfaces. We first construct a static model to compare reproductive fitness in terms of the numbers of eggs oviposited by SS and RR mosquitoes. We then extend this to a dynamic mosquito population model with proportions of SS, SR and RR mosquitoes changing over time. We adapt a previously published model [28,29] of the behavioural interactions between host seeking *Anopheles gambiae* mosquitoes and pyrethroid treated surfaces to estimate a mosquito's daily mortality

and feeding frequency. ITNs and IRS are assumed to have three effects on susceptible mosquitoes: (i) directly killing mosquitoes that land on treated surfaces; (ii) repelling and possibly diverting mosquitoes to an animal blood host due to either insecticide irritation or the physical barrier of the net; and (iii) lengthening the duration of the gonotrophic cycle leading to a reduced oviposition rate (by denying a blood meal). It is assumed that when kdr resistant mosquitoes encounter a pyrethroid treated net or surface, they (i) have a lower probability of being killed by pyrethroids; (ii) have a higher probability of successful feeding; and (iii) tend to be diverted by the physical barrier of the net as opposed to the irritant effect of the pyrethroids.

The key model parameters and the literature from which they are derived are defined in Table 1. The probabilities of a pyrethroid-susceptible mosquito feeding successfully ($s = 0.03$), being repelled ($r = 0.56$) or dying ($d = 0.41$) on exposure to an ITN are derived from empirical observations in experimental huts (26, 27). Resistant mosquitoes either die ($d = 0.10$) or are thwarted by the physical barrier of the net ($r = 0.24$) (27). The remainder is assumed to feed successfully. See Text S1, Tables S1 and S2 in Text S1, and Figure S1 for further explanation and illustration.

The fitness of susceptible or resistant phenotypes is recorded as the expected number of eggs that a female mosquito will oviposit in her lifetime. A susceptible mosquito with daily mortality $\mu_{M,ITN}^{SS}$, ovipositing ε eggs every δ_{ITN}^{SS} days, will oviposit an expected E^{SS} eggs over her lifetime, where

$$E^{SS} = \left(\varepsilon e^{-\mu_{M,ITN}^{SS} \delta_{ITN}^{SS}} + \varepsilon e^{-2\mu_{M,ITN}^{SS} \delta_{ITN}^{SS}} + ... \right)$$

$$= \frac{\varepsilon}{e^{\mu_{M,ITN}^{SS} \delta_{ITN}^{SS}} - 1}$$

Without contacting PPF, pyrethroid-resistant mosquitoes will oviposit ε eggs every δ_{ITN}^{RR} days and experience daily mortality $\mu_{M,ITN}^{RR}$. In the presence of PPF/pyrethroid co-treated nets at coverage C, resistant mosquitoes are exposed to PPF while attempting to feed with probability $p_{PPF} = C_{ITN} Q_0 \phi (s_{ITN} + r_{ITN})$. See SI text, section 2.2 for more detail. When exposed to PPF at co-treated surfaces, resistant mosquitoes will oviposit $\varepsilon_{PPF} \leq \varepsilon$ eggs and be subject to daily mortality $\mu_{M,PPF-ITN}^{RR} \geq \mu_{M,ITN}^{RR}$. The expected number of eggs oviposited over the mosquito's lifetime will be

$$E^{RR} = \underbrace{(1 - p_{PPF}) \frac{\varepsilon}{e^{\mu_{M,ITN}^{RR} \delta_{ITN}^{RR}} - 1 + p_{PPF}}}_{\text{eggs oviposited before PPF contact}}$$

$$+ \underbrace{p_{PPF} \frac{\varepsilon_{PPF}}{e^{\mu_{M,ITN}^{RR} \delta_{ITN}^{RR}} - 1 + p_{PPF}} \frac{1}{1 - e^{-\mu_{M,ITN-PPF}^{RR} \delta_{ITN}^{RR}}}}_{\text{eggs oviposited after PPF contact}}$$

These equations describe the comparative reproductive fitness of homozygous pyrethroid-susceptible and resistant mosquitoes in terms of the numbers of eggs oviposited. See SI text section 3 and Figure S2 for more detail. The numbers and ratios of homozygous susceptible (SS) and resistant (RR) eggs that result from the presence of co-treated nets are illustrated in Figure 1. In situations where pyrethroid resistance is emerging, there will be a dynamic mix of SS, SR and RR mosquitoes. The model can also be

Table 1. Parameters for reproduction and interaction with pyrethroid/PPF co-treated surfaces.

Parameter	Description	Value pyrethroid resistance		Reference
		susceptible	resistant	
C	ITN coverage (proportion of people under nets)	fixed	fixed	
μ_M	daily non-insecticide mosquito mortality (day^{-1})	0.096	0.096	[30,39]
ε	eggs per oviposition	74	74	[30]
δ	duration of gonotrophic cycle (days)	3	3	[40]
Q_0	human blood index	0.90	0.90	[41]
ϕ	proportion of bites taken on humans while in bed	0.89	0.89	[42]
s	successful feeding with ITN	0.03	0.66	[43,44]
r	cycle repeating probability for ITN	0.56	0.24	[43,44]
d	insecticide mortality probability for ITN	0.41	0.10	[43,44]
p_{PPF}	probability of surviving contact with PPF treated surfaces $p_{PPF} = C\, Q_0 \varphi(s+r)$	0*	model estimate	
$\mu_{M,ITN}\,(C)$	daily mosquito mortality in the presence of ITNs (day^{-1}) – see SI for details	model estimate	model estimate	
$f_{ITN} = 1/\delta_{ITN}$	blood feeding frequency in the presence of ITNs (day^{-1}) – see SI for details	model estimate	model estimate	
	reduction in eggs: ITNs–0.001% w/v PPF	68%	68%	[9]
	reduction in eggs: ITNs–0.01 or 0.1% w/v PPF	100%	100%	[9]
	reduction in lifespan: ITNs–0.001% w/v PPF	38%	38%	[9]
	reduction in lifespan: ITNs–0.01% w/v PPF	55%	55%	[9]
	reduction in lifespan: ITNs–0.1% w/v PPF	75%	75%	[9]
	reduction in eggs: PPF treated surfaces	60–100%	60–100%	[10]
	reduction in lifespan: PPF treated surfaces	0%	0%	[10]

*Pyrethroid susceptible mosquitoes that contact a pyrethroid/PPF co-treated surface will be killed by the pyrethroid component. The survival of susceptible insects that avoid contact with the net (described by the terms *s*, *r* and *d*) is independent of this parameter.

extended to incorporate the number of eggs oviposited by heterozygous resistant mosquitoes and track the mixing of genotypes using a model of *An. gambiae* s.l. population dynamics [30]. Resistance is assumed (as is the case for kdr) to reflect a single locus, incompletely recessive allele [17] and we assume that SR mosquitoes have phenotypic properties intermediate between those of SS and RR. See SI text, Table S3 and Figure S4 for more detail on the dynamic model.

Results

Increasing coverage of ITNs treated only with pyrethroids imparts a fitness advantage to pyrethroid-resistant mosquitoes. These are more likely to survive, blood-feed and oviposit. The consequent ratios of resistant: susceptible eggs will be large (Figure 1A). Co-treatment with PPF can reverse this advantage if the reduction in fecundity in pyrethroid-resistant mosquitoes contacting PPF is sufficiently large (Figure 1B).

At 50% coverage of co-treated ITNs, a 65% reduction in fecundity in exposed mosquitoes will reverse resistance selection by pyrethroids (Figure 1B). Higher levels of ITN coverage require increased impact of PPF to reverse that increased selection for resistance by pyrethroids (Figure 1C). Contact with PPF-treated surfaces may also shorten a mosquito's lifespan and reduce the number of gonotrophic cycles and oviposition events [9]. This can affect disease transmission by reducing the time available for the incubation of viruses and parasites but, in this model we examine

its additive effects on fitness costs in pyrethroid-resistant mosquitoes exposed to co-treated ITNs. The reductions in fecundity and life expectancy of mosquitoes exposed to nets treated with the formulations of pyriproxyfen used by Ohashi *et al* [9] and Harris *et al* [10] are highlighted in Figure 1D. These scenarios are not unrealistic: recent data shows that PPF can induce total sterilisation of mosquitoes using just 0.01% w/w on nets [9] or 3 mg/m^2 on other substrates [10]. There is considerable potential to increase those doses.

Figure S3 extends the results of Figure 1 by illustrating the reversal of resistance selection at 30% and 80% ITN coverage. At low levels of ITN coverage, the emergence of pyrethroid resistance can be prevented either by modest reductions in fecundity or life expectancy. At higher levels of ITN coverage, reductions in lifespan alone are not sufficient to prevent the emergence of resistance, and large reductions in fecundity (>80%) are required.

The emergence of pyrethroid resistance is likely to be a complex stochastic event, with unpredictable evolutionary scales. The deterministic model implemented here does not account for the emergence of novel resistance mechanisms or chance immigration of resistant mosquitoes, but it does illustrate the evolutionary outcomes that eventuate from selective pressure due to combinations of pyrethroids and PPF. It demonstrates a strong advantage to pyrethroid-susceptible genotypes. Figure 2A shows the emergence of pyrethroid resistance after the introduction of ITNs at 50% coverage and the subsequent reversal in allele frequency following the introduction of a PPF co-treatment that imposes a

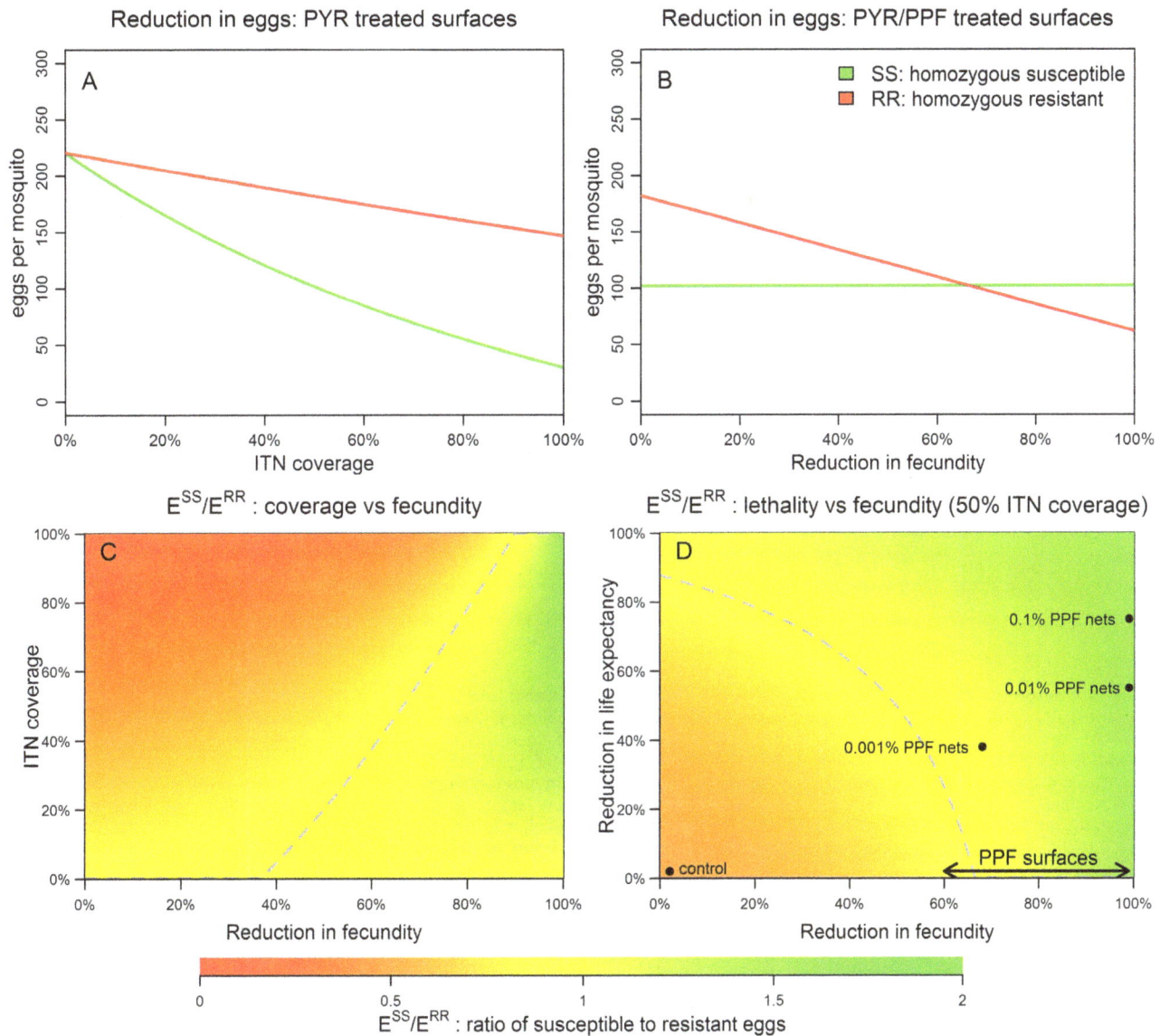

Figure 1. Reproductive fitness of pyrethroid susceptible (green) and resistant (red) mosquitoes in the presence of co-treated nets. Reduction in fecundity is defined as the proportional decrease in the number of eggs per oviposition. PYR = pyrethroid, PPF = pyriproxyfen. (**A**) Reduction in number of oviposited eggs with increasing coverage of ITNs. (**B**) Reduction in the number of oviposited eggs in presence of co-treated nets at 50% coverage. No reduction in life expectancy following PPF exposure is assumed. (**C**) Regions in parameter space where more eggs are oviposited by susceptible (green) than resistant (red) mosquitoes. No reduction in life expectancy following PPF exposure is assumed. (**D**) Regions in parameter space where more eggs are oviposited by susceptible (green) or resistant (red) mosquitoes at 50% ITN coverage. Reductions in fecundity and life expectancy observed by different concentration of PPF on bed nets by Ohashi *et al* [9] are represented as points. The range of reduction in fecundity seen by Harris *et al* [10] is represented by the black arrowed line. The dashed grey lines divide the parameter space into regions where susceptible mosquitoes are fitter than resistant mosquitoes ($E^{SS}>E^{RR}$), and where resistant mosquitoes are fitter than susceptible mosquitoes ($E^{RR}>E^{SS}$). The R code used to derive this figure is available as part of the supporting information (R code S1).

modest 68% decrease in fecundity (the reduction caused by exposure to 0.001% PPF exposure on nets [9]). Figure 2B shows the corresponding change in mosquito densities. In these scenarios, heterozygous resistant mosquitoes (SR) that encounter pyrethroid-treated surfaces display intermediate phenotypic behaviours in comparison to homozygous resistant (RR) or susceptible (SS) forms, i.e. the dominance co-efficient is $h=0.5$. Figures S5 ($h=1$) and S6 ($h=0.01$) illustrate that the dominance co-efficient has relatively little impact on model outcomes at these levels of bed net coverage and imposed fitness cost.

Discussion

The model-based investigations undertaken here suggest that the co-application of pyriproxyfen to pyrethroid treated nets or surfaces constitutes a plausible, practicable strategy for selecting against kdr resistant alleles. The technique that we exploit is distinct from the conventional use of binary mixtures and rotations where there is no hyper-sensitivity of resistant alleles and little advantage in terms of resistance management if the target pest has already developed resistance to either insecticide [5].

Prevalence of resistance

Mosquito numbers

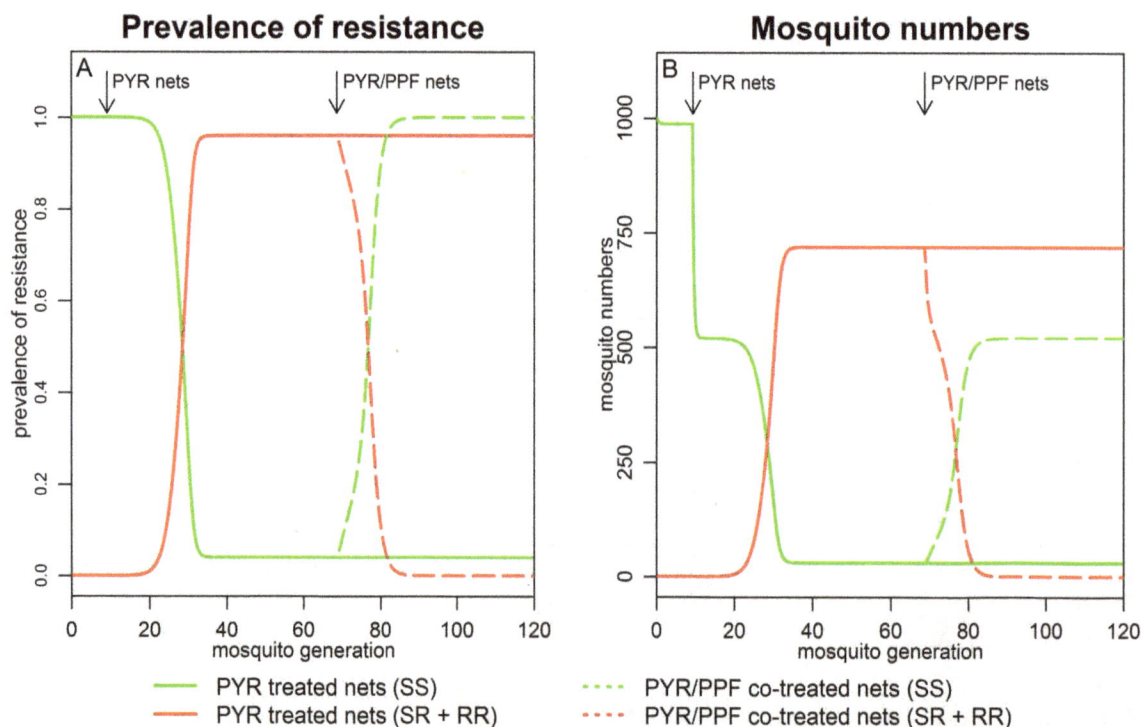

Figure 2. Emergence of pyrethroid resistance in the absence (solid lines) and presence of co-treated nets (dashed lines) at 50%
coverage. PYR = pyrethroid, PPF = pyriproxyfen. Heterozygous resistant mosquitoes (SR) display behaviours intermediate to SS or RR genotypes, i.e.
h = 0.5. (A) The introduction of ITNs treated with pyrethroids alone leads to the emergence of pyrethroid resistance but this is reversed by co-treating
nets with PPF. The rate of reversal will depend on the percentage reduction in fecundity. (B) The introduction of ITNs causes a rapid decline in
mosquito numbers, followed by the emergence of resistance and an increase in mosquito numbers. When resistance is reversed by the introduction
of PPF, numbers remain suppressed as a consequence of mortality in the now largely pyrethroid-susceptible population. The initial frequency of
homozygous resistant mosquitoes is assumed to be 10^{-5}. A mosquito generation is assumed to be the expected lifespan of the aquatic plus adult
stages. The R code used to derive this figure is available as part of the supporting information (R code S2).

Although the sterilising effects of PPF might be used in a number of ways to suppress mosquito populations [31] we stress that, in this instance, the pyrethroid component of our proposed strategy is essential: it is the immediate, lethal impact of pyrethroids that permits the co-treated net to remain a successful disease intervention. In contrast, and unlike conventional toxins, PPF has little impact on mosquito longevity and exposed but infected mosquitoes will retain their capacity to survive the extrinsic incubation period and transmit disease. The purpose of the PPF component is to impose a cost on pyrethroid resistance, regain pyrethroid-susceptibility, and restore the overall effectiveness of ITNs.

Our model does not consider immigration of resistant alleles and makes the assumption that resistance is selected solely through interactions with treated bed nets. This reflects some empirical systems [32–34] but ignores the potential role of selection by pyrethroids used in agriculture and livestock [35]. It is expected, however, that resistant immigrants that encounter co-treated nets will be subject to the same fitness differential as resident insects. Other challenges to the feasibility of this resistance management approach might include avoidance of co-treated surfaces by RR or SR insects or recovery of fecundity with age. Neither scenario is likely. There is no evidence that PPF is repellent, even at high doses [36] and changes in fecundity are thought likely to be life-long following exposure to extremely practicable PPF concentrations [9,10]. One other tangible threat to this chemically-based vector control solution is the appearance of novel resistance mechanisms (i.e. ones that reduce or negate the chemosterilant

effect of PPF). In our modelled scenario, co-treatment offers some protection against that possibility: the pyrethroid-resistant individuals that encounter PPF will be strongly selected to evolve an additional PPF-resistance mechanism but pyrethroid-susceptible mosquitoes will be protected from PPF exposure, and hence from selection for PPF resistance, because of their responses to pyrethroids. Assuming random mating between genotypes, selection for PPF-resistant alleles should be constantly diluted by this pool of fully susceptible insects.

The modelled pyrethroid resistant mosquito population carries the incompletely recessive kdr resistance mechanism. This target-site mutation is an excellent diagnostic of pyrethroid-resistance [11] but additional mechanisms such as cytochrome P450 (CYP) are increasingly commonly described. Like kdr, these metabolic mechanisms are intermediately dominant in their heterozygote form [37] and individuals of species that exhibit mixtures of target site and metabolic mechanisms are observed to be spend a great deal of time in contact with pyrethroid-treated nets [23,38]. It is likely therefore, that the behavioural differentials that we apply in our model are valid for most pyrethroid resistance mechanisms. Importantly, there is no evidence that CYP mechanisms alter the impact of pyriproxyfen's chemosterilant effect.

An additional impact of PPF exposure, which we do not model, is PPF's potential to be transferred from co-treated surfaces and to lethally affect juveniles developing in aquatic habitats. This phenomenon of "autodissemination" [26] may have profound impacts on population size but it will target aquatic environments irrespective of the juvenile phenotypes therein.

The co-application of pyrethroids and PPF may offer a powerful resistance management tool that complements the essential impacts of pyrethroids on mosquito population suppression and disease transmission. We offer an entirely different approach to the development of "resistance breaking" chemistries, which are simply new molecules as yet unresisted, or old molecules in new, more efficient guises. Solutions involving physiological NCR (in which pyrethroid-resistant populations are hyper-sensitive to a second insecticide but pyrethroid-susceptible populations are not) have no candidate molecules. Although we focus on a strategy where PPF is co-applied to pyrethroid treated nets, the model is broadly applicable to the same chemical combination deployed as an indoor residual spray. Our proposed strategy of "behaviourally-mediated NCR" utilises extant, registered and safe chemistries and merits urgent empirical investigation. Considerably more experimental data are needed to evaluate its practicality.

Supporting Information

Figure S1 Flow chart of mosquito life cycle based on the diagram from Le Menach et al [7] and Griffin et al [6].

Figure S2 Flow chart depicting the life history and expected number of oviposited eggs of a pyrethroid-resistant mosquito. M_n denotes a mosquito having completed n gonotrophic cycles. $M_{PPF,n}$ denotes a mosquito that has completed n gonotrophic cycles and also been exposed to PPF. p_{PPF} is the probability that a mosquito contacts PPF at each feeding attempt. $q_{PPF} = 1 - p_{PPF}$ is the probability that a mosquito avoids contact with PPF during a feeding attempt.

Figure S3 Comparison of the reproductive fitness of pyrethroid-susceptible and pyrethroid-resistant mosquitoes in the presence of co-treated pyrethroid/PPF nets at 30% and 80% coverage. Reduction in fecundity is defined as the proportional reduction in the number of eggs per oviposition. Red regions of parameter space represent scenarios where more eggs are oviposited by pyrethroid-resistant mosquitoes than pyrethroid-susceptible mosquitoes. Green regions of parameter space represent scenarios where more eggs are oviposited by pyrethroid-susceptible mosquitoes than pyrethroid- resistant mosquitoes. Yellow regions of parameter space represent scenarios where approximately the same number of eggs is oviposited by pyrethroid-susceptible mosquitoes and pyrethroid- resistant mosquitoes. Reductions in fecundity and life expectancy observed by different concentration of PPF on bed nets by Ohashi et al [11] are represented as points.

The range of reduction in fecundity seen by Harris et al [12] is represented by the black arrowed line. The dashed grey lines divide the parameter space into regions where susceptible mosquitoes are fitter than resistant mosquitoes ($E^{SS} > E^{RR}$), and resistant mosquitoes are fitter than susceptible mosquitoes ($E^{RR} > E^{SS}$). The R code for generating this figure is included as a supporting file (R code S1).

Figure S4 Flow chart for the numbers of aquatic stages (early and late larval instars and pupae) and adult mosquitoes stratified by gonotrophic cycle and PPF exposure status.

Figure S5 Emergence of pyrethroid resistance in the absence (solid lines) and presence of co-treated nets (dashed lines) at 50% coverage. PYR = pyrethroid, PPF = pyriproxyfen. It is assumed that heterozygous resistant mosquitoes (SR) have the same phenotypic behaviour as homozygous resistant mosquitoes (RR), i.e. $h = 1$.

Figure S6 Emergence of pyrethroid resistance in the absence (solid lines) and presence of co-treated nets (dashed lines) at 50% coverage. PYR = pyrethroid, PPF = pyriproxyfen. It is assumed that heterozygous resistant mosquitoes (SR) have the similar phenotypic behaviour as homozygous susceptible mosquitoes (SS), i.e. $h = 0.1$.

R Code S1 Code is for Figure 1 and Figure S3.

R Code S2 Code is for Figure 2.

Text S1 Contains Table S1, parameters describing the behaviour and life history of An. gambiae s. s. mosquitoes. Table S2, parameters describing interactions between a mosquito and an insecticide-treated net. Table S3, notation, definition and values of the variables and parameters for the model of A. gambiae population dynamics. All parameter values are taken from White et al [2].

Author Contributions

Wrote the paper: GJD MTW. Conceived and designed the model: MTW ID GJD JM GCL DAB DL. Executed and interpreted the model: MTW JM GJD ID.

References

1. Hemingway J, Vontas J, Poupardin R, Raman J, Lines J, et al. (2013) Country-level operational implementation of the Global Plan for Insecticide Resistance Management. Proc Natl Acad Sci U S A 110: 9397–9402.
2. Maharaj R, Mthembu DJ, Sharp BL (2005) Impact of DDT re-introuction on malaria transmission in KwaZulu-Natal. Samj South African Medical Journal 95: 871–874.
3. Trape JF, Tall A, Diagne N, Ndiath O, Ly AB, et al. (2011) Malaria morbidity and pyrethroid resistance after the introduction of insecticide-treated bed nets and artemisinin-based combination therapies: a longitudinal study. Lancet Infect Dis 11: 925–932.
4. Ranson H, N'Guessan R, Lines J, Moiroux N, Nkuni Z, et al. (2011) Pyrethroid resistance in African anopheline mosquitoes: what are the implications for malaria control? Trends in Parasitology 27: 91–98.
5. Tabashnik BE (1990) Modeling and Evaluation of Resistance Management Tactics. In: Roush RT, Tabashnik BE, editors. Pesticide Resistance in Arthropods. New York: Chapman and Hall. 153–182.
6. Khambay BPS, Denholm I, Carlson GR, Jacobson RM, Dhadialla TS (2001) Negative cross-resistance between dihydropyrazole insecticides and pyrethroids in houseflies, Musca domestica. Pest Management Science 57: 761–763.
7. Yamamoto I, Kyomura N, Takahashi Y (1993) Negatively correlated cross resistance - combinations of N-methylcarbamate with N-propylcarbamate or oxadiazolone for green rice leafhopper. Archives of Insect Biochemistry and Physiology 22: 277–288.
8. Kolaczinski JH, Curtis CF (2004) Investigation of negative cross-resistance as a resistance-management tool for insecticide-treated nets. Journal of Medical Entomology 41: 930–934.
9. Ohashi O, Nakada K, Ishiwatari T, Miyaguchi J, Shono Y, et al. (2012) Efficacy of pyriproxyfen treated nets in sterilizing and shortening the longevity of Anopheles gambaie (Diptera: Culicidae). Journal of Medical Entomology 49: 1052–1058.
10. Harris C, Lwetoijera DW, Dongus S, Matowo NS, Lorenz LM, et al. (2013) Sterilising effects of pyriproxyfen on Anopheles arabiensis and its potential use in malaria control. Parasites & Vectors 6.
11. Donnelly MJ, Corbel V, Weetman D, Wilding CS, Williamson MS, et al. (2009) Does kdr genotype predict insecticide-resistance phenotype in mosquitoes? Trends in Parasitology 25: 213–219.

12. Chandre F, Darriet F, Duchon S, Finot L, Manguin S, et al. (2000) Modifications of pyrethroid effects associated with kdr mutation in Anopheles gambiae. Medical And Veterinary Entomology 14: 81–88.

13. Corbel V, Chandre F, Brengues C, Akogbeto M, Lardeux F, et al. (2004) Dosage-dependent effects of permethrin-treated nets on the behaviour of Anopheles gambiae and the selection of pyrethroid resistance. Malaria Journal 3.

14. Diabate A, Brengues C, Baldet T, Dabire KR, Hougard JM, et al. (2004) The spread of the Leu-Phe kdr mutation through Anopheles gambiae complex in Burkina Faso: genetic introgression and de novo phenomena. Tropical Medicine & International Health 9: 1267–1273.

15. Kolaczinski JH, Fanello C, Herve JP, Conway DJ, Carnevale P, et al. (2000) Experimental and molecular genetic analysis of the impact of pyrethroid and non-pyrethroid insecticide impregnated bed nets for mosquito control in an area of pyrethroid resistance. Bulletin Of Entomological Research 90: 125–132.

16. Mourou JR, Coffinet T, Jarjaval F, Pradines B, Amalvict R, et al. (2010) Malaria transmission and insecticide resistance of Anopheles gambiae in Libreville and Port-Gentil, Gabon. Malaria Journal 9.

17. Martinez Torres D, Devonshire AL, Williamson MS (1997) Molecular studies of knockdown resistance to pyrethroids: Cloning of domain II sodium channel gene sequences from insects. Pesticide Science 51: 265–270.

18. Hougard JM, Corbel V, N'Guessan R, Darriet F, Chandre F, et al. (2003) Efficacy of mosquito nets treated with insecticide mixtures or mosaics against insecticide resistant Anopheles gambiae and Culex quinquefasciatus (Diptera: Culicidae) in Cote d'Ivoire. Bulletin Of Entomological Research 93: 491–498.

19. Pennetier C, Corbel V, Hougard JM (2005) Combination of a non-pyrethroid insecticide and a repellent: A new approach for controlling knockdown-resistant mosquitoes. American Journal Of Tropical Medicine and Hygiene 72: 739–744.

20. Asidi A, N'Guessan R, Akogbeto M, Curtis C, Rowland M (2012) Loss of household protection from use of insecticide-treated nets against pyrethroid-resistant mosquitoes, benin. Emerg Infect Dis 18: 1101–1106.

21. Corbel V, Chabi J, Dabire RK, Etang J, Nwane P, et al. (2010) Field efficacy of a new mosaic long-lasting mosquito net (PermaNet (R) 3.0) against pyrethroid-resistant malaria vectors: a multi centre study in Western and Central Africa. Malaria Journal 9.

22. Dabire RK, Diabate A, Baldet T, Pare-Toe L, Guiguemde RT, et al. (2006) Personal protection of long lasting insecticide-treated nets in areas of Anopheles gambiae s.s. resistance to pyrethroids. Malaria Journal 5.

23. Irish SR, Guessan RN, Boko PM, Metonnou C, Odjo A, et al. (2008) Loss of protection with insecticide-treated nets against pyrethroid-resistant Culex quinquefasciatus mosquitoes once nets become holed: an experimental hut study. Parasites & Vectors 1.

24. N'Guessan R, Corbel V, Akogbeto M, Rowland M (2007) Reduced efficacy of insecticide-treated nets and indoor residual spraying for malaria control in pyrethroid resistance area, Benin. Emerging Infectious Diseases 13: 199–206.

25. Dhadialla TS, Carlson GR, Le DP (1998) New insecticides with ecdysteroidal and juvenile hormone activity. Annual Review of Entomology 43: 545–569.

26. Devine GJ, Perea EZ, Killeen GF, Stancil JD, Clark SJ, et al. (2009) Using adult mosquitoes to transfer insecticides to Aedes aegypti larval habitats. Proceedings of the National Academy of Sciences of the United States of America 106: 11530–11534.

27. Ohba S, Ohashi K, Pujiyati E, Higa Y, Kawada H, et al. (2013) The effect of pyriproxyfen as a "Population Growth Regulator" against Aedes albopictus under semi-field conditions. PLoS ONE 8(7): e67045.

28. Griffin JT, Hollingsworth TD, Okell LC, Churcher TS, White M, et al. (2010) Reducing Plasmodium falciparum malaria transmission in Africa: a model-based evaluation of intervention strategies. PLoS Med 7.

29. Le Menach A, Takala S, McKenzie FE, Perisse A, Harris A, et al. (2007) An elaborated feeding cycle model for reductions in vectorial capacity of night-biting mosquitoes by insecticide-treated nets. Malaria Journal 6.

30. White MT, Griffin JT, Churcher TS, Ferguson NM, Basáñez M, et al. (2011) Modelling the impact of vector control interventions on Anopheles gambiae population dynamics. Parasites and Vectors 4.

31. Lwetoijera DW, Harris C, Kiware SS, Killeen GF, Dongus S, et al. (2014) Short Report: Comprehensive Sterilization of Malaria Vectors Using Pyriproxyfen: A Step Closer to Malaria Elimination Am J Trop Med Hyg.

32. Czeher C, Labbo R, Arzika I, Duchemin J-B (2008) Evidence of increasing Leu-Phe knockdown resistance mutation in Anopheles gambiae from Niger following a nationwide long-lasting insecticide-treated nets implementation. Malaria Journal 7.

33. Vulule JM, Beach RF, Atieli FK, Roberts JM, Mount DL, et al. (1994) Reduced susceptibility of Anopheles gambiae to permethrin associated with the use of permethrin-impregnated bed nets and curtains in Kenya. Medical and veterinary entomology 8: 71–75.

34. Ndiath MO, Sougoufara S, Gaye A, Mazenot C, Konate L, et al. (2012) Resistance to DDT and pyrethroids and increased kdr mutation frequency in An. gambiae after the implementation of permethrin-treated nets in Senegal. Plos One 7.

35. Diabate A, Baldet T, Chandre F, Akogbeto M, Guiguemde TR, et al. (2002) The role of agricultural use of insecticides in resistance to pyrethroids in Anopheles gambiae SL in Burkina Faso. American Journal of Tropical Medicine and Hygiene 67: 617–622.

36. Sihuincha M, Zamora-Perea E, Orellana-Rios W, Stancil JD, Lopez-Sifuentes V, et al. (2005) Potential use of pyriproxyfen for control of Aedes aegypti (Diptera: Culicidae) in Iquitos, Peru. Journal of Medical Entomology 42: 620–630.

37. Witzig C, Parry M, Morgan JC, Irving H, Steven A, et al. (2013) Genetic mapping identifies a major locus spanning P450 clusters associated with pyrethroid resistance in kdr-free Anopheles arabiensis from Chad. Heredity 110: 389–397.

38. Norris LC, Norris DE (2011) Insecticide resistance in Culex quinquefasciatus mosquitoes after the introduction of insecticide-treated bed nets in Macha, Zambia. J Vector Ecol 36: 411–420.

39. Molineaux L, Gramiccia G (1980) The Garki project: Research on the Epidemiology and Control of Malaria in the Sudan Savanna of West Africa. Geneva: WHO. 311 p.

40. Killeen GF, McKenzie FE, Foy BD, Schieffelin C, Billingsley PF, et al. (2000) A simplified model for predicting malaria entomologic inoculation rates based on entomologic and parasitologic parameters relevant to control. American Journal Of Tropical Medicine And Hygiene 62: 535–544.

41. Dia I, Diop T, Rakotoarivony I, Kengne P, Fontenille D (2003) Bionomics of Anopheles gambiae Giles, An. arabiensis Patton, An. funestus Giles and An. nili (Theobald) (Diptera: Culicidae) and transmission of Plasmodium falciparum in a Sudano-Guinean zone (Ngari, Senegal). J Med Entomol 40: 279–283.

42. Killeen GF, Kihonda J, Lyimo E, Oketch FR, Kotas ME, et al. (2006) Quantifying behavioural interactions between humans and mosquitoes: Evaluating the protective efficacy of insecticidal nets against malaria transmission in rural Tanzania. BMC Infectious Diseases 6.

43. Curtis CF, Myamba J, Wilkes TJ (1996) Comparison of different insecticides and fabrics for anti-mosquito bed nets and curtains. Medical And Veterinary Entomology 10: 1–11.

44. Lines JD, Myamba J, Curtis CF (1987) Experimental hut trials of permethrin-impregnated mosquito nets and eave curtains against malaria vectors in Tanzania. Medical And Veterinary Entomology 1: 37–51.

8

Pesticide Residues and Bees – A Risk Assessment

Francisco Sanchez-Bayo[1]*, Koichi Goka[2]

1 Faculty of Agriculture and Environment, The University of Sydney, Eveleigh, New South Wales, Australia, **2** National Institute for Environmental Sciences, Tsukuba, Ibaraki, Japan

Abstract

Bees are essential pollinators of many plants in natural ecosystems and agricultural crops alike. In recent years the decline and disappearance of bee species in the wild and the collapse of honey bee colonies have concerned ecologists and apiculturalists, who search for causes and solutions to this problem. Whilst biological factors such as viral diseases, mite and parasite infections are undoubtedly involved, it is also evident that pesticides applied to agricultural crops have a negative impact on bees. Most risk assessments have focused on direct acute exposure of bees to agrochemicals from spray drift. However, the large number of pesticide residues found in pollen and honey demand a thorough evaluation of all residual compounds so as to identify those of highest risk to bees. Using data from recent residue surveys and toxicity of pesticides to honey and bumble bees, a comprehensive evaluation of risks under current exposure conditions is presented here. Standard risk assessments are complemented with new approaches that take into account time-cumulative effects over time, especially with dietary exposures. Whilst overall risks appear to be low, our analysis indicates that residues of pyrethroid and neonicotinoid insecticides pose the highest risk by contact exposure of bees with contaminated pollen. However, the synergism of ergosterol inhibiting fungicides with those two classes of insecticides results in much higher risks in spite of the low prevalence of their combined residues. Risks by ingestion of contaminated pollen and honey are of some concern for systemic insecticides, particularly imidacloprid and thiamethoxam, chlorpyrifos and the mixtures of cyhalothrin and ergosterol inhibiting fungicides. More attention should be paid to specific residue mixtures that may result in synergistic toxicity to bees.

Editor: Raul Narciso Carvalho Guedes, Federal University of Viçosa, Brazil

Funding: These authors have no support or funding to report.

Competing Interests: The authors have declared that no competing interests exist.

* E-mail: sanchezbayo@mac.com

Introduction

Growing concern about the impact of pesticides on pollinators is reflected in the enormous literature on the topic in the past few years [1]. In response to this concern, considerable amounts of new data on toxic effects of pesticides on wild bees, in particular bumble bees, have been obtained from laboratory and semi-field experiments [2,3].

A number of reviews on the topic have highlighted the importance of bees as natural pollinators not only for our crops but also for wildflowers and plants of forests and tropical ecosystems [4,5]. That is why the current declining trend of pollinators is worrying [6]. For example, it has been estimated that without bees, some 60 species of crop plants would fail to produce fruit [7]; the economic consequences of this impact are obvious. Importation of bumble bees to make up for the losses of pollinators in the areas affected not only does not solve the issue but also creates more problems by exporting parasites to other regions or countries [8,9] or competing with native species [10].

Of particular importance is the collapse of honey bee (*Apis mellifera*) colonies (CCD) in America and other developed countries, because they provide honey and wax commodities to our society. Attempts to explain the CCD have focussed on two main fronts: i) biological diseases, which includes virus [11] *Nosema* infections [12], parasites such as mites [13,14] and hive beetles [15]; and ii) pesticides, including not only insecticides and acaricides but also fungicides and herbicides [16,17]. Naturally,

low levels of pesticides may act as stressors that make bees more prone to biological infections [3,18,19]. Among the pesticides, newly developed systemic insecticides such as fipronil and neonicotinoids have been targeted as the main culprits involved in the collapses since they were launched to the market in the mid-1990s [20,21,22,23].

Biological factors have been responsible for many of the problems that beekeepers have with their bee hives [24], but they are unlikely to be the main cause of disappearance of a number of wild bee species, or the decline of bumble bees in North America and Europe in recent years [12,25]. Although there are scant data on bee populations from other parts of the world to make a proper evaluation, the fact that bee declines have been observed in countries that have a long history of using pesticides in agriculture points to these agrochemicals as one of the important factors underlying wild bee and honey bee colony losses. To resolve this issue, several surveys have been carried out in recent years in North America [26,27,28], France [29,30], Spain [31] and India [32] among others, to find out the amounts and prevalence of pesticide residues present in pollen, honey, wax and other matrices of the bee hives (e.g. combs). They constitute a useful dataset to evaluate the impact that current pesticide residue levels have on honey bees and, possibly, wild bees as well; this risk is different to the risk of being affected by spray drift of these plant-protection products [33,34].

Typical risk assessments consider only acute toxicity of the chemicals either by topical or oral exposure in 24 or 48 hours, ignoring thus the negative effects derived from constant exposure to pesticide residues over longer periods. Some assessments have focused on environmental fate of pesticides and their application rates to estimate Toxicity Exposure Ratios (TERs) that were then used as indicators of the risk for honeybees due to particular exposure routes, e.g. ingestion of pollen or contact with it [35]. Recently, an individual study on pollen residues evaluates the possible risk of such residues to honey bees by both contact with and ingestion of contaminated pollen [28]. Neither study, however, includes the frequency of contaminated pollen among the risk parameters, while they also ignore the residues in honey or nectar. This we consider a serious flaw, as risk assessments should be based on the probability of exposure to actual residue levels. Indeed, none of the frequency data from the surveys mentioned above have been used to assess the impact that individual chemical residues and their combinations may or may not have on bees.

Some authors have tried to link the residue levels to the CCD in America [36], but by and large no risk assessment that includes residue levels, their prevalence and toxicity has been carried out to date. The handicap here is not insufficient residue data or acute toxicity data, but rather a lack of understanding as to how chronic toxicity by constant dietary exposure to residues found in pollen and honey affect the mortality of individual bees and the growth and reproduction of their colonies. Such effects include not only sublethal impairments but also delayed mortality [37]. Experiments with bumble bees have demonstrated that the lethal effects of new insecticidal compounds, including insect growth regulators and neonicotinoids, cannot be assessed based on acute toxicity data alone [22]. To understand the impact of small but constant doses of toxic residues on bee colonies it is necessary to apply different approaches where the time of exposure is taken into account [38].

Here, we attempt to provide a comprehensive risk assessment for all pesticide residues found in pollen and honey, or nectar, to bumble and honey bees using all residue and toxicity information available to date in the open literature and databases. Residue data originate from application of pesticides in accordance with standard agricultural practices in the countries surveyed, not from worst case, theoretical scenarios. Bees rely on nectar and pollen to meet the majority of their nutritional requirements, and therefore our risk assessment is focused on these two plant materials; honey is just concentrated nectar. Residues in wax are not included in this assessment since their availability to the bees was considered to be negligible compared to the direct exposure by contact with or dietary intake of pollen and honey [39]. However, recent research indicates that wax residues may also have an impact higher than expected until now [40], so available residue data in wax is presented for comparison only. Inhalation of volatile pesticides near treated crops is also excluded, since this is considered a minor route of exposure for most pesticides [41]. Traditional as well as novel methods of risk assessment will be used and compared in their predictions.

This assessment differs from those intended for regulatory purposes in several aspects: i) our focus is on the actual exposure of bees to the current pesticide residues found in the environment of developed and developing countries, not on the predicted exposure levels determined by models used in the tiered process of pesticide registration; ii) our assessment does not consider the particular application method of individual chemicals to their specific crops (e.g. foliar spray, granular, seed treatment, etc.) as it is based on the residue levels that are actually found in pollen and honey, regardless of the way they get there; iii) our assessment considers bee larvae and two castes of worker bees with different food requirements: nurses that feed on pollen, and nectar foragers.

While the viability of the bee colonies depends largely on the queen's health and her reproductive output, at present there is insufficient knowledge to assess the impact that pesticides have on the queen's performance – the exception being recent studies with honey bees [42] and bumble bees [43].

The aim of this risk assessment is to identify the main chemicals that may pose a threat to the life of bees in their natural environment, which is currently contaminated with a large array of pesticides and other chemicals. By highlighting the compounds with higher risk to bees, we hope that apiculturists, beekeepers and policy makers involved in agricultural production will be able to screen the products most harmful to bees and find the appropriate remedies to avoid further damage.

Materials and Methods

This assessment is restricted to honey bees (*Apis mellifera*) and bumble bees (*Bombus* spp.), which are very important pollinators and have been well studied. Information on ecotoxicity of a few pesticides to other wild bees exists [44,45,46], and their assessment can be inferred from the risk to the most common bee species presented here.

Residues Data

Data on pesticide residues in pollen, honey and wax from bee hives were taken from several sources, including recent pesticide surveys in the USA [26,27], France [47] and Spain [31] as well as a survey of neonicotinoids in Poland [48]. The review by Johnson et al. [17] provided further data on maximum residues in all these matrices. Residues in honey include additional data from surveys in Greece [49], Spain [50], Brazil [51] and India [52], complemented with sparse data from other sources as well as with residues in nectar from treated plants [44,47,53,54,55,56]. Residues in wax also include other data from Spain [57] and the USA [26,58]. The data were compiled to obtain average and maximum residue loads for each compound, and their frequency, in pollen, wax, honey or nectar (see Table S1).

Toxicity Data

Acute oral and contact toxicity of pesticides to honey bees are available for the majority of pesticides as either median lethal doses per bee (LD50) or median lethal concentrations (LC50) in the tested media. Median values are preferred to no-observed effect level (NOEL) or the lowest-observed effect level (LOEL) values, which are only available for a small number of compounds and which relevance for risk assessment has been questioned on statistical grounds [59] and inaccuracy [60].

Toxicity data for honey bees were obtained from the Pesticide Manual [61], the ECOTOX database of the U.S. Environment Protection Agency (http://cfpub.epa.gov/ecotox/) and the Agri-Tox Database of the Agence Nationale de Sécurité Sanitaire de l'Alimentation, de l'Environnement et du Travail in France (http://www.agritox.anses.fr/index.php). Toxicity of 29 insecticides to bumble bees was obtained from ECOTOX and the open literature [2].

Agreement between the toxicity data sources was remarkably high (>95% of all compounds), with only a handful of compounds (8) showing obvious discrepancies. It is concerning, however, that LD50 values for 30% of the most highly toxic compounds to bees are not reported in the Pesticide Manual, since this is the database most commonly used by consultants in the agricultural business. Notable among these omissions are imidacloprid, emamectin benzoate, etofenprox, flumethrin, prallethrin and several organophosphorus compounds: dicrotophos, parathion (ethyl), omethoate and acephate. Surprisingly, toxicity data for coumaphos – which is widely used in apiaries for mite control – were absent from the Pesticide Manual and Agritox databases, as noted also by other researchers [28].

Oral toxicities were available for 221 compounds of the 322 pesticides compiled (69%), whereas contact toxicity (topical) covered 96% of the pesticides (see Table S2). Included in the data are 76% of existing insecticides and fungicides and 83% of acaricides registered for use in agriculture. Herbicides were excluded since they are non-toxic to bees, i.e. LD50 values above 100 or 200 µg bee^{-1}. When more than one oral or contact LD50 value was available, a geometric mean was calculated. For data reported as "more than a given value", that value was used in the calculations. Oral toxicities were referred in almost all cases to 48-h exposures, whereas contact exposures varied between less than a day and 96-h, with a median of 48-h, so the average LD50s or LC50s used here can be considered representative of acute exposures to bees in about 2 days.

Unfortunately, no toxicity data for larvae are available (but see [40]), so here we assume the same LD50 values for larvae as for adult bees. Chronic data for bees are extremely rare and only reported for 1 systemic insecticide [62,63], and 6 insecticide growth regulators [22], as indicated in Table S2.

A regression of insecticides' LD50s (µg bee^{-1}) between honey and bumble bees reveals that the sensitivity of honey bees by oral exposure is similar to that of bumble bees (slope = 0.34, $r^2 = 0.94$, p<0.001, n = 13), whereas bumble bees are 28 times less sensitive than honey bees in regard to contact exposure with insecticides (slope = 28.3, $r^2 = 0.93$, p<0.001, n = 16) (Fig. 1). Even after correcting for weight between species, bumble bees are about 7 times less sensitive to insecticides by contact than honey bees. Because such difference varies from chemical to chemical, extrapolations of toxicity from honey bee to bumble bee have been avoided in this study, even if they may be useful in some situations [33,64].

Data Analysis

Standard risk approach. It should be recognised that the standard hazard quotient (i.e. HQ = PEC/LD50, where PEC is the predicted environmental concentration) is not a measure of risk because it does not indicate the probability of a hazard to occur. And yet, previous studies on pesticides and bees used HQs in their evaluations [28,35]. To estimate the risk of bees being affected by contaminated pollen or nectar it is necessary to consider the frequency of detection of pesticides residues in such matrices, because prevalence indicates the probability of exposure to the contaminants. Therefore, a simple risk assessment should incorporate this probability as follows

$$Risk = \frac{frequency\ [\%] \times residue\ dose\ [\mu g]}{LD50\ [\mu g\ bee^{-1}]} \quad (1)$$

This expression indicates that a given pesticide residue has a certain probability of causing 50% mortality among the bees that come into contact with or ingest the contaminated pollen or nectar. Since we only use here LD50 values to estimate risks, our assessment should be considered very conservative. For estimation of risks at the lowest effect level, approximate estimates of LD10 can be calculated as 0.1 × LD50. Such approximation is based on previous field studies that determined the lowest effect levels of many pesticides on aquatic organisms [65].

For expression (1) to represent the actual probability of risk, residue loads should be first converted to the actual doses of

residue that come in contact with (topical exposure) or are ingested (oral exposure) by the bees. Having data on average and maximum residue loads allows us establish a range of possible risks to bees. For average loads, the frequency of detection shown in Table S1 was used in the calculations; for maximum loads it should be noted that their frequency of detection is 1/total number of samples analysed in each survey. As the number of samples per survey varies between 99 and 845, the frequency of appearance of maximum residues is in the range 0.1–1.0%, i.e. much lower than the average prevalence of residues.

In the case of exposure by contact with pollen, topical LD50s shown in Table S2 were used to calculate the risk for a worker bee that comes in contact with 1 g of contaminated pollen per day. For oral exposure we focus on nurses, which feed exclusively on pollen for 10 days, and nectar foragers, which feed on nectar/honey for another 20 days during the summer season (Table 1). These types are considered representative of the bee colonies during the summer, when most pesticides are applied to crops, although winter bees can be exposed to residues in nectar for up to 100 days or more [66]. Lack of specific data on intake by bumble bees obliged us to scale the same rates as honey bees multiplied by a factor of 5, based on average intake of syrup by workers of *Bombus terrestris* and *Apis mellifera* - see File SI for estimation of contact doses and daily intake of residues.

Risk of synergistic mixtures. Because of the known synergism between ergosterol inhibiting fungicides with pyrethroids and cyano-substituted neonicotinoids (i.e. thiacloprid and acetamiprid), risks of residue mixtures were also included in this assessment for both contact and dietary exposures. These fungicides disable the monooxygenase detoxification system in honey bees, thus increasing the lethal effects several fold [67,68,69]. Synergistic factors vary for each combination and are reported only for topical exposure, but here we assume the same factor applies to oral exposures. In any case, the LD50 of the mixture was estimated as

$$LD50_{mixture} = \frac{LD50_{insecticide}}{synergistic\ factor} \quad (2)$$

Risks of mixtures are estimated using equation (1), with residue loads of the insecticide and the combined frequency of the compounds. Since the probability of finding residues of both insecticide and synergist in the same pollen or honey cannot be estimated here we considered the lowest frequency of either compound only.

New approaches to risk. The above expressions of risk indicate probabilities of causing serious effects (e.g. 20% risk of resulting in 50% mortality) within short periods of exposure, i.e. about 2 days. They suit well the assessment of risks by contact exposures. However, they may not be appropriate to assess risks by chronic, dietary exposure because the bees constantly consume pollen, nectar and honey. Assuming the residues ingested remain in the body, the median lethal dose may be reached after some time; in practice, there is some elimination and metabolism for most compounds [70], so the cumulative residue amounts estimated this way represent a worse case scenario. As the residue loads in pollen and honey are already known, the only limitation is the life-span of the individual bees, which varies from 5 days in worker larvae to 100 or more days in winter worker bees (Table 1).

Consequently, a simple way to assess the dietary risk of pesticide residues is by estimating the time to reach their corresponding LD50s, and compare those times with the actual life-span of each stage of development. Only times which are shorter than the life-

Insecticides

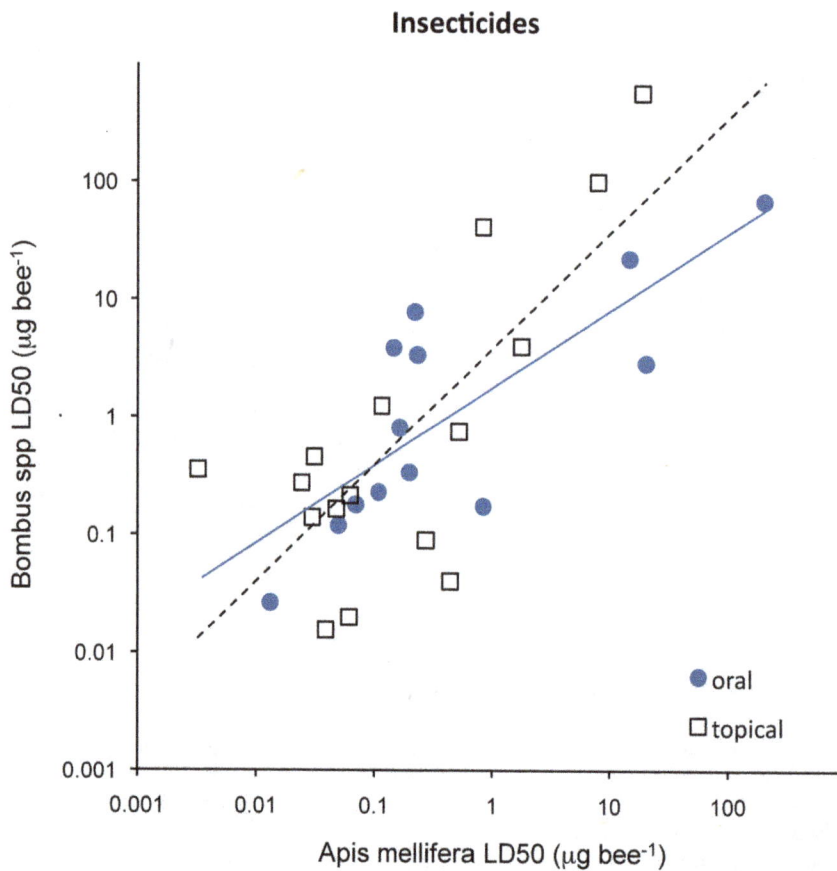

Figure 1. Comparison of the sensitivity of honey bees (*Apis mellifera*) and bumble bees (*Bombus* spp.) to 29 insecticides, as expressed by their contact or oral LD50s (μg bee⁻¹). Susceptibility of both species by oral exposure is similar (line, slope = 0.34, p<0.001), whereas on average bumble bees are 7 times less sensitive than honey bees by contact exposure, after correcting for weight (stippled line, slope = 28.3, p<0.001).

span would represent a serious risk, as they indicate that surely more than 50% of the bees exposed would die. We use two distinct new approaches for assessing dietary exposure:

i) Fixed dose approach. Assuming that acute LD50 values are constant for each pesticide, estimates of the time to reach the

dietary LD50 (henceforth T50) of each pesticide were calculated as follows

$$T50\ [days] = \frac{LD50\ [\mu g\ bee^{-1}]}{daily\ dose\ [\mu g]} \tag{3}$$

Table 1. Life-span of larvae and worker bees and their consumption rates of pollen and honey (After [76]).

	Apis mellifera			Bombus spp.[1]	
	Daily rate (mg/day)		Life-span (days)	Daily rate (mg/day)	
	Honey	Pollen		Honey	Pollen
Drone larvae	15.1	1.1	6.5	75.5	5.5
Worker larvae	28.9	1.1	5	144.5	5.5
Brood attendant	34–50		8	170–250	
Nectar forager	80.2		30	401	
Nurse worker		6.5	10		32.5
Pollen forager	13		30	65	
Wax-bees	18		6	90	
Winter bees	8.8		91+	44	

[1]Assuming 5 times the consumption of *Apis mellifera* in the same proportion.

As with the standard risk assessment, T50s were estimated for intake of average residues as well as maximum residues, so as to provide a range rather than an exact number of days. This approach may be valid for most pesticides, but there are some exceptions that justify another way of assessing risks.

ii) Time-cumulative effects. This approach is based on the experimental observation that dietary LD50s for certain compounds decrease with exposure time [37]. Consequently, the estimated T50s will be reached earlier than expected. The rate of change of LD50 with time can be estimated experimentally by a simple log-to-log regression of the LD50s on the exposure times

$$Ln\,T50\,[days] = a + b \cdot Ln\,LD50\,[\mu g\,bee^{-1}] \qquad (4)$$

where a (intercept) and b (slope) are empirical parameters specific to each chemical and species tested [71]. Slope values <1 result in an exponential increase of effects over time, according to the Druckrey-Küpfmüller equation (D Tn = constant, where the exponent n = 1/slope, D = dose and T = time) [37]. To date, there is empirical evidence of time-cumulative toxicity for some carcinogenic substances, neonicotinoid insecticides, rodenticides and methylmercury, and its underlying mechanism is thought to be the irreversible binding of the toxicant to specific receptors [72]. In the case of bees, the only data available are for the neonicotinoids imidacloprid [62,63] and thiamethoxam [73], so this new approach will be used here only for these two compounds.

Results

Residue Data

A total of 161 pesticides have been found so far in bee hives, of which 124 appeared in pollen, 95 in wax and 77 in honey or nectar. The majority were insecticides (83), with fungicides (40), herbicides (27) and acaricides (10) making up the remainder; only one insecticide synergist (piperonyl butoxide) was found [17]. Among the survey's data, 15 metabolites were reported due to their toxicity and persistence (e.g. aldicarb sulfoxide, endosulfan sulfate, fipronil sulfone, 5-hydroxy-imidacloprid, DDE, etc.). Whenever metabolites are reported, total loads for a compound were calculated as the sum of loads of the individual metabolites and their respective parent compound, weighted for the frequency of appearance in the respective surveys. Some persistent compounds are no longer used in one or several of the countries surveyed (e.g. dieldrin, DDT, HCB) but their residues are still present in their environment and need to be taken into account for risk assessment of bees. Obviously not all chemicals appeared in all surveys, and their concentrations and frequency differed markedly among surveys, reflecting the usage pattern of agrochemicals in each country or region.

The highest residue concentrations were found in wax and pollen (average 126 and 66 $\mu g\,kg^{-1}$ respectively), whereas the highest frequency of detection corresponds to wax (over 50% for chlorfenvinphos, tau-fluvalinate, bromopropylate, coumaphos and chlorothalonil) and honey (over 50% for thiacloprid, thiamethoxam and acetamiprid [48]). It is unclear whether the residues detected in pollen collected from apiaries (Fig. 2a) originated from sprayed fields or from hives treated with pesticides; they are probably a mixture of both. Whatever their source, such pollen feeds the nurse bees and larvae. Among the residues in honey, systemic insecticides stand out for their high prevalence: neonicotinoids are the most commonly found, while

phorate, dimethoate and carbofuran are typically present in more than 5% of nectar collected from treated plants (Fig. 2b). The presence of hydrophobic pesticides such as coumaphos or vinclozolin, and to a lesser extent tau-fluvalinate, in honey suggests contamination from the comb, since honey bee colonies are commonly treated with these pesticides for mite and fungal control [74].

Risk by Contact Exposure

A total of 92 individual compounds could be assessed for risk to contaminated pollen by contact exposure after matching residue and toxicity data. To these were added the synergistic combinations of cyhalothrin, thiacloprid and acetamiprid with three ergosterol inhibiting fungicides: propiconazole, penconazole and myclobutanil. Table 2 shows the risk for honey bees and bumble bees exposed to average and maximum residue levels of each compound, after taking into account their average prevalence in Europe, America and Asia. Only 33 compounds and 5 mixtures that have some relevance (i.e. risk >0.1) are shown, since 65% of the compounds have negligible or no risk to the bees. Risks above 5% can be considered high, as they correspond to T50s of 2 days or less; between 1 and 5% the risk is moderate, usually corresponding to T50s between 2 and 7 days, which are within the life-span of larvae and adult workers; risk below 1% can be regarded as low, for which T50s range from 7 to 60 days and more, covering the life-span of nectar foragers in summer and most of the life-span of winter bees.

Not surprisingly, the bulk of chemicals posing contact risk to bees are insecticides (20) or insecticide-acaricides (10), with only 2 acaricides, 1 fungicide and 5 fungicide mixtures appearing in that list. The risk of being seriously affected by contact with pollen residues is generally low, with only 5 compounds showing high risks (>5%): thiamethoxam (3.7–29.6% for honey bees), phosmet (14.6–23.9% for honey bees), chlorpyrifos (8.3–12.9% for both bees), imidacloprid (10.3–16% for honey bees but 31.8–49% for bumble bees) and clothianidin (1.0–5.3% for honey bees and 2.5–13.3% for bumble bees). It should be noted that the risk of these neonicotinoids to bumble bees is about two to three times as high as for honey bees, due to the different sensitivity among the species (Fig. 1). These compounds pose high risk to bees on account of their extremely high toxicity to both honey and bumble bees, with topical LD50s in the range 0.02–0.09 $\mu g\,bee^{-1}$, and also because their average residues (12–35 ppb) were present in 11 to 16% of the pollen surveys worldwide. By contrast, the high risk of phosmet is mainly due to average residues of 339 ppb (highest 16.5 ppm) in spite of its moderate toxicity to honey bees (topical LD50 = 0.62 $\mu g\,bee^{-1}$). While six other compounds were more common among the residues (coumaphos, tau-fluvalinate, chlorothalonil, acetamiprid, amitraz and thiacloprid), their toxicity to bees is 100–5000 times lower than that of thiamethoxam or chlorpyrifos.

Mixtures of fungicides with cyhalothrin or thiacloprid pose also high risks for honey bees (3.7–8.8%) and a moderate risk to bumble bees (1.1–2.6%), even if the prevalence of the three fungicides is relatively low (1.8–5.5%). Attention should be paid to the synergism of propiconazole with such insecticides, as it changes markedly the risk of the individual compounds from being moderate (0.2–1.8% cyhalothrin) or negligible (<0.1% thiacloprid) to a high risk. The synergistic factor of propiconazole for thiacloprid is 560 [67,75] and for cyhalothrin 16.2 [67,68]. Only the mixtures acetamiprid with propiconazole and fenuconazole showed low risk for honey bees (0.1–0.7%) and negligible risk for bumble bees (0.01–0.07%) based on synergistic factors of 100-fold (propiconazole) or 4.5-fold (fenuconazole) and the low frequency of such fungicides in pollen (1.8–3.3%).

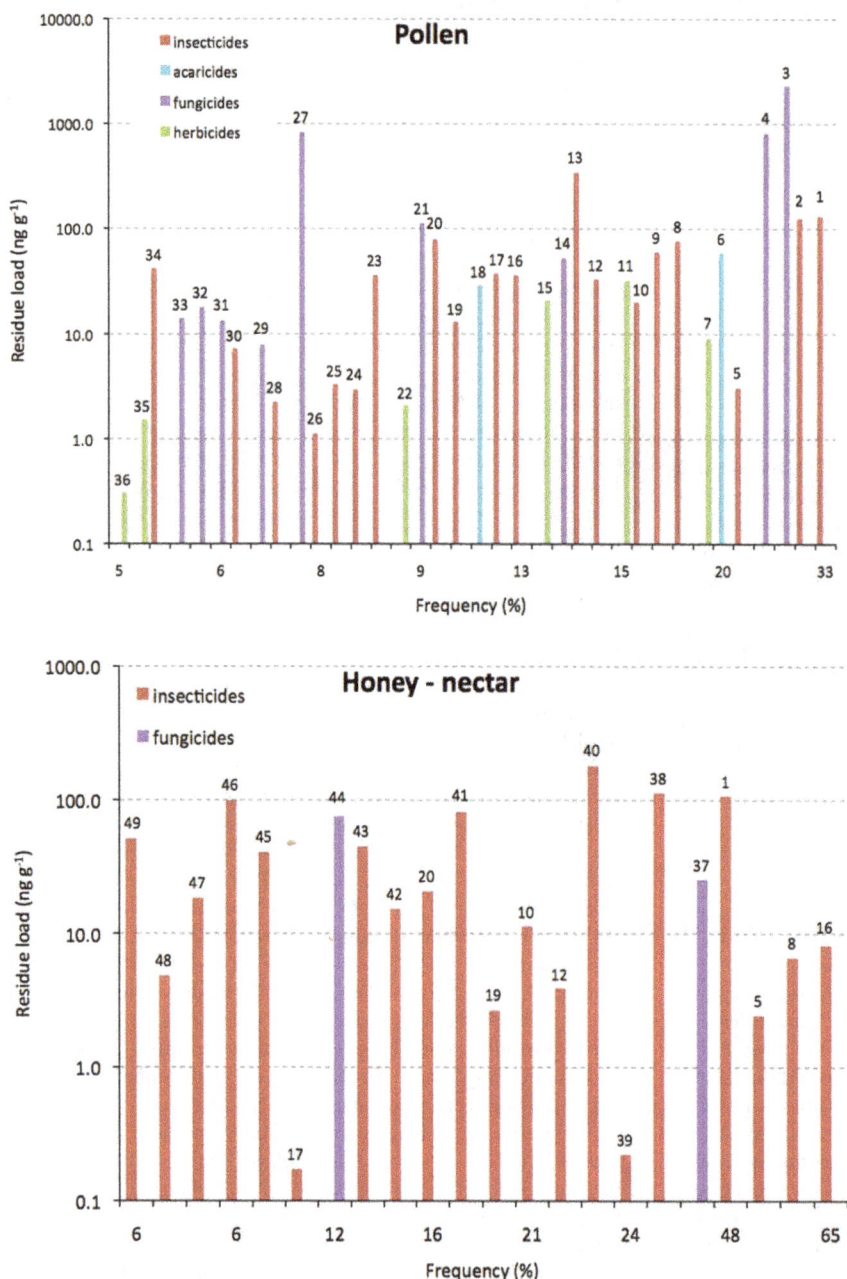

Figure 2. Residue loads of the most common pesticides plotted against their prevalence (frequency) in: A) pollen; B) honey or nectar. Key: 1 coumaphos (total); 2 tau-fluvalinate; 3 thymol; 4 chlorothalonil; 5 acetamiprid; 6 amitraz (total); 7 dithiopyr; 8 thiacloprid; 9 carbaryl; 10 imidacloprid (total); 11 pendimethalin; 12 chlorpyrifos; 13 phosmet; 14 carbendazim; 15 atrazine; 16 thiamethoxam; 17 chlorfenvinphos; 18 fenpyroximate; 19 clothianidin; 20 endosulfan (total); 21 thiophanate-methyl; 22 metolachlor; 23 fenpropathrin; 24 methoxyfenozide; 25 esfenvalerate; 26 tebufenozide; 27 captan (total); 28 bifenthrin; 29 azoxystrobin; 30 lambda-cyhalothrin; 31 diphenylamine; 32 penconazole; 33 trifloxystrobin; 34 fenthion; 35 norflurazon; 36 metribuzin; 37 hexachlorobenzene; 38 HCH (alpha and beta); 39 phorate; 40 gamma-HCH (lindane); 41 heptenofos; 42 methiocarb; 43 DDT (total); 44 vinclozolin; 45 methidathion; 46 malathion; 47 cypermethrin; 48 dimethoate; 49 carbofuran (total).

Moderate risk by contact exposure (i.e. 1–5%) includes 6 pyrethroids (acrinathrin, deltamethrin, cypermethrin, bifenthrin, esfenvalerate and lambda-cyhalothrin, in that order), the carbamate carbaryl, the organophosphorus fenthion, the neonicotinoid dinetofuran, the pyrazole fipronil and the acaricide pyridaben. However, risks of these compounds to bumble bees are below 1%, because they are less toxic to the large pollinators (Table S2). Notice that, despite fipronil and bifenthrin being among the most

toxic insecticides to honey bees (topical LD50 0.007 and 0.015 µg bee^{-1} respectively), their risk by contact exposure is reduced because of their low residue loads (1.6–29 ppb and 2.2–13 ppb respectively) and low prevalence in pollen residues (average 2.8 and 6.6% respectively).

The remaining 17 compounds pose a low risk (<1%) for being less toxic (chlorothalonil, coumaphos, tau-fluvalinate, endosulfan, chlorfenvinphos), appearing rarely (phenothrin, prallethrin, ace-

Table 2. Risk (% probability) and time to reach topical LD50 (T50 in days) for worker bees, estimated as contact exposure with 1 g of contaminated pollen, at average or maximum levels, during 2 days.

Use[1]	Chemical	Honey bee					Bumble bee				
		Topical LD50 (µg bee^{-1})	Risk (%)		TD50 (days)		Topical LD50 (µg bee^{-1})	Risk (%)		TD50 (days)	
			Average	Max	Average	Max		Average	Max	Average	Max
I	thiamethoxam	0.02	29.58	3.66	1	0.2					
I	phosmet	0.62	14.56	23.89	2	0.04					
I	chlorpyrifos	0.07	12.92	10.33	2	0.1	0.09	10.32	8.26	3	0.1
I	imidacloprid (total)	0.06	10.34	16.00	3	0.1	0.02	31.77	49.15	1	0.02
I+F	cyhalothrin+propiconazole	0.003[2]	8.79	5.90	0.4	0.1	0.01[2]	2.56	1.71	1.4	0.3
I+F	cyhalothrin+myclobutanil	0.004[2]	7.86	5.52	0.6	0.1	0.02[2]	2.28	1.60	2.1	0.4
I+F	cyhalothrin+penconazole	0.011[2]	7.23	3.68	2	0.3	0.04[2]	2.10	1.07	5	1.0
I	clothianidin	0.04	5.28	0.99	4	1	0.02	13.26	2.49	2	0.4
I+F	thiacloprid+propiconazole	0.065[2]	4.21	7.43	1	0.1					
A	acrinathrin (total)	0.17	3.44	1.21	1	0.2					
I	deltamethrin	0.02	3.27	2.26	1	0.3	0.28	0.29	0.20	11	3
I-A	cypermethrin	0.03	2.40	1.75	2	1					
I	carbaryl	0.84	2.32	0.95	14	1	41.16	0.05	0.02	699	39
I	bifenthrin	0.01	1.97	1.14	7	1					
I	esfenvalerate	0.03	1.96	2.99	8	0.4					
I	fenthion	0.22	1.89	0.57	5	1					
I	dinotefuran	0.05	1.84	2.18	1	0.3					
I	lambda-cyhalothrin	0.05	1.83	0.71	7	1	0.17	0.53	0.21	23	5
I	fipronil (total)	0.007	1.19	2.49	5	0.3					
A	pyridaben	0.05	1.00	0.29	3	2					
I-A	tau-fluvalinate	8.66	0.92	0.28	70	3					
I	indoxacarb	0.59	0.81	0.43	5	1					
I	permethrin	0.06	0.60	1.90	6	1	0.22	0.18	0.55	20	2
I	beta-cyfluthrin	0.03	0.56	1.41	14	1	0.46	0.04	0.10	209	14
I	prallethrin	0.03	0.42	0.16	4	4					
I-A	coumaphos (total)	20.29	0.41	0.28	158	3					
I	phenothrin	0.13	0.38	0.37	2	2					
F	chlorothalonil	135.32	0.32	0.35	169	1					
I-A	endosulfan (total)	6.35	0.26	0.48	82	2					
I-A	carbofuran (total)	0.16	0.22	0.55	13	1					
I-A	chlorfenvinphos	4.10	0.22	0.05	112	10					
I-A	aldicarb (total)	0.38	0.17	1.61	29	0.3					

Table 2. Cont.

Use[1]	Chemical	Honey bee					Bumble bee				
		Topical LD50 (µg bee⁻¹)	Risk (%)		TD50 (days)		Topical LD50 (µg bee⁻¹)	Risk (%)		TD50 (days)	
			Average	Max	Average	Max		Average	Max	Average	Max
I-A	methomyl	0.49	0.16	0.03	48	21					
I+F	acetamiprid+propiconazole	0.076[2]	0.14	0.85	25	0.6	0.95[2]	0.01	0.07	318	7.1
I-A	diazinon	0.38	0.14	0.05	45	9					
I	acephate	1.78	0.11	0.06	20	11	3.99	0.05	0.02	44	24
I-A	malathion	0.47	0.11	0.07	28	7					
I+F	acetamiprid+fenbuconazole	1.76[2]	0.01	0.09	587	13	22.2[2]	0.00	0.01	7407	166
IGR	diflubenzuron	114.80	0.00	0.00	1441	897	0.10[3]	1.68	1.66	1	1

[1]A = acaricide; F = fungicide; I = insecticide; IGR = insect growth regulator; IS = insecticide synergist.
[2]Mixture LD50 estimated in accordance with known synergistic ratios [67,68,69].
[3]Chronic LD50 for 77 days exposure.

phate, carbofuran, malathion and permethrin) or both (diazinon, methonyl, aldicarb, beta-cyfluthrin and indoxacarb). Except for pyrethroid residues, which can have almost immediate effects by contact exposure, the average T50 for all other pesticides in this group is above 60 days for honey bees, denoting a very low risk by contact with pollen. Obviously, maximum residues of these pesticides would result in serious effects in very few days or even less (Table 2). Presumably, a similar risk would apply to contact with residues in wax.

Risk by Dietary Exposure

Information on oral toxicity to bees is less comprehensive than that of topical toxicity (see Table S2), so only 77 compounds could be evaluated here. Average and maximum daily doses of residues ingested (Table S3) were calculated first to assess the risk of worker larvae, nurses and nectar foragers when exposed to the array of pesticides found in pollen and nectar (Table S1). Considering the life spans of each type of bee, the risk of consuming contaminated food during their lifetime and the T50 were assessed using the standard risk method. Results for 25 pesticides and 1 mixture that pose some risks (i.e. >0.1%) to honey bees are shown in Table 3; the remaining 67% of pesticides pose a negligible or no dietary risks to these bees.

Dietary risk to honey bees. Extremely high risks were found for thiamethoxam and lindane residues in honey, which affect primarily nectar foragers and secondarily the larvae. Daily consumption of nectar or honey contaminated with these compounds at the average residue levels found worldwide would cause nectar forager mortalities of 50% or above within 3 days in the case of lindane, or a week for thiamethoxam (Table 3). The risk of these two insecticides to larvae is moderate (0.6–4.0% lindane, 0.2–2.8% thiamethoxam), since larvae consume less amounts and their exposure is only during 5 days. In addition, two other neonicotinoid insecticides found in honey pose high risks to foragers (3–22% clothianidin, 6–23% imidacloprid) and moderate risks to larvae (0.2–1.2% for either compound).

Residues of the pyrethroid cypermethrin in honey pose a moderate risk to nectar foragers (4.0–6.8%) but a low risk to larvae (0.1%). Moderate risks (1–5%) are also found for the organophosphorus coumaphos and quinalphos, the neonicotinoid dinetofuran and the carbamate methiocarb, but only coumaphos and dinetofuran present some risk to larvae. Nectar foragers are at low risk (0.1–1%) when feeding on honey contaminated with 9 more insecticides: the organophosphorus chlorpyrifos, dimethoate, pirimiphos ethyl, diazinon and malathion, the carbamates carbaryl and pirimicarb, the pyrethroid beta-cyfluthrin and total residues of DDT (i.e. DDT and its metabolites). Among these, only carbaryl seems to pose a minor risk to bee larvae (0.03–0.4%) and foragers (0.5–0.8%) alike, but daily consumption of its residues would only inflict some mortality among the adult foragers (T50 of 45–80 days). Residues of the synergistic fungicides, myclobutanil, penconazole and propiconazole have so far not been detected in honey, and therefore nectar foragers are exempt of higher risks in this regard.

The residual composition of pollen is different from that of honey, with 70 out of the 124 pesticides found only in pollen (Table S1). Among the worker bees, only nurses depend entirely on this kind of food, but the queen and larvae are fed substantial amounts of pollen as well ([76], Table 1). Moderate risks of pollen residues (1–5%) to both nurses and larvae were found for thiamethoxam, clothianidin, imidacloprid and phosmet. Estimated T50s for thiomethoxam are 6–27 days for nurses and 8–23 days for larvae, depending on the residue load. Obviously, the high toxicity of this insecticide to honey bees (oral LD50 0.005 µg

Table 3. Risk (% probability) and time to reach oral LD50 (T50 in days) for larvae and workers of honey bees feeding on contaminated pollen and/or nectar at average or maximum residue levels (see Table S3).

Chemical	Oral LD50 (µg bee⁻¹)	Risk (%)						T50 (days)					
		Worker larvae[1]		Nurses[2]		Nectar forager[3]		Worker larvae		Nurses		Nectar forager	
		Average	Max	Average	Max	Average	Max	Average	Max	Average	Max	Average	Max
thiamethoxam	0.005	2.77	0.23	4.80	0.59	200.18	3.86	23	8	27	6	10	4
gamma-HCH (lindane)	0.05	0.62	4.02	0.01	0.01	200.40	313.09	9	0.4	979	323	3	0.1
imidacloprid (total)	0.013	1.19	0.64	1.57	2.43	23.33	5.93	68	4	103	2	28	2
clothianidin	0.004	1.02	0.23	1.91	0.36	22.04	3.25	54	10	58	13	23	4
cypermethrin	0.06	0.13	0.10	0.04	0.03	4.00	6.77	119	23	711	154	44	9
coumaphos (total)	4.61	0.11	0.03	0.06	0.04	2.62	4.37	1444	71	5524	120	545	28
dinotefuran	0.02	0.10	0.06	0.13	0.16	1.50	2.37	49	27	74	20	20	13
quinalphos	0.07	0.00	0.00			1.29	0.69	253	236			91	85
methiocarb	0.47	0.00	0.00	0.00	0.00	1.08	0.28	1080	601	51648	51648	391	217
chlorpyrifos	0.24	0.04	0.01	0.13	0.10	0.86	0.30	1605	176	1118	44	764	197
carbaryl	0.15	0.41	0.03	0.42	0.17	0.54	0.82	202	63	392	22	80	45
beta-cyfluthrin	0.05	0.10	0.03	0.01	0.03	0.43	0.62	190	123	3497	226	69	49
dimethoate	0.17	0.01	0.00	0.00	0.00	0.40	0.24	1198	662	11303	6190	440	243
DDT (total)	5.08	0.00	0.00	0.00	0.00	0.29	0.62	3871	266	25061	7108	1432	96
pirimiphos ethyl	0.22	0.00	0.00			0.21	0.24	401	346			144	125
diazinon	0.21	0.04	0.01	0.01	0.01	0.19	0.39	426	202	3869	781	156	76
malathion	9.17	0.00	0.00	0.00	0.00	0.15	0.12	3218	1292	82696	20162	1167	471
pirimicarb	3.84	0.00	0.00			0.10	0.09	3500	1873			1261	675
thiacloprid+propiconazole	0.03[4]	0.08	0.29	0.30	0.53	0.00	0.00	109	4	61	5	57	2
phosmet	0.37	0.07	0.11	0.79	1.29			991	20	168	3		
fipronil (total)	0.001	0.02	0.05	0.27	0.57			596	33	101	6		
acrinathrin (total)	0.12	0.01	0.36	0.17	0.06		49.74	719	2	122	20		1
acephate	0.23	0.00	0.01	0.03	0.01	0.55	0.55	2264	135	383	214		54
permethrin	0.13	0.00	0.02	0.01	0.03	0.52	0.52	10842	142	1835	209		58
methoxychlor	5.02	0.00	0.00			0.28	0.28		293				106
dichlorvos	0.29	0.00	0.00	0.00	0.00	0.07	0.11	1215	735	6561	4746	452	272

[1]Exposure period of 5 days.
[2]Exposure period of 10 days.
[3]Exposure period of 30 days.
[4]Mixture LD50 estimated in accordance with known synergistic ratios [67].

Table 4. Risk (% probability) and time to reach oral LD50 (T50 in days) for larvae and workers of bumble bees feeding on contaminated pollen and/or nectar at average or maximum residue levels (see Table S3).

Chemical	Oral LD50 (µg bee⁻¹)	Risk (%)						T50 (days)					
		Worker larvae[1]		Nurses[2]		Nectar forager[3]		Worker larvae		Nurses		Nectar forager	
		Average	Max	Average	Max	Average	Max	Average	Max	Average	Max	Average	Max
imidacloprid (total)	0.03	2.93	1.57	3.86	5.98	57.39	14.58	28	2	42	1	11	1
heptenophos	0.53	0.00	0.00			28.98	10.42	46	16			17	6
chlorpyrifos	0.23	0.23	0.07	0.65	0.52	4.44	1.57	313	34	218	9	149	38
quinalphos	0.18	0.00	0.00			2.51	1.34	130	122			47	44
beta-cyfluthrin	0.12	0.22	0.05	0.02	0.06	0.90	1.28	92	59	1683	109	33	23
dimethoate	0.82	0.01	0.00	0.00	0.00	0.41	0.25	1159	641	10930	5986	425	235
lambda-cyhalothrin	0.18	0.02	0.00	0.08	0.03	0.19	0.10	1248	567	757	149	624	572
carbaryl	3.88	0.08	0.01	0.08	0.03	0.10	0.16	1050	328	2034	114	415	231
cyhalothrin+propiconazole	0.01[4]	0.12	0.03	0.11	0.13	0.19	0.10	77	35	47	9	624	572
cyhalothrin+myclobutanil	0.02[4]	0.11	0.03	0.07	0.12	0.19	0.10	114	52	69	14	624	572
cyhalothrin+penconazole	0.04[4]	0.10	0.02	0.03	0.08	0.19	0.10	282	128	171	34	624	572
acetamiprid+propiconazole	0.21[4]	0.02	0.02	0.00	0.02	0.07	0.00	585	80	2179	49	23125	4173
acetamiprid	22.20	0.00	0.00	0.00	0.00	0.07	0.00	61259	8370	228262	5110	23125	4173
diflubenzuron	1.46	0.00	0.00	0.02	0.02		0.02	3349	2084	567	353		
bifenthrin	0.34	0.00	0.00	0.01	0.01		0.11	28186	675	4770	807	284	284
rotenone	0.83	0.00	0.00	0.00	0.00		0.00	6538	6538	1106	1106		
acephate	7.87	0.00	0.00	0.00	0.00		0.08	15762	937	2667	1488	378	378
phosalone	3.98	0.00	0.00	0.00	0.00		0.00	23177	23177	3922	3922		
methomyl	3.38	0.00	0.00	0.00	0.00		0.00	59814	25670	10122	4344		

[1]Exposure period of 5 days.
[2]Exposure period of 10 days.
[3]Exposure period of 30 days.
[4]Mixture LD50 estimated in accordance with known synergistic ratios for honeybees [67,68].

bee^{-1}), together with its relatively high residue loads (29 ppb) and worldwide prevalence (12.8%) are the main reasons behind this risk. Although clothianidin is 4 times more toxic than imidacloprid, average residues of the latter insecticide are slightly higher and more frequently found in pollen than those of the former, so their overall risk is very similar. Nevertheless, only the highest residues of imidacloprid would seriously affect nurses and larvae alike, with T50s of 2 and 4 days respectively, whereas the highest clothianidin residues would have a smaller impact because the T50s are longer than the life spans of the bees (Table 3). In addition to neonicotinoids, the highest residues of phosmet and fipronil in pollen can result in 50% mortality of nurses in 3 and 6 days, respectively; but the risk can be considered low due to their low average residues (0.8% phosmet and 0.3% fipronil).

Residues in pollen of 4 other insecticides (carbaryl, acrinathrin, dinotefuran and chlorpyrifos) have low risk to honey bees (0.1–1%), as their T50s exceed by a long margin the life spans of nurses and larvae (Table 3). Also, the mixture of thiacloprid+propiconazole may pose some risk to larvae and nurses (T50s of 4 and 5 days respectively) only when residues of thiacloprid in pollen are at the highest recorded levels (1 ppm); otherwise, under normal circumstances the average residues of this neonicotinoid in pollen (75 ppb) wouldn't be of concern to either forager bees (0.5%, and T50 57 days) or larvae (0.08%, and T50 109 days). Risk of the remaining compounds found in pollen is considered negligible.

Dietary risk to bumble bees. In the case of bumble bees, estimates of risks for 15 compounds and 4 mixtures for which toxicity data are available are shown in Table 4. Contrasting with honey bees, the dietary risk of imidacloprid to bumble bees is very high: 14.5–57.4% for nectar foragers that consume honey or nectar and 3.8–6% for nurses that feed on pollen, while a moderate risk (1.6–2.9%) was found for worker larvae that consume both types of food (Table 4). Moreover, the maximum imidacloprid residues ingested by the different types of bumble bee would reach the oral LD50 within their respective life spans, while average residues in honey result in T50 of 11 days for nectar foragers, indicating than half of them would probably die before reaching the end of their lives. In addition, residues of heptenophos in honey present high risk to forager bumble bees (10.4–29% and T50 6–17 days) but not to their larvae. Lack of toxicity data for thiamethoxam on bumble bees prevents us from a

further assessment of this toxic compound using standard methods, even if some researchers have proven its negative effects on experimental bee colonies [42,43].

Moderate risks to nectar foragers (1–5%) were determined for the organophosphorus chlorpyrifos and quinalphos as well as the pyrethroid beta-cyfluthrin, but once again the risk of theses insecticides to larvae are low or negligible (Table 4). Based on the estimated T50s, only the highest residues of chlorpyrifos in pollen and beta-cyfluthrin in honey may represent a considerable risk to nurses and foragers respectively.

Residues of dimethoate, lambda-cyhalothrin, bifenthrin and carbaryl in honey pose a low risk to forager bumble bees. The risk of fungicides mixtures with cyhalothrin and acetamiprid is estimated as low as well. Since the three fungicides are only present in pollen, their synergism only affects the larvae and nurses, and even then the resulting risks are low: <0.12 for larvae and <0.13% for nurses, with T50s above their life span (Table 4). The only exception is with the highest residues of cyhalothrin in pollen (36 ppb), for which a T50 of 9 days was estimated.

Risks by Cumulative Toxicity

Estimates of T50s for imidacloprid and thiamethoxam in honey bees were also carried out using the time-cumulative approach (Table 5). Indeed, the time to reach their oral LD50 is shorter than the T50s estimated by the standard method. For example, nurse bees feeding on pollen contaminated with imidacloprid would reach their LD50 in 7 to 9 days (within their life-span), and those ingesting thiamethoxam would die in large numbers after one day, no matter what the residue loads are. The same applies to nectar foragers, which would be at serious risk when feeding on nectar or honey contaminated with either chemical, and so would be the larvae consuming thiamethoxam.

This approach should apply to all insecticides that exhibit time-cumulative toxicity, which requires the binding to specific receptors to be persistent [72]. However, it would not apply to the majority of chemical residues found in pollen or nectar. Although fipronil and its toxic metabolites have systemic properties and high toxicity to bees [77], so far there is no evidence of time-cumulative effects of fipronil on bees or other organisms. Also, residues of pyrethroid insecticides have little effect when ingested by bees, as they are either metabolised or quickly

Table 5. Comparison of estimated times to LD50 (T50 range in days) for dietary exposure of honey bees to two neonicotinoid insecticides, using standard and cumulative risk approaches.

Risk approach	Imidacloprid			Thiamethoxam		
	T50 (days)			T50 (days)		
	Worker larvae	Nurse	Nectar forager	Worker larvae	Nurse	Nectar forager
Cumulative	6–10	7–9	4–8	<0.1–1	0.2–1	<0.1–0.5
Standard	4–68	2–103	2–28	8–23	6–27	4–10
	Experimental data[1]			Experimental data[2]		
Exposure (days)	2	4	31	1	5.2	8.04
Oral LD50 (ng bee^{-1})	28.5	10.79	0.18	0.109	0.057	0.009
Equation	Ln T50 = 2.55 − 0.53 *Ln LD50			Ln T50 = −1.03 − 0.70 *Ln LD50		
Power exponent (n = 1/slope)	1.89			1.44		
r2	1.00			0.68		

[1]Sources: [62,63].
[2]Source : [73].

eliminated; consequently, their oral toxicity is on average 3 times lower than their topical toxicity (Table S2). Organophosphorus and carbamate compounds undergo a similar fate, so only a handful of persistent compounds (i.e chlorpyrifos, coumaphos and chlorfenvinphos) may last long enough to cause time-dependent toxicity, if any, during chronic ingestion of their residues.

Discussion

Bees can be exposed to plant protection products in two ways:

i) by direct exposure to either drift droplets, which are scattered during the foliar spraying of crops [33], dust from seed drilling at planting [78], or inhalation of volatile pesticides during or after application to the crops; and

ii) by exposure to residues present in pollen, wax, nectar, honey and guttation drops, which may result either from direct spray contamination of flowers, translocation through the treated plants or soil [20,54], or direct contamination during treatment of the combs (for honey bees only). Bees also drink water [79], and we have observed them drinking from paddy field waters contaminated with pesticides.

Our risk assessment in this paper deals only with residues in pollen and honey or nectar, because these constitute the essential food of bees [39]. Exposure to residues in these matrices may be by contact, while gathering pollen in the field and in the storages of the comb, or most likely by dietary and chronic ingestion of contaminated nectar, honey and pollen. Foragers presumably feed on nectar from flowers, rather than consuming the honey stores. Nectar is carried in the insect's honey stomach, and then processed by the bees before it becomes honey. Foragers carry and process far more nectar than they consume. We do not know how much of the active ingredients enter the insects during these processes, but it can be assumed that the exposure of foragers could thus be much higher than estimated here. Exposure to guttation drops was not considered here, as it is unlikely to affect most bees since such drops only appear in the early hours of the day [80]. For risk during agricultural operations the reader can consult [81,82].

The large number of agricultural chemicals found in pollen demands a rigorous evaluation of their risk to bee pollinators. Of the 124 parent compounds found in pollen from honey bee apiaries, about half of them appear with a frequency of 2% or more, 20 are present in more than 10%, and two insecticide-acaricides (coumaphos and tau-fluvalinate) appear regularly in more than 30%, particularly in North America [26,27]. It is also worrying that residues of the four most common compounds (tau-fluvalinate, coumaphos, thymol and chlorothalonil) are present at average concentrations above 100 ng g^{-1} of pollen (Fig. 2a). Highest residues can be up to 20 times higher (see Table S1), although they only appear occasionally.

Some 77 compounds have also been found in honey, with 23 of them being exclusive to this matrix. Residues in honey include mainly systemic compounds, among which the most commonly found are neonicotinoid insecticides – up to 65% prevalence (Fig. 2b). Systemic insecticides can move from the soil, where they are applied as granules or seed-coatings, through the sap of the plants and reach the nectar glands at the time of pollination, when the bees are attracted to the flowers [83]. It is no surprise, therefore, that many residues found in honey are of hydrophilic herbicides (5) and fungicides (15), as they are known to translocate within the various parts of the treated plants [84,85]. The highest residue loads in honey, however, correspond to hydrophobic compounds such as lindane and coumaphos, the latter being used to treat the combs for mite control [86].

Traditional risk assessments have considered only the residue loads in pollen and the acute oral or contact toxicity of the compounds [28,33]. We draw attention here to this important distinction, as the toxicity of hydrophobic insecticides and acaricides is mostly by contact exposure whereas the toxicity of hydrophilic fungicides and systemic insecticides is mainly by oral ingestion of residues in pollen and honey. It should be noted that pyrethroids, which are highly hydrophobic compounds, are on average 3 times more toxic to bees by contact than by oral exposure. By contrast, 60% of the systemic (hydrophilic) pesticides have oral toxicities higher than their contact toxicities, up to 11 and 13 times higher in the case of clothianidin and phorate, respectively (Table S2). It follows that regulators should pay more attention to dietary toxicities of any hydrophilic compound suspected of getting into the food chain.

Also essential to any risk assessment are not just the actual residue loads, but the frequency with which they appear in pollen and/or nectar. This is because the risk of bees being affected by pesticide residues is directly proportional to the prevalence of such residues in the environment (see equation (1)). For example, chlorpyrifos and methomyl have equal oral toxicities to honey bees (0.24 ng bee^{-1}), so assuming equal residue loads in pollen their hazard quotient is the same. However, because chlorpyrifos is present in 14.3% of pollen and methomyl only in 3.8%, a nurse bee is more likely to be intoxicated with the former compound while feeding on pollen, and hence the risk of chlorpyrifos to the bees is greater than that of methomyl.

When considering the relative weight of these three factors, residue loads, prevalence and toxicity, in the estimation of risks, it is evident that toxicity is the most important factor. Thus, risks above 1% by contact exposure are obtained for compounds with topical LD50s of 1 µg bee^{-1} or below, and this includes fipronil, four neonicotinoids (clothianidin, dinetofuran, imidacloprid and thiamethoxam), all synthetic pyrethroid insecticides (except tau-fluvalinate), pyridaben and six cholinesterase inhibitors: phosmet, chlorpyrifos, carbaryl, carbofuran, fenthion and aldicarb (Table 2). Risks by contact with residues in pollen or wax are likely overestimated here because not all residues are bioavailable by this exposure route [39]. High acute toxicity also determines the risk through dietary exposure, with neonicotinoids, cypermethrin, lindane and three cholinesterase inhibitors (coumaphos, quinalphos and methiocarb) posing the highest risks for honey bees (Table 3). In stark contrast, low risks were determined for three acaricides used in apiaries to control mites, tau-fluvalinate (0.3–1%), coumaphos (0.3–0.4%) and chlorfenvinphos (0.05–0.2%), even if their residues loads (36–128 ppb) are above the 60 ppb average and appear in pollen with a frequency of 12 to 32% ([26], Table S1). All of them present little risk to the bees because their toxicities by contact are low (4, 8 and 20 ng bee^{-1}). They may be of concern, however, when present in high concentrations, and it is only then that they can reach the topical LD50 in 2 to 4 days (Table 2).

The second factor used in the estimation of risk is a combination of both residue loads and prevalence. Indeed, the risk of 10 ppb of residues of a compound appearing in 10% of the pollen is equivalent to the risk of 100 ppb of the same compound appearing only in 1% of pollen. For example, among the highest average residue loads found in pollen are the fungicides captan (821 ppb) and chlorothalonil (802 ppb); chlorothalonil is of greater concern (risk 0.32%) not only because is twice as toxic as captan (135 µg bee^{-1} vs 215 µg bee^{-1}) but also its residues are found in 27% of the pollen, whereas captan is present only in 7% of the pollen.

Since toxicity is the main factor affecting risk, the synergistic combinations of ergosterol inhibiting fungicides with pyrethroids and cyano-substituted neonicotinoids are of great concern for one reason: the intrinsic toxicity of the individual compounds is already very high in the case of pyrethroids, and it is boosted up to 16-fold when propiconazole is present among the residues [68]. For thiacloprid the synergistic factor is as high as 560-fold, and for acetamiprid 100-fold [67], so their safety features [87] become all of a sudden hazardous. However, combinations of anilinopyrimidine fungicides with the same neonicotinoids do not show synergism in bees [75], perhaps because they do not interfere with the P-450 detoxification system. The relatively low prevalence of these fungicides among pollen residues (1.8–5.5%) could be an ameliorating factor for nurses, queens and larvae, while nectar foragers would not be affected by the synergism as honey appears to be free of these fungicides. Not considered here is the synergism of chlorothalonil with fluvalinate and coumpahos (which only occurs at high concentrations of fluvalinate), because the presence of coumaphos significantly reduces the toxicity of the fluvalinate and chlorothalonil mixture [40]. The risks of fungicide-insecticide mixtures calculated here are based on the lowest prevalence among the fungicide-insecticide pairs, but even then they may be overestimated: it is obvious that not all pollen contaminated with the insecticides (e.g. 6.2% lambda-cyhalothrin and 17.7% thiacloprid) contains at the same time residues of the synergistic fungicides. Also, risks of some residue mixtures are high for contact exposures (Table 2), but low or negligible for dietary exposures (Tables 3 and 4). Experimental evidence has shown that mixtures of imidacloprid and lambda-cyhalothrin increase mortality of bumble bees (*B. terrestris*) and reduce brood production in their colonies more than when fed on pollen contaminated with only one insecticide [88]. However, the effects of insecticide mixtures are additive, not synergistic.

The risk of dietary exposure was estimated for representatives of three different types of bees (larvae, nurses and nectar foragers) in the assumption that ingested residues and/or toxic metabolites [89] remain in their bodies. As mentioned above, this is a worst-case scenario, since elimination and metabolism over time are not taken into account; therefore, dietary risks may be overestimated for some compounds in this assessment. Although only a handful of individual pesticides appear to pose a serious threat to the bees (Tables 3 and 4), we should not forget that our evaluation considered oral LD50 values, not NOEL or LOEL values. For example, nurse honey bees feeding on pollen contaminated with imidacloprid may never reach the oral LD50 for that insecticide during their short life-time of 10 days, but toxic effects will be felt among those bees well before reaching the median dose, including some mortality. There is ample evidence that honey bees and bumble bees exposed to relevant sublethal doses of imidacloprid suffer motor and learning difficulties [90,91,92] and may even die in small proportions [88]. More meaningful risk assessments can be done using 1/10 of the oral LD50 values, which represent the lowest doses required for causing toxic effects. In this way, the probabilities of risk shown in Tables 3 and 4 would increase 10 times, and the T50s will be reduced correspondingly.

What is clear from the dietary assessment shown here is that systemic insecticides rank at the top of the list of risky chemicals: thiamethoxam, clothianidin, imidacloprid, dinetofuran, and to a lesser extent methiocarb, dimethoate and carbaryl [93]. These are more likely than any other pesticide to produce long-term toxic effects in workers and larvae of bumble and honey bees. However, while the systemic aldicarb is known to translocate to nectar and affect bees in the first four weeks after treatment of plants [94] its residues have not been found in honey from apiaries in recent years. In view of these findings, banning of some neonicotinoids by the European Community seems to be justified alone on the

grounds of residues in the food of bees, apart from other considerations [20] and side-effects that these compounds may have [95]. Surely, the high prevalence of neonicotinoids in honey (17–65%) is of great concern not only for worker bees but also for larvae (Fig. 2b and Table 3). Presumably, queens would be affected in a similar way as larvae, because both consume royal jelly and pollen, with the queens consuming larger quantities. Some experimental evidence indicates that the reproductive output of bumble bee queens is seriously curtailed when fed on pollen contaminated with imidacloprid [88] or thiamethoxam [43].

Moreover, the risk of neonicotinoids by dietary exposure above appears to be underestimated because it is known that these insecticides have chronic toxicities that exceed the known acute toxicities [62,73,96]. Time-cumulative effects justify a new approach to calculate T50s based on the exponential effects with time during dietary exposure. Indeed, mortality of bees increases by a power factor of 1.5 to 2, so the LD50s are reached sooner than expected. Consequently, average residues of thiamethoxam found in honey and pollen would approach the oral LD50 within the life span of larvae and worker honey bees, while average residues of imidacloprid would cause more than 50% mortality among nectar foragers and nurses and substantial mortality among larvae (Table 5). The latter predictions are deemed more realistic than the standard risks as they are in agreement with the negative effects of these insecticides observed in laboratory and semi-field experiments [63,97]. They contrast, however, with the conclusions of a recent report, funded by the chemical industry, suggesting that residues of thiamethoxam in pollen and nectar of oilseed rape and maize do not reduce the performance of honey bee colonies [98]. It should also be noted that sublethal and side-effects of neonicotinoids, such as immune suppression [99], have not been taken into account in our assessment.

Among the hydrophobic pesticides, the highest risks by dietary exposure correspond to four organophosphorus compounds (coumaphos, chlorpyrifos, heptenophos and quinalphos) on account of their high toxicity, residue loads and average prevalence in pollen (14–32%) and/or honey (4–47%). Other highly toxic insecticides such as fipronil, and pyrethroids could also have some impact on larvae and nurse bees, but low prevalence of their residues in pollen (usually <5%) and their absence in nectar or honey ensures their risks are low compared to that of neonicotinoids and cholinesterase inhibitors. Despite being designed to stop moulting in insects, average residues of diflubenzuron in pollen (80 ppb) pose little risk to bumble bees under chronic exposure because they are rarely found in that matrix (1% prevalence).

Conclusions

The large number and frequency of pesticide residues found in pollen and nectar of crop plants pose a clear risk to bee pollinators. Based solely on contact exposure, some 18 compounds (mostly pyrethroids and neonicotinoids) pose a threat to worker bees, but only five insecticides, namely thiamethoxam, phosmet, imidacloprid, chlorpyrifos and clothianidin, and four insecticide-fungicide mixtures pose risks with probabilities above 5%.

Those three neonicotinoids plus the organochlorine lindane pose the highest risk to worker bees and larvae when feeding on contaminated honey or nectar, but only thiamethoxam is of great concern when they feed on contaminated pollen, honey or nectar. In addition, risks of systemic neonicotinoids are probably underestimated because of their time-cumulative toxicity, synergistic effects with ergosterol inhibiting fungicides, and additive effects in combination with pyrethroids. Further research on the combined effects of such mixtures is needed to fully understand the reasons behind the collapse of honey bee and bumble bee colonies.

Supporting Information

Table S1 Pesticide residues (µg kg^{-1} or ppb) found in pollen, honey or nectar and wax together with their average prevalence (%) in Europe, the Americas and Asia.

Table S2 Acute toxicity (LD50 µg bee^{-1}) of pesticides to honey bees and bumble bees.

Table S3 Estimated average and maximum daily doses (ng bee^{-1}) of pesticide residues ingested by bees – herbicides excluded.

File SI Estimation of doses and risks by contact and dietary exposure.

Acknowledgments

We thank P. Mineau for making useful comments to earlier drafts of this paper, S. Kegley for providing data on neonicotinoids and bees, and J. Chapman for editing the manuscript.

Author Contributions

Conceived and designed the experiments: FSB. Performed the experiments: FSB. Analyzed the data: FSB. Contributed reagents/materials/analysis tools: FSB KG. Wrote the paper: FSB KG.

References

1. Osborne JL (2012) Ecology: Bumblebees and pesticides. Nature 491: 43–45.
2. Marletto F, Patetta A, Manino A (2003) Laboratory assessment of pesticide toxicity to bumble bees. Bulletin of Insectology 56: 155–158.
3. Mommaerts V, Smagghe G (2011) Side-effects of pesticides on the pollinator *Bombus*: An overview. In: Stoytcheva M, editor. Pesticides in the Modern World – Pests Control and Pesticides Exposure and Toxicity Assessment. Rijeka, Croatia: InTech. pp. 508–552.
4. Klein A-M, Vaissile BE, Cane JH, Steffan-Dewenter I, Cunningham SA, et al. (2007) Importance of pollinators in changing landscapes for world crops. Proceedings of the Royal Society B: Biological Sciences 274: 303–313.
5. Allen-Wardell G, Bernhardt P, Bitner R, Burquez A, Buchmann S, et al. (1998) The potential consequences of pollinator declines on the conservation of biodiversity and stability of food crop yields. Conservation Biology 12: 8–17.
6. Potts SG, Biesmeijer JC, Kremen C, Neumann P, Schweiger O, et al. (2010) Global pollinator declines: trends, impacts and drivers. Trends in Ecology & Evolution 25: 345–353.
7. Heard TA (1999) The role of stingless bees in crop pollination. Annual Review of Entomology 44: 183–206.
8. Graystock P, Yates K, Evison SEF, Darvill B, Goulson D, et al. (2013) The Trojan hives: pollinator pathogens, imported and distributed in bumblebee colonies. Journal of Applied Ecology 50: 1207–1215.
9. Goka K, Okabe K, Yoneda M (2006) Worldwide migration of parasitic mites as a result of bumblebee comercialization. Population Ecology 48: 285–291.
10. Kenta T, Inari N, Nagamitsu T, Goka K, Hiura T (2007) Commercialized European bumblebee can cause pollination disturbance: an experiment on seven native plant species in Japan. Biological Conservation 134: 298–309.
11. Cox-Foster DL, Conlan S, Holmes EC, Palacios G, Evans JD, et al. (2007) A metagenomic survey of microbes in honey bee Colony Collapse Disorder. Science 318: 283–287.
12. Cameron SA, Lozier JD, Strange JP, Koch JB, Cordes N, et al. (2011) Patterns of widespread decline in North American bumble bees. Proceedings of the National Academy of Sciences USA 108: 662–667.
13. Thompson H, Ball R, Brown M, Bew M (2003) *Varroa destructor* resistance to pyrethroid treatments in the United Kingdom. Bulletin of Insectology 56: 175–181.
14. Underwood RM, vanEngelsdorp D (2007) Colony Collapse Disorder: have we seen this before? Bee Culture 135: 13–15.
15. Buczek K (2009) Honey bee colony collapse disorder (CCD) [Zespo masowego giniecia pszczoy miodnej (CCD)]. Annales Universitatis Mariae Curie-Skodowska Sectio DD, Medicina Veterinaria 64: 1–6.
16. Maini S, Medrzycki P, Porrini C (2010) The puzzle of honey bee losses: a brief review. Bulletin of Insectology 63: 153–160.
17. Johnson RM, Ellis MD, Mullin CA, Frazier M (2010) Pesticides and honey bee toxicity - USA. Apidologie 41: 312–331.
18. Pettis J, vanEngelsdorp D, Johnson J, Dively G (2012) Pesticide exposure in honey bees results in increased levels of the gut pathogen *Nosema*. Naturwissenschaften 99: 153–158.
19. Vidau C, Diogon M, Aufauvre J, Fontbonne R, Vigues B, et al. (2011) Exposure to sublethal doses of fipronil and thiacloprid highly increases mortality of honeybees previously infected by *Nosema ceranae*. PLoS One: e21550.
20. Goulson D (2013) An overview of the environmental risks posed by neonicotinoid insecticides. Journal of Applied Ecology 50: 977–987.
21. Blacquière T, Smagghe G, van Gestel C, Mommaerts V (2012) Neonicotinoids in bees: a review on concentrations, side-effects and risk assessment. Ecotoxicology 21: 973–992.

22. Mommaerts V, Reynders S, Boulet J, Besard L, Sterk G, et al. (2010) Risk assessment for side-effects of neonicotinoids against bumblebees with and without impairing foraging behavior. Ecotoxicology 19: 207–215.
23. Aliouane Y, El-Hassani AK, Gary V, Armengaud C, Lambin M, et al. (2009) Subchronic exposure of honeybees to sublethal doses of pesticides: effects on behavior. Environmental Toxicology and Chemistry 28: 113–122.
24. Williams GR, Tarpy DR, vanEngelsdorp D, Chauzat M-P, Cox-Foster DL, et al. (2010) Colony Collapse Disorder in context. BioEssays 32: 845–846.
25. Goulson D, Lye GC, Darvill B (2008) Decline and conservation of bumble bees. Annual Review of Entomology 53: 191–208.
26. Mullin CA, Frazier M, Frazier JL, Ashcraft S, Simonds R, et al. (2010) High levels of miticides and agrochemicals in North American apiaries: implications for honey bee health. PLoS One 5: e9754.
27. Rennich K, Pettis J, vanEngelsdorp D, Bozarth R, Eversole H, et al. (2012) 2011–2012 National Honey Bee Pests and Diseases Survey Report. USDA. 17 p.
28. Stoner KA, Eitzer BD (2013) Using a hazard quotient to evaluate pesticide residues detected in pollen trapped from honey bees (*Apis mellifera*) in Connecticut. PLoS One 8: e77550.
29. Chauzat MP, Carpentier P, Martel AC, Bougeard S, Cougoule N, et al. (2009) Influence of pesticide residues on honey bee (Hymenoptera: Apidae) colony health in France. Environmental Entomology 38: 514–523.
30. Chauzat M-P, Faucon J-P, Martel A-C, Lachaize J, Cougoule N, et al. (2006) A survey of pesticide residues in pollen loads collected by honey bees in France. Journal of Economic Entomology 99: 253–262.
31. Bernal J, Garrido-Bailon E, Nozal MJd, Gonzalez-Porto AV, Martin-Hernandez R, et al. (2010) Overview of pesticide residues in stored pollen and their potential effect on bee colony (*Apis mellifera*) losses in Spain. Journal of Economic Entomology 103: 1964–1971.
32. Choudhary A, Sharma DC (2008) Dynamics of pesticide residues in nectar and pollen of mustard (*Brassica juncea* (L.) Czern.) grown in Himachal Pradesh (India). Environmental Monitoring and Assessment 144: 143–150.
33. Thompson HM (2001) Assessing the exposure and toxicity of pesticides to bumblebees (*Bombus* sp.). Apidologie 32: 305–321.
34. Greig-Smith PW, Thompson HM, Hardy AR, Bew MH, Findlay E, et al. (1994) Incidents of poisoning of honeybees (*Apis mellifera*) by agricultural pesticides in Great Britain 1981–1991. Crop Protection 13: 567–581.
35. Villa S, Vighi M, Finizio A, Bolchi Serini G (2000) Risk assessment for honeybees from pesticide-exposed pollen. Ecotoxicology 9: 287–297.
36. VanEngelsdorp D, Speybroeck N, Evans JD, Nguyen BK, Mullin C, et al. (2010) Weighing risk factors associated with Bee Colony Collapse Disorder by classification and regression tree analysis. Journal of Economic Entomology 103: 1517–1523.
37. Tennekes HA, Sánchez-Bayo F (2012) Time-dependent toxicity of neonicotinoids and other toxicants: implications for a new approach to risk assessment. Journal of Environmental & Analytical Toxicology S4: S4–001.
38. Halm M-P, Rortais A, Arnold G, Taséi JN, Rault S (2006) New risk assessment approach for systemic insecticides: the case of honey bees and imidacloprid (Gaucho). Environmental Science & Technology 40: 2448–2454.
39. EPA (2012) White Paper in Support of the Proposed Risk Assessment Process for Bees. In: Office of Chemical Safety and Pollution Prevention OoPP, Environmental Fate and Effects Division, editor. Washington, D. C. pp. 275.
40. Zhu W, Schmehl DR, Mullin CA, Frazier JL (2014) Four common pesticides, their mixtures and a formulation solvent in the hive environment have high oral toxicity to honey bee larvae. PLoS One 9: e77547.
41. Geoghegan TS, Kimberly JH, Scheringer M (2013) Predicting honeybee exposure to pesticides from vapour drift using a combined pesticide emission and

atmospheric transport model. SETAC Australasia - Multidisciplinary approaches to managing environmental pollution. Melbourne 1–3 October. pp. 174.

42. Henry Ml, Beguin M, Requier F, Rollin O, Odoux JFo, et al. (2012) A common pesticide decreases foraging success and survival in honey bees. Science 336: 348–350.

43. Whitehorn PR, O'Connor S, Wackers FL, Goulson D (2012) Neonicotinoid pesticide reduces bumble bee colony growth and queen production. Science 336: 351–352.

44. Scott-Dupree CD, Conroy L, Harris CR (2009) Impact of currently used or potentially useful insecticides for canola agroecosystems on *Bombus impatiens* (Hymenoptera: Apidae), *Megachile rotundata* (Hymentoptera: Megachilidae), and *Osmia lignaria* (Hymenoptera: Megachilidae). Journal of Economic Entomology 102: 177–182.

45. Helson BV, Barber KN, Kingsbury PD (1994) Laboratory toxicology of six forestry insecticides to four species of bee (Hymenoptera: Apoidea). Archives of Environmental Contamination and Toxicology 27: 107–114.

46. Tasei JN (2002) Impact of agrochemicals on non-*Apis* bees. In: Devillers J, Pham-Delegue MH, editors. Honey bees: estimating the environmental impact of chemicals. pp. 101–131.

47. Chauzat M-P, Martel A-C, Cougoule N, Porta P, Lachaize J, et al. (2011) An assessment of honeybee colony matrices, *Apis mellifera* (Hymenoptera: Apidae) to monitor pesticide presence in continental France. Environmental Toxicology and Chemistry 30: 103–111.

48. Pohorecka K, Skubida P, Miszczak A, Semkiw P, Sikorski P, et al. (2012) Residues of neonicotinoid insecticides in bee collected plant materials from oilseed rape crops and their effect on bee colonies. Journal of Apicultural Science 56: 115–134.

49. Balayiannis G, Balayiannis P (2008) Bee honey as an environmental bioindicator of pesticides' occurrence in six agricultural areas of Greece. Archives of Environmental Contamination and Toxicology 55: 462–470.

50. Blasco C, Fernandez M, Pena A, Lino C, Silveira MI, et al. (2003) Assessment of pesticide residues in honey samples from Portugal and Spain. Journal of Agricultural and Food Chemistry 51: 8132–8138.

51. Rissato SR, Galhiane MS, de Almeida MV, Gerenutti M, Apon BM (2007) Multiresidue determination of pesticides in honey samples by gas chromatography-mass spectrometry and application in environmental contamination. Food Chemistry 101: 1719–1726.

52. Amit C, Sharma DC (2008) Pesticide residues in honey samples from Himachal Pradesh (India). Bulletin of Environmental Contamination and Toxicology 80: 417–422.

53. Schmuck R, Schöning R, Stork A, Schramel O (2001) Risk posed to honeybees (*Apis mellifera* L, Hymenoptera) by an imidacloprid seed dressing of sunflowers. Pest Management Science 57: 225–238.

54. Stoner KA, Eitzer BD (2012) Movement of soil-applied imidacloprid and thiamethoxam into nectar and pollen of squash (*Cucurbita pepo*). PLoS One 7: e39114.

55. Dively GP, Kamel A (2012) Insecticide residues in pollen and nectar of a cucurbit crop and their potential exposure to pollinators. Journal of Agricultural and Food Chemistry 60: 4449–4456.

56. Byrne FJ, Visscher PK, Leimkuehler B, Fischer D, Grafton-Cardwell EE, et al. (2014) Determination of exposure levels of honey bees foraging on flowers of mature citrus trees previously treated with imidacloprid. Pest Management Science 70: 470–482.

57. Serra-Bonvehí J, Orantes-Bermejo J (2010) Acaricides and their residues in Spanish commercial beeswax. Pest Management Science 66: 1230–1235.

58. Wu JY, Anelli CM, Sheppard WS (2011) Sub-lethal effects of pesticide residues in brood comb on worker honey bee (*Apis mellifera*) development and longevity. PLoS One 6: e14720.

59. Landis WG, Chapman PM (2011) Well past time to stop using NOELs and LOELs. Integrated Environmental Assessment and Management 7: vi–viii.

60. Fox DR, Billoir E, Charles S, Delignette-Muller ML, Lopes C (2012) What to do with NOECS/NOELS–prohibition or innovation? Integrated Environmental Assessment and Management 8: 764–766.

61. Tomlin CDS (2009) The e-Pesticide Manual. In: Tomlin CDS, editor. 12 ed. Surrey, U.K.: British Crop Protection Council.

62. Suchail S, Guez D, Belzunces LP (2001) Discrepancy between acute and chronic toxicity induced by imidacloprid and its metabolites in *Apis mellifera*. Environmental Toxicology and Chemistry 20: 2482–2486.

63. Dechaume-Moncharmont F-X, Decourtye A, Hennequet-Hantier C, Pons O, Pham-Delegue M-H (2003) Statistical analysis of honeybee survival after chronic exposure to insecticides. Environmental Toxicology and Chemistry 22: 3088–3094.

64. Thompson HM, Hunt LV (1999) Extrapolating from honeybees to bumblebees in pesticide risk assessment. Ecotoxicology 8: 147–166.

65. Wijngaarden RPAV, Brock TCM, van den Brink PJ (2005) Threshold levels for effects of insecticides in freshwater ecosystems: a review. Ecotoxicology 14: 355–380.

66. Mattila HR, Harris JL, Otis GW (2001) Timing of production of winter bees in honey bee (*Apis mellifera*) colonies. Insectes Sociaux 48: 88–93.

67. Iwasa T, Motoyama N, Ambrose JT, Roe RM (2004) Mechanism for the differential toxicity of neonicotinoid insecticides in the honey bee, *Apis mellifera*. Crop Protection 23: 371–378.

68. Pilling ED, Jepson PC (1993) Synergism between EBI fungicides and a pyrethroid insecticide in the honeybee (*Apis mellifera*). Pesticide Science 39: 293–297.

69. Biddinger DJ, Robertson JL, Mullin C, Frazier J, Ashcraft SA, et al. (2013) Comparative toxicities and synergism of apple orchard pesticides to *Apis mellifera* (L.) and *Osmia cornifrons* (Radoszkowski). PLoS One 8: e72587.

70. Suchail S, Debrauwer L, Belzunces LP (2004) Metabolism of imidacloprid in *Apis mellifera*. Pest Management Science 60: 291–296.

71. Sánchez-Bayo F (2009) From simple toxicological models to prediction of toxic effects in time. Ecotoxicology 18: 343–354.

72. Tennekes HA, Sánchez-Bayo F (2013) The molecular basis of simple relationships between exposure concentration and toxic effects with time. Toxicology 309: 39–51.

73. Oliveira RA, Roat TC, Carvalho SM, Malaspina O (2013) Side-effects of thiamethoxam on the brain and midgut of the africanized honeybee *Apis mellifera* (Hymenopptera: Apidae). Environmental Toxicology 28: (in press) DOI: 10.1002/tox.21842.

74. Chauzat M-P, Faucon J-P (2007) Pesticide residues in beeswax samples collected from honey bee colonies (*Apis mellifera* L.) in France. Pest Management Science 63: 1100–1106.

75. Schmuck R, Stadler T, Schmidt HW (2003) Field relevance of a synergistic effect observed in the laboratory between an EBI fungicide and a chloronicotinyl insecticide in the honeybee (*Apis mellifera* L, Hymenoptera). Pest Management Science 59: 279–286.

76. Rortais A, Arnold G, Halm MP, Touffet-Briens F (2005) Modes of honeybees exposure to systemic insecticides: estimated amounts of contaminated pollen and nectar consumed by different categories of bees. Apidologie 36: 71–83.

77. Jacob CRO, Soares HM, Carvalho SM, Nocelli RCF, Malaspina O (2013) Acute toxicity of fipronil to the stingless bee *Scaptotrigona postica* Latreille. Bulletin of Environmental Contamination and Toxicology 90: 69–72.

78. Girolami V, Marzaro M, Vivan L, Mazzon L, Greatti M, et al. (2012) Fatal powdering of bees in flight with particulates of neonicotinoids seed coating and humidity implication. Journal of Applied Entomology 136: 17–26.

79. Kovac H, Stabentheiner A, Schmaranzer S (2010) Thermoregulation of water foraging honeybees - Balancing of endothermic activity with radiative heat gain and functional requirements. Journal of Insect Physiology 56: 1834–1845.

80. Thompson HM (2010) Risk assessment for honey bees and pesticides–recent developments and 'new issues'. Pest Management Science 66: 1157–1162.

81. Tapparo A, Marton D, Giorio C, Zanella A, Solda L, et al. (2012) Assessment of the environmental exposure of honeybees to particulate matter containing neonicotinoid insecticides coming from corn coated seeds. Environmental Science & Technology 46: 2592–2599.

82. Mineau P, Harding KM, Whiteside M, Fletcher MR, Garthwaite D, et al. (2008) Using reports of bee mortality in the field to calibrate laboratory-derived pesticide risk indices. Environmental Entomology 37: 546–554.

83. Tasei JN, Ripault G, Rivault E (2001) Hazards of imidacloprid seed coating to *Bombus terrestris* (Hymenoptera: Apidae) when applied to sunflower. Journal of Economic Entomology 94: 623–627.

84. Hsieh Y-N, Liu L-F, Wang Y-S (1998) Uptake, translocation and metabolism of the herbicide molinate in tobacco and rice. Pesticide Science 53: 149–154.

85. Vieira RF, Sumner DR (1999) Application of fungicides to foliage through overhead sprinkler irrigation – a review. Pesticide Science 55: 412–422.

86. Higes M, Martín-Hernández R, Martínez-Salvador A, Garrido-Bailón E, González-Porto AV, et al. (2010) A preliminary study of the epidemiological factors related to honey bee colony loss in Spain. Environmental Microbiology Reports 2: 243–250.

87. Schmuck R (2001) Ecotoxicological profile of the insecticide thiacloprid. Pflanzenschutz-Nachrichten Bayer 54: 161–184.

88. Gill RJ, Ramos-Rodriguez O, Raine NE (2012) Combined pesticide exposure severely affects individual- and colony-level traits in bees. Nature 491: 105–108.

89. Suchail S, Sousa Gd, Rahmani R, Belzunces LP (2004) *In vivo* distribution and metabolisation of 14C-imidacloprid in different compartments of *Apis mellifera* L. Pest Management Science 60: 1056–1062.

90. Decourtye A, Armengaud C, Renou M, Devillers J, Cluzeau S, et al. (2004) Imidacloprid impairs memory and brain metabolism in the honeybee (*Apis mellifera* L.). Pesticide Biochemistry and Physiology 78: 83–92.

91. Desneux N, Decourtye A, Delpuech J-M (2007) The sublethal effects of pesticides on beneficial arthropods. Annual Review of Entomology 52: 81–106.

92. Teeters BS, Johnson RM, Ellis MD, Siegfried BD (2012) Using video-tracking to assess sublethal effects of pesticides on honey bees (*Apis mellifera* L.). Environmental Toxicology and Chemistry 31: 1349–1354.

93. Moffett JO, Macdonald RH, Levin MD (1970) Toxicity of carbaryl-contaminated pollen to adult honey bees. Journal of Economic Entomology 63: 475–476.

94. Johansen CA, Rincker CM, George DA, Mayer DF, Kious CW (1984) Effects of aldicarb and its biologically active metabolites on bees. Environmental Entomology 13: 1386–1398.

95. van der Sluijs JP, Simon-Delso N, Goulson D, Maxim L, Bonmatin J-M, et al. (2013) Neonicotinoids, bee disorders and the sustainability of pollinator services. Current Opinion in Environmental Sustainability 5: 293–305.

96. Laurino D, Manino A, Patetta A, Porporato M (2013) Toxicity of neonicotinoid insecticides on different honey bee genotypes. Bulletin of Insectology 66: 119–126.

97. Cresswell J (2011) A meta-analysis of experiments testing the effects of a neonicotinoid insecticide (imidacloprid) on honey bees. Ecotoxicology 20: 149–157.

98. Pilling E, Campbell P, Coulson M, Ruddle N, Tornier I (2014) A four-year field program investigating long-term effects of repeated exposure of honey bee colonies to flowering crops treated with thiamethoxam. PLoS One 8: e77193.

99. Di Prisco G, Cavaliere V, Annoscia D, Varricchio P, Caprio E, et al. (2013) Neonicotinoid clothianidin adversely affects insect immunity and promotes replication of a viral pathogen in honey bees. Proceedings of the National Academy of Sciences 110: 18466–18471.

9

Geometric Morphometrics of Nine Field Isolates of *Aedes aegypti* with Different Resistance Levels to Lambda-Cyhalothrin and Relative Fitness of One Artificially Selected for Resistance

Nicolás Jaramillo-O.*, Idalyd Fonseca-González, Duverney Chaverra-Rodríguez

Instituto de Biología, Facultad de Ciencias Exactas y Naturales, Universidad de Antioquia UdeA, Medellín, Colombia

Abstract

Aedes aegypti, a mosquito closely associated with humans, is the principal vector of dengue virus which currently infects about 400 million people worldwide. Because there is no way to prevent infection, public health policies focus on vector control; but insecticide-resistance threatens them. However, most insecticide-resistant mosquito populations exhibit fitness costs in absence of insecticides, although these costs vary. Research on components of fitness that vary with insecticide-resistance can help to develop policies for effective integrated management and control. We investigated the relationships in wing size, wing shape, and natural resistance levels to lambda-cyhalothrin of nine field isolates. Also we chose one of these isolates to select in lab for resistance to the insecticide. The main life-traits parameters were assessed to investigate the possible fitness cost and its association with wing size and shape. We found that wing shape, more than wing size, was strongly correlated with resistance levels to lambda-cyhalothrin in field isolates, but founder effects of culture in the laboratory seem to change wing shape (and also wing size) more easily than artificial selection for resistance to that insecticide. Moreover, significant fitness costs were observed in response to insecticide-resistance as proved by the diminished fecundity and survival of females in the selected line and the reversion to susceptibility in 20 generations of the non-selected line. As a practical consequence, we think, mosquito control programs could benefit from this knowledge in implementing efficient strategies to prevent the evolution of resistance. In particular, the knowledge of reversion to susceptibility is important because it can help in planning better strategies of insecticide use to keep useful the few insecticide-molecules currently available.

Editor: Immo A. Hansen, New Mexico State University, United States of America

Funding: This research was supported by Colciencias (www.colciencias.gov.co) and Universidad de Antioquia (www.udea.edu.co), Colombia, project No. 1115-343-19131, Contract No. 360-2006. The funders had no role in studying design, data collection and analysis, decision to publish, or preparation of the manuscript.

Competing Interests: The authors have declared that no competing interests exist.

* E-mail: nicolas.jaramillo@udea.edu.co

Introduction

Aedes aegypti (Linnaeus) is an urban mosquito which transmits several viruses, mainly the four serotypes of dengue virus, of increasing concern to public health policies. The global burden of dengue is huge; Bhatt et al. [1] estimates there are 390 million dengue infections per year, with only 96 million of which are apparent.

Currently, there are no available licensed vaccines or specific drugs to prevent infection, so the more effective way to reduce transmission is vector control or interruption of vector-human contact [2]. However, so far all efforts to control *Ae. aegypti* are threatened by the rapid and widespread emergence of insecticide-resistance [2,3]. Insecticide resistance is a heritable trait and thus subject to natural selection [4]. In the absence of insecticides, mosquitoes carrying resistant alleles frequently exhibit a diminished fitness relative to susceptible ones; but that fitness costs vary between species and even between populations of the same species in time and space [5,6]. Brown et al. [7] demonstrated that this natural variation of fitness costs can qualitatively affect the economic prescriptions of optimal control models of mosquitoes.

Seizing on fitness costs, an effective integrated management of insecticide-resistance requires monitoring resistance and understanding the underlying evolutionary biology of mosquitoes [2,7].

It is therefore essential to perform both periodic surveillance of insecticide-resistance and research on the factors that promote or prevent the emergence of resistance. Research on the variation of components of fitness can help to develop policies for effective integrated management and control, identifying when and how agencies must make changes in the types of insecticides to keep susceptible vector populations and preserve the utility of insecticides molecules [3,6,7].

Size and shape could be strongly associated with fitness [8,9]. Geometric morphometrics is a powerful analytical tool for studying size and shape variability [10]. In turn, life tables provide the basis for studying the integrated life history characteristics that we would expect to be connected with fitness [11]. Then, combining geometric morphometrics with construction of life tables might improve our understanding of the trade-offs behind the outcomes of relative fitness.

In this work, we investigated the variation in wing size and shape of nine *Ae. aegypti* field isolates, which showed different resistance ratios 50 (RR_{50}) to lambda-cyhalothrin, a widely used adulticide pyrethroid. To deepen the relationship between morphology, insecticide resistance, and fitness, we chose one of those isolates for laboratory selection of resistance because it exhibited the highest natural resistance ratio against lambda-cyhalothrin. Relative fitness cost and morphometric variation of the selected line was measured and compared with a non-selected line from the same isolate and with the susceptible reference Rockefeller strain (ROCK). The selected and the non-selected lines shared the same genetic background, which solves one of the most important weaknesses of many works: comparisons of unrelated resistant and susceptible strains. Resistant and susceptible strains may differ in many other genes than those involved in resistance, which is particularly relevant because populations from different geographical origins often differ in life histories [12].

Materials and Methods

Insects

The first level of comparison included nine field isolates. These were from two very distant Colombian municipalities, Cúcuta and Quibdó, located at the northeast and at the west of the country, respectively (Table 1, Fig. S1). We used the ROCK strain as reference of susceptibility to lambda-cyhalothrin [13]. The different levels of natural resistance were calculated as RR_{50} for all isolates, a procedure that we explained in detail here under the section: *Bioassays*.

The second level of comparison included both the non-selected and selected lines for resistance to lambda-cyhalothrin from Comuneros (COM) isolate, which were followed in laboratory for 20 generations.

Ethics Statement

Neither humans nor endangered or protected animals were involved in this study. The procedure of feeding mosquitoes on mice was adjusted to the guidelines established by the Ethics Committee for Animal Experiments of the University of Antioquia, which approved the study as stated in Act 30 of the April 6, 2006.

Mosquito rearing

Larvae and pupae from each isolate were collected from different types of natural and man-made container habitats, using standardized protocols to determine aedic indexes. Within each municipality, isolates were separated by 1.5 to 5 km, and within each isolate we sampled at random 15–20 breeding-sites, thus avoiding the collection of larvae and pupae descendants from the same female. Samplings were coordinated with the "Instituto Departamental de Salud, Dirección de Vigilancia en Salud Pública" from municipality of Cúcuta, and with the "Departamento Administrativo de Salud, Unidad de Control de Vectores" from municipality of Quibdó, which helped with trained field-assistants and with transport.

The collected larvae and pupae were taken to the laboratory using containers filled with water from their habitats and afterward they were transferred to plastic pans of 2 L filled with dechlorinated water. Larvae were fed with fish food and the emerged adults were transferred to cages ($25 \times 25 \times 25$ cm) where plastic cups covered internally with paper towels and filled with water were placed for oviposition. Mosquitoes were fed *ad libitum* with a 10% (w/v) sugar solution and, additionally, females were fed each two days on mice Balb/c 3–4 months old. Insectary colonies were raised to adults at $28 \pm 2°C$, $70 \pm 5\%$ RH and a photoperiod of 12 hours light/dark.

Biossays

The resistance ratios 50 (RR_{50}) were evaluated in larvae from nine field isolates and each generation for the selected and the non-selected lines of mosquitoes from the parental COM population and the ROCK strain using the WHO protocols [14,15]. Groups of 20 fourth instar larvae were exposed for 24 h at least to five different concentrations of lambda-cyhalothrin (purchased from Chem Service Inc., West Chester, PA, USA) producing mortalities between 2 and 90%. The insecticide was prepared from a 1 mL stock solution (10 ug dissolved in 1 ml of acetone) in 99 mL of dechlorinated water; the control sample contained a stock solution of acetone instead of insecticide. Each insecticide concentration was tested in triplicate and each bioassay was performed twice for each field isolate and generation.

Larval mortality data obtained after 24 h of exposure to each insecticide-concentration and that of the controls were used to calculate lethal concentration 50 (LC_{50}) through Log-Probit linear

Table 1. Origin and resistance status to lambda-cyhalothrin of *Ae. aegypti* isolates.

MUNIC.	NEIGHB.	COORD. (north, west)	RR_{50}-LAMB	DATE OF COLLEC.	CODE	N
Cúcuta	Belén	7°52'18.42", 72°32'4.02"	8.08	06/28/08	BEL	52
Cúcuta	Comuneros	7°54'54.42", 72°31'44.82"	24.23	07/09/07	COM	51
Cúcuta	El Contento	7°53'9.66", 72°30'43.56"	8.75	04/04/08	CONT	54
Cúcuta	Guaimaral	7°54'56.7", 72°29'48.96"	23.00	08/14/08	GUAI	51
Cúcuta	La Libertad	7°53'23.88", 72°28'49.38"	17.50	12/21/07	LIBE	50
Quibdó	Cristo Rey	5°41'39", 76°39'31"	9.50	04/26/08	CREY	50
Quibdó	Jardín	5°41'01", 76°38'42"	8.00	04/26/08	JARD	52
Quibdó	Playita	5°40'44", 76°39'17"	10.25	04/26/08	PLAY	50
Quibdó	Porvenir	5°41'49", 76°38'58"	1.60	04/26/08	PORV	51
Reference	The Rockefeller strain		1.00		ROCK	49
TOTAL						510

MUNIC.: Municipality, NEIGHB.: Neighborhood, COORD.: Geographical coordinates, RR_{50}-LAMB: Resistance Ratios 50 to lambda-cyhalothrin, DATE OF COLLEC.: date of collection. N: sample size; CODE: isolate code; Ref.: Reference of susceptibility.

regression [16]. Mortality data were corrected by the Abbot's formula [17]. The RR_{50} was calculated for each generation by dividing the LC_{50} of the field isolate and both the selected and the non-selected lines by the LC_{50} of the ROCK strain. The RR_{50} values were interpreted according to Mazzarri and Georghiou [18] thus: a value <5-fold is taken as low or "susceptible"; a value between 5-fold and 10-fold is taken as moderate or "tolerant", and a value >10-fold is taken as high or "resistant".

The resistance status was confirmed in adults using the CDC bottle bioassays [19]. Each field isolate and generation of the selected and non-selected lines from COM isolate and the ROCK strain were evaluated relative to diagnostic doses (DD) and times previously known to kill 100% of mosquitoes from the ROCK strain: 6.25 ug/15 min [20]. For this, we exposed to lambda-cyhalothrin groups of 20 three days old females, fed only with a 10% (w/v) sugar solution, into 250 mL glass bottles internally coated with the insecticide. Each bioassay, tested in triplicate, consisted of four treated bottles (impregnated with DD of the insecticide) and one control bottle (impregnated only with acetone). Mortality was checked each 5 min for 2 h. The mortality criteria included mosquitoes with difficulty in flying or as resting on the bottle's surface.

Selection of resistance to lambda-cyhalothrin

The COM isolate from the municipality of Cúcuta was chosen as the parental for selection of insecticide-resistance because it showed the highest natural RR_{50} to lambda-cyhalothrin (24.23X) (Table 1). The first offspring (F1) was randomly separated into two batches, one to be exposed to lambda-cyhalothrin, each generation; the other batch was kept in parallel but without exposure to insecticide. From 1000 to 2000 larvae (divided in groups of 250) were exposed during 24 h to the LC_{50} of lambda-cyhalothrin estimated by dose-response assays. The emerged adults from survivor larvae were reared at the same laboratory at the above mentioned conditions. This procedure was iterated for 20 generations. The LC_{50} increased in each generation as the larval and adult bioassays showed a progressive increase in the insecticide-resistance.

The same procedure was performed for the other batch of 1000 to 2000 larvae from COM isolate and for the ROCK strain; but using a solution of acetone and water in the same proportion of LC_{50} instead of lambda-cyhalothrin.

We took care to keep an equal proportion of parental males and females (500 to 700 mosquitoes) for each generation in all the treatments.

Geometrics morphometrics

We studied a total of 758 females chosen at random from each field isolate and the ROCK strain (n = 510), and the selected and non-selected lines at F9 and F20 generations (n = 248). The following procedures were performed for the isolates from field, the selected and non-selected lines and the ROCK strain. Right and left wings were carefully dissected and mounted between microscope slides and cover slips, and afterward were photographed taking care to avoid optical distortion of the microscope's peripheral visual field. We selected 14 type I landmarks (sensu [21]) on vein intersections of hemelytra (Fig. 1). Landmarks were digitized to Cartesian coordinates and those corresponding to right and left wings were averaged to eliminate possible bias because of fluctuating asymmetry. The measurement error was estimated by the Pearson correlation coefficient in two set of coordinates taken on the same subsample of 30 specimens chosen at random from one of the isolates [22].

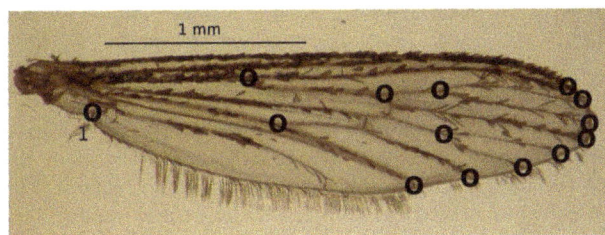

Figure 1. Arrangement of landmarks digitized on right and left wings of 758 *Aedes aegypti* females. The landmarks were digitized in a clockwise sequence from the number one.

To analyze wing size variation, we used an isometric estimator, the centroid-size (CS), which was calculated as the square root of the sum of the squared distances between the center of the configuration of landmarks and each landmark [21]. The CS variation was analysed with the Kruskal-Wallis rank sum tests and the post hoc pairwise Wilcoxon rank sum tests, and their significance was tested with a permutation procedure (10,000 permutations), with p-values corrected with the Bonferroni method. The relationships in wing size were visualized through a violin plot [23] and a dendrogram built with the UPGMA method. The cophenetic correlation coefficient [24] was used to measure the correlation between the distances used to construct the UPGMA dendrogram and the distances implied by the dendrogram itself. To analyse the association between wing size and resistance, we performed a linear regression of CS on RR_{50}.

To verify the correlation between wing centroid-size and body size, we performed a linear regression of the dry weight of a sample of 150 females on the CS. The sample of 150 females was chosen randomly from the ROCK strain and the selected and the non-selected lines at F20 generation. They were first dried at 37°C in the oven with silica gel bags for 48 h and then weighed in groups of five on an analytical balance (OHAUS Explorer mod. AV114, 110 g×0.1 mg).

To analyze wing shape variation, the raw coordinates matrix was transformed with the generalized Procrustes analysis [25]. The algorithm consists of a series of iterative superimpositions of homologous landmarks to adjust them by scale, translation and rotation. The iterations allowed computing the mean wing shape which becomes a reference to compute the distances between individuals. Then, the resulting variables are shape variables which can be analysed with multivariate statistical methods. We analysed shape variables by relative warps analysis which corresponds to a principal component analysis of wing shape variation [21]. The relationships in wing shape were visualized by a scatter plot of the individuals on the two first relative warps and by a dendrogram built by the UPGMA method. The dendrogram fidelity was verified by the cophenetic correlation coefficient [24]. Mahalanobis distances between isolates means were computed and their significance was tested with a permutation procedure (10,000 permutations). Furthermore, the Mahalanobis distances were used to perform a jack-knife cross-validated classification [26].

Allometry

Allometry was tested using a multivariate regression of wing shape on wing size, and their significance was tested with a permutation procedure (10,000 permutations). MANCOVA was used to test for a common allometric slope. When a common slope was deemed appropriate, residuals of the regression of shape on centroid size were used for downstream analyses. The regression was scaled by size to test for differences in shape without the size effect.

Life tables and fitness cost

Life history was evaluated for the following groups of insects: parental COM population, the ROCK strain, and both the selected and non-selected lines at F9 and F10 generations. The model that we studied for life histories assumed ideal conditions for laboratory with discrete, non-overlapping generations. For each group, five cohorts (five replicates) each of 100 larvae of 12-h old were kept at the same conditions mentioned above, but in plastic pans of 500 ml. Daily register of larval mortality and molting were recorded until pupation. Pupae were transferred to new pans with freshly dechlorinated water and female and male emergence was daily recorded. For each group, four cohorts (four replicates) each of 25 females and 25 males of 12 hours old were kept in boxes of $25 \times 25 \times 25$ cm, and fed *ad libitum* with a 10% sugar solution (w/v) and daily on mice Balb/c aged 3–4 months old at the same environmental conditions previously mentioned. One 30 mL plastic cup per cage covered internally with paper towels and filled with water were disposed for oviposition. Daily mortality was recorded at 7:00 h and 19:00 h, until the last mosquito was dead. Oviposition paper towels were daily removed one time per day, and all the eggs were counted under the microscope. This experiment was repeated for the 200 females and 200 males used.

The vital parameters studied were: average time in days to develop to pupae, average time in days to develop to adults, proportion of male and female emergence, hatchability, age-specific survival of females (l_x: number of females alive at beginning of stage x) and fecundity (m_x: average number of eggs produced per female). Kruskal-Wallis rank sum tests or ANOVA were used for comparisons, and when significance was found, pairwise comparisons were performed with Tukey tests corrected by Bonferroni method. Age-specific survival comparisons were performed by log-rank analysis and the Kaplan-Meier method was used to build survival curves.

These parameters were used to prepare life tables from which the following demographic growth estimates were calculated: generation time $(G = \sum_{x=0}^{n} x l_x m_x / R_0)$, where x is the proportion of individuals in day x), net reproductive rate per generation $(R_0 = \sum_{x=0}^{n} l_x m_x)$, and intrinsic rate of increase in days $(r = \ln R_0 / G)$. Both, R_0 and r can be estimators of fitness, but here we used r to estimate the relative fitness because it combines information from both survivorship and fecundity, in terms of R_0 and G. The r value is the most comprehensive life table parameter and therefore the most useful to explain evolutionary changes in generation time [27]. We use the Euler-Lotka equation $(1 = \sum_{x=0}^{n} l_x m_x e^{-rx})$ as a more accurate way to calculate r [28] and thus the relative fitness. The R_0 and r demographic parameters were compared across groups by using ANOVA and post hoc pair-wise Tukey's studentized range tests.

Software

Most of the statistical analyses were performed using the R language [29]. Geometric morphometrics and multivariate analyses of shape were performed using the CLIC package [26]. The tpsRelw program [30] was used to make deformations grids for the first and second relative warps on the mean configuration. The Probit program [31] was used for calculating LC_{50} through Log-Probit linear regression.

Results

Repeatability of landmarks

Size and shape showed fairly good precision in the digitalization of landmarks in two sets of repeated wings (0.998 for centroid-size, and 0.902, 0.868, 0.929 for the first three relative warps).

Geometric morphometrics and resistance to lambda-cyhalothrin of field isolates

Kruskal-Wallis rank sum test showed significant differences in wing size among isolates from the cities of Cúcuta and Quibdó and the susceptible reference ROCK strain ($\chi2 = 187.4598$, df = 9, $P = 2.2 \times 10^{-16}$). COM isolate displayed the highest resistance against lambda-cyhalothrin (Table 1) and was the most different in size (Fig. 2) and included the smallest individuals (Fig. 3). Additionally, most of the isolates that exhibited higher resistance included the smaller individuals. However, the ROCK strain contained the biggest individuals (Fig. 3), but the correlation between resistance and size was not perfect (Fig. 2). The ROCK strain ($RR_{50} = 1$) was not statistically different from the other three isolates (Wilcoxon rank sum test, $P > 0.3$), which exhibited RR_{50} of 8.0, 8.8, and 10.3, but it was statistically different (Wilcoxon rank sum test, $P = 0.005$) from an isolate with a RR_{50} of 1.6 (Fig. 2). Linear regression of wing size (CS) on RR_{50} was significant, but the coefficient of determination was moderate (adjusted $R^2 = 0.2307$, F = 153.6, df = 1 and 508, $P < 2.2 \times 10^{-16}$). Furthermore, the dendrogram (Fig. 2) shows that similar wing sizes are found in isolates belonging to different municipalities far between (Cúcuta and Quibdó) or in isolates from different climates.

Wing shape was also different among isolates. Outcomes of relative warp analysis visualized by a dendrogram showed two levels of similarity in wing shape; the first one coincides to geographical origin (three clusters corresponding to the cities of Cúcuta and Quibdó, and to the ROCK strain, respectively), and the second one to RR_{50} against lambda-cyhalothrin (Fig. 4). The dendrogram also shows a stronger relationship of wing shape with RR_{50}, than that of wing size (Fig. 4). Mahalanobis distances were significantly different between the ROCK strain, and the cities of Cúcuta, and Quibdó (from 10,000 permutations, no Mahalanobis distance was equal or higher than observed: $P < 0.0001$). Isolates within cities of Cúcuta and Quibdó were also different except for two of them (Table 2). Cross-validated classification of each individual to its original isolate shows fairly good correct classification rates of the three clusters (ROCK: 85%, Cúcuta: 75%, Quibdó: 77%) but moderate within each cluster (27% to 66%).

Geometric morphometrics and resistance to lambda-cyhalothrin of artificially lambda-cyhalothrin selected and non-selected lines

The development of resistance was confirmed using bioassays on larvae and adults of COM isolate, on the selected and non-selected lines for lambda-cyhalothrin, and on the ROCK strain. Artificial selection increased the RR_{50} in each generation by a factor of 1.25X on average. The RR_{50} of larvae of parental COM was 24.23X, and the selected line at F9, F10, and F20 generations showed a RR_{50} of 225.4X, 91.52X, and 114.23X, respectively; However, the non-selected COM line showed for F9, F10, and F20 generations a RR_{50} of 10.0X, 12.13X, and 1.95X respectively. Adults followed in parallel way: parental COM was susceptible; but the selected line at F9, F10, and F20 generations showed a mortality of 25, 32, and 8%, respectively. However, the

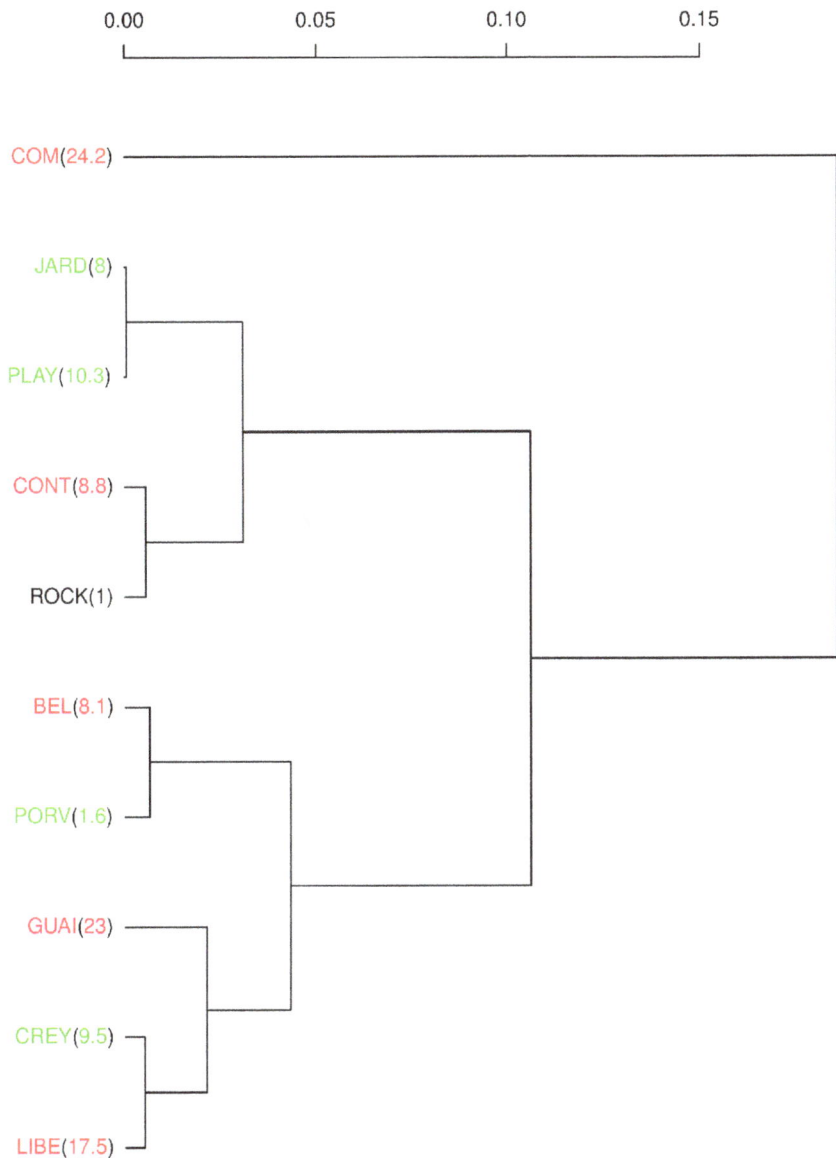

Figure 2. Dendrogram showing the relationships in wing size of isolates from Cúcuta and Quibdó. Between parentheses is the RR_{50} to lambda-cyhalothrin. Cophenetic correlation coefficient: 0.892522. isolates from Cúcuta are colored in red, and those from Quibdó in green ROCK: the Rockefeller strain.

non-selected line at F9, F10, and F20 generations showed a mortality of 100, 100, and 97%, respectively (Table S1 and S2).

Wing centroid-size (CS) as estimator of global size was confirmed by a lineal regression of dry weight of body on CS (Adjusted $R^2 = 0.2625$, $F = 11.32$, $df = 1$ and 28, $P = 0.002236$).

Kruskal-Wallis rank sum test showed differences in wing size of the selected and the non-selected lines, their parents, and the ROCK strain ($\chi2 = 111.1079$, $df = 4$, $P < 2.2 \times 10^{-16}$). Wing size did not show an appreciable change between the parental and the F9 generation of the selected and the non-selected lines (Fig. 5), but there was a significant increase in wing size of the F20 generation of both lines relative to its ascendants (Fig. 5). Notably, the selected and the non-selected lines at F9 generation did not show differences in wing size between them, but the non-selected F20 generation was bigger than the selected (Bonferroni corrected Wilcoxon $P = 0.023$).

In similar way, wing shape becomes progressively different between the selected line and the non-selected one from the parental to the F9 and to the F20 generation (Fig. 6). Mahalanobis distances were statistically significant between the parental isolate and all of its descendants at F9 and F20 generations for both the selected and the non-selected lines and they were also significant between lines and generations (from 10,000 permutations, no Mahalanobis distance was equal or higher than observed, $P < 0.0001$). Cross-validated classification of each individual to its original isolate shows fairly good correct classification rates of the three clusters, being the best classified the parental isolate, followed by the selected line at F20 generation, and after by the non-selected line of the same F20 generation (parental $= 82\%$, F20-S $= 80\%$, F20-NS $= 78\%$, F9-S $= 67\%$, F9-NS $= 59\%$).

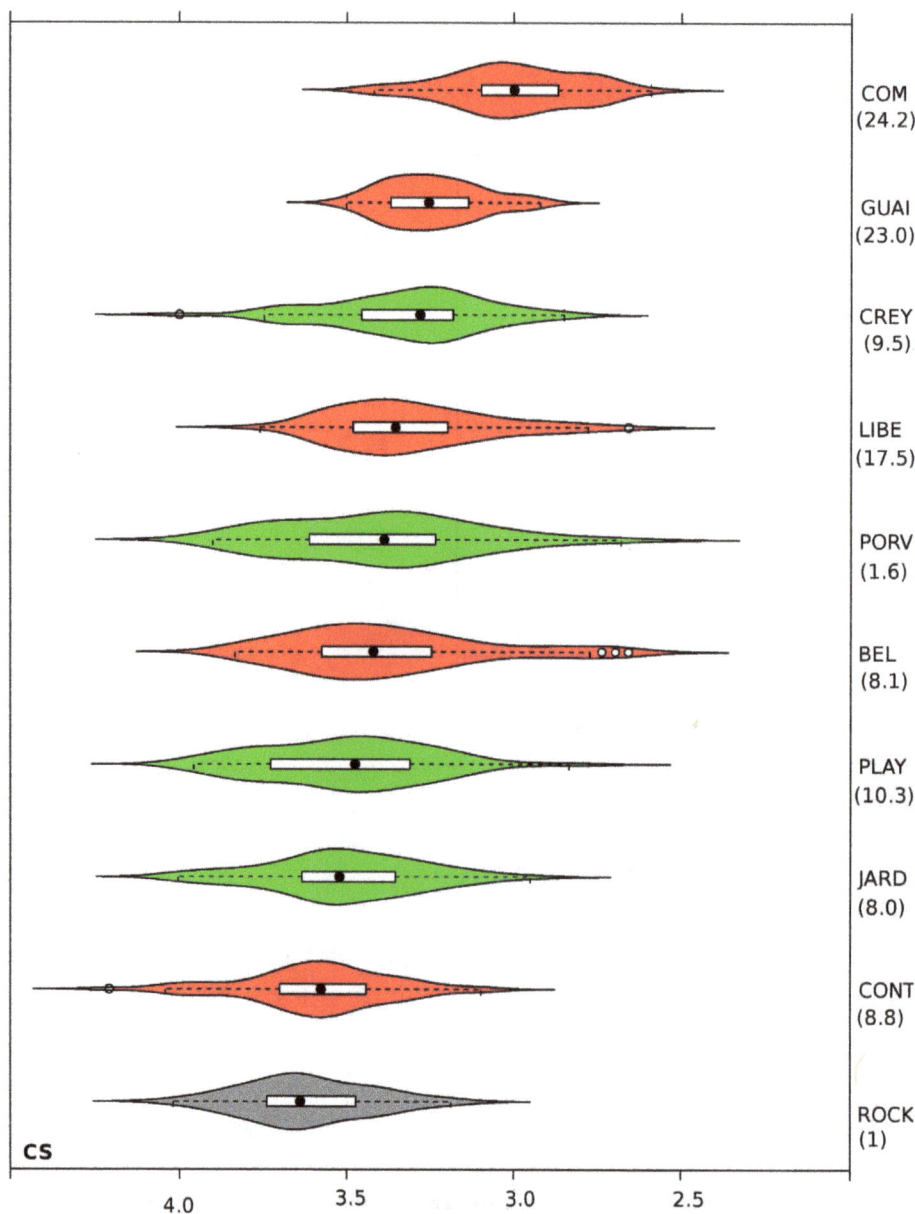

Figure 3. Wing size variation among *Aedes aegypti* isolates from field. Violin plots enclose box plots. Each box is divided by the median (black circles), which top and bottom correspond to 25th and 75th quartiles, respectively. Isolates from Cúcuta are colored in red, and those from Quibdó in green. Between parentheses is the RR_{50} to lambda-cyhalothrin. CS: centroid-size.

Allometry

For the field isolates, differences in wing shape were affected by wing size (multivariate linear regression of relative warps on CS after 10,000 permutations shows 0 values found equal or higher than those observed, $P<0.0001$); and the test for common slopes shows significance (Wilks lambda = 0.548, F = 1.354, df = 216 and 3970.166, $P=0.0006$), which means that a common allometric slope was not found. We repeated the MANCOVA, but excluding the ROCK strain, and the model shows significance as well (Wilks lambda = 0.567, F = 1.295, df = 192 and 3215.4, $P=0.0048$).

For the selected and non-selected lines, differences in wing shape were affected by the wing size (multivariate linear regression of relative warps on CS after 10,000 permutations shows 11 values found equal or higher than observed, $P=0.001$); and the model of common slopes could be accepted (Wilks lambda = 0.636,

F = 1.079, df = 96 and 854.24, $P=0.294$). Size-free analysis of wing shape confirmed statistic differences among parental isolate and its descendants of F9 and F20 generations of the selected and the non-selected lines but with higher differences at F20 generation than at F9 (after 10,000 permutations no Euclidian distances among lines and isolates were equal or higher than observed, $P<0.0001$).

Life tables and fitness costs

The following parameters did not show statistical differences at P-level of 0.05 between the selected and the non-selected lines at F9 and F10 generations: average time to pupation, average time to eclosion, proportion of male and female emergence, and hatchability (Table S3).

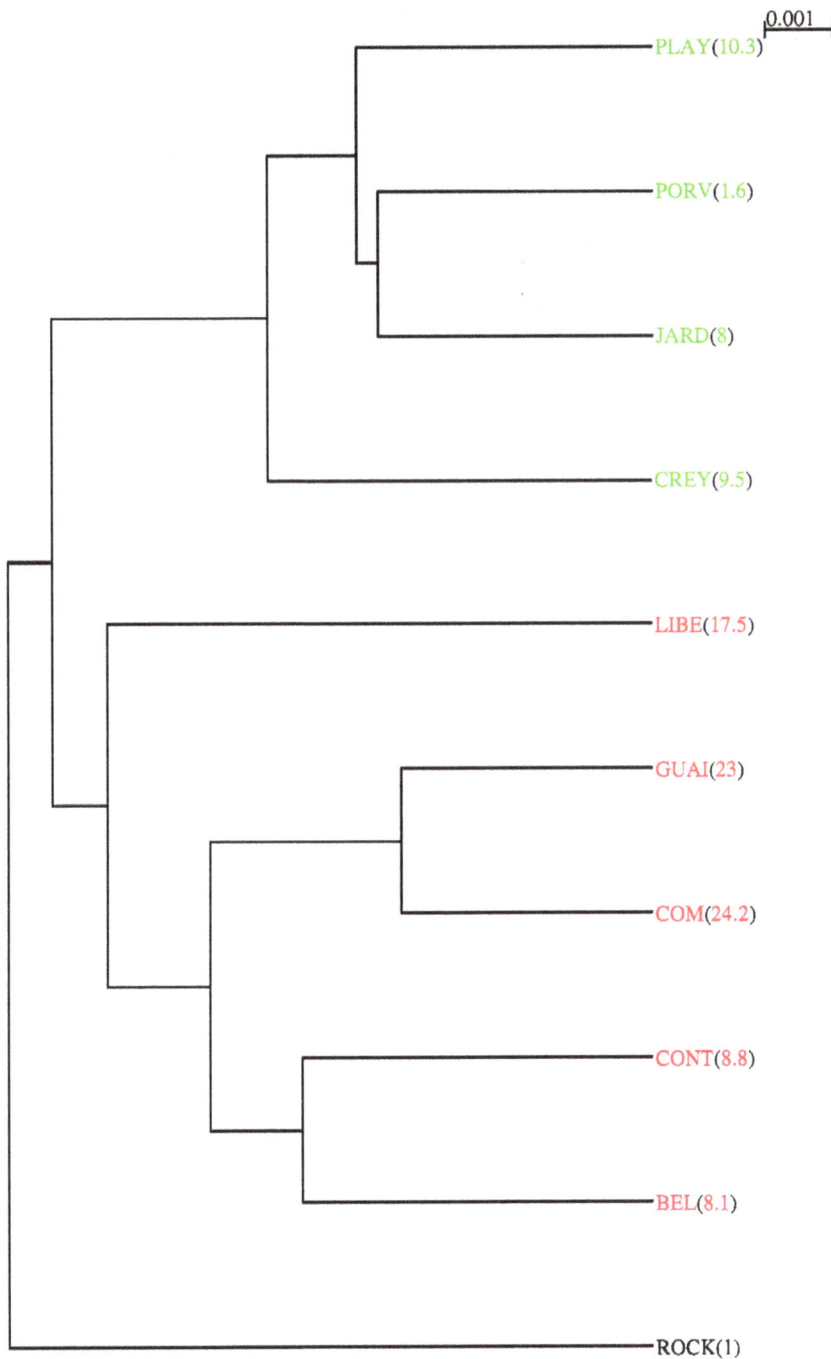

Figure 4. Dendrogram showing the relationships in wing shape of isolates from Cúcuta and Quibdó. Cophenetic correlation coefficient: 0.8935737. Isolates from Cúcuta are colored in red, and those from Quibdó in green. Between parentheses is the RR_{50} to lambda-cyhalothrin. ROCK: the Rockefeller strain.

However, the probability of survival was very different among tested groups (Log-Rank test: $\chi^2 = 171$, df $= 4$, $P<0.0001$; Fig. 7). The median survival of the selected line at F9 and F10 generations was 15 days for both of them; survival of the non-selected line was significantly higher (28 days for F9 and 22 for F10) as it was also for the ROCK strain (29 days). Fecundity was different among tested groups (ANOVA: F $= 21.36$, df $= 4$ and 15, $P=4.73\times10^{-6}$). Pair-wise comparisons showed differences between the ROCK strain and the selected line at F9 and F10 generations (Tukey test

$= 9.679$, $P = 0.0001816$, and Tukey test $= 9.68$, P: 0.0001815, respectively (Fig. 8) but was similar between the ROCK strain and the non-selected line at F9 and F10 generations (Tukey test $= 1.575$, P: 0.7971 and Tukey test $= 2.705$, P: 0.3527, respectively).

Likewise, the net reproductive rate per generation (R_0) and the intrinsic rate of increase in days (r) showed differences among the tested groups (ANOVA R_0: F $= 18.56$, df $= 4$ and 15, $P=1.126\times10^{-5}$; ANOVA r: F $= 12.99$, df $= 4$ and 15,

Table 2. Differences in wing shape among *Ae. aegypti* isolates.

	ROCK	BEL	COM	CONT	GUAI	LIBE	CREY	JARD	PLAY	PORV
ROCK	0	0.0000	0.0000	0.0000	0.0000	0.0000	0.0000	0.0000	0.0000	0.0000
BEL	3.5900	0	0.0000	0.0000	0.0000	0.0000	0.0000	0.0000	0.0000	0.0000
COM	4.7200	3.2600	0	0.0000	0.0000	0.0000	0.0000	0.0000	0.0000	0.0000
CONT	3.6000	2.7200	4.2500	0	0.0000	0.0000	0.0000	0.0000	0.0000	0.0000
GUAI	3.5300	2.0500	1.9800	2.8800	0	0.0000	0.0000	0.0000	0.0000	0.0000
LIBE	3.1700	2.7500	3.6900	3.0800	2.3500	0	0.0000	0.0000	0.0000	0.0000
CREY	3.8000	2.5300	3.4100	2.8000	2.4600	3.3400	0	0.0000	0.0000	**0.0615**
JARD	3.6600	2.3800	3.8600	2.9600	2.6700	3.2500	1.8000	0	0.0000	0.0000
PLAY	3.7500	2.8800	3.8800	3.2200	2.7800	3.5000	1.7500	1.7400	0	0.0002
PORV	3.3000	2.1300	3.3400	2.8300	2.3100	2.9800	1.3100	1.8300	1.7000	0

Mahalanobis distances are shown below diagonal and its corresponding P values are above diagonal. Using Bonferroni correction, a significant P value is <0.00111.

$P = 9.139 \times 10^{-5}$). Both parameters were different between the ROCK strain and the selected line at F9 and F10 generations but were similar between the ROCK strain and the non-selected line at F9 and F10 generations (Table 3).

Finally, relative to the ROCK strain, the fitness, which was calculated from R_0 and r, was 0.79 and 0.82 for the selected line at F9 and F10 generations, respectively; meanwhile, 0.95 and 0.97 for the non-selected line at F9 and F10 generations, respectively (Table 4).

Discussion

Variation in body size affects fitness as it has been suggested by the strong correlation to one or several of its components, like fecundity or survival age specific [32–37] Wing length, is often used in mosquitoes as an estimator of body size [32,36,38], but it cannot always be demonstrated [39,40]. We used wing centroid-size (CS) as estimator of global size, instead of wing length, because we found a positive correlation with dry body weight and because CS detects size changes in all directions, not only along the largest axis of wing [8].

Furthermore, we observed that field isolate with highest lambda-cyhalothrin resistance contained the smallest insects and the susceptible reference strain, the biggest one (Fig. 3). But the correspondence between lambda-cyhalothrin resistance to wing size was not perfect (Fig. 2). We observed that insects similar in wing size were found in isolates separated by 520 km, which covers three branches of the Andes mountain range and at least two of the largest rivers of the country (Fig. S1). Those municipalities have different altitudes, climates and history of *Ae. aegypti* colonization [41,42]. Moreover, here we studied adults which emerged in the laboratory from larvae and pupae collected in the field. Our method of collection controlled maternal effects, but water containers where larvae and pupae were collected could have different microclimates. Then, besides lambda-cyhalothrin resistance, other factors like density, nutrition, temperature, and relative humidity could affect the wing size [43,44]; thus, unmeasured environmental variables in addition to their own genetic factors could also affect the lambda-cyhalothrin resistance to wing size relationship. Related to possible seasonal effects due to the fact that collections of isolate spanned over a 13-month period (Table 1), we discarded them because this Neotropical region does not have significant oscillations along the year (see World Meteorological Organization at http://www.wmo.int/, for climate parameters of Quibdó and Cúcuta).

CS of the selected and the non-selected lines seem hard to change, because they did not change during nine generations relative to their parents; but in the F20 generation, individuals were significantly bigger, with the non-selected line contained the biggest ones (Fig. 5). This result seems to follow the same pattern of those observed in field where most of the isolates containing the smaller individuals exhibited the higher RR_{50} (Fig. 3), and the biggest individuals belonged to the ROCK susceptible strain.

Smaller size associated with increased resistance could be a disadvantage, if smaller individuals have decreased fecundity because of reduced ability to ingest blood or having fewer ovarioles [36]. In *Culex pipiens*, the organophosphate resistance was associated with a smaller adult size [12,45]; but *Ae. albopictus* resistance to permethrin was associated with larger size when compared with the susceptible strain [46,47]. *Aedes aegypti* females exposed to sublethal doses of Spinosad were significantly larger than control females; male survivors, in contrast, were significantly smaller than controls [48]; although, resistance level to Spinosad

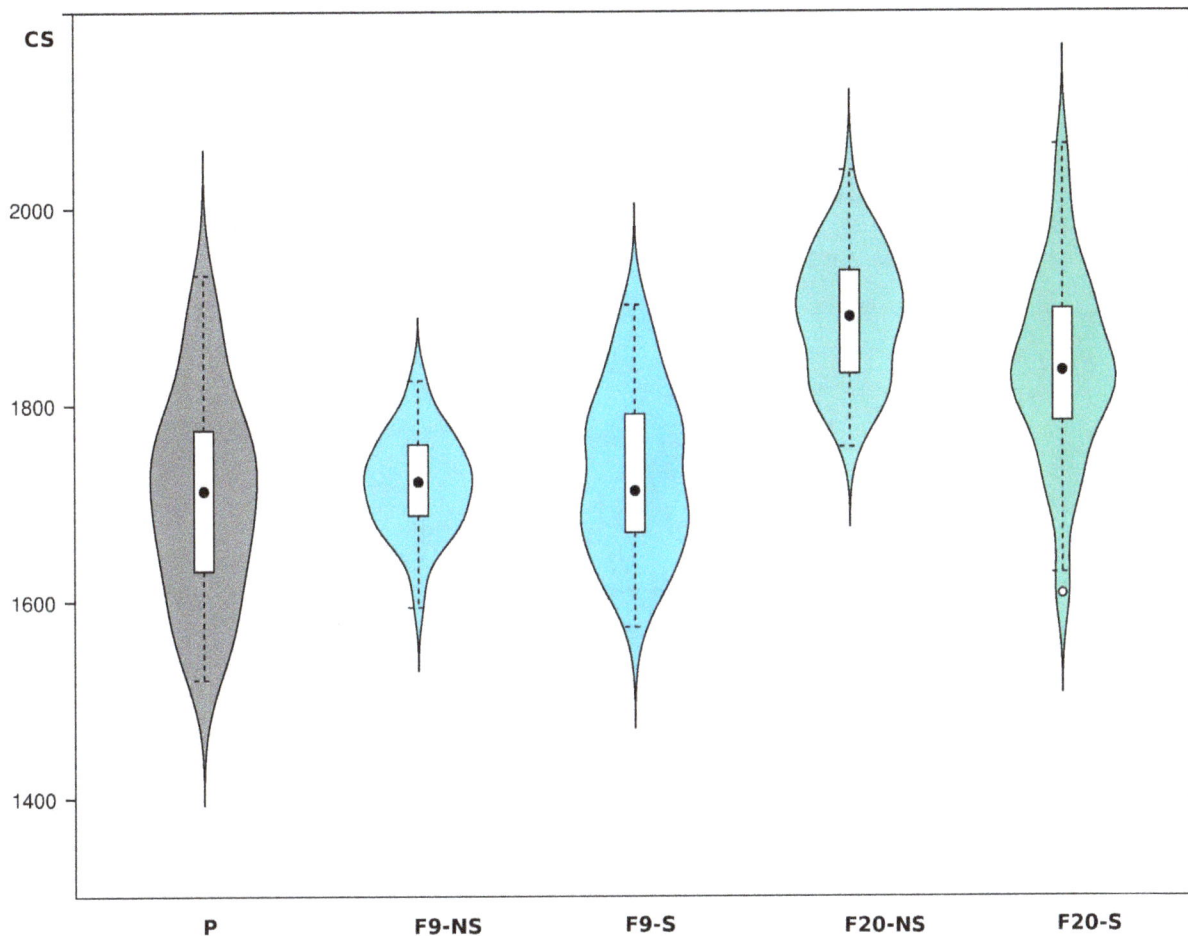

Figure 5. Wing size variation of both lambda-cyhalothrin selected and non-selected lines. Violin plots enclose box plots. Each box is divided by the median (black circles), which top and bottom correspond to 25th and 75th quartiles, respectively. CS: centroid-size; P: parental Comuneros isolate; F9-NS and F20-NS: ninth and twenty generations non-selected for lambda-cyhalothrin resistance; F9-S and F20-S: ninth and twenty generations selected for lambda-cyhalothrin resistance. Sample-sizes were 51 parental females, 42 F9-NS, 49 F9-S, 50 F20-NS, and 56 F20-S.

was not reported. Thus, it seems that the family Culicidae has not a stable response in body size related to insecticide resistance.

The significant increase in wing size of both the selected and the non-selected lines at F20 generation relative to their parent population (Fig. 5) may reflect the ideal conditions of growth in the laboratory. However, Jirakanjanakit and Dujardin [49] observed a progressively significant reduction in size of four *Ae. aegypti* lines with the lifetime spent in laboratory from 17 to 564 generations. We found it difficult to explain these conflicting results, but perhaps the most likely explanation rests in genetic drift and inbreeding effects and/or different micro-environmental conditions of different laboratories.

As far as we know, a lot of attention has been paid to size of mosquitoes and its relationship with insecticide resistance and fitness, but nothing between their shape, resistance and fitness. However, shape has proven to be more stable, less labile than size and expresses more about genetics and evolution of organisms [9,50]. In particular, wing shape of *Ae. aegypti* has shown more consistent evidence for genetic determinism than wing size [42,51].

In this work, wing shape matched very well with geographic origins of isolates and with lambda-cyhalothrin resistance (Fig. 4). Although differentiation in wing shape across large distances occurs, it has been shown [42] that different populations of *Ae.*

aegypti belong to the same species, and are distinct from the closely related species *Ae. albopictus*. Wing shape differentiation observed here could be related to the history of colonization of *Ae. aegypti* in Colombia. Before 1952, *Ae. aegypti* was widely distributed throughout the country and by the 1960 s it was eradicated completely from Colombia, except for the city of Cúcuta in the east, bordering Venezuela. Unfortunately, in 1971 there was a new and intensive re-infestation from the northern port cities of Colombia [41].

In addition to the association between wing shape and geography, we observe a second level of significant relationship between wing shape and resistance to lambda-cyhalothrin (Fig. 4). As far as we know, there are no current studies about mosquitoes' shape changes in response to insecticides resistance in these biological vectors; and in other insects there are few [52]. Our results also demonstrate the importance of focusing on shape more than in size when studying the morphological response to insecticide resistance in mosquitoes.

We observed in the F9 generation that the selected and the non-selected lines had a very different wing shape than that of their parents, but it was not different between the two lines (Fig. 6). However, in the F20 generation, the selected and the non-selected lines were significantly separated between them and with the F9 generation and its parents. We saw in this behavior the joint effects

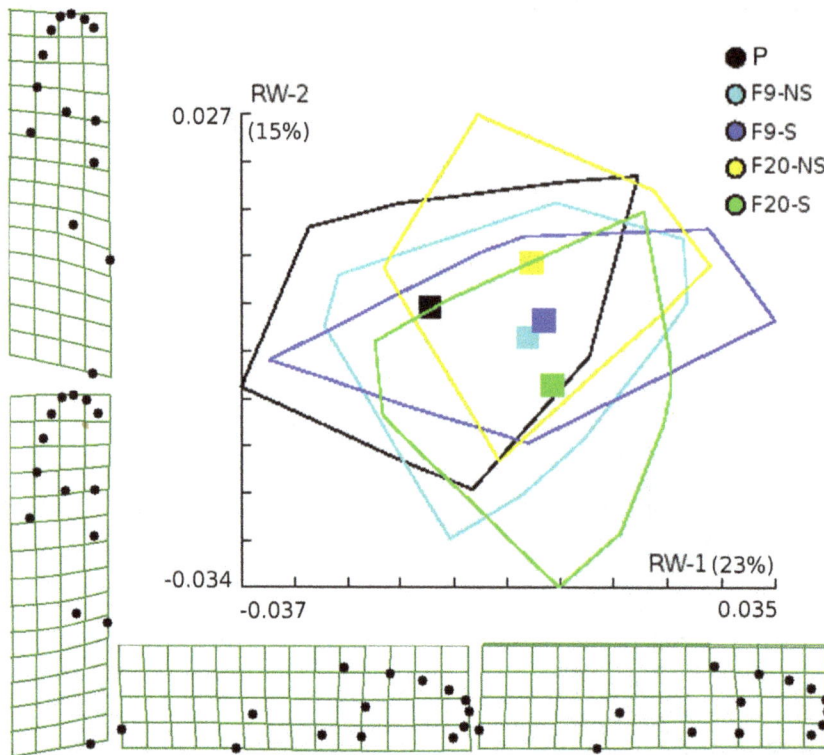

Figure 6. Scatterplot of the scores along the first two principal components (relative warps), the convex hull for each group and the average scores for each group. Convex hulls enclose individuals from parental and its descendants at ninth and twenty generations which were ordered on the two first relative warps. Squares represent the centroids (average scores for each group). To easy viewing of the group's centroids, the individual positions are not shown. Deformation grids for the extreme values of the first and the second relative warps on the mean configuration, are also shown. P: parental Comuneros isolate; F9-NS and F20-NS: ninth and twenty generations non-selected for lambda-cyhalothrin resistance; F9-S and F20-S: ninth and twenty generations selected for lambda-cyhalothrin resistance. Sample-sizes were 51 parental females, 42 F9-NS, 49 F9-S, 50 F20-NS, and 56 F20-S.

of laboratory colonization and the response to selection. We think that wing shape changes are easier when an organism suffers a bottleneck and founder effects (likely at initial conditions of laboratory colonization) and are more difficult in response to selection to insecticide-resistance.

Wing shape in each isolate was statistically different from others, with one exception (Table 2), but the wing size effects on wing shape (allometry) could not be discarded. Because a common allometric slope to all groups was not found, we could not verify whether wing shape variation was not simply the passive consequence of wing size variation [8]. We repeated the analysis excluding the ROCK strain because we think that the peculiar history of this reference strain could distort the model. But again, we could not find a common allometric slope. Therefore, we interpreted that isolates studied here did not share a common allometric growth. The different histories between Cúcuta and Quibdó isolates could also account their different allometries.

When we examined the allometries of the non-selected and the selected lines, we found that wing size significantly contributed to wing shape variation and that a common allometric slope could be found. Then, we could examine the variation in wing shape, scaling it by wing size, and we concluded that the lab's selection for resistance to lambda-cyhalothrin generated differences in wing shape that were not a passive consequence of variation in wing size.

Generally, insecticide-resistance is associated with negative effects on fitness [53,54]. It is not easy to measure all components of fitness, but we measured probably the most important for

quantifying the costs of lambda-cyhalothrin resistance in *Ae. aegypti*. The parental COM isolate showed an initial RR_{50} of 24.23 times relative to the susceptible ROCK strain, which was interpreted as a high level of natural resistance to this pyrethroid. Throughout the lab's selection we could raise this RR_{50} to 225.4X, 91.53X and 114.25X for F9, F10 and F20 generations, respectively. However, the parallel line not subjected to selection showed a progressive decrease in resistance until 1.95X. Such a change from a moderate resistance to susceptibility suggests that resistance to lambda-cyhalothrin carries a cost in absence of the insecticide.

We observed different levels of resistance to lambda-cyhalothrin in field isolate from the two municipalities evaluated (Table 1). The schemes of the use of insecticides in Cúcuta and Quibdó are different. Cúcuta employs mainly lambda-cyhalothrin, periodically spraying throughout all the year and covering the entire city. Quibdó employs mainly malathion, an organophosphate, which is only used when and where cases are detected. In this municipality, lambda-cyhalothrin is also used, but with less frequency than malathion. On the other hand, previous work showed that the selected line had cross-resistance to other pyrethroid (permethrin) but not to the organophosphates temephos and malathion [55], a situation which suggests that rotation of both families of insecticides is an appropriate alternative for managing the insecticide-resistance evolution.

Even the proximate collections within each municipality exhibited distinct resistance levels (Table 1). The scheme of collections could help to understand this phenomena. Given that the flight range of *Ae. eagypti* has been calculated in 840 m [56], we

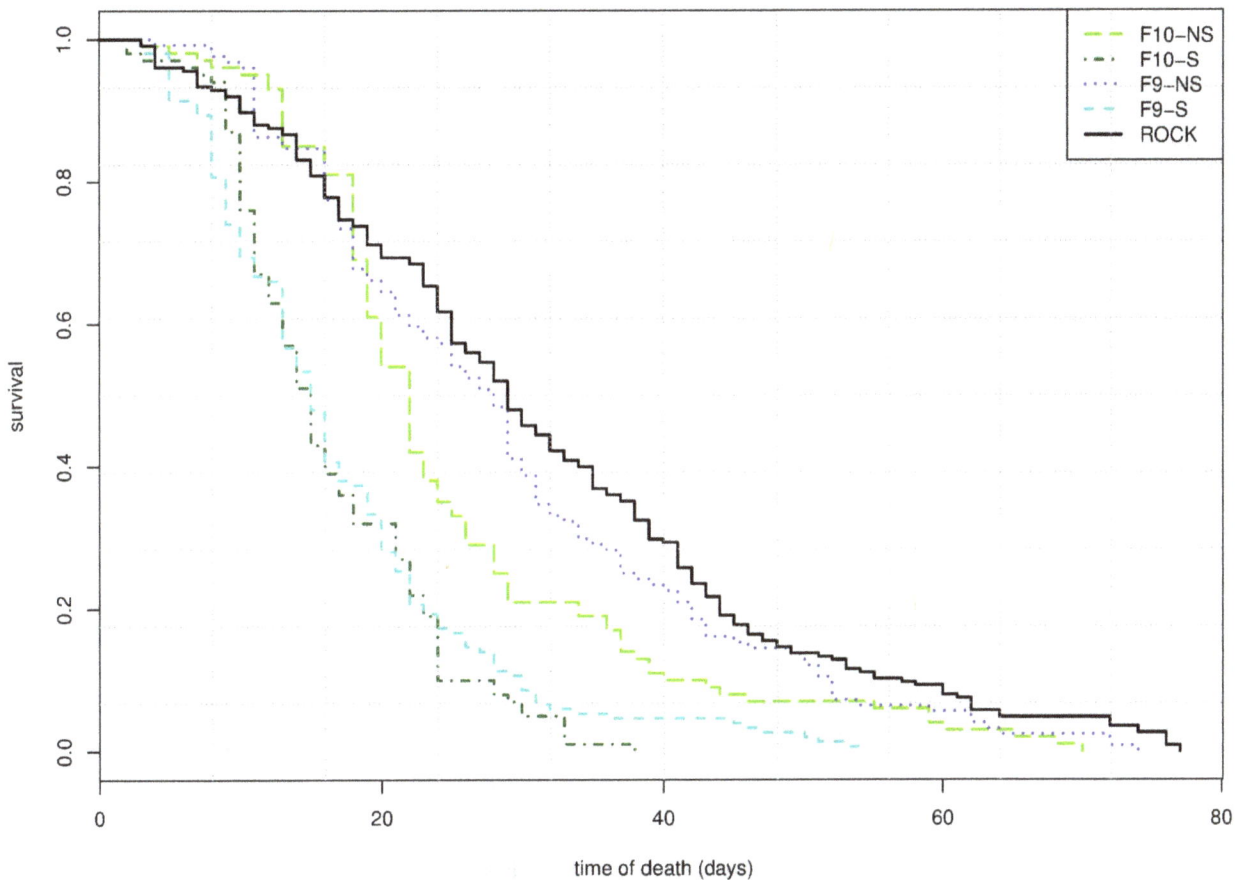

Figure 7. Survival analysis using Kaplan-Meier estimates. Curves represent daily survival of the Rockefeller strain (ROCK) females, and the ninth (F9-S) and tenth (F10-S) generations of the lambda-cyhalothrin selected line from COM isolate, and the ninth (F9-NS) and tenth (F10-NS) generations of the non-selected line from the same isolate. Sample-sizes were 100 females for both F10-S and F10-NS, 124 for F9-NS, 150 for F9-S, and 225 for the ROCK strain.

collected the insects in isolates separated by 1.5 to 5 km. Then each isolate could have been experienced genetic drift when they were founded by few adult survivors of lambda-cyhalothrin control. On the other hand, microarray analyses showed that laboratory adaptation to the pyrethroid permethrin in *Ae. aegypti* is genetically complex, largely conditioned by pre-existing target site insensitivity in the *para* gene and dependent on the geographic origin of the isolates [57]. Maybe, lab adaptation to the pyrethroid lambda-cyhalothrin behaves similarly. Furthermore, differential pressures from non-planned use of pyrethroids at domestic level, the non-standardized use of lambda-cyaholothrin impregnated bednets, and indirect exposure to xenobiotics could also influence such different resistance levels [20].

Related to the dramatic decrease in RR_{50} of the selected line from F9 to F10, and then increased in F20 to nearly half of the F9, Saavedra-Rodríguez et al [57] found a similar trend. They selected six field isolates of *Ae. aegypti* with permethrin for five generations in laboratory and registered a progressive increase in resistance levels except for two isolates, in which the resistance levels suddenly decrease in the fourth generation, and then raised again in the fifth generation. The authors observed an important amount of uniquely transcribed genes in individual isolates and thought that the differential response to selection of insecticide resistance could be associated with lethal or deleterious recessive alleles or even with genes related to metabolic resistance.

On the other hand, we do not discard a human error in the experimental protocols, but in spite the notorious oscillation, the selected F10 showed a RR_{50} of nearly nine times higher than the non-selected line, and the selected F20 of nearly 60 times. It is important remember that the selected and the non-selected lines were derived from the same parental insects, and that demographic parameters were studied in parallel on the F9 generation when RR_{50} reached the peak. The F10 and F20 generations were used to verify that under the selective regime, resistance was kept significantly different from the non-selected relatives, and to assess the corresponding morphometric changes.

To examine more deeply the fitness costs, we measured six recognized components of fitness, but only age-specific survival of females and fecundity showed significant differences between the selected and the non-selected lines (Table S3). We measured the components of fitness under ideal laboratory's conditions, which, although it did not reflect entirely the isolate dynamics on the field, is a way to study the biological potential of this species. Comparisons between non-related strains are one of the more common weaknesses in studies about fitness costs of resistant mosquitoes [12]; however, our work studied two lines of mosquitoes selected and non-selected for insecticide-resistance, which shared the same genetic background, which is fundamental to understand the evolutionary capabilities of *Ae. aegypti* and then planning optimal policies of vector control [7].

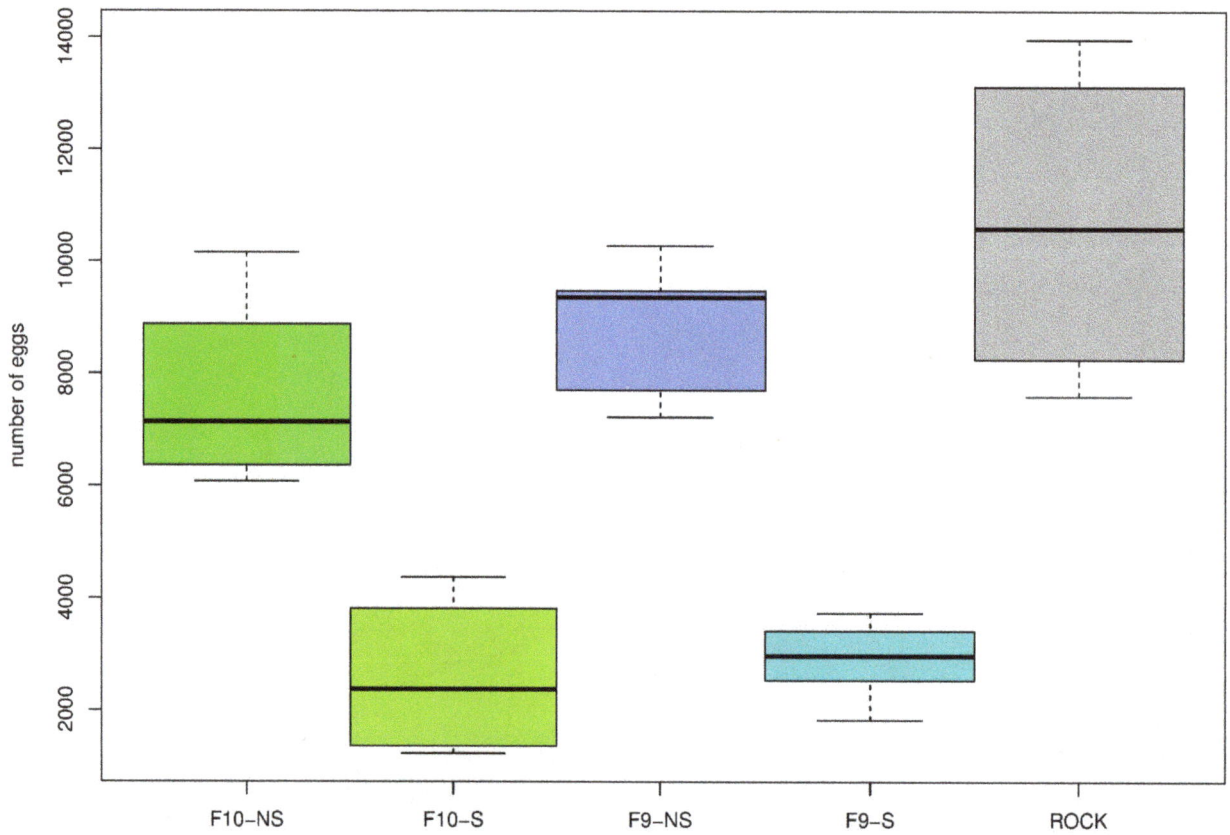

Figure 8. Box-plots showing the fecundity variation of both lambda-cyhalothrin selected and non-selected lines. Each box is divided by the median, which top and bottom correspond to 25th and 75th quartiles, respectively. ROCK: the Rockefeller strain, F9-S and F10-S: the selected line at ninth and tenth generations, and F9-NS and F10-NS: the non-selected line at ninth and tenth generations. Sample-sizes were 100 females for both F10-S and F10-NS, 124 for F9-NS, 150 for F9-S, and 225 for the ROCK strain.

The survival probability of females, which were selected for resistance to lambda-cyhalothrin, quickly decreased in youngest individuals (10–30 days old); but afterwards, it progressively slowed its decline and some individuals could reach older ages (Fig. 7). The susceptible ROCK strain and the non-selected females displayed similar curves; meanwhile, the fecundity was

Table 3. Comparisons of the net reproductive rate per generation (R_0) and the intrinsic rate of increase in days (r).

R_0					
	ROCK	F9-NS	F9-S	F10-NS	F10-S
ROCK	0	0.9919	**0.0002812**	0.4515	**0.0002809**
F9-NS	1.6132	0	**0.0004547**	0.7027	**0.0004541**
F9-S	8.708	8.095	0	**0.003833**	1.00
F10-NS	2.433	1.82	6.275	0	**0.003825**
F10-S	8.71	8.096	0.001557	6.276	0
r					
ROCK	0	0.9991	**0.0008817**	0.9726	**0.003287**
F9-NS	0.355	0	**0.001329**	0.9964	0.005271
F9-S	7.456	7.101	0	**0.002482**	0.9401
F10-NS	0.8521	0.4971	6.604	0	**0.01034**
F10-S	6.391	6.036	1.065	5.539	0

Q-values of paired Tukey tests are shown below the diagonal, and P values above the diagonal. Using Bonferroni correction, a significant P value is <0.005.
ROCK: the Rockefeller strain; F9-S and F10-S: the selected line at ninth and tenth generations, respectively; F9-NS and F10-NS: the non-selected line at ninth and tenth generations, respectively.

Table 4. Relative fitness and parameters used to calculate it. Fitness was calculated relative to the ROCK strain.

	net reproductive rate per generation: R_0	intrinsic rate of increase in days: r	generation time in days: G	r derived from Euler-Lotka equation	relative fitness
ROCK	203.65	26.23	0.203	0.27	1.00
F9-NS	176.16	27.49	0.188	0.26	0.95
F10-NS	152.50	25.46	0.198	0.26	0.97
F9-S	57.66	23.27	0.175	0.21	0.79
F10-S	51.70	18.80	0.209	0.22	0.82

l_x: age specific survival, m_x: fecundity. ROCK: the Rockefeller strain, F9-S and F10-S: the selected line at ninth and tenth generations, and F9-NS and F10-NS: the non-selected line at ninth and tenth generations.

significantly lower for the selected line (Fig. 8). It is important to notice that the ROCK strain and the non-selected line did not share a genetic background, and the analysis of life history components were performed at the same time and in the same environment. Moreover, both the selected and non-selected lines shared the same genetic background. Because of this, we think there were no other intrinsic triggers of mortality and fecundity between the selected and the non-selected lines than insecticide-resistance, and consequently it suggests costs in age-specific survival of females, and fecundity for Ae. aegypti resistant to lambda-cyhalothrin.

Fitness costs due to lower survival and fecundity in Ae. aegypti females associated with insecticide-resistance has been reported in several works [58–62], but this is the first time that two estimators of the per capita rate of increase are calculated (R_0 and r) and from them the relative fitness.

The statistical analyses of R_0 (net reproductive rate per generation) and r (intrinsic rate of increase in days) showed significant differences between the selected and the non-selected lines; besides, the r parameter did not show differences with the ROCK strain (Table 3). Setting a relative fitness value of 1.0 to the ROCK strain, the selected line reduced its fitness by 21%, meanwhile the non-selected line by only 4% (Table 4).

In conclusion, we were able to increase 10-fold the RR_{50} to lambda-cyhalothrin of a mosquito-line from COM isolate and held it in laboratory conditions, at least for two years. Furthermore, we observed that the parallel line, related to the selected one, and not subjected to insecticide pressure, reverted to its susceptibility status in just 20 generations. This information tells us about the evolutionary capacities of Ae. aegypti and strongly suggests that there are costs in fitness in absence of insecticides. Beserra et al. [63,64] estimated in average between 21 to 31 the number of generations per year under natural conditions for several populations in Brazil. Thus in absence of lambda-cyhalothrin, reversion to susceptibility in the field could occur in less of one year, if we take in account that natural conditions are probably not as optimal as laboratory conditions. As a practical consequence the knowledge of reversion to insecticide-susceptibility in relative short time is useful to preserve the utility of the few current insecticides molecules currently available for vector control of tropical diseases.

Moreover, the present study showed the fitness disadvantage focused in less fecundity and survival, but not in the average hatchability, average time to pupation, average time to eclosion adults, and proportion of males and females emerging. Such fitness costs could negatively affect the extrinsic incubation period of the dengue virus, altering the dynamic of the virus transmission by mosquitoes [65].

Wing shape is a more reliable marker of the resistance status of Ae. aegypti natural populations, at least for lambda-cyhalothrin, than wing size.

Mosquito control programs could benefit from the knowledge of natural resistance status of populations and from the fitness costs. If we know fitness costs of resistance to insecticides and time to revert to susceptibility in absence of insecticides, it might be possible to implement efficient strategies to combine insecticides over time and/or space to delay or prevent the evolution of resistance and supplement this with biological control and education of communities [6].

Supporting Information

Figure S1 Map of Colombia showing the locations of municipalities of Cúcuta and Quibdó and the locations of isolates within both municipalities. The map of Colombia was downloaded from the public domain at the CIA (https://www.cia.gov/library/publications/the-world-factbook/index.html). The maps of Cúcuta and Quibdó were opened in the OpenStreetMap's viewer and exported as PNG image files (http://www.openstreetmap.org/). The points and labels of locations were edited with the GIMP software ver. 2.8 (http://www.gimp.org/).

Figure S2 Scatter plot of the scores along the first two principal components (relative warps), the convex hull for each group and the average scores for each group. Convex hulls enclose individuals from nine natural isolates which were ordered on the two first relative warps. To easy viewing of the group's centroids, the individual positions are not shown. Deformation grids for the extreme values of the first and the second relative warps on the mean configuration, are also shown. The isolates of the municipality of Cúcuta are represented by red tones and the isolates of the municipality of Quibdó are represented by green tones.

Table S1 Changes in susceptibility to lambda-cyhalothrin of the selected line relative to the susceptible reference ROCK strain.

Table S2 Changes in susceptibility to lambda-cyhalothrin of the non-selected line relative to the susceptible reference ROCK strain.

Table S3 Summary of parameters which did not show statistical differences ($P>0.05$) among the reference

susceptible ROCK strain, and the selected and the non-selected lines at F9 and F10 generations. ROCK: reference susceptible ROCK strain; F9-S and F10-S: the selected line at F9 and F10 generations, respectively; F9-NS and F10-NS: the non-selected line at F9 and F10 generations, respectively. Between parenthesis: standard deviation.

Acknowledgments

We thank Dr. Henry R. Rupp, for his kind comments and suggestions to the text, and for help to improving the use of English in the manuscript.

References

1. Bhatt S, Gething PW, Brady OJ, Messina JP, Farlow AW, et al. (2013) The global distribution and burden of dengue. Nature 496: 504–507.
2. WHO (2012) Global strategy for dengue prevention and control 2012–2020. Geneva: WHO, World Health Organization. WHO/HTM/NTD/VEM/2012.5.
3. Hoy M A (1998) Myths, models and mitigation of resistance to pesticides. Philos Trans R Soc Lond B, Biol Sci 353: 1787–1795.
4. Labbé P, Berticat C, Berthomieu A, Unal S, Bernard C, et al. (2007) Forty years of erratic insecticide resistance evolution in the mosquito Culex pipiens. PLoS Genet 3: e205.
5. Berticat C, Bonnet J, Duchon S, Agnew P, Weill M, et al. (2008) Costs and benefits of multiple resistance to insecticides for Culex quinquefasciatus mosquitoes. BMC Evol Biol 8: 104.
6. REX Consortium (2013) Heterogeneity of selection and the evolution of resistance. Trends Ecol Evol 28: 110–118.
7. Brown ZS, Dickinson KL, Kramer RA (2013) Insecticide resistance and malaria vector control: The importance of fitness cost mechanisms in determining economically optimal control trajectories. J Econ Entomol 106: 366–374.
8. Dujardin J-P (2008) Morphometrics applied to medical entomology. Infect genet Evol 8: 875–890.
9. Klingenberg CP (2010) Evolution and development of shape: integrating quantitative approaches. Nature Rev Genet 11: 623–635.
10. Adams DC, Rohlf FJ, Slice DE (2004) Geometric morphometrics: Ten years of progress following the "revolution". Ital J Zool 71: 5–16.
11. Deevey ES (1947) Life tables for natural populations of animals. Q Rev Biol 22: 283–314.
12. Bourguet D, Guillemaud T, Chevillon C, Raymond M (2004) Fitness costs of insecticide resistance in natural breeding sites of the mosquito Culex pipiens. Evolution 58: 128–135.
13. Kuno G (2010) Early history of laboratory breeding of Aedes aegypti (Diptera: Culicidae) focusing on the origins and use of selected strains. J Med Entomol 47: 957–971.
14. WHO (1998) Test procedures for insecticide resistance monitoring in malaria vectors, bio-efficacy and persistence of insecticides on treated surfaces. Geneva: WHO, World Health Organization. WHO/CDS/CPC/MAL/98.12.
15. WHO (1981) Instructions for determining the susceptibility or resistance of mosquitoes larvae to insecticide. Geneva: WHO, World Health Organization. WHO/VBC/81807: 6.
16. Fitnney DJ (1971) Probit Analysis. A Statistical Treatment of the Sigmoid Response Curve. 3rd ed. Cambridge, UK: Cambridge University Press. 333 p.
17. Abbott WS (1925) A Method of computing the effectiveness of an insecticide. J Econ Entomol 18: 265–267.
18. Mazzarri MB, Georghiou GP (1995) Characterization of resistance to organophosphate, carbamate, and pyrethroid insecticides in field populations of Aedes aegypti from Venezuela. J Am Mosq Control Assoc 11: 315–322.
19. Brogdon WG, McAllister JC (1998) Simplification of adult mosquito bioassays through use of time-mortality determinations in glass bottles. J Am Mosq Control Assoc 14: 159–164.
20. Fonseca-González I, Quiñones ML, Lenhart A, Brogdon WG (2011) Insecticide resistance status of Aedes aegypti (L.) from Colombia. Pest Manag Sci 67: 430–437.
21. Bookstein FL (1991) Morphometric tools for landmark data, Geometry and Biology. 1st ed. Cambridge, NY: University Press. 435 p.
22. Arnqvist G, Martensson T (1998) Measurement error in geometric morphometrics: empirical strategies to assess and reduce its impact on measure of shape. Acta Zool Academ Sci Hung 44: 73–96.
23. Hintze JL, Nelson RD (1998) Violin plots: a box plot-density trace synergism. Am Stat. 52: 181–184.
24. Sokal RR, Rohlf FJ (1962). The comparison of dendrograms by objective methods. Taxon 11: 33–40.
25. Rohlf FJ (1999) Shape statistics: Procrustes superimpositions and tangent spaces. J Classif 16: 197–223.
26. Dujardin JP, Kaba D, Henry AB (2010) The exchangeability of shape. BMC Res Notes 3: 266.
27. Futuyma DJ (1998) The Evolution of Life Histories. Evolutionary Biology. Sunderland, Massachusetts: Sinauer Associates.pp. 561–578.
28. Sibly RM, Hone J (2002) Population growth rate and its determinants: an overview. Philos Trans R Soc Lond B Sci 357: 1153–1170.
29. R Core Team (2012) R: A language and environment for statistical computing. Vienna, Austria: R Foundation for Statistical Computing.
30. Rohlf FJ (2013) tpsRelw, relative warps analysis, version 1.53. Department of Ecology and Evolution, State University of New York at Stony Brook. Software available at: http://life.bio.sunysb.edu/morph/soft-tps.html.
31. Raymond M (1985) Présentation d'un programme Basic d'analyse log-probit pour micro-ordinateur. Cahiers ORSTOM, série Entomologie médicale et Parasitologie 23: 117–121.
32. Armbruster P, Hutchinson RA (2002) Pupal Mass and Wing Length as Indicators of Fecundity in Aedes albopictus and Aedes geniculatus (Diptera: Culicidae). J Med Entomol 39: 699–704.
33. Berger D, Olofsson M, Friberg M, Karlsson B, Wiklund C, et al. (2012) Intraspecific variation in body size and the rate of reproduction in female insects - adaptive allometry or biophysical constraint? J Anim Ecol 81: 1244–1258.
34. Blackmore MS, Lord CC (2000) The relationship between size and fecundity in Aedes albopictus. J Vector Ecol 25: 212–217.
35. Honek A (1993) Intraspecific variation in body size and fecundity in insects: A general relationship. Oikos 66: 483–492.
36. Packer MJ, Corbet PS (1989) Size variation and reproductive success of female Aedes punctor (Diptera: Culicidae). Ecol Entomol 14: 297–309.
37. Xue R-D, Barnard DR, Muller GC (2010) Effects of body size and nutritional regimen on survival in adult Aedes albopictus (Diptera: Culicidae). J Med Entomol 47: 778–782.
38. Koella JC, Lyimo EO (1996) Variability in the relationship between weight and wing length of Anopheles gambiae (Diptera: Culicidae). J Med Entomol 33: 261–264.
39. Urbanski J, Mogi M, O'Donnell D, DeCotiis M, Toma T, et al. (2012) Rapid adaptive evolution of photoperiodic response during invasion and range expansion across a climatic gradient. Am Nat 179: 490–500.
40. Reiskind MH, Zarrabi AA (2012) Is bigger really bigger? Differential responses to temperature in measures of body size of the mosquito, Aedes albopictus. J Insect Physiol 58: 911–917.
41. Groot H (1980) The reinvasion of Colombia by Aedes aegypti: aspects to remember. Am J Trop Med Hyg 29: 330–338.
42. Henry A, Thongsripong P, Fonseca-González I, Jaramillo-Ocampo N, Dujardin J-P (2010) Wing shape of dengue vectors from around the world. Infect Genet Evol 10: 207–214.
43. Jirakanjanakit N, Leemingsawat S, Thongrungkiat S, Apiwathnasorn C, Singhaniyom S, et al. (2007) Influence of larval density or food variation on the geometry of the wing of Aedes (Stegomyia) aegypti. Trop Med Int Health 12: 1354–1360.
44. Morales-Vargas RE, Ya-Umphan P, Phumala-Morales N, Komalamisra N, Dujardin J-P (2010) Climate associated size and shape changes in Aedes aegypti (Diptera: Culicidae) populations from Thailand. Infect Genet Evol 10: 580–585.
45. Duron O, Labbé P, Berticat C, Rousset F, Guillot S, et al. (2006) High Wolbachia density correlates with cost of infection for insecticide resistant Culex pipiens mosquitoes. Evolution 60: 303–314.
46. Gómez A, Seccacini E, Zerba E, Licastro S (2011) Comparison of the insecticide susceptibilities of laboratory strains of Aedes aegypti and Aedes albopictus. Mem Inst Oswaldo Cruz 106: 993–996.
47. Chan HH, Zairi J (2013) Permethrin Resistance in Aedes albopictus (Diptera: Culicidae) and associated fitness costs. J Med Entomol 50: 362–370.
48. Antonio GE, Sánchez J, Williams T, Marina CF (2009) Paradoxical effects of sublethal exposure to the naturally derived insecticide spinosad in the dengue vector mosquito, Aedes aegypti. Pest Manag Sci 65: 323–326.
49. Jirakanjanakit N, Dujardin J-P (2005) Discrimination of Aedes aegypti (Diptera: Culicidae) laboratory lines based on wing geometry. Southeast Asian J Trop Med Public Health 36: 858–861.
50. Flint J, Mackay TF (2009) Genetic architecture of quantitative traits in mice, flies, and humans. Genome Res 19: 723–733.
51. Jirakanjanakit N, Leemingsawat S, Dujardin JP (2008) The geometry of the wing of Aedes (Stegomyia) aegypti in isofemale lines through successive generations. Infect Genet Evol 8: 414–421.

Author Contributions

Conceived and designed the experiments: IFG NJO. Performed the experiments: DCR NJO. Analyzed the data: NJO DCR IFG. Contributed reagents/materials/analysis tools: IFG NJO. Wrote the paper: NJO.

We specially thank Dr. Hector Anaya R. of the "Instituto Departamental de Salud, Dirección de Vigilancia en Salud Pública" from municipality of Cúcuta, and Dr. Zulma Bejarano M. of the "Departamento Administrativo de Salud, Unidad de Control de Vectores" from municipality of Quibdó, by providing logistical support to insect sampling. We thank to Dr. Carmelo Fruciano and the other three reviewers by their careful revision which certainly helped us to improve the manuscript.

52. Gaspe MS, Gurevitz JM, Gürtler RE, Dujardin J-P (2013) Origins of house reinfestation with *Triatoma infestans* after insecticide spraying in the Argentine Chaco using wing geometric morphometry. Infect Genet Evol 17: 93–100.

53. Hemingway J, Ranson H (2000) Insecticide resistance in insect vectors of human disease. Ann Rev Entomol 45: 371–391.

54. Kliot A, Ghanim M (2012) Fitness costs associated with insecticide resistance. Pest Manag Sci 68: 1431–1437.

55. Chaverra-Rodríguez D, Jaramillo-Ocampo N, Fonseca-Gonzalez I (2012) Selección artificial de resistencia a lambda-cialotrina en *Aedes aegypti* y resistencia cruzada a otros insecticidas. Rev Colomb Entomol 38: 100–107.

56. Reiter P, Amador MA, Anderson RA, Clark GG (1995) Short report: dispersal of *Aedes aegypti* in an urban area after blood feeding as demonstrated by rubidium-marked eggs. Am J Trop Med Hyg 52: 177–179.

57. Saavedra-Rodriguez K, Suarez AF, Salas IF, Strode C, Ranson H, et al. (2012) Transcription of detoxification genes after permethrin selection in the mosquito *Aedes aegypti*. Insect Mol Biol 21: 61–77.

58. Kumar S, Thomas A, Samuel T, Sahgal A, Verma A, et al. (2009) Diminished reproductive fitness associated with the deltamethrin resistance in an Indian strain of dengue vector mosquito, *Aedes aegypti* L. Trop Biomed 26: 155–164.

59. Paris M, David J-P, Despres L (2011) Fitness costs of resistance to Bti toxins in the dengue vector *Aedes aegypti*. Ecotoxicology 20: 1184–1194.

60. Belinato TA, Martins AJ, Valle D (2012) Fitness evaluation of two Brazilian *Aedes aegypti* field populations with distinct levels of resistance to the organophosphate temephos. Mem Inst Oswaldo Cruz 107: 916–922.

61. Martins AJ, Ribeiro CD, Bellinato DF, Peixoto AA, Valle D, et al. (2012) Effect of insecticide resistance on development, longevity and reproduction of field or laboratory selected *Aedes aegypti* populations. PloS ONE 7: e31889.

62. Brito LP, Linss JG, Lima-Camara TN, Belinato TA, Peixoto AA, et al. (2013) Assessing the effects of *Aedes aegypti* kdr mutations on pyrethroid resistance and its fitness cost. PLoS ONE 8: e60878.

63. Beserra EB, Castro Jr. FP de, Santos JW dos, Santos TdeS, Fernandez CRM (2006) Biologia e Exigências Térmicas de *Aedes aegypti* (L.) (Diptera: Culicidae) Provenientes de Quatro Regiões Bioclimáticas da Paraíba. Neotrop Entomol 35: 853–860.

64. Beserra EB, Fernandes CRM, Silva SA de O, Silva LA da, Santos JW dos (2009) Efeitos da temperatura no ciclo de vida, exigências térmicas e estimativas do número de gerações anuais de *Aedes aegypti* (Diptera, Culicidae). Iheringia Série Zool 99: 142–148.

65. Bellan SE (2010) The importance of age dependent mortality and the extrinsic incubation period in models of mosquito-borne disease transmission and control. PloS ONE 5: e10165.

10

Lipotropes Protect against Pathogen-Aggravated Stress and Mortality in Low Dose Pesticide-Exposed Fish

Neeraj Kumar, Subodh Gupta, Nitish Kumar Chandan, Md. Aklakur, Asim Kumar Pal, Sanjay Balkrishna Jadhao*

Department of Fish Nutrition, Biochemistry and Physiology, Central Institute of Fisheries Education, Versova, Mumbai, Maharashtra, India

Abstract

The decline of freshwater fish biodiversity corroborates the trends of unsustainable pesticide usage and increase of disease incidence in the last few decades. Little is known about the role of nonlethal exposure to pesticide, which is not uncommon, and concurrent infection of opportunistic pathogens in species decline. Moreover, preventative measures based on current knowledge of stress biology and an emerging role for epigenetic (especially methylation) dysregulation in toxicity in fish are lacking. We herein report the protective role of lipotropes/methyl donors (like choline, betaine and lecithin) in eliciting primary (endocrine), secondary (cellular and hemato-immunological and histoarchitectural changes) and tertiary (whole animal) stress responses including mortality (50%) in pesticide-exposed (nonlethal dose) and pathogen-challenged fish. The relative survival with betaine and lecithin was 10 and 20 percent higher. This proof of cause-and-effect relation and physiological basis under simulated controlled conditions indicate that sustained stress even due to nonlethal exposure to single pollutant enhances pathogenic infectivity in already nutritionally-stressed fish, which may be a driver for freshwater aquatic species decline in nature. Dietary lipotropes can be used as one of the tools in resurrecting the aquatic species decline.

Editor: Jonathan H. Freedman, NIEHS/NIH, United States of America

Funding: SBJ acknowledges the support for institutional project [CIFE (M)-2007/T6-1]. NK and NKC received an institutional research fellowship from the Central Institute of Fisheries Education (Indian Council of Agricultural Research). The funders had no role in study design, data collection and analysis, decision to publish, or preparation of the manuscript.

Competing Interests: The authors have declared that no competing interests exist.

* E-mail: sbjadhao@hotmail.com

Introduction

Freshwater species decline and endangerment around the world has been receiving increased attention, but there are missing spots in linking these patterns to physiological mechanisms and in finding possible remedies. It is a well-recognized multifactorial problem related to habitat destruction, climate change, and pesticide use, among which, little is known about the latter. Unsustainable trends in the use of synthetic fertilizers and pesticides are evident from their respective seven-fold and three-fold increases in last four decades [1]; application of pesticide is expected to increase by 170 percent by 2050 [2]. These trends' negative effects on ecosystem health are corroborated by an increased outbreak of diseases and greater loss of biodiversity. Among the various ecosystems being radically altered, freshwater ecosystems are the most endangered by these anthropogenic activities, because unlike in marine systems, the likelihood of pollutant dilution is rare in freshwaters systems, resulting in the suppression of fish immune systems and increase in mortality. Some 34 percent of fish species, mostly from fresh water, are threatened with extinction [3] and the number of threatened fish species in the Red List [4] (version 2011.2) has increased from 734 in 1996 to 2028 in 2011, and the list is growing every year. In India, out of a total 46 percent evaluated from 700 total freshwater fish species, 70 percent are threatened [5]. Of anthropogenic stressors, pesticides are of prime concern as they aggravate the effects of other stressors, which has negative implications for

biodiversity and the aquaculture industry. In the United States, pesticides were found to pollute every stream and over 90 percent of wells sampled in a study by the US Geological Survey [6] and the situation around the world is similar.

This is an enormous challenge, and effective tools are necessary for the successful conservation of threatened species, as well as for producing an extra 37 million tons of fish and aquatic food by 2030 in order to feed the world's burgeoning population under even mildly polluted aquatic systems. From simple aquaculture producers' utilitarian perspectives, the cost of cleaning up pollution when resources are scarce could favor secondary prevention strategies, such as nutritional strategies, for mitigating this environmental insult and restoring endangered species. Nutritional strategies are handy and appropriate for combating various stressors. Available evidence indicates that stress alters metabolism, causes hypo/demethylation of DNA [7] and also changes requirement of variety of nutrients [8,9]. Methyl groups are of vital importance as animals cannot synthesize them and thus need to receive them through diet [10]. Despite the established link between stress and methylation, direct studies on methyl donor (lipotropic compound) supplementation in an ecological context are scarce.

Although pesticide regulations are designed to protect human and wildlife communities from large-dose exposures to pesticides and prevent acute disease symptoms and mortality, nothing exists in regulations to prevent low-level exposures and sublethal effects

[11], which should be a concern. Even sublethal/nonlethal environmental pollutants and pesticides like endosulfan (organochlorine lipophilic insecticide) and others at ecologically relevant doses cause immunosuppressive effects in several species [12,13] including fish [14], affect recruitment (and reproduction), increase pathogen virulence [15] and delay development, decrease longevity, decrease foraging success and cause decline in species populations [16]. While acute toxicity aspects have been extensively investigated and there are a few discrete studies on sublethal toxicity effects in fish on cortisol secretion and glutathione-s-transferase [17] and immunological parameters [14,18] along with an indication of increased susceptibility of juvenile Chinook salmon to vibriosis after exposure to chlorinated and aromatic compounds found in contaminated urban estuaries [19], comprehensive studies elucidating the physiological mechanisms of the nonlethal toxicity and concurrent pathogenic infections under experimental conditions along with counteractive measures are rarely attempted. This study was done in *Labeo rohita*, a commercially important freshwater carp species in the Indian subcontinent [20] contributing 80–90 percent of the carp polyculture, and almost a dozen species of this genus are endangered/threatened [4]. These experiments were carried out with the aim of studying the effects of low dose endosulfan exposure on comprehensive stress responses, including mortality in pesticide-exposed (nonlethal dose) and pathogen-challenged fish, and whether lipotropes can counteract these responses elicited by exposure of fish to pesticide.

Material and Methods

Ethics Statement

The use of animals conforms to the existing laws in India. The care and treatment of animals used in this study were in accordance with the guidelines of the CPCSEA [(Committee for the Purpose of Control and Supervision of Experiments on Animals), Ministry of Environment & Forests (Animal Welfare Division), Government of India] on the care and use of animals in scientific research. The study protocol and experimental endpoints were approved by the Advisory Committee of this research work, Board of Studies and authorities of the Central Institute of Fisheries Education (Deemed University), Mumbai (India). As the experimental fish *L. rohita* is a commercially important and non-endangered fish, the provisions of the Government of India's Wildlife Protection Act of 1972 are not applicable for experiments on this fish.

Fish and Experimental Design

Fingerlings of *Labeo rohita* (average weight 7.95±0.04 g) were procured from Prem Fisheries Consultancy, Gujarat, India, and transported to the experimental facilities at the Institute in a circular container (500 L) with sufficient aeration. The animals were acclimatized to the experimental rearing conditions for 15 days. Fifteen fish of uniform size (average weight 7.96±0.04 g) per container were stocked in five distinct groups with three replicates for each treatment in plastic containers of 150 L capacity (80×57×42 cm) each following a completely randomized design. Control fish were reared in normal water and fed basal (control) feed. Fish in the four experimental groups were exposed to low dose endosulfan (1/10th dose of LC_{50}) for 37 days and fed with control feed or given feed supplemented with 0.1 percent choline, or 0.5 percent betaine, or 2 percent lecithin twice daily (09:00 and 17:00 h) to approximate satiation. The LC50 of endosulfan used in this study was based on our study in same species [21]. The quality and preparation details of endosulfan have been described

earlier [21,22]. Growth was monitored on the 17th and 37th day by collectively weighing each group of fish. Fishes were starved overnight before taking the weight. Round-the-clock aeration was provided to all the containers from a compressed air pump and manual water exchange (two third) was carried out daily. Water quality parameters were checked every week using the methods of APHA [23] and the water quality conformed to carp-rearing standards (**Table S1**). As the residue determination required large quantity of fish mass, another set of fishes (average weight 262 g) procured from the same source were used in a setup similar to that given above for th same duration. This additional set allowed determination of serum caspase, HSP70, methyl transferase (MT) and vitellogenin in these fishes along with endosulfan residue in muscle tissue.

Experimental Diets and Proximate Analysis

Four isoproteinous (35% crude protein) and isocaloric (410 kcal/100 g) practical diets were formulated: control basal feed, and basal feed containing: choline as choline chloride (SD Fine Chemicals Ltd, Mumbai) (0.1 g/kg); betaine as betaine hydrochloride (HIMEDIA, Mumbai, India) (0.5 g/kg); and lecithin (20 g/kg) as soylecithin (HIMEDIA, Mumbai, India), respectively. Choline and betaine were included at the expense of wheat flour, while lecithin (being lipid) was included at the expense of oil components in the diet. The composition of the basal diet is given in **Table 1**. Betaine and choline chloride were first dissolved in water and incorporated into a vitamin mineral premix, whereas lecithin was mixed in oil. For formulation of the pelleted diet, a manually prepared vitamin and mineral mixture along with ascorbyl phosphate (SRL Ltd., Mumbai, India) as the source of vitamin C was used. The dough was mixed properly and was pelleted, air dried and kept in a hot air oven at 60°C until dry and was subsequently stored at 4°C until required for feeding. The proximate composition of all the experimental diets and fish were analyzed as per the methods of AOAC [24].

Sampling

At the end of the of 37-day feeding trial, the first sampling was carried out for analysis of the different blood parameters, respiratory burst and serum lysozyme activity. Three fish from each replicate with a total of nine fish from each treatment were anaesthetized with clove oil (50 µl/1) and blood was collected from the caudal vein. For serum, another six fish from each treatment were anaesthetized and blood was collected without anticoagulant and allowed to clot for 2 h followed by collection of serum with a micropipette and stored at −20°C until use. The procedures used for blood collection, estimation of blood glucose, haemogobin, total erythrocyte and total leucocyte count in Neubauer's chamber, total serum protein, albumin and globulin, the respiratory burst activity of phagocytes (as measured by intracellular superoxide radical-induced reduction of nitro blue tetrazolium (NBT), tissue homogenate preparation and analysis of enzyme (superoxide dismutase (SOD) (E.C.1.15.1.1), catalase (E.C.1.11.1.6) and glutathione-s-transferase (EC 2.5.1.18) are described earlier [22]. Serum cortisol was determined via radioimmunoassay as described [25]. For removal of organs, the required numbers of fish were anesthetized with an overdose of clove oil and cessation of heartbeat was observed.

Quantification of Markers of Stress and Endocrine Disruption and Sex Steroids

The expression of HSP-70 (EIA kit, catalog o. EKS-700B), and caspase-3 (colorimetric detection kit, catalog no. ADI-907-013) in

Table 1. Composition of the basal diet.

Ingredients	Percent
Soybean meal[a]	45.5
Fish meal[a]	10.00
Sunflower meal[a]	10.00
Wheat flour[a]	14.97
Wheat bran[a]	10.00
Sunflower oil[a]	4.00
Cod liver oil[a]	2.00
CMC[b] (Carboxyl methylcellulose)	1.00
Vitamin + mineral mix[c]	2.00
Vitamin C[d]	0.030
Chromic oxide	0.50

[a, b, c, d]Sources: [a]procured from local market, [b]HiMedia (JTJ Enterprises, Mumbai, India), [c]Prepared manually and all components from HiMedia Ltd and [d]SD Fine-Chemicals Ltd (Mumbai, India).
[c]Composition of vitamin mineral mix (quantity/250 g starch powder): Vitamin A 550,000 IU; Vitamin D_3 110,000 IU; Vitamin B_1 20 mg; Vitamin B_2 200 mg; Vitamin E 75 mg; Vitamin K 100 mg; Vitamin B_{12} 0.6 µg; Calcium Pantothenate 250 mg; Nicotinamide 1000 mg; Pyridoxine 100 mg; Mn 2,700 mg; I 100 mg; Fe 750 mg; Zn 500 mg; Cu 200 mg; Co 45 mg; Ca 50 g; P 30 g; Selenium 5 ppm.

the gill and liver samples and serum methyl transferase (MT) (MT detection kit, catalog no. ADI-907-025) were determined as per the manufacturer's instructions (Bioguenix/Enzo Life Science, Mumbai, India). Yolk precursor protein, vitellogenin and 11-keto testosterone (11-KT) were quantified in the male and female serum using ELISA with a vitellogenin EIA kit (Catalog No. V01003402) (Biosense, Bergen, Norway) and an EIA kit (Catalog no. 582751) (Cayman Chemical Co. Ann Arbor, MI), respectively, as per the manufacturer's instructions. The absorbance was read in an ELISA plate reader (Biotek India Pvt. Ltd, Mumbai, India).

Serum Lysozyme Activity

Serum samples were diluted with phosphate buffer (pH 7.4) to a final concentration of 0.33 mg ml^{-1}. In a cuvette, 3 ml of *Micrococcus luteus* (Bangalore Genei, India) suspension in phosphate buffer ($A_{450} = 0.5-0.7$) and 50 µl of diluted serum sample were mixed well for 15 sec, and the reading was taken in a spectrophotometer at 450 nm exactly after 60 sec of addition of serum sample. This absorbance was compared with a standard lysozyme (Bangalore Genei, India) of known activity following the same procedure as above. The activity was expressed as U min^{-1} mg^{-1} protein of serum.

Post-challenge Protection and Agglutinating Antibody Titre

A compound is said to have immunostimulatory properties if it meets the classical definition of an immunostimulant, which requires the administered compound to boost animal immune response to a level at which they pass a survival test following a challenge with a pathogenic microorganism. Thus, to determine if the lipotropic compounds used in the experiment meet the definition of immunostimulant in fish, a challenge test was carried out using the pathogenic bacteria *Aeromonas hydrophila*, which was cultured and prepared as described [25]. After 37 days of feeding, 24 fish per group were challenged with an intraperitoneal injection of *A. hydrophila*, and survival was monitored over a seven-day

period. As it was logistically difficult to watch the challenged fish 24 hrs a day for seven days, periodic observations were carried out at a minimum interval of every six hours, and more often during the mornings. At these times dead fish were noticed more often than severely morbid (about to die) fish. As survival of fish over a period of seven days was the experimental endpoint for a pathogen challenge test, every attempt was made to keep good hygiene in the experimental tub, and any dead fish were removed as soon as noticed. Severely morbid fish, when discovered, were anaesthesized with clove oil (50 µl/1) and blood was withdrawn from the caudal vein followed by overdosing with clove oil until cessation of heartbeat was observed, and the organs were harvested to provide a better histology picture than already-dead fish could provide. At the end of the challenge, surviving fish from each group were anaesthetized, and sera samples were collected for antibody titre determination by agglutination assay as per Plumb and Areechon [26]. To prevent the spread of infection, after the end of each procedure, surviving fish were overdosed with clove oil and the cessation of heart beat was monitored.

Determination of Bioaccumulation of Pesticide

For pesticide extraction, fish muscle (250 g) was ground in a high speed blender with excess anhydrous sodium sulfate (100 g) [24]. The lipid fraction was extracted using petroleum ether, passed through an anhydrous sodium sulfate column, and filtered. The elute was made up to a known volume. Acetonitrile and saturated petroleum ether were used for pesticide fractionation. Petroleum ether was then collected and evaporated to 10 mL. The concentrate (2 µL) was injected into GC (Shimazdu 14 B) using a capillary column of 1.85- m length, 4 mm internal diameter, made of glass, packed with 10% D.C. 200 (w/w) on solid support 80-mesh chromatosorb WHP, and measured with an electron capture detector (63 Ni). Nitrogen was the carrier gas (flow rate 30 mL/min). The column temperature was increased from 170–240°C at the rate of 10°C/min. The temperature of the detector and injector was 270°C. Residue quantification was done using appropriate standards.

Histology

For histopathological studies, immediately after fish dissection, liver tissue was stored in 10% neutral buffered formalin (Na_2HPO_4: 0.6 g, NaH_2PO_4: 4 g, distilled water: 100 ml and formalin: 10 ml). The samples were processed and embedded in paraffin, and after blocking and cooling, sectioning (5 µm) was done using a rotatory microtome. Mounted sections were dewaxed in xylene and dehydrated serially in alcohol after embedding in paraffin wax, cut into, and stained by Haematoxylin and Eosin (H&E) as described by Roberts [27] and examined under a light microscope (Olympus CX-31, Japan).

Statistical Analysis

The main effect among five different groups was analyzed by one way ANOVA. The comparison of any two mean values was done by Duncan's multiple range test (DMRT). The mean values for pre- and post-challenge attributes were compared by Student's t-test. The statistical analysis was performed using SPSS (version 16).

Results

Nonlethal Low Dose Endosulfan Exposure Elicits Primary Stress Response in Fish but Lipotropes Counteract It

Primary stress response is the immediate effect on endocrine hormones following extraneous stressors. Primary stress responses in fish were quantified by measuring serum markers of stress (cortisol) and endocrine disruption (vitellogenin induction) and the male sex steroid hormone 11- keto-testosterone (11-KT). The level of serum cortisol (P<0.01) was significantly increased by exposure to endosulfan, as was the vitellogenin. Exposure decreased the level of 11-KT in the animals. Dietary lecithin, betaine and choline supplementation prevented the effects of endosulfan on these hormones from being invoked and while serum levels of cortisol were lower, those of 11-KT and vitellogenin were on par with control (**Figure 1**).

Nonlethal Low Dose Endosulfan Exposure Elicits Cellular (Secondary) Stress Responses in Fish but Lipotropes Counteract Them

Metabolic stress responses in terms of blood glucose, liver glycogen and body protein content were not influenced (P>0.05) by exposure to low dose endosulfan and by lecithin, betaine and choline supplementation in exposed fish (**Table 2**). However, nonlethal exposure to low dose endosulfan significantly (P<0.01) enhanced cellular stress indicators like antioxidant enzymes (superoxide dismutase and catalase), phase II enzymes in xenobiotic metabolism (glutathione-s-transferase, GST, and methyl transferase, MT), protein such as caspase-3 involved in apoptosis and heat shock proteins in gill and liver tissue. While choline, betaine and lecithin supplementations were able to prevent the effects of endosulfan (values either comparable or lower than control) on these parameters, the effects of betaine and especially lecithin were more pronounced leading to lowered values of SOD, catalase and GST in liver and gill tissue (**Figure 2**).

Pathogen Infection Aggravates Low Dose Endosulfan Exposure-induced Secondary Stress (Hemato-immunological and Histological) Responses but Lipotropes Counteract Them

In field conditions, fishes are exposed to multiple stressors. To evaluate whether nonlethal exposure to low dose endosulfan aggravates pathogenic infectivity, fishes were exposed to a nonlethal dose of endosulfan for 37 days (pre-challenge) and subsequently injected intraperitoneally with the pathogenic bacteria *Aeromonas hydrophila* (post-challenge), and secondary stress responses were studied in fish fed with or unfed with lipotropic nutritional compounds. Successful experimental infection of *A. hydrophila* resulted in typical symptoms (**Figure 3**) such as hemorrhagia, shallow to deep necrotizing ulcers, and abdominal distension with sero-hemorrhagic fluids exuding from the vent. Dietary lipotropes potentiated the hematological profile (**Table 3**). Compared to control, RBC count was not affected by nonlethal exposure to endosulfan in pathogen-unchallenged fish. But in the same fish, lecithin supplementation was found to elevate (P<0.01) RBC count even under conditions of nonlethal endosulfan exposure. Following bacterial challenge, while the RBC count was significantly decreased (P<0.01) in endosulfan-exposed groups, there was no effect on WBC count. Choline, betaine and lecithin supplementation significantly (P<0.01) elevated the RBC and WBC count, which was even higher than control. The RBC and WBC count in post-challenge groups was higher than corresponding pre-challenge groups (P<0.05). Hemoglobin content was unaffected by treatments or bacterial challenge.

Among serum proteins (**Table 4**), only albumin was significantly affected (P<0.05) by treatments, with no effects on serum total protein and globulin (both pre- and post-challenge). Compared to control, exposure to nonlethal levels of endosulfan had no effect on serum albumin, but albumin levels during the

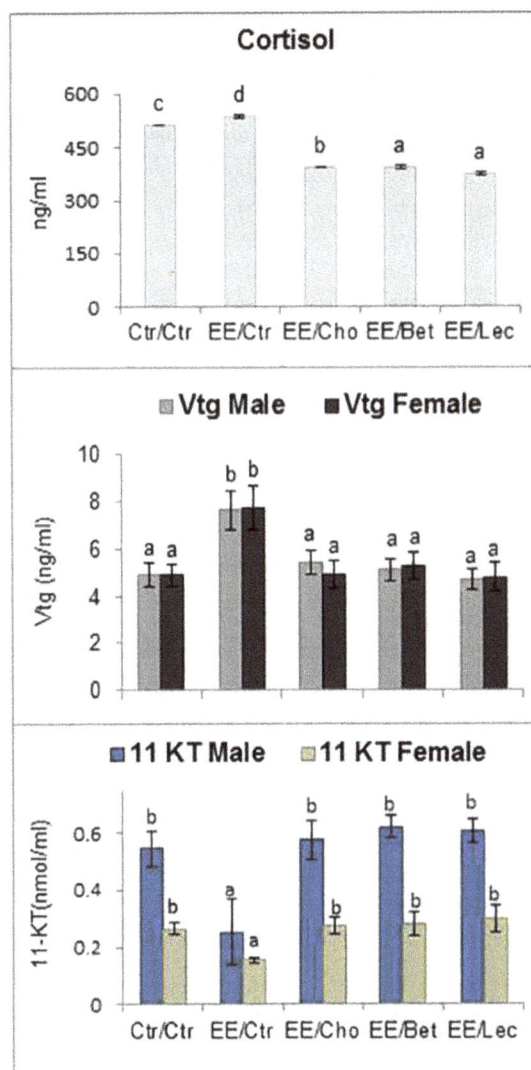

Figure 1. Primary stress response to low (nonlethal) dose endosulfan exposure in fish unfed or fed with lipotropes for 37 days: Steroid hormones in serum. Abbreviations for exposure/diet treatments of fish: Ctr/Ctr, fish group reared in normal/control (Ctr) water and fed control (Ctr) feed; EE/Ctr, low dose endosulfan-exposed (EE) and control feed (Ctr) fed group; EE/Cho, low dose endosulfan-exposed and supplemental choline (Cho) fed group; EE/Bet, low dose endosulfan-exposed and supplemental betaine (Bet) fed group; EE/Lec, low dose endosulfan-exposed and supplemental lecithin (Lec) fed group. The values reported in bar charts represent the mean±SE. Bars bearing different letters (a, b, c, d) indicate significant differences between treatment means for that parameter and for that particular sex. Probability (P) values: cortisol (P=0.001), vitellogenin in males (P=0.013), vitellogenin in females (P=0.030), 11-Ketotestosterone (KT) in males (P=0.02) and 11-KT in females (P=0.01). Number of observations (n): n=6 for cortisol and n=7 for vitellogenin and for 11-KT in males and females. Levels of 11-KT in females were lower (P<0.05) by student's t-test than for males in respective groups.

pre-challenge period were lower in betaine- and lecithin-supplemented (P<0.05) groups than in the control and endosulfan-exposed non-supplemented groups in the post-challenge state. Serum protein levels during post-challenge were lower than in corresponding pre-challenge groups (P<0.05).

Lipotropes boosted the immunological response of fish exposed to pesticide and pathogenic stressors (**Figure 4**). Most significant-

Table 2. Secondary stress response to low dose endosulfan-exposure in *L. rohita* fish fingerlings unfed or fed with lipotropes for 37 days: Effect on body composition and some metabolites.

Exposure/Diet	Control/Control	Endosulfan/Control	Endosulfan/Choline	Endosulfan/Betaine	Endosulfan/Lecithin	P-Value
Body OM[1]	86.57±0.41	85.74±0.46	84.62±1.58	86.46±0.10	86.46±0.21	0.41
Body CP[2]	64.42±3.57	58.11±0.68	61.42±6.56	59.52±1.28	60.63±4.94	0.84
Body ash[3]	13.43±0.41	15.38±1.58	13.54±0.10	13.73±0.21	14.26±0.46	0.32
Liver glycogen	0.01±0.003	0.03±0.006	0.01±0.0016	0.10±0.078	0.02±0.006	0.24
Blood glucose (mg/dL)	70.03±4.28	76.89±4.93	67.35±2.29	67.76±6.28	65.77±2.4	0.27

OM[1], Organic Matter, CP[2], Crude Protein and [3]Ash expressed as % DM.
Glycogen expressed as mg glycogen/g tissue. Data expressed as Mean ± SE (n = 6).

ly, nonlethal exposure to pesticide was found to have no effect on different nonspecific blood immune parameters such as the A: G ratio, lysozyme and NBT score during pre- or- post-challenge state. There were no differences between treatments for pre-challenge A: G ratio and lysozyme activity. None of the treatments were significantly different from control for post-challenge A: G ratios or lysozyme levels. Among the endosulfan-exposed but supplemented groups, betaine- and lecithin-fed groups exhibited higher lysozyme activity and the betaine-fed group exhibited a higher NBT score than in the control and endosulfan-exposed groups. The post-challenge bacterial agglutination titre was significantly enhanced (P<0.01) in endsosulfan-exposed groups. While the A: G ratio during post-challenge was not different from corresponding pre-challenge groups (P>0.05), post-challenge lysozyme in the control, endosulfan-exposed unsupplemented and choline-supplemented groups were lower and NBT in the choline-fed group was higher than in corresponding pre-challenge groups.

Methyl donors provided histoarchitectural protection (**Figure 5**). Control liver histology showed normal polygonal hepatocytes with distinct nuclei and normal sinusoids (Figure 5A). In *A. hydrophila*-infected fish livers, slight degenerative changes with minimal hepatocellular hypertrophy and slight cytoplasmic vacuolation were observed (Figure 5B). In fish exposed only to a nonlethal dose of endosulfan, minimal vacuolation and hepato-cellular hypertrophy were observed (Figure 5C). Exposure to a nonlethal dose of endosulfan aggravated *A. hydrophila*-induced pathological lesions in the livers of fish as evidenced by hepatocytocellular cloudy swelling/hypertrophy, more pronounced pyknotic/karyorrhectic nuclei, moderate cytoplasmic vacuolation and focal hepatocellular necrosis (Figure 5D). In the livers of fish exposed to a nonlethal dose of endosulfan and injected with *A. hydrophila* but fed with the methyl donors choline (Figure 5E), betaine (Figure 5F) or lecithin (Figure 5G), there were clear-cut signs of protection as revealed by reduced pathognomic histological lesions. Thus the protective efficacy of nutritional supplements on nonlethal low dose endosulfan-aggravated secondary stress profiles was evident.

Low dose Endosulfan-exposed Fish Accumulate Residue, Have Enhanced Disease Susceptibility and Mortality (Tertiary Stress Responses) but Lipotropes Lower/Counteract It

Nonlethal exposure of fish to endosulfan had no effect on the initial and final body weights of the fish among different treatment groups (P>0.05) (**Table 5**), however, there was a slight variation in feed efficiency (weight gained per unit feed) (P = 0.06)

(**Figure 6**). Importantly, the residue level in endosulfan-exposed but choline-fed fish was significantly reduced, while endosulfan was undetectable in other lipotrope-fed fish (P<0.01). The survival rate with nonlethal endosulfan exposure was 50 percent less (55.56% in control vs. 27.78% in endosulfan exposure). Choline supplementation prevented mortality as relative survival was similar to control, while survival in betaine- and lecithin-fed groups was 61.11 and 66.67 percent (i.e. 10% and 20% higher than control), respectively (Figure 6).

Discussion

The study, in addition to providing greater ecophysiological insights into the already demonstrated [19,28–30] synergistic effects of concurrent exposure to low dose pesticides and other stressors as a cause for species decline and reduced productivity, also prescribes practical preventive or mitigation strategies that would be useful to aquaculturists and biodiversity conservators. The integrated biological response to stressors is also a result of methylation of genes in stress-modulating LHPA circuitry (such as the neuropeptide corticotrophin-releasing factor [31], oxytocin and brain-derived neurotrophic factor [32], vitellogenin-1 [33] and the immune system [34]. This study under simulated conditions attempted to correct possible epigenetic dysregulation as noticed in other toxicities [7] through supplementation of methyl donor compounds like choline, betaine and lecithin with metabolic interrelationship. Lipotropes are used in synthesis of useful compounds such as creatine and phosphatidycholine (PC) [35] and for maintenance of epigenetic methylation. Our selection of the most appropriate compounds to protect fish from very low dose endosulfan-induced oxidative stress is also supported by recent ^{1}H-NMR-based metabolic fingerprint report [36], which showed a decrease in choline content and lipid LDL in mice exposed to low doses of endosulfan.

The major sources of methyl groups for practical diets are betaine, choline, methionine and the choline derivative lecithin, which is a nutritionally superior source of choline. Aside from containing phospholipids like PC, phosphatidyl inositol, phosphatidyl ethanolamine and phosphatidic acid, soylecithin contains oil, some sterols and B group vitamins. About 40–50 percent of total phospholipids in eukaryotic cell membranes are PC. As even low dose endosulfan is known to damage cell membranes [37], lecithin may be involved in cellular homeostasis maintenance by keeping the cell membrane intact or providing components for repair. Betaine, being a compatible osmolyte, increases the water retention of cells, replaces inorganic salts, and protects intracellular enzymes against osmotically- or temperature-induced inactivation [38]. Choline is also an important component of some plasmal-

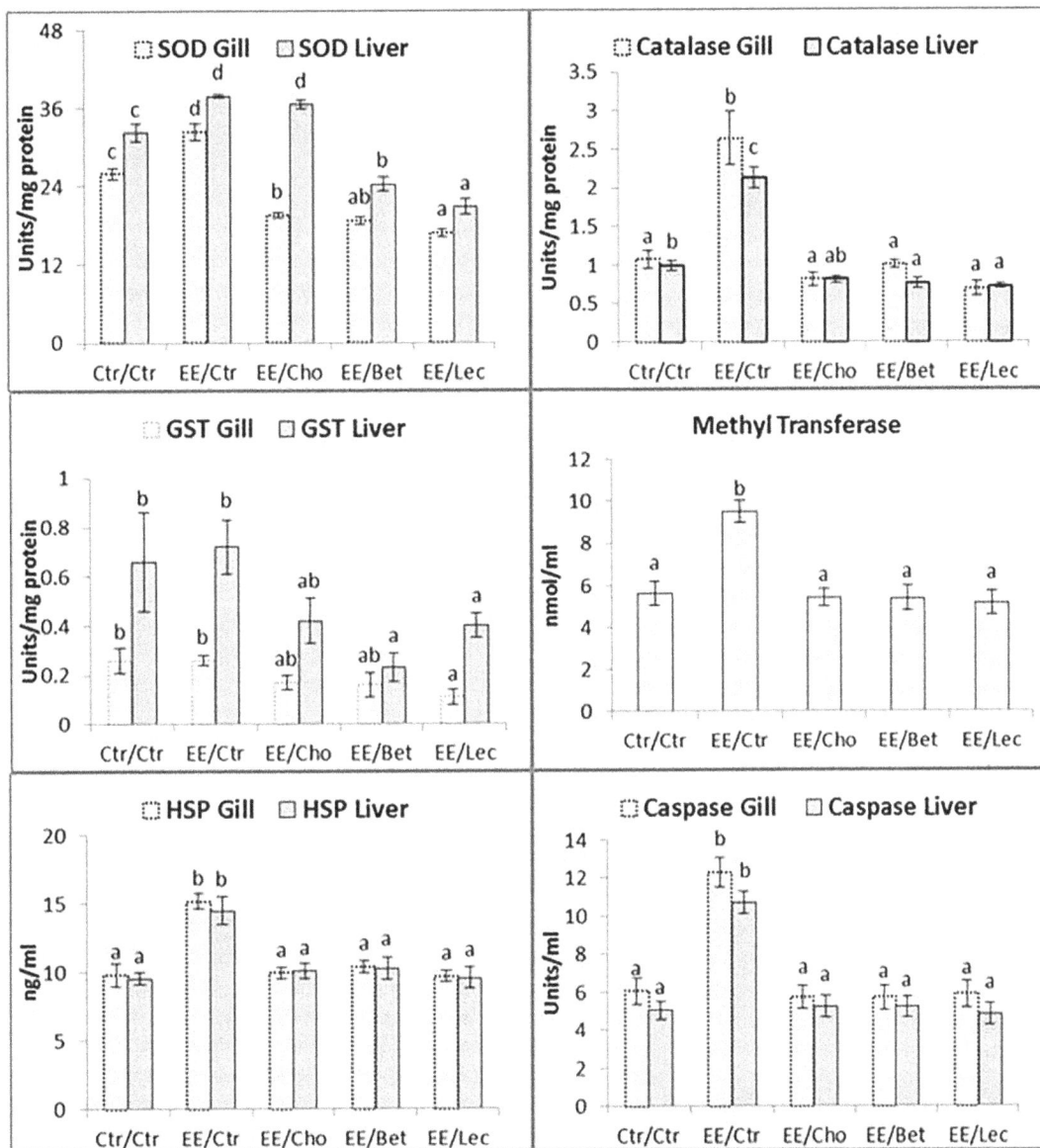

Figure 2. Secondary stress response to low dose endosulfan exposure in fish unfed or fed with lipotropes for 37 days: Cellular responses. Secondary cellular stress responses included activities/levels of: antioxidant enzymes superoxide dismutase (SOD) and catalase; phase II metabolism enzymes glutathione-s-transferase (GST) and SAM-dependent methyl transferase (MT); heat shock protein (HSP70); and caspase. While MT was measured in serum, all other attributes were quantified in the liver and gills. Abbreviations for exposure/diet treatments of fish are the same as used in Figure 1: Ctr, control; EE, endosulfan-exposed; Cho, choline; Bet, betaine and Lec, lecithin. The values reported in bar charts represent the mean±SE. Bars bearing different letters (a, b, c) indicate significant differences between treatment means for the level/activity of a marker in respective tissue or serum. Probability (P) values: SOD liver (P = 0.002), SOD gill (P = 0.005), catalase gill (P = 0.003), catalase liver (P = 0.005), GST gill (P = 0.02), GST liver (P = 0.03), serum MT (P = 0.001), HSP70 and caspase (P = 0.001). Number of observations (n): n = 6 for SOD, GST, catalase, HSP70 and caspase, and n = 7 for MT.

ogens, sphingomyelins and lecithin and acts as a source of methyl groups, via betaine, for the synthesis of various methylated metabolites.

Physiological methods and concepts can be useful in conservation biology [39], and the link between diet quantity or quality [40] and the state of fasting and feeding and concentration of dietary macronutrients like dietary protein and lipid [41] and to some extent micronutrients has been studied, but the role of many other micronutrients in facilitating detoxification is not well delineated. Lipotropes have important functions in health and disease and the literature is centered around hepatotoxicity and

carcinogenicity [42,43]. As the liver is the principal organ for detoxification of endosulfan in fish [44] and pesticides also damage cellular membranes (composed of phospholipids), we used lipotropes for potentiating the detoxification capacity of fish. This is the first paper showing mechanistic effects of lipotrope compounds in ecotoxicity.

In a quest to regain homeostasis after low dose toxicity-induced physiological perturbations, the foremost stress response is a primary response consisting of catecholamine and glucocorticoid (cortisol) hormone, which is dependent on the duration and strength of the stressor [45]. Significantly elevated cortisol level was noted in the same fish species from the same stock (as used in the study) exposed to the nonlethal endosulfan from the same batch [25] at the end of a sixty day study. However, groups fed lipotropes on a preventive basis had these values significantly lower. Endosulfan has been known to damage the endocrine system and reproductive system [46] and among the negative effects of organochlorines includes lower plasma concentrations of gonadotropin, testosterone and 11-ketotestosterone (11-KT) [47]. In this study, a high amount of vitellogenin (Vtg) and decreased 11-KT suggests that nonlethal exposure to endosulfan is estrogenic in nature [48]. It is well known that Vtg cannot normally be detected in male fish. But this study employed soybean meal and fish meal as practical protein sources, and phytoestrogens (genestein, daidzein, coumestrol and equol) [49] from soybean meal and sex steroids (estradiol and estrone) from fish meal [50] are well known and potent Vtg inducers. Reported plasma levels of <10 ng/ml for male minnows, bream, gudgeon and zebrafish and 15 ng/ml for roach fish [51] agree with those found in this study. The fish used in our experiment were immature and possibly no sexual distinctions existed in the liver (where vitellogenin is synthesized) at this stage, as is also noted for other species [52]. The lack of significant differences in Vtg induction in males and females was similar to reports on immature fish (barfin plaice, *Liopsetta pinnifasciata*) from the moderately contaminated area of Amursky Bay in the Sea of Japan [53]. Abnormal Vtg induction in male summer flounder correlates with depression of plasma testosterone and with gonadal abnormalities [54]. The 11-Keto-testosterone in male *H. fossilis* was significantly decreased in response to hexachlorocyclohexane exposure [55]. It is likely that decreased sex steroids might be indicative of a change in the biosynthetic pathway of steroid hormones in the stressor group. The intervention might have been at the step where 17α-OH progesterone is converted into testosterone and then later to 11-keto testosterone. Hence, it would result in the depression of testosterone and 11-keto testosterone levels in the serum of males. However, low-dose endosulfan- exposed but choline-, betaine- and lecithin-fed groups had normal vitellogenin and 11-KT.

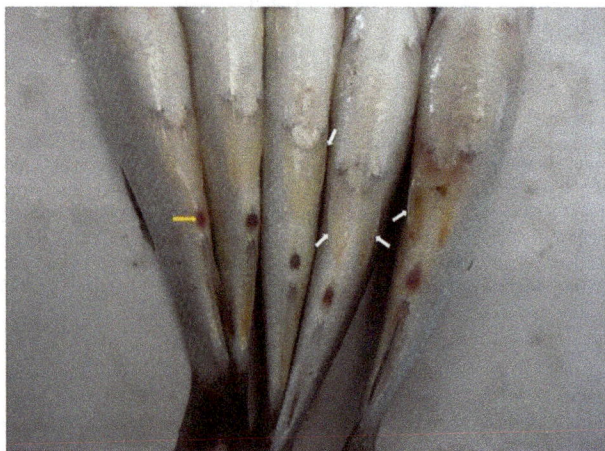

Figure 3. Fish injected with pathogenic bacteria *Aeromonas hydrophila* showing one or more typical signs of infection according to the stage of disease. Signs included hemorrhagia (large and irregular hemorrahages), shallow to deep necrotizing ulcers, and abdominal distension with sero-hemorrhagic fluids exuded from the inflamed vent. White arrows indicate edges of the ulcers. A yellow arrow indicates vent. Abdominal distension is clear in the right side fish. All fish shown are infected.

The primary hormonal response stimulates secondary responses, which are typically of short duration (up to hours) [56]; however, the stress response may persist during extended contaminant exposures [57]. We also studied secondary responses, such as changes in plasma and tissue metabolite levels, hematological features, and HSPs, which relate to physiological homeostasis [58].

Body organic matter, crude protein and ash and liver glycogen were not affected by a nonlethal level of endosulfan exposure for 37 days. In studies by Sarvanan et al. [59], glycogen content in the liver was decreased up to the 10th day, after that it gradually increased (p<0.05) from day 15 to day 25 of sublethal lindane (organochlorine) exposure. No change in body composition (or weight) in this study may indicate that the fish could satisfy their energy requirements and fish could accommodate stress due to nonlethal exposure to endosulfan. Endosulfan is a potent stimulator of the nervous system, and upon entering a fish's body it brings out several physiological alterations. Blood glucose levels increase upon exposure to organochlorine pesticides like endosulfan and lindane [25,59]. In this study, the blood glucose levels (before challenge) were unchanged by treatments, but after challenge with pathogen blood glucose in the endosulfan-exposed group was increased, but was decreased in lipotrope-fed groups. Glucose levels in lipotrope-supplemented groups were on par with control, indicating efficient utilization of glucose from the blood, also attested by cortisol levels.

It is well established that exposure to organochlorine pesticides is associated with increased cytochrome P-450 (CYP) 1A1, a source of reactive oxygen species (ROS), as measured by EROD activity [60,61]. Further, anti-oxidative enzyme (superoxide dismutase and catalase) activities, which protect cells against oxygen radical damage, and activities of phase II conjugating enzymes like GST (which helps in conjugation of products of phase I xenobiotic metabolism) also increase under pesticide stress [60,61], as was also observed in this experiment and our earlier work [22]. Similarly, the activity of S-adenosyl methionine- (SAM) dependent methyl transferases (MTases) was also increased due to endosulfan exposure. The MTases use different substrates (e.g. DNA, RNA, protein, lipid and small molecules such as arsenic) and atoms for methylation (like C or S), and are involved in small biomolecule synthesis, elimination of small molecules and xenobiotics, stabilization of DNA, RNA and proteins, cellular signaling pathways, and protein synthesis. Most of the methylation reactions (about 85 percent) and 50 percent of all methionine metabolism take place in a single organ: the liver. Increased anxiety in female catechol-O-methyltransferase (COMT) knockout animals with increased cortisol levels and a role for COMT in modulating stress-related hormonal and immune parameters in a manner that depends on chronicity of the stressor has been demonstrated [62]. The liver is the principal organ that detoxifies endosulfan [63]. Lowered activity of antioxidative enzymes and phase II enzymes in xenobiotic metabolism in the treatment groups suggests that the supplementation of dietary methyl groups helps in detoxification in the liver. Improvement in antioxidative status with nutritional supplementation [64], even in sublethal exposure with pesticide or with stress [8,41], has been reported.

Endosulfan generates reactive oxygen species (ROS) [65] and ROS-induced oxidative damage to mitochondria is a preliminary step to caspase-3 activation leading to apoptosis and necrosis [66]. The caspase-3 activity is induced by pesticides [67], and increased activity of this enzyme in liver and gill tissue of fishes exposed to nonlethal doses of endosulfan indicates its involvement in apoptosis. The HSPs affect cell survival by interacting with various components of the programmed cell death machinery,

Table 3. Secondary stress response to low dose endosulfan-exposure in *L. rohita* fish fingerlings unfed or fed with lipotropes for 37 days: Pre-challange[+] and post-challenge[++] hematological profile.

Exposure/Diet	Control/Control	Endosulfan/Control	Endosulfan/Choline	Endosulfan/Betaine	Endosulfan/Lecithin	P-Value
RBC-Pre	1.66[ab]±0.08	1.35[a]±0.13	1.66[ab]±0.10	1.54[ab]±0.13	1.96[b]±0.18	0.001
RBC-Post	2.23[b,**]±0.03	2.08[a,**]±0.02	2.33[c,**]±0.03	2.41[d,**]±0.02	2.43[d,*]±0.02	0.001
WBC-Pre	110[a]±1.15	105[a]±3.76	119[b]±1.33	129[c]±0.88	132[c]±2.03	0.001
WBC-Post	153[a,**]±3.46	148[a,**]±2.65	170[b,**]±1.67	181[c,**]±1.76	185[c,**]±0.58	0.001
Hb-pre	9.07±0.67	7.67±0.62	10.30±0.65	9.20±0.93	10.23±0.45	0.330
Hb-post	8.67±0.32	8.13±0.09	9.10±0.81	9.07±0.35	9.10±0.17	0.460

[+]Pre-challange blood samples were taken after 37 days of experiment. Subsequently, fish were challenged with the infectious bacteria, *A. hydrophila*, injected intraperitoneally and [++]post-challange samples were taken at the end of 7 days in surviving fish or just before sacrificing severely morbid fish.
Units: RBC count (x 10[6] cells/mm[3]), WBC count (x 10[3] cells/mm[3]) and Hemoglobin (Hb) g/dL.
**Indicates significant difference from pre-challenge values (P<0.01) with in a group by student's t-test.
[a, b, c, d]Means bearing different superscript letters in a row differ significantly against the P value indicated in the last column. Data expressed as Mean ± SE (n = 6).

both upstream and downstream of the mitochondrial events [68]. Under stressed conditions, increased intracellular levels of HSP play an essential role in maintaining cellular homeostasis by assisting with the correct folding of nascent and stress-accumulated misfolded proteins, preventing protein aggregation or promoting selective degradation of misfolded or denatured proteins [69]. Induction of stress proteins is highly tissue-specific in animals [70]. Significantly higher (P<0.05) induction of HSP was observed in the gill and liver of the endosulfan-exposed group. Swimming Chinook salmon exposed for 30 days to sublethal levels of bleached kraft pulp mill effluent or sodium dodecylsulphate (100 percent survival in 100 percent effluent) showed significantly higher total HSP70 expression in the liver at all concentrations compared with the control group [71]. However, supplementation of choline, betaine and lecithin prevented induction of cellular HSP70 and caspase-3 in low dose endosulfan-exposed fish in this study.

The fish showed typical clinical signs of *A. hydrophila* infection and responded to the infection through changes in hematology, serum protein profile and immunology, as also noticed by Misra et al. [72]. While low dose endosulfan exposure showed no significant (P>0.05) negative effects on the majority of haemato-immuno-logical parameters except decreased post-challenge RBC count and increased agglutination titer, significant (P<0.01) positive modulation of haemato-immunological profiles with dietary

methyl donor compounds in *A. hydrophila*-challenged and endosulfan-exposed fish, as revealed by strong nonspecific or innate immune enhancement indicators such as enhanced WBC count [73], a lower A:G ratio [74], increased respiratory burst activity of phagocytes (measured by reduction of NBT by intracellular superoxide radicals produced by leucocytes) [75], increased lysosomal activity (except with choline) and restored agglutination titer [76]. A 60-day study from our laboratory utlizing *L. rohita* from the same source and the same endosulfan stock as this experiment observed differential effects of low dose endosulfan on varying immune parameters of the fish, which were ameliorated by dietary pyridoxine [25]. Fish are able to maintain their integrity through an innate immune system based on cell phagocytosis and secretion of soluble antimicrobial molecules. The innate system is characterized by being non-specific and therefore not dependent upon previous recognition of the surface structures of the invader. Immunostimulants can increase serum lysozyme activity, due to either an increase in the number of phagocytes secreting lysozyme, or to an increase in the amount of lysozyme synthesized per cell [77]. Elevation of lysozyme following immunostimulation has been demonstrated in a number of fish species [78]. Lysozyme has been found in mucus, serum and ova of fish [79]. Lysozyme may also act as an opsonin [80] and thereby help induce lysis of bacterial cell walls and stimulate the phagocytosis of bacteria and improve the innate immune response. A higher antibody agglutination titer

Table 4. Secondary stress response to low dose endosulfan-exposure in *L. rohita* fish fingerlings unfed or fed with lipotropes for 37 days: Pre-challange[+] and post-challenge[++] serum protein profile.

Exposure/Diet	Control/Control	Endosulfan/Control	Endosulfan/Choline	Endosulfan/Betaine	Endosulfan/Lecithin	P-Value
TP-pre	8.45±0.30	8.40±1.58	8.66±1.42	9.06±0.68	9.25±1.22	0.97
TP-post	1.80*±0.08	1.54*±0.06	1.85*±0.06	1.84*±0.12	1.74*±0.12	0.38
Albumin-pre	1.77[bc]±0.05	1.91[c]±0.06	1.74[bc]±0.09	1.54[ab]±0.05	1.40[a]±0.11	0.006
Albumin-post	0.46[bc,*]±0.03	0.48[c,*]±0.01	0.34[ab,*]±0.01	0.31[a,*]±0.00	0.30[a,*]±0.08	0.02
Globulin-pre	6.68±0.26	6.49±1.63	6.93±1.48	7.52±0.73	7.85±1.33	0.91
Globulin-post	1.35*±0.10	1.06*±0.15	1.51*±0.16	1.53*±0.12	1.44*±0.09	0.24

[+]Pre-challange blood samples were taken after 37 days of experiment. Subsequently, fish were challenged with the infectious bacteria, *A. hydrophila*, injected intraperitoneally and [++]post-challange samples were taken at the end of 7 days in surviving fish or just before sacrificing severely morbid fish.
TP indicates Total Protein. Serum proteins expressed as g/dL.
*Indicates significant difference from pre-challenge value (P<0.01) with in a group by student's t-test.
[a, b, c]Means bearing different superscript letters in a row differ significantly against P value indicated in the last column. Data expressed as Mean ± SE (n = 6).

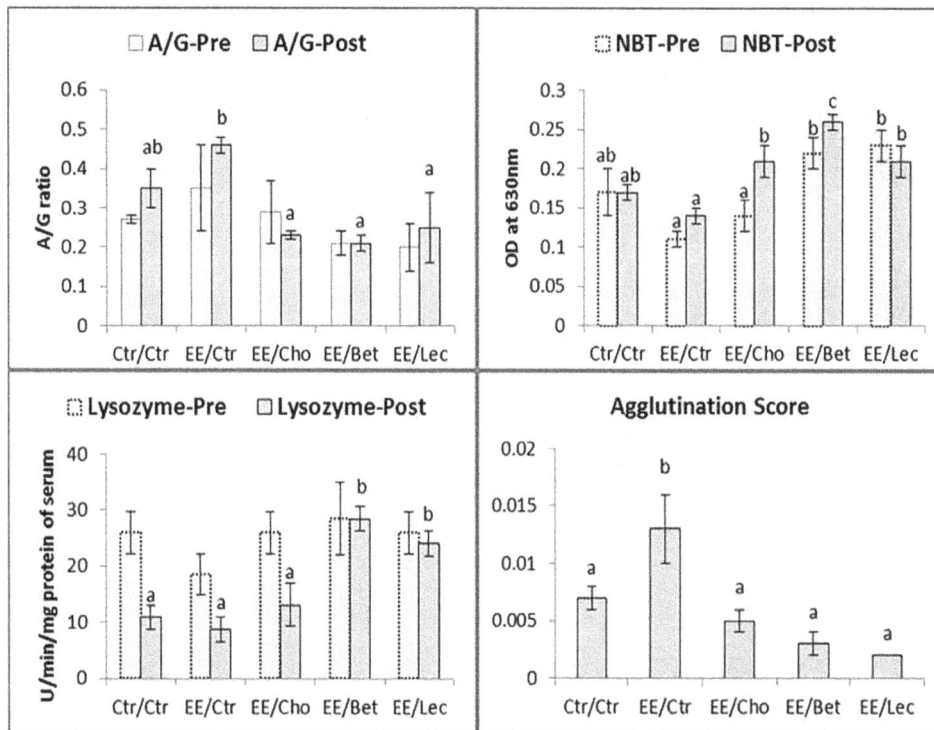

Figure 4. Secondary stress response to low dose endosulfan exposure in fish unfed or fed with lipotropes for 37 days: Pre- and post-challenge immunological responses. Pre-challenge samples were collected after 37 days of experiment. Fish were challenged with a pathogen, *A. hydrophila*, injected intraperitoneally and samples were collected at the end of 7 days post-challenge in surviving fish, or just before sacrificing severely morbid fish. Abbreviations for exposure/diet treatments of fish are the same as used in Figure 1: Ctr, control; EE, endosulfan exposed; Cho, choline; Bet, betaine; and Lec, lecithin. The values reported in bar charts represent the mean±SE. Bars bearing different letters (a, b, c) indicate significant differences between treatment means for respective attributes during a separate comparison of pre- or post-challenge data. Probability (P) values during pre-challenge: A: G ratio (P = 0.53), NBT (P = 0.013), lysozyme (P = 0.46). P values during post-challenge: A: G (P = 0.018), NBT (P = 0.002), lysozyme (P = 0.02) and agglutination score (P = 0.002). Number of observations (n): n = 6 for A: G, lysozyme and agglutination score, and n = 3 for NBT. Comparisons by t-test indicated significantly reduced (P<0.05) lysozyme post-challenge compared to pre-challenge within these respective groups: Ctr/Ctr, EE/Ctr and EE/Cho.

was noticed in fingerlings inoculated with *Aeromonas hydrophila* and exposed to atrazine compared to inoculated nonexposed fish [81]. The erythropoiesis-stimulating effect of lipotropes noticed in this study corroborates with the observations of Rehulka and Minarik [82].

Protection of the histoarchitecture of the liver for effective detoxification and survival of fish exposed to multiple stressors is important. While the livers of fish exposed to a nonlethal dose of endosulfan had hypertrophied cells and very mild vacuolation and more vacuoles were noticed in the liver of fish that were only infected with *A. hydrophila*, the concurrent effect of both nonlethal exposure and bacterial infection was clearly additive in terms of hypertrophied cells, vacuolation and degenerative and necrotic changes. Vacuolation is caused by the protoxin of *A. hydrophila* which has been shown to be inserted into cell membranes, where its activation causes pore formation through a process of oligomerisation. This alters cell membrane K+ permeability, but the endoplasmic reticulum membranes are also altered, causing considerable 'vacuolar' distension [83]. Vacuolation in cells is generally seen as an adaptive physiological response for damage limitation, but very little is known about the intracellular homeostatic mechanisms which operate to restore the *status quo*. Where damage limitation fails, cells usually die quickly [84]. Provision of a good diet containing lipotropes thus appears to be a factor in fish attempts to maintain cellular homeostasis and histoarchitecure. While similar pathological lesions were observed

in the liver of *A. hydrophila*-infected tilapia [85] and in Japanese flounder (*Paralichthys olivaceus*) caused by another infectious bacteria, *Edwardsiella tarda* [86], and independent reports also exist for similar liver pathology in fish exposed to sublethal endosulfan [87], there appears to be no literature that demonstrates the aggravating effect of low dose pesticide on liver lesions due to infectious bacteria or possible protective measures.

Choline, lecithin and betaine were found to have immunostimulatory role in fish in this study. The literature on the use of lipotropes in fish under nonlethal exposure in general and on their anti-oxidative and immunostimulatory properties in particular is scanty. We recently reported that these lipotropes promote immunobiochemical plasticity and protect fish against low-dose pesticide-induced oxidative stress during a 21-day experiment [88]. Unlike in earlier reports [88], in the present experiments the fish were challenged with pathogenic bacteria (*A. hydrophila*), a mandatory test to declare a compound to be an immunostimulant. Earlier, Klasing et al. [89] also demonstrated a modulatory effect of dietary betaine on the pathogenesis of *E. acervulina* infection in chicks and attributed the protective effect of betaine to enhancement of monocyte chemotaxis and nitrous oxide production by heterophils and macrophages. Methionine and betaine have shown immunomodulation in chicks [89,90]. Their importance in immune responses may be due to their role in DNA-methylation occurring during immune recognition and antibody production [91]. In addition, the metabolic product of betaine,

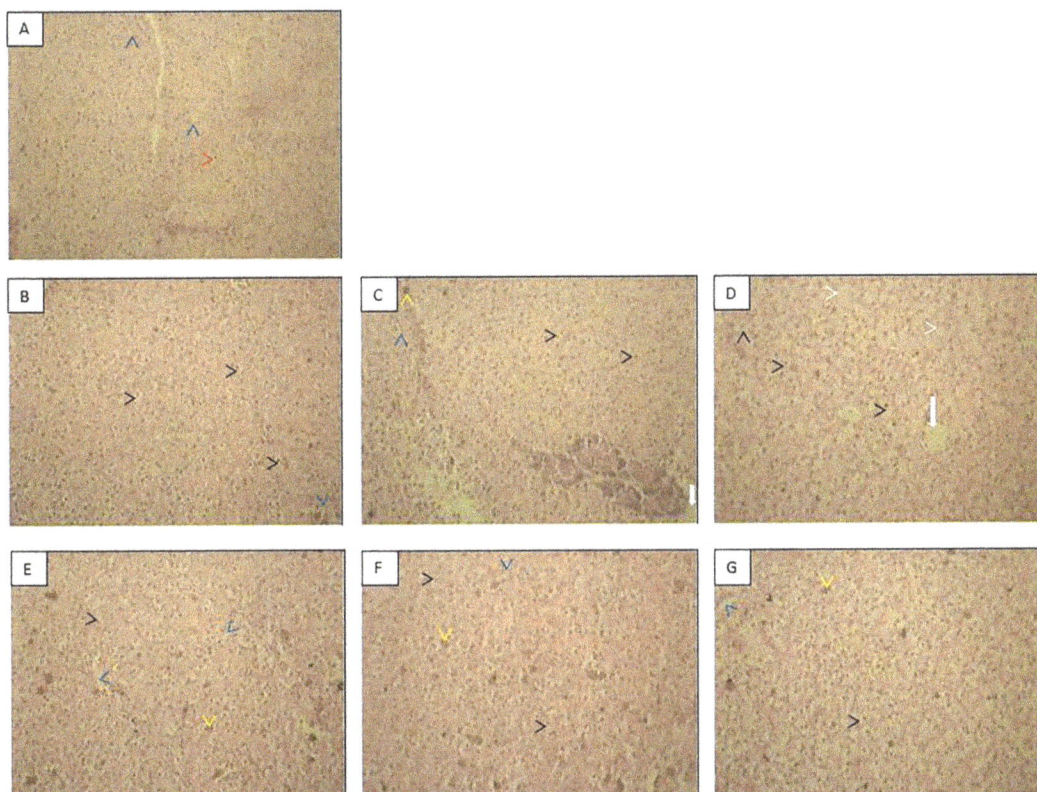

Figure 5. Secondary stress response to low dose endosulfan exposure in fish unfed or fed with lipotropes for 37 days: Histoarchitectural response. Histoarchitecture of the liver revealed a protective role for lipotropes in fish exposed to a nonlethal dose of endosulfan and intraperitoneally injected with the infectious bacteria, *A. hydrophila*. Pre-challenge samples were collected after 37 days of experiment. Post-challenge samples were collected at the end of 7 days in surviving fish or just before sacrificing severely morbid fish. Section from control fish (A), *A. hydrophila* injected fish (B), fish exposed to nonlethal dose of endosulfan (C) and also injected with *A. hydrophila* but either fed with no supplements (D), or fed with choline (E), betaine (F) or lecithin (G). Blue arrowhead: sinusoids, Red arrowhead: hepatocyte with nucleus, Black arrowhead: vacuole in hepatocyte, White arrow: central vein, White arrowhead: ghost cell without nucleus (due to karyolysis), Yellow arrowhead: focal inflammatory infiltrate. Histological changes are described in detail in the text.

dimethylglycine, has been shown to enhance both the humoral and cell-mediated immune responses in humans [92] and in mice [93], although the mechanisms of the effect are unknown. Choline is a precursor of betaine, acetylcholine (neurotransmitter) and phosphatidylcholine (PC). Mustafa et al. [94] could not find immunostimulation with 0.25 percent phosphatidylcholine in Nile Tilapia, *Oreochromis niloticus*, reared at cold temperature. With around 20 percent PC in lecithin [95], the PC in this study (i.e. 0.4 percent) was higher than that of Mustafa et al. [94], in addition to other components provided by lecithin. Phospholipids are the key players in apoptosis and immune regulation [96] and the strong responses obtained indicate this fact.

Under extended exposure, as in this study, secondary stress responses may give rise to tertiary stress responses that will be detrimental to the organism's survival and reproduction [97]. Tertiary (organismal) reponses like initial and final body weight changes and feed efficiency showed non-significant differences among different treatment groups (P>0.05), which is in agreement with earlier reports on sublethal phenol [98] and sublethal dimethoate and malathion [99].

The primary stress response (endocrine hormones), secondary cellular stress responses such as activities of anti-oxidant enzymes (SOD and catalase), phase II (conjugation of xenobiotic metabolites) enzymes involved in detoxification (GST and methyltransferase), HSP70 and caspase, immune responses, disease suscepti-

bility and survival and pathological histoarchitecture corresponded with the bioaccumulation of pesticide. The residue level in endosulfan-exposed but choline-fed fish was significantly reduced, while endosulfan was not detectable in betaine- and lecithin-fed fish (P<0.01). This is expected in a short duration study with low nonlethal dose of endosulfan as this using an otherwise balanced diet supplemented with lipotropes. An earlier report from our laboratory [8] indicated a potential role of nutritional intervention such as high protein (also means higher methionine, a methyl donor) and ascorbic acid in mitigating endosulfan toxicity through enhancing the liver's detoxification ability, leading to decreased residue accumulation in spotted murrel fish, *Channa punctatus*. Similarly, less toxicity but no mortality and less residue accumulation was reported when methionine was increased from 0.96 to 2.2 percent in the diet of rainbow trout subjected to dieldrin (an organochlorine) toxicity [100]. This is consistent with advances in the understanding of mechanisms that govern detoxification of foreign compounds, which revealed that diets (especially micronutrients) can have important impact on the efficacy of phase I and II enzymes [101]. The survival rate under a nonlethal dose decreased by 50 percent (55.56% for controls vs. 27.78% for endosulfan exposure). Choline supplementation prevented mortality as relative survival was similar to control, while survival in betaine and lecithin groups was 61.11 and 66.67 percent (i.e. 10% and 20% higher than control). Relyea and Mills [28] reported that

Table 5. Tertiary stress response to low dose endosulfan-exposure in *L. rohita* fish fingerlings unfed or fed with lipotropes for 37 days: Body weights.

Exposure/Diet	Control/Control	Endosulfan/Control	Endosulfan/Choline	Endosulfan/Betaine	Endosulfan/Lecithin	P-Value
Initial weight (g)	7.96±0.04	8.19±0.01	8.08±0.04	8.08±0.08	8.03±0.10	0.18
Final weight (g)	13.35±0.45	12.58±0.23	13.88±0.74	13.64±0.42	14.37±0.30	0.15

Data from three tubs expressed as Mean ± SE.

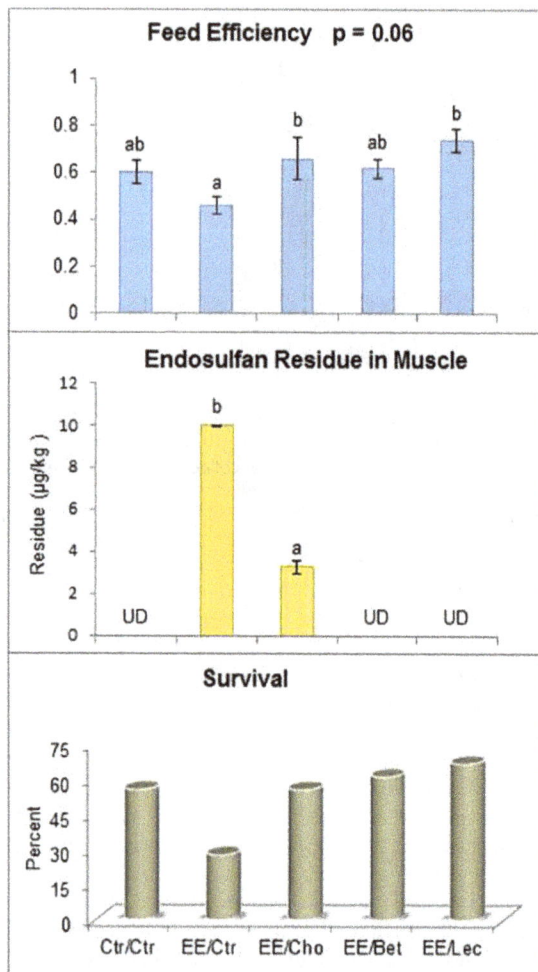

Figure 6. Tertiary stress response to low dose endosulfan exposure in fish unfed or fish fed with lipotropes for 37 days: Feed efficiency, residue accumulation and disease resistance against pathogen. Abbreviations for exposure/diet treatments of fish are the same as used in Figure 1: Ctr, control; EE, endosulfan-exposed; Cho, choline; Bet, betaine and Lec, lecithin. UD indicates undetectable. The values reported in bar charts represent the mean±SE. Bars bearing different letters (a, b) indicate significant differences between treatment means for that attribute. Probability (P) values: feed efficiency (P = 0.06), endosulfan residue in muscle (P≤0.01). Number of observations (n): feed efficiency data from 3 containers, n = 5 for endosulfan residue and n = 18 for survival. Fish were challenged with *A. hydrophila* after 37 days and post-challenge mortality/survival as a disease resistance indicator was recorded during a seven-day period.

predator-induced stress makes the pesticide carbaryl more deadly to gray treefrog tadpoles (*Hyla versicolor*) and is a reason for species decline. Increased susceptibility of baby salmon [19], and silver catfish fingerlings [22] to pathogens after exposure to harmful compounds at ecologically relevant levels and a decrease in the mean time to death after infection [23] along with the results of the current study sheds light on the synergistic effects of pesticides and pathogens on species disappearance. However, preventative measures as proposed can be used for conservation and resurrection of a declining population with a caveat that requirements of these nutrients for specific functions under specific conditions would be different.

Conclusions

The comprehensive stress response in fish studied in this experiment realistically resonates with the state of aquatic animal health worldwide. Although, we used just one organo-chlorine pesticide at a nonlethal level, the severity of implications to health of concurrent exposure to a cocktail of low dose pesticides and biotic/abiotic stressors may be far greater, as can be correlated with the noticeable enormous fish biodiversity loss worldwide. We found synergistic effects of exposures to very low concentrations of a pesticide and pathogen infection, leading to further decreased immunocompetence and enhanced mortality in fish with already compromised stress responses, which can be counteracted with dietary lipotropic compounds that enhance immunity and the detoxification efficiency of the liver. This physiological basis indicates that nonlethal pesticide toxicities, along with induced nutritional deficiency stress, may be a driver for aquatic species decline or extinction, and appropriate strategies like dietary lipotropes may be used for resurrecting the endangered and declining aquatic species.

Acknowledgments

The authors are grateful to the Director, Central Institute of Fisheries Education, Mumbai, for providing necessary facilities for carrying out the experiments. We are thankful to M/s Reliable Laboratories, Thane for their help in analyzing pesticide residue from biological samples.

Author Contributions

Conceived and designed the experiments: SBJ. Performed the experiments: NK. Analyzed the data: NK SBJ. Contributed reagents/materials/analysis tools: SG NKC MA AKP SBJ. Wrote the paper: SBJ.

References

1. Tilman D, Cassman KG, Matson PA, Naylor R, Polasky S (2002) Agricultural sustainability and intensive production practices. Nature 418: 671–677.

2. Runge CF, Senauer B, Pardey PG, Rosegrant MW (2003) Ending hunger in our lifetime: food security and globalization. Baltimore and London: The Johns Hopkins University Press for International Food Policy Research Institute. 39 p.

3. IUCN (1996) IUCN Red List of Threatened Animals. Gland, Switzerland and Cambridge, UK: IUCN. 24 p.

4. IUCN (2011) IUCN Red List of Threatened Animals Version 2011.2 available at www.iucnredlist.org. Accessed 1/14/2012.

5. Kumar A, Walker S, Molur S (2000) Prioritisation of endangered species. In: Singh S, Sastry ARK, Mehta R, Uppal V, editors. Setting biodiversity priorities for India. New Delhi: World Wide Fund for Nature, India. pp 341–425.

6. Gilliom RJ (2007) The Quality of our nation's waters: Pesticides in the nation's streams and ground water, 1992–2001, Chapter 1, US Geological Survey; at: http://pubs.usgs.gov/circ/2005/1291/pdf/circ1291_chapter1.pdf pp 4. Accessed 9/9/2012.

7. Ren X, McHale CM, Skibola CF, Smith AH, Smith MT, et al. (2011) An emerging role for epigenetic dysregulation in arsenic toxicity and carcinogenesis. Environ Health Perspect 119: 11–19.

8. Sarma K, Pal AK, Sahu NP, Ayyappan S, Baruah K (2009) Dietary high protein and vitamin C mitigates endosulfan toxicity in the spotted murrel, Channa punctatus (Bloch, 1793). Sci Total Environ 407: 3668–3673.

9. Akhtar MS, Pal AK, Sahu NP, Alexander C, Gupta SK (2012) Effects of dietary pyridoxine on growth and biochemical responses of Labeo rohita fingerlings exposed to endosulfan. Pestic Biochem Physiol 103: 23–30.

10. Kidd MT, Ferket PR, Garlick JD (1997) Nutritional osmoregulatory functions of betaine. World Poultry Sci J 53: 125–139.

11. Resolve (1994) Assessing pesticide impacts on birds: final report of the Avian Effects Dialogue Group. Washington, DC: Center for Environmental Dispute Resolution. 156pp.

12. Pushpanjali, Pal AK, Prasad SK, Prasad A, Kumar SA, et al. (2005) In ovo embryotoxicity of α-endosulfan adversely influences liver and brain metabolism and the immune system in chickens. Pestic Biochem Physiol 82: 103–114.

13. Garg UK, Pal AK, Jha GJ, Jadhao SB (2004) Pathophysiological effects of chronic toxicity with synthetic pyrethroid, organophosphate and chlorinated pesticides on bone health of broiler chicks. Toxicol Pathol 32: 364–369.

14. Girón-Pérez MI, Montes-Lopez M, Gracia-Ramirez LA, Romero-Banuelos CA, Robledo-Marenco ML, et al. (2008) Effect of sub-lethal concentrations of endosulfan on phagocytic and hematological parameters in Nile tilapia (Oreochromis niloticus). Bull Environ Contam Toxicol 80: 266–269.

15. Hayes TB, Falso P, Gallipeau S, Stice M (2010) The cause of global amphibian declines: a developmental endocrinologist's perspective. J Exp Biol 213: 921–933.

16. Henry M (2012) A common pesticide decreases foraging success and survival in honey bees. Science 336: 348–350.

17. Ezemonye LIN, Ikpesu TO (2011) Evaluation of sub-lethal effects of endosulfan on cortisol secretion, glutathione S-transferase and acetylcholinesterase activities in Clarias gariepinus. Food Chem Toxic 49: 1898–1903.

18. DaCuna RH, Rey Vázquez G, Piol MN, Guerrero NV, Maggese MC, et al. (2011) Assessment of the acute toxicity of the organochlorine pesticide endosulfan in Cichlasoma dimerus (Teleostei: Perciformes). Ecotoxicol Environ Saf 74: 1065–1073.

19. Arkoosh MR, Clemons E, Huffman P, Kagley AN, Casillas E, et al. (2001) Increased susceptibility of juvenile chinook salmon (Oncorhynchus tshawytscha) to vibriosis after exposure to chlorinated and aromatic compounds found in contaminated urban estuaries. J Aquat Anim Health 13: 257–268.

20. Chondar SL (1999) Biology of Fish and Shellfish. Howrah: SCSC Publishers (India). 514 p.

21. Muthappa NA, Jadhao SB, Gupta S (2011) Acute toxicity of endosulfan and its effect on activity of some enzymes in Labeo rohita fingerlings. Natl Acad Sci Lett-India 34: 211–218.

22. Kumar N, Prabhu PAJ, Pal AK, Remya S, Aklakur M, et al. (2011) Antioxidative and immuno-hematological status of Tilapia (Oreochromis mossambicus) during acute toxicity test of endosulfan. Pestic Biochem Physiol 99: 45–52

23. APHA-AWWA-WEF (1998) Standard methods for the estimation of water and waste water, 20th edn.In: Clesceri LS, . Greenberg AE, . Eaton AD, editors. Washington DC: American Public Health Association, American Water Works Association, Water Environment Federation.

24. AOAC International (2000) Official Methods of Analysis of AOAC International, 17th Edn, Gaithersburg, :MD, USA AOAC International.

25. Akhtar MS, Pal AK, Sahu NP, Alexander C, Gupta SK, et al. (2009) Stress mitigating and immunomodulatory effect of dietary pyridoxine in Labeo rohita (Hamilton) fingerlings. Aquac Res 41: 991–1002.

26. Plumb JA, Areechon N (1990) Effect of malathion on humoral immune response of channel catfish. Dev Comp Immunol 14: 355–358.

27. Roberts RJ (1989) Nutritional pathology of teleosts. In: Roberts RJ, editor. Fish pathology. London: Bailliere Tindall. pp 337–62.

28. Relyea RA, Mills N (2001) Predator-induced stress makes the pesticide carbaryl more deadly to gray treefrog tadpoles (Hyla versicolor). Proc Natl Acad Sci USA 98: 2491–2496.

29. Kreutz LC, Barcellos LJ, Marteninghe A, Dos SED, Zanatta R (2010) Exposure to sublethal concentration of glyphosate or atrazine-based herbicides alters the phagocytic function and increases the susceptibility of silver catfish fingerlings (Rhamdia quelen) to Aeromonas hydrophila challenge. Fish Shellfish Immunol 29: 694–697.

30. Danion M, Le Floch S, Castric J, Lamour F, Cabon J, et al. (2012) Effect of chronic exposure to pendimethalin on the susceptibility of rainbow trout, Oncorhynchus mykiss L., to viral hemorrhagic septicemia virus (VHSV). Ecotoxicol Environ Saf 79: 28–34.

31. Elliott E, Ezra NG, Regev L, Neufeld CA, Chen A (2010) Resilience to social stress coincides with functional DNA methylation of the Crf gene in adult mice. Nat Neurosci 13: 1351–1353.

32. Unternaehrer E, Luers P, Mill J, Dempster E, Meyer AH, et al. (2012) Dynamic changes in DNA methylation of stress-associated genes (OXTR, BDNF) after acute psychosocial stress. Translational Psychiatry 2: e150; DOI: 10.1038/tp.2012.77.

33. Strömqvist M, Tooke N, Brunström B (2010) DNA methylation levels in the 5' flanking region of the vitellogenin I gene in liver and brain of adult zebrafish (Danio rerio)-Sex and tissue differences and effects of 17alpha-ethinylestradiol exposure. Aquat Toxicol 98: 275–281.

34. Suarez-Alvarez B, Rodriguez RM, Fraga MF, López-Larrea C (2012) DNA methylation: a promising landscape for immune system-related diseases. Trends Genet 28: 506–514

35. National Research Council (1993) Nutrient Requirements of Fish. Washington, DC: National Academies Press.

36. Canlet C, Tremblay-Franco M, Gautier R, Molina J, et al. (2013) Specific metabolic fingerprint of a dietary exposure to a very low dose of endosulfan. J Toxicol Article ID doi: 10.1155/2013/545802.

37. Daniel CS, Agarwal S, Agarwal SS (1986) Human red blood cell membrane damage by endosulfan. Toxicol Lett 32: 113–118.

38. Yancey PH, Clar ME, Hand SC, Bowlus RD, Somero GN (1982) Living with water stress: Evalution of osmolyte system. Science 217: 1214–23.

39. Carey C (2005) How Physiological Methods and Concepts Can Be Useful in Conservation Biology. Integr Comp Biol 45: 4–11.

40. Kennedy CJ, Tierney KB (2008) Energy intake affects the biotransformation rate, scope for induction, and metabolite profile of benzo[a]pyrene in rainbow trout. Aqua Toxic 90: 172–181.

41. Morrow G, Battistini S, Zhang P, Tanguay RM (2004) Decreased lifespan in the absence of expression of the mitochondrial small heat shock protein Hsp22 in Drosophila. J Biol Chem 279: 43382–43385.

42. Rogers AE (1991) Diet and toxicity of chemicals J Nutr Biochem 2: 579–593.

43. Ghyczy M, Boros M (2001) Electrophilic methyl groups present in the diet ameliorate pathological states induced by reductive and oxidative stress: a hypothesis. Br J Nutr 85: 409–414.

44. Dutta HM, Adhikari NK, Singh PK, Munshi JS (1993) Histopathological changes induced by malathion in the liver of a freshwater catfish, Heteropneustes fossilis (Bloch). Bull Environ Contam Toxicol 51: 895–900.

45. Barton BA, Iwama GK (1991) Physiological changes in fish from stress in aquaculture with emphasis on the response and effects of corticosteroids. Ann Rev Fish Dis 1: 3–26.

46. Sang S, Petrovic S (1999) Endosulfan - A Review of its Toxicity and its Effects on the Endocrine System. WWF (World Wild Life Fund-Canada).

47. Toppari J, Larsen JC, Christiansen P, Giwercman A, Grandjean P, et al. (1996) Male reproductive health and environmental xenoestrogens. Environ Health Perspect 104: 741–776.

48. Varayoud J, Monje L, Bernhardt T, Munoz-de-Toro M, Luque EH, et al. (2008) Endosulfan modulates estrogen-dependent genes like a non-uterotrophic dose of 17β-estradiol. Repro Toxicol 26: 138–145.

49. Pelissero C, Le Menn F, Kaushick S (1991) Estrogenic effect of dietary soya bean meal on vitellogenesis in cultured Siberian sturgeon Acipenser baeri. Gen Comp Endocrinol 83: 447–457.

50. Pelissero C, Sumpter JP (1992) Steroids and "steroid-like" substances in fish diets. Aquaculture 107: 283–301.

51. Tyler CR, Vandereerden B, Jobling S, Panter G, Sumpter JP (1996) Measurement of vitellogenin, a biomarker for exposure to oestrogenic chemicals, in a wide variety of cyprinid fish. J Comp Physiol B 166: 418–426.

52. Morales MH, Osuna R, Sánchez E (1991) Vitellogenesis in Anolis pulchellus: Induction of VTG-like protein in liver explants from male and immature lizards. J Exp Zool 260: 50–58.

53. Shved N, Kumeiko V, Syasina I (2011) Enzyme-linked immunosorbent assay (ELISA) measurement of vitellogenin in plasma and liver histopathology in barfin plaice Liopsetta pinnifasciata from Amursky Bay, Sea of Japan. Fish Physiol Biochem 37: 781–799

54. Mills LJ, Gutjahr-Gobell RE, Haebler RA, Horowitz DJB, Jayaraman S, et al. (2001) Effects of estrogenic (o, p9-DDT; octylphenol) and anti-androgenic (p,p9-DDE) chemicals on indicators of endocrine status in juvenile male summer flounder (Paralichthys dentatus). Aquat Toxicol 52: 157–176.

55. Singh PB, Canario AVM (2004) Reproductive endocrine disruption in the freshwater catfish, Heteropneustes fossilis, in response to pesticide hexachlorocyclohexane exposure. Ecotoxicol Environ Saf 58: 77–83.

56. Wendelaar Bonga SE (1997) The stress response in fish. Physiol Rev 77: 591–625.

57. Bennett RO, Wolke RE (1987) The effect of sublethal endrin exposure on rainbow trout, *Salmo gairdneri* Richardson. I. Evaluation of serum cortisol concentrations and immune responsiveness. J Fish Biol 31: 375–385.

58. Iwama GK, Thomas PT, Forsyth RB, Vijayan MM (1998) Heat shock protein expression in fish. Rev Fish Biol 8: 35–56.

59. Saravanan M, Prabhu K, Ramesh M (2011) Haematological and biochemical responses of freshwater teleost fish *Cyprinus carpio* (Actinopterygii: Cypriniformes) during acute and chronic sublethal exposure to lindane. Pestic Biochem Phys 100: 206–211.

60. Lemaire B, Priede IG, Collins MA Bailey DM, Schtickzelle N, et al. (2010) Effects of organochlorines on cytochrome P450 activity and antioxidant enzymes in liver of roundnose grenadier *Coryphaenoides rupestris*. Aquat Biol 8: 161–168.

61. Salvo LM, Bainy AC, Ventura EC, Marques MR, Silva JR, et al. (2012) Assessment of the sublethal toxicity of organochlorine pesticide endosulfan in juvenile common carp (*Cyprinus carpio*). J Environ Sci Health A Tox Hazard Subst Environ Eng 47: 1652–1658.

62. Desbonnet L, Tighe O, Karayiorgou M, Gogos JA, Waddington JL, et al. (2012) Physiological and behavioural responsivity to stress and anxiogenic stimuli in COMT-deficient mice. Behav Brain Res 228: 351–358.

63. Rao DMR, Devi AP, Murty AS (1980) Relative toxicity of endosulfan its isomers, and formulated products to the freshwater fish, *Labeo rohita*. J Toxicol Environ Health 6: 825–834.

64. Jane JC, Yuan MD, Yuan MD, Chen PY, Chien SW (2002) Vitamin C and E supplements improve the impaired antioxidant status and decrease plasma lipid peroxides in hemodialysis patients. J Nutr Biochem 13: 653–663.

65. Tellez-Bañuelos MC, Santerre A, Casas-Solis J, Bravo-Cuellar A, Zaitseva G (2009) Oxidative stress in macrophages from spleen of Nile tilapia (*Oreochromis niloticus*) exposed to sublethal concentration of endosulfan. Fish Shellfish Immunol 27: 105–111.

66. Anuradha CD, Kanno S, Hirano S (2001) Oxidative damage to mitochondria is a preliminary step to caspase-3 activation in fluoride-induced apoptosis in HL-60 cells. Free Radic Biol Med 31: 367–373.

67. Jia Z, Misra HP (2007) Exposure to mixtures of endosulfan and zineb induces apoptotic and necrotic cell death in SH-SY5Y neuroblastoma cells, in vitro. J Appl Toxicol 27: 434–446.

68. Gupta SC, Sharma A, Mishra M, Mishra RK, Chowdhuri DK (2010) Heat shock proteins in toxicology: how close and how far. Life Sci 86: 377–84.

69. Morimoto RI (1993) Cells in stress: transcriptional activation of heat shock genes. Science 259: 1409–1410.

70. Sanders BM (1993) Stress proteins in aquatic organisms: an environmental perspective. Crit Rev Toxicol 23: 49–75.

71. Vijayan MM. Pereira C, Kruzynski G, Iwama GK (1998) Sublethal concentrations of contaminant induce the expression of hepatic heat shock protein 70 in two salmonids. Aquat Toxicol 40: 101–108.

72. Misra CK, Das BK, Mukherjee SC, Pattnaik P (2006) Effect of multiple injections of beta-glucan on non-specific immune response and disease resistance in *Labeo rohita* fingerlings. Fish Shellfish Immunol 20: 305–19.

73. Roberts RJ (1978) The pathophysiology and systemic pathology of teleosts. In: Roberts RJ, editor. Fish pathology. London: Bailliere Tindal. pp. 55–91

74. Wiegertjes GF, Stet RJM, Parmentier HK, Muiswinkel WB (1996) Immuno-genetics of Disease Resistance in Fish: A Comparative Approach. Dev Comp Immunol 20: 365–381.

75. Sharp GJE, Secombes CJ (1993) The role of reactive oxygen species in the killing of the bacterial fish pathogen *Aeromonas salmonicida* by rainbow trout macrophages. Fish Shellfish Immunol 3: 119–129.

76. Misra S, Sahu NP, Pal AK, Xavier B, Kumar S, et al. (2006) Pre- and post-challenge immuno-haematological changes in *Labeo rohita* juveniles fed gelatinised or non-gelatinised carbohydrate with n-3 PUFA. Fish Shellfish Immunol 21: 346–56.

77. Engstad RE, Robertsen B, Frivold E (1992) Yeast glucan induces increase in lysozyme and complement mediated haemolytic activity in Atlantic salmon blood. Fish Shell Immunol 2: 287–297.

78. Paulsen SM, Lunde H, Engstad RE, Robertsen B (2003) In vivo effects of glucan and LPS on regulation of lysozyme activity and mRNA expression in Atlantic salmon (*Salmo salar L.*). Fish Shellfish Immunol 14: 39–54.

79. Murray CK, Fletcher TC (1976) The immunohistochemical localization of lysozyme in plaice (*Pleuronectes platessa* L.) tissues. J Fish Biol 9: 32–34.

80. Ellis AE (1990) Lysozyme assays. In: Stolen JS, Fletcher TC, Anderson BS, Van Muiswinkel WB, editors. Techniques in fish immunol. Fair Haven (NJ, USA): SOS Publications. 1013 p.

81. Kreutz LC, Gil Barcellos LJ, de Faria Valle S, de Oliveira Silva T, Anziliero D, et al. (2011) Altered immunological parameters in silver catfish (*Rhamdia quelen*) exposed to sublethal concentration of an atrazinebased herbicide. 38th Congress of Brazilian Veterinary Medicine, available at http://www.sovergs.com.br/site/38conbravet/resumos/102.pdf. Accessed 9/9/2012.

82. Rehulka J, Minarik B (203) Effect of lecithin on the haematological and condition indices of the rainbow trout *Oncorhynchus mykiss* (Walbaum). Aquacult Res 34: 617–27.

83. Abrami L, Marc Fivaz, Pierre-Etienne G, Robert GP, van der GF, et al. (1998) A pore-forming toxin interacts with a GPIanchored protein and causes vacuolation of the endoplasmic reticulum. J Cell Biol 140: 525–540.

84. Henics T, Wheatley DN (1999) Cytoplasmic vacuolation, adaptation and cell death: a view on new perspectives and features. Biol Cell 91: 485–98.

85. Yardimci B, Aydin Y (2011) Pathological findings of experimental *Aeromonas hydrophila* infection in Nile tilapia (*Oreochromis niloticus*). Ankara Üniv Vet Fak Derg, 58: 47–54.

86. Miwa S, Mano N (2000) Infection with *Edwardsiella tarda* causes hypertrophy of liver cells in the Japanese flounder *Paralichthys olivaceus*. Dis Aquat Organ 42: 227–231.

87. Altinok I, Capkin E (2007) Histopathology of rainbow trout exposed to sublethal concentrations of methiocarb or endosulfan. Toxicol Pathol 35: 405–410.

88. Muthappa NA, Gupta S, Yengkokpam S, Debnath D, Kumar N, et al. (2014) Lipotropes promote immunobiochemical plasticity and protect fish against low-dose pesticide- induced oxidative stress. Cell Stress Chaperon 19: 61–81.

89. Klasing KC, Adler KL, Remus JC, Calvert CC (2002) Dietary betaine increases intraepithelial lymphocytes in the duodenum of coccidia-infected chicks and increases functional properties of phagocytes. J Nutr 132: 2274–2282.

90. Hess JB, Eckman MK, Bilgili SF (1998) Influence of betaine of broilers challenged with two levels of *Eimeria acervulina*. Poultry Sci 77: 43–43.

91. Sano IM, Sager R (1998) Detection of heavy methylation in human repetitive DNA subsets by monoclonal antibody against 5-methylcytosine. Biochim Biophys Acta 951: 157–165.

92. Graber CD, Goust JM, Glassman AD, Kendall R, Loadhot CB (1981) Immunomodulating properties of dimethylglycine in human. J Infect Dis 143: 101–105.

93. Reap EA, Lawson JW (1990) Stimulation of the immune response by dimethylglycine, a nontoxic metabolite. J Lab Clin Med 115: 481–486.

94. Mustafa A, Randolph L, Dhawale S (2011) Effect of phosphatidylcholine and beta-carotene supplementation on growth and immune response of Nile Tilapia, *Oreochromis niloticus*, in cool water. J App Aqua 23: 136–146.

95. Scholfield CR (1981) Composition of Soybean Lecithin. J Am Oil Chem Soc 58: 889–892.

96. Chaurio RA, Janko C, Muñoz LE, Frey B, Herrmann M, et al. (2009) Phospholipids: key players in apoptosis and immune regulation. Molecules 14: 4892–4894.

97. Barton BA (2002) Stress in fishes: a diversity of responses with particular reference to changes in circulating corticosteroids. Integr Comp Biol 42: 517–525.

98. Nair JR, Sherief PM (1998) Acute toxicity of phenol and long-term effects on food consumption and growth of juvenile rohu *Labeo rohita* (Ham.) under tropical conditions. Asian Fish Sci 10: 179–187.

99. Sweilum MA (2006) Effect of sublethal toxicity of some pesticides on growth parameters, haematological properties and total production of Nile tilapia (*Oreochromis niloticus L*) and water quality of ponds. Aquac Res 37: 1079–1089.

100. Mehrle PM, Mayer FL, Johnson WW (1977) Diet quality in fish toxicology: effects on acute and chronic toxicity. In: Mayer FL, . Hamelink JL, Editors, Aquatic Toxicology and Hazard Evaluation ASTM STP 634. Philadelphia, PA: American Society for Testing and Materials. Pp.269–280.

101. Chen CH (2012) Activation and Detoxification Enzymes: Functions and Implications. New York: Springer Science. 4 p.

11

Macro-Invertebrate Decline in Surface Water Polluted with Imidacloprid: A Rebuttal and Some New Analyses

Martina G. Vijver[1]*, **Paul J. van den Brink**[2,3]

1 Institute of Environmental Sciences (CML), Leiden University, Leiden, The Netherlands, **2** Alterra, Wageningen University and Research centre, Wageningen, The Netherlands, **3** Wageningen University, Wageningen University and Research centre, Wageningen, The Netherlands

Abstract

Imidacloprid, the largest selling insecticide in the world, has received particular attention from scientists, policymakers and industries due to its potential toxicity to bees and aquatic organisms. The decline of aquatic macro-invertebrates due to imidacloprid concentrations in the Dutch surface waters was hypothesised in a recent paper by Van Dijk, Van Staalduinen and Van der Sluijs (PLOS ONE, May 2013). Although we do not disagree with imidacloprid's inherent toxicity to aquatic organisms, we have fundamental concerns regarding the way the data were analysed and interpreted. Here, we demonstrate that the underlying toxicity of imidacloprid in the field situation cannot be understood except in the context of other co-occurring pesticides. Although we agree with Van Dijk and co-workers that effects of imidacloprid can emerge between 13 and 67 ng/L we use a different line of evidence. We present an alternative approach to link imidacloprid concentrations and biological data. We analysed the national set of chemical monitoring data of the year 2009 to estimate the relative contribution of imidacloprid compared to other pesticides in relation to environmental quality target and chronic ecotoxicity threshold exceedances. Moreover, we assessed the relative impact of imidacloprid on the pesticide-induced potential affected fractions of the aquatic communities. We conclude that by choosing to test a starting hypothesis using insufficient data on chemistry and biology that are difficult to link, and by ignoring potential collinear effects of other pesticides present in Dutch surface waters Van Dijk and co-workers do not provide direct evidence that reduced taxon richness and abundance of macroinvertebrates can be attributed to the presence of imidacloprid only. Using a different line of evidence we expect ecological effects of imidacloprid at some of the exposure profiles measured in 2009 in the surface waters of the Netherlands.

Editor: Christopher Joseph Salice, Texas Tech University, United States of America

Funding: These authors have no support or funding to report.

Competing Interests: For transparency reasons, we mentioning the following: PvdB's chair was cofunded between 2008 and 2011 by the following pesticide producers, Bayer, which produces imidacloprid and Syngenta. We feel that this cofunding provides no compete of interest since we don't claim that imidacloprid poses less risks or toxicity than stated in the Van Dijk et al. (2013) as in the current paper we only criticized their methodology. This current work has not been funded. Sponsors thus had no role in study design, data collection and analysis, decision to publish, or preparation of the manuscript.

* E-mail: vijver@cml.leidenuniv.nl

Introduction

The Netherlands is one of the world's foremost agricultural producers, with 2/3 of the total land mass devoted to agriculture or horticulture. Land use is highly intensive in terms of output per hectare or head of livestock [2]. To achieve such high outputs a vast range of agricultural chemicals are used, including fertilizers, veterinary drugs, pesticides and biocides. Different pesticides are used depending on the crop that is grown on the land. There are several routes that pesticides may enter surface waters. Pesticides may be washed into ditches and rivers by rainfall; surface waters can be contaminated by direct overspray or via runoff and leaching from agricultural fields [3]. Emission to surface waters (and thus pesticide residue concentrations) is dictated by many factors such as distance of the crop from the ditch and the mode of application, weather conditions and so on.

Neonicotinoids are the first new class of insecticides to be introduced in the last 50 years. The neonicotinoid imidacloprid is currently one of the most widely used insecticides in the world [4]. Recently, imidacloprid has received much negative attention: The use of certain neonicotoids has been restricted in some countries due to evidence of an unacceptably high risk of toxicity to bees, but this restriction was not in effect in the Netherlands at the time of writing this paper. On April 29, 2013, the European Union passed a two-year ban on the use of three neonicotinoids: European law restricts the use of imidacloprid, clothianidin, and thiamethoxam on flowering plants for two years unless compelling evidence comes out that proves that the use of the chemicals is environmentally safe [5]. This ban is partially, restricted to some applications in specific crops and likely covers 15% of the total use of the three neonicotinoids in the Netherlands [6]. Temporary suspensions had previously been enacted in countries such as France, Germany, Switzerland and Italy. In March 2013, a review of 200 studies on neonicotinoids was published by Mineau and Palmer [7], calling for a ban on neonicotinoid use as seed treatments because of their toxicity to birds, aquatic invertebrates, and other wildlife. The EPA – USA is now re-evaluating the safety of neonicotinoids.

Van Dijk and co-workers [1] aimed to assess the specific relationship between imidacloprid residues in Dutch surface waters, and the abundance of non-target macro-invertebrate taxa.

As also stated by the authors, finding a statistical relationship between those two datasets does not necessarily reflect causality, because there could be other factors (e.g. other pesticide residues, other local habitat factors) which drive observed patterns of abundance. We have some fundamental criticisms on the way the data were analysed and the results were interpreted, and we feel that this can be challenged by existing data. Therefore as a response to the paper of Van Dijk et al [1], and by using additional data, we explore their two key assumptions: 1) residues of pesticides other than imidacloprid, that are collinear with imidacloprid exposure either do not exist or have negligible effects on macroinvertebrate abundance and 2) that imidacloprid concentrations can be extrapolated successfully over 160 days and at a 1 km^2 spatial scale.

Materials and methods

Data collection and treatment

Data on pesticides concentrations in surface water in the Netherlands were obtained from the Dutch Pesticides Atlas. [8]. This is an online tool from which Dutch monitoring data can be collected and processed into a graphic format. Here, data of all pesticide active ingredients and metabolites (n = 634) collected in 2009 were used, since this data set is contiguous with the data used by Van Dijk et al. [1]. Only one year was selected since it can be expected that the correlations between pesticide occurrences will be year-specific, so this correlation should also be assessed for each year specifically. The 2009 dataset covered 302111 individual measurement records of which 19693 measurements exceeded the reporting limit (LOR). The measurements were performed on 4816 samples obtained from 723 different locations. The sample by pesticide dataset is characterised by missing values (90% of entries) and below LOR values (9% of all entries). This is a result of the fact that every water manager has his own suite of pesticides that is sampled, measured and evaluated. The selection of this suite of pesticides is based on the crops and land-use in their region. This selection of pesticides to be monitored improves the efficiency of the monitoring efforts of the individual water managers but yields a data set that has missing values and with many < LOR values when the data of multiple water managers are combined into one. To obtain frequency distributions of the imidacloprid concentrations, data from 2010 and 2011 have also been used.

Environmental quality standards (EQS) of all pesticides were as follows: for imidacloprid the annual average-EQS value (AA-EQS) is 0.067 µg/L (database value set 2-6-2010), and the maximum allowable concentration (MAC-EQS) is 0.2 µg/L (database value set 2-6-2010) as specified by the European Water Framework Directive. In addition, in the Netherlands, the maximum permissible concentration (MPC) of 0.013 µg/L is an important additional criterion (database value set 8-10-2008).

For all samples in which a pesticide could not be detected or quantified, the database substitutes a value of lower than the LOR. The values of reporting limits vary across samples (unique location x time). In our calculations these measurements below LOR are set as zero. We chose to do so, as choosing any other value below LOR would be arbitrary. Moreover, if not taking zero as a value, any other chosen value will result in relatively high toxicity at intensively measured surface waters even if the pesticides are not applied in that area since all measurements results in a lowest value possible of being below the LOR. These types of assumptions are inherent when working with data sets based on monitoring efforts.

Collinearity of imidacloprid concentrations with concentrations of other pesticides

Collinearity refers to a linear relationship between two explanatory variables, meaning that one can be linearly predicted from the others with a non-trivial degree of accuracy. Collinearity was determined on the data set of 2009 measurements restricted to all samples with at least one measurement above the LOR. The reduced data contained measured values for 18% of the samples, of which 8% of the total were measurements above the LOR. In order to assess the correlation between the concentrations of different pesticides we needed a sample by pesticide matrix with as little missing values as possible. From this gappy database, the largest closed data sets were extracted using Principal Component Analysis [9]. For this, measured values in the database were coded as one and missing data by zero. After running the PCA, the species-by-substance matrix was sorted, based on the scores of the substances and samples on the first principal component. Using this approach, it was possible to extract closed data sets by extracting groups of samples with the same score on the first principal component. Four data sets could be extracted that contained more than 100 samples in which the same pesticides were measured. One data set did not include imidacloprid and was not taken into account. The remaining three matrices contained 114, 108 and 191 samples, 27, 51 and 54 pesticides, with 11, 11 and 13% of the measurements above the LOR for data set 1, 2 and 3, respectively. All sampling points of data set 1 were within the provinces of Utrecht and Gelderland while all sampling points of data set 2 and 3 were located in the province of South Holland.

The log((1000 * conc) +1) transformed pesticide concentration values were analysed with Principal Component Analysis (PCA) using the Canoco5 computer programme [10], (see Zafar et al. [11] for the rationale of the transformation]. The pesticide data were centred and standardised for each pesticide. The graphical pictures based on orthogonal coordinate systems describe optimal variance in a dataset. Points that are clustered near each other have a strong correlation. PCA [9] transforms data to a new coordinate system such that the greatest variance by any projection of the data comes to lie on the first coordinate (called the first principal component), the second greatest variance on the second coordinate [12].

Calculating multi substance PAF

The potential affected fraction (PAF) is a common way to express ecotoxicological risks [13]. Following this approach, measured pesticides concentrations were translated into PAF using the species sensitivity distribution (SSD) approach. Toxicity data for each pesticide was obtained from De Zwart [14], and based on acute median effect concentrations (EC50) as derived in the laboratory (database eTox, RIVM as described in [14]). The eTox database consists mainly of data entries from the ECOTOX EPA database. The SSD for imidacloprid is given in Figure 1, and includes 41 different species from 7 different taxonomic groups. Underlying data including references are given in Table S1 of the Supplementary Information. The full database used for the multi substance PAF (msPAF) calculations contained data of 496 different pesticides with 75 different modes of action. To quantify the ecological impacts due to imidacloprid concentrations amongst all other pesticide concentrations as measured in the surface waters, the msPAF was calculated. Firstly, all concentrations of individual pesticides measured over one month per location were aggregated using the maximum measured value. Secondly individual pesticide concentrations were compared to the toxicity data resulting in the PAF. Thirdly, pesticides were grouped based on their mode of action. The PAF's of the pesticides with a similar mode-of-action were added using a concentration addition equation. In this equation, each substance concentration is divided by its effect concentration, EC_{xa}, i.e., the concentration of a that represents a standard effect expressed as EC50 for endpoint x. This gives: $Emix (Cmix) = (Ca/EC_{xa}) + (Cb/EC_{xb}) + \ldots$ In which Emix(Cmix) is the summed ratio of the mixture components at the exposure concentration of each chemical (Cx). Fourthly, the different pesticides groups with dissimilar mode-of-action were added using a response addition equation. In response addition, the toxicity of the substances in the mixture can be predicted from the product of the fractional effects of the mixture components. This gives $Emix (Cmix) = 1 - ((1 - E(Ca)) * (1 - E(Cb)) * \ldots$ In which Emix(Cmix) is the calculated effect of the mixture, Ca the exposure concentration of substance a, and E(ca) the effect of substance a at concentration Ca.

Both models for mixture toxicity are described in Hewlett and Plackett [15]. Chemicals with an unknown mode-of-action were treated according to a unique mode-of-action. As a result an msPAF value per month per monitoring location was derived. In this study we reported the maximum msPAF of the year 2009. The quantification of the relative contribution of imidacloprid on the total chemical pressure as expressed by msPAF was based on acute toxicity data as insufficient chronic toxicity data were available in the literature.

Pairwise combinations of samples taken within 1 km and 160 days

Datasets on imidacloprid concentrations and abundances of macroinvertebrates were linked to each other by Van Dijk and co-workers [1] by using the criteria ≤1 km distance and ≤ 160 days

of time difference. We performed pairwise comparisons of imidacloprid measurements to determine whether imidacloprid concentrations at sites that meet these criteria, matched successfully. Therefore, all imidacloprid measurements were extracted from the 2009 data set. All sampling sites were first ranked on their x coordinate and the difference in distance with the next sample was assessed (using Pythagoras theorem). All site combinations which yielded a difference less than 1 km were extracted. The same procedure was performed using a ranking based on the y-coordinate. The site combinations from both queries were combined. This procedure is not exhaustive since two sites that are not ranked next to each other can also be closer to 1 km from each other, but is likely to find most combinations. The imidacloprid concentrations of all samples taken at the paired sites were compared to each other when the samples were taken within 160 days. The result of the comparison were categorised into: 1) two measurements below the LOR, 2) one measurement below and one above the LOR (0% matching), 3) two measurements above the LOR, of which the number of sample pairs that matched 100% (based on one decimal) was also noted. The analysis resulted in 37 pairs of sites containing a total of 260 observations and 584 concentration measurement pairs being evaluated.

Time series of imidacloprid exposure

For each sampling site it was determined how often imidacloprid samples were analysed. For 34 sampling sites 10 or more samples were analysed, of which imidacloprid was not detected in any of the samples at 14 sites (41%), and in less than half of the samples at 28 sites (82%). The concentration dynamics of the

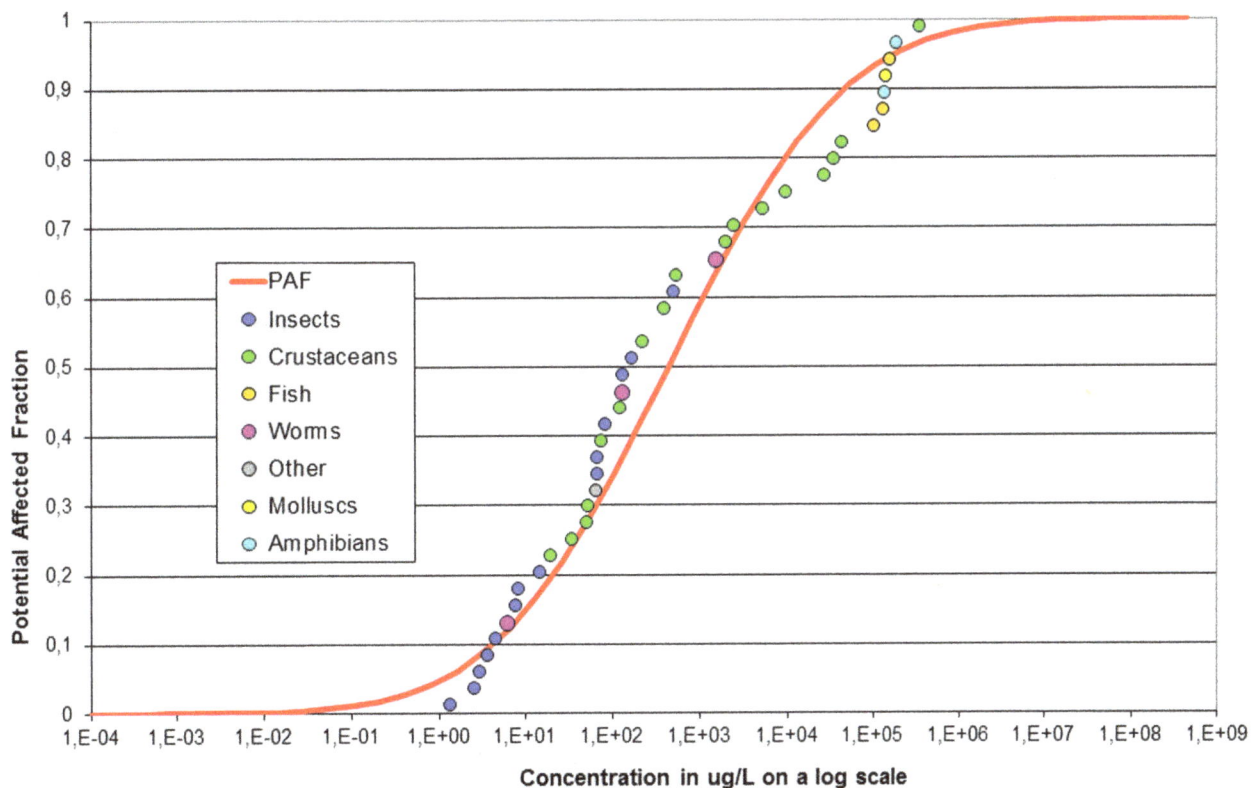

Figure 1. The Species Sensitivity Distribution of imidacloprid based on acute toxicity data. The data consist of 7 different taxomonic groups and 41 species. EPA database downloaded at Oct 23th 2013.

remaining 18% of the sites were plotted to evaluate whether chronic concentrations of imidacloprid may be expected.

Cumulative frequency of maximum imidacloprid concentrations

The measured maximum concentration of each site was compared with threshold concentrations based on the findings of Roessink et al. [16], i.e. the chronic EC10 of the mayfly species *Caenis horaria* and *Cloeon dipterum* (≈ 0.03 μg/L) and the different environmental quality standards. In order to remove the within-site sample dependency, for each sampling site the maximum imidacloprid concentration was extracted. The analysis resulted in 225 negative measurements (below the LOR) and 226 positive measurements (above the LOR).

MPC exceedances of imidacloprid compared to other pesticides

Since only for a restricted number of pesticide AA-EQS and MAC-EQS values have been set in the WFD, we used the (Dutch) MPC standard to compare exceedance frequencies between pesticides. For this comparison, both the magnitude of exceedance as well as the frequency of exceedance was incorporated. Firstly, the exceedance of the MPC of an individual pesticide concentration was derived per measuring location. Secondly, the degree of standard exceedance was weighted according to the following classes: 0 ($\leq$MPC); 1 ($>$ MPC and $\leq$ 2 x MPC), 2 ($>$ 2 x MPC and $\leq$ 5 x MPC) and 5 ($>$ 5 x MPC exceedance). Thirdly, the exceedance classes were summed over all measuring locations per year. Fourthly, pesticides were ranked on the basis of the weighted number of monitoring sites at which the MPC for the compound was exceeded, i.e. corrected for the number of monitoring sites by taking the percentage of sites that show an exceedance of the MPC. Compounds monitored at fewer than ten sites were ignored.

Results and Discussion

For many locations pesticide concentrations have been found to exceed the MPC in 2009 (see Fig. 2). Figure 2 shows that throughout the entire country more than one pesticide exceeds their respective quality standard, so this exceedance is not a common regionally problem. The maximum amount of pesticides exceeding their MPC in one sample is 35. From this it can be concluded that a single pesticide is not likely to drive solely the macro-invertebrate quality, rather all pesticides exceeding the quality standards should be considered.

Collinearity of imidacloprid concentrations with other pesticides

Figure 3A clearly shows that imidacloprid exposure is highly correlated with all chemicals placed on the right, lower side of the diagram, like carbendazim and DEET and to a lesser extend with the large group of chemicals which have a high loading with the horizontal axis, which explains almost double the amount of variance compared to the vertical axis. The results of the second data set (Fig. 3B) show that imidacloprid is placed in the centre of a large group of pesticides placed in the middle of the diagram, since it was measured only in a few samples (7% of the total). The results of the third data set shows a high occurrence of imidacloprid above the LOR (78% of all samples), with concentrations strongly collinear with those that have a high loading on the horizontal axis which explains almost triple the amount of variance of the vertical one (Fig. 3C). The results of the first and third data set show that the contribution of imidacloprid toxicity in surface waters cannot

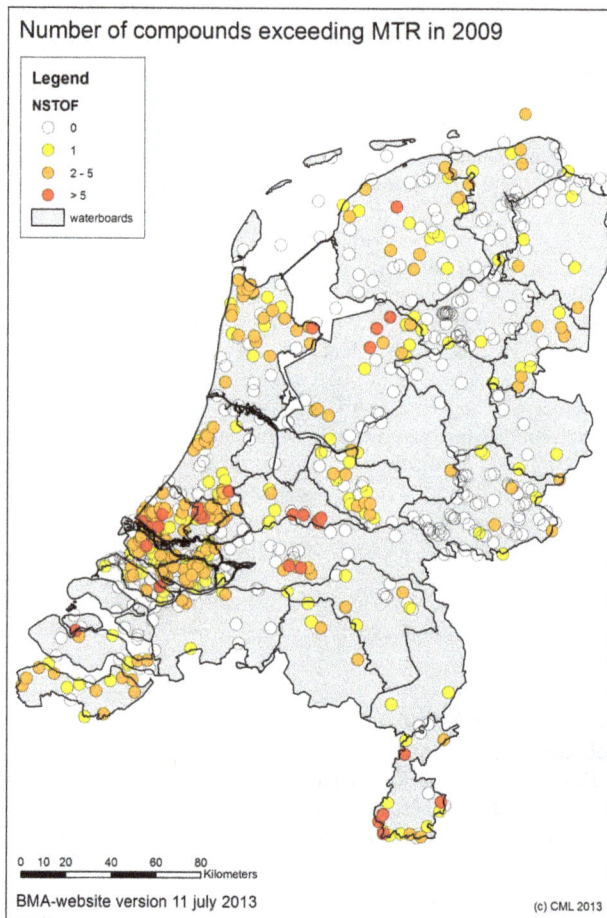

Figure 2. Number of pesticides exceeding the MPC in 2009. All monitoring locations in the Dutch surface waters with one (yellow); two till five pesticides concentrations (orange); and > five different pesticides (red) exceeding their MPC-values are depicted. Locations were measurements were performed but no exceedances were found are depicted in white.

easily be separated from the toxicity arising from other co-occurring pesticides, or indeed any other co-occurring chemical or physical stressing agent.

The correlations derived from the PCA-plots (Fig. 3) can also be explained from the fact that the active ingredient imidacloprid currently has several authorizations in 38 different products (database ctgb.nl [17], accessed 21-5-2013). The professional use ranges from the use in crops grown in glasshouses such as all different vegetables and in open systems for different bulbs of flowers, potatoes and sugarbeets. Imidacloprid is also registered for use in fruit trees including apple and pear trees. Generally, more than one pesticide is used to protect a specific crop from pest attack. Thus, depending on the land use type, imidacloprid is invariably emitted to surface waters in combination with other pesticides that are authorized to be used on those crops.

Imidacloprid contribution in the msPAF

The potentially affected fraction of the aquatic species by the measured pesticides is higher than 5% in 11 locations (reflecting 1.2 % of all monitoring sites) in the Netherlands in the year 2009. The maximum level that we determined based on the msPAF was 23% in the province of South-Holland. Imidacloprid contributed

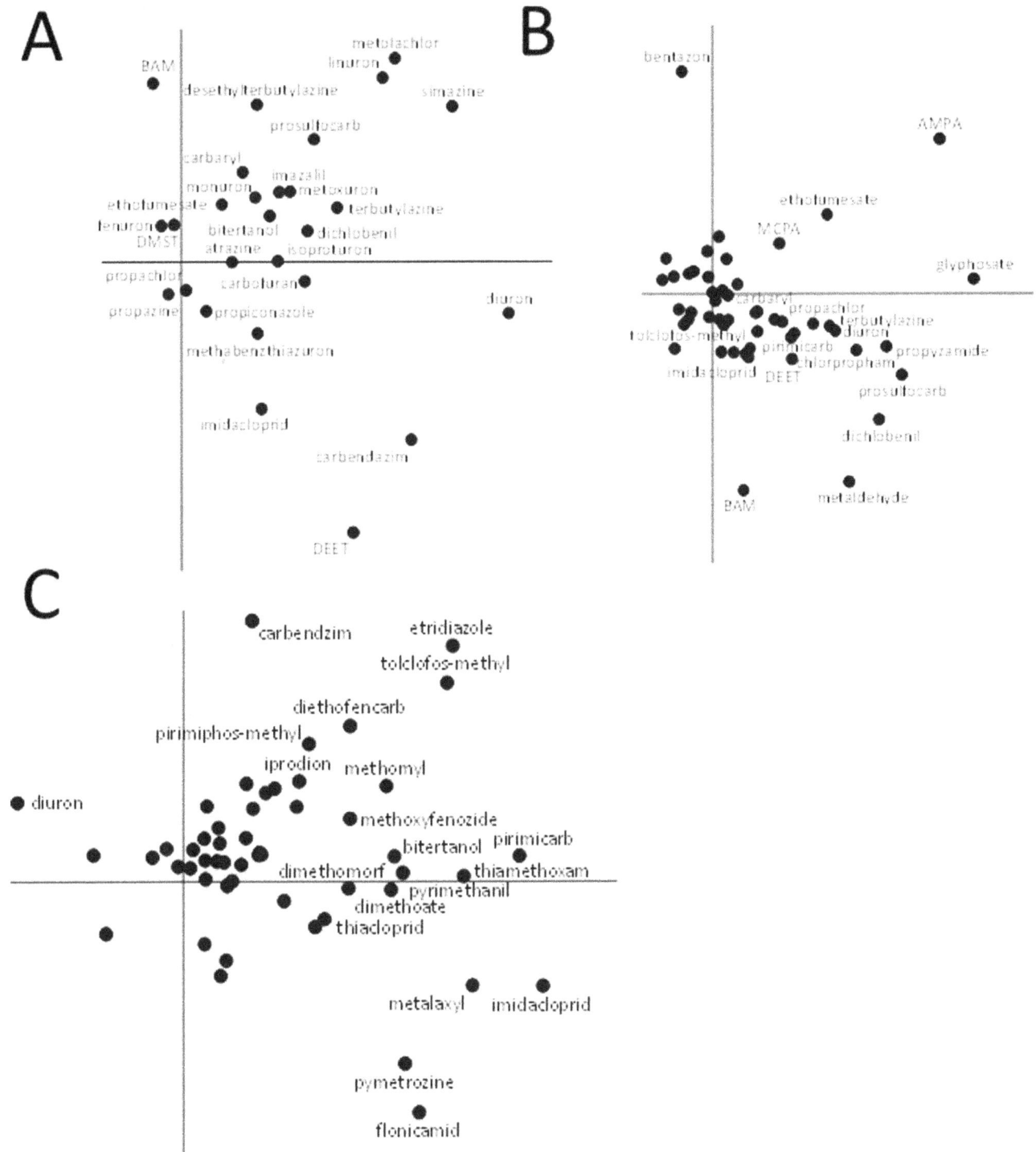

Figure 3. Results of the PCA analysis on data set 1 (A), 2 (B) and 3 (C). The PCA diagram of data set 1 displays 51% (33% on horizontal axis and 18% on vertical one) of the variation in chemical concentrations between the sites while 34% is displayed for data set 2 (21% on horizontal axis and 13% on vertical one) and 38% for data set 3 (28% on horizontal axis and 10% on vertical one).

in 8 out of 11 cases to this potential risk (Table 1). The relative contribution compared to other pesticides as measured at the same location at the same sampling time is rather modest and varied with a maximum of 21% at one location. Note that this calculation was based on acute toxicity data only, so likely is an underestimation of the potential risks that include both acute as chronic effects. From Table 1, it can be deduced that depending on

location, the contribution of specific individual active ingredients differs.

Pairwise combinations of samples

Imidacloprid measurements performed within a time window of 160 days which were taken at sites closer than 1 km from each other were compared. By this pairwise analysis we investigate if

Table 1. Contribution of imidacloprid to the msPAF at locations where msPAF > 5%.

x-coordinate	y-coordinate	Province	Total msPAF of measured pesticides (%)	Relative contribution of imidacloprid to the total msPAF of measured pesticides (%)
N 51 46 39.9	E 4 16 36.7	South Holland	22.53	0
N 52 1 29.6	E 4 30 24.7	South Holland	13.85	7.59
N 51 43 11.8	E 4 16 1.5	Zealand	12.48	0.002
N 51 52 33.5	E 4 10 26.2	South Holland	10.11	0
N 51 46 38.6	E 4 33 19.3	South Holland	9.91	0.009
N 51 45 0.4	E 4 25 46.2	South Holland	9.44	0
N 52 31 7.8	E 4 40 36.5	North Holland	9.25	0.014
N 51 57 10.2	E 4 15 8.8	South Holland	7.09	21.04
N 51 50 20	E 4 35 16.7	South Holland	6.61	0.001
N 51 21 52	E 4 2 10.1	Zealand	6.36	11.49
N 52 41 42.6	E 6 53 54.9	Drenthe	5.64	0.011

selected pairs of imidacloprid concentrations match with each other, and subsequently can be used to accurately link biological effect data and imidacloprid concentrations. Table 2 shows that in 39% of the comparisons there was no match in the presence of imidacloprid above the LOR, while only in 23% of the cases imidacloprid was present above the LOR in both samples. The remaining 38% of comparisons showed two measurements below the LOR. So when imidacloprid is found in at least one of the samples there is a large probability (62%) of not finding imidacloprid in the other site, which hampers the extrapolation of imidacloprid over a time window of 160 day and over a distance of 1 km (Table 2). We, therefore, conclude that the criteria used by Van Dijk et al. [1] to link chemical with biological observations result in a large probability (46%) of linking a site where imidacloprid was detected with a site, where the biological sample was taken, where actually no imidacloprid could be detected. The alternative, i.e. the first measurement being below the LOR and the second one above also has a relatively high probability (34%) (Table 2). Especially in a water-rich country such as the Netherlands, that has more than 350.000 km of ditch systems [18], it should be noted that sampling locations taken within 1 km, not necessarily have a hydrological connection with each other.

Imidacloprid dynamics

The concentration dynamics of imidacloprid (reflecting the concentrations of imidacloprid at the sampling locations with 10 or more samples taken in 2009 and with detection above the LOR in

at least 50% of those samples) are shown in Figure 4. In all but two (Fig. 4B and 4C) of these sampling sites the 28d, EC10 values for *C. horaria* and *C. dipterum* are exceeded for a period longer than 28 days, so at these sites chronic effects of imidacloprid exposure on mayflies can be expected. Also all standards are exceeded for some time in most of the sampling sites, with Fig. 4G showing the largest exeedence for a site near Boskoop in the province of South Holland. It should be noted that these 7 sites only constitute a small percentage (18%) of the total number of sites with 10 or more observations, so likely these exposure patterns represent the worst-cases of the exposure patterns at sites with 10 or more observations. Since we don't know whether there is a bias to measure imidacloprid more intensively at sites where exposure is expected we cannot extrapolate this to the whole population of sites.

Maximum concentrations of imidacloprid

Figure 5 shows the cumulative frequency of the all concentration measurements on the maximum level of imidacloprid for the years 2009, 2010 and 2011. The below LOR measurements are indicated at the 0.001 µg/L level and constituted 50, 53 and 55% of the maximum concentrations in 2009, 2010 and 2011, respectively. The results in Figure 4 show that peak concentrations of imidacloprid in the Dutch surface waters often exceeds the chronic effect concentrations of mayfly as determined in the chronic single species studies by Roessink et al. [16], as well as the three standards. In 2011 the MPC, 28d, EC10, AA-EQS and

Table 2. Result of the comparison of imidacloprid concentrations in samples taken in 2009 at sampling sites closer than 1 km and within 160 days.

Category	# sample pairs	% of total comparisons	% when 1st observation is above LOR	% when 1st observation is below LOR
Two below LOR	217	38		66
One below and above LOR	223	39	46	34
Two above LOR	134	23	54	
100% matching measurements	10	1.7		

LOR = analytical reporting limit.

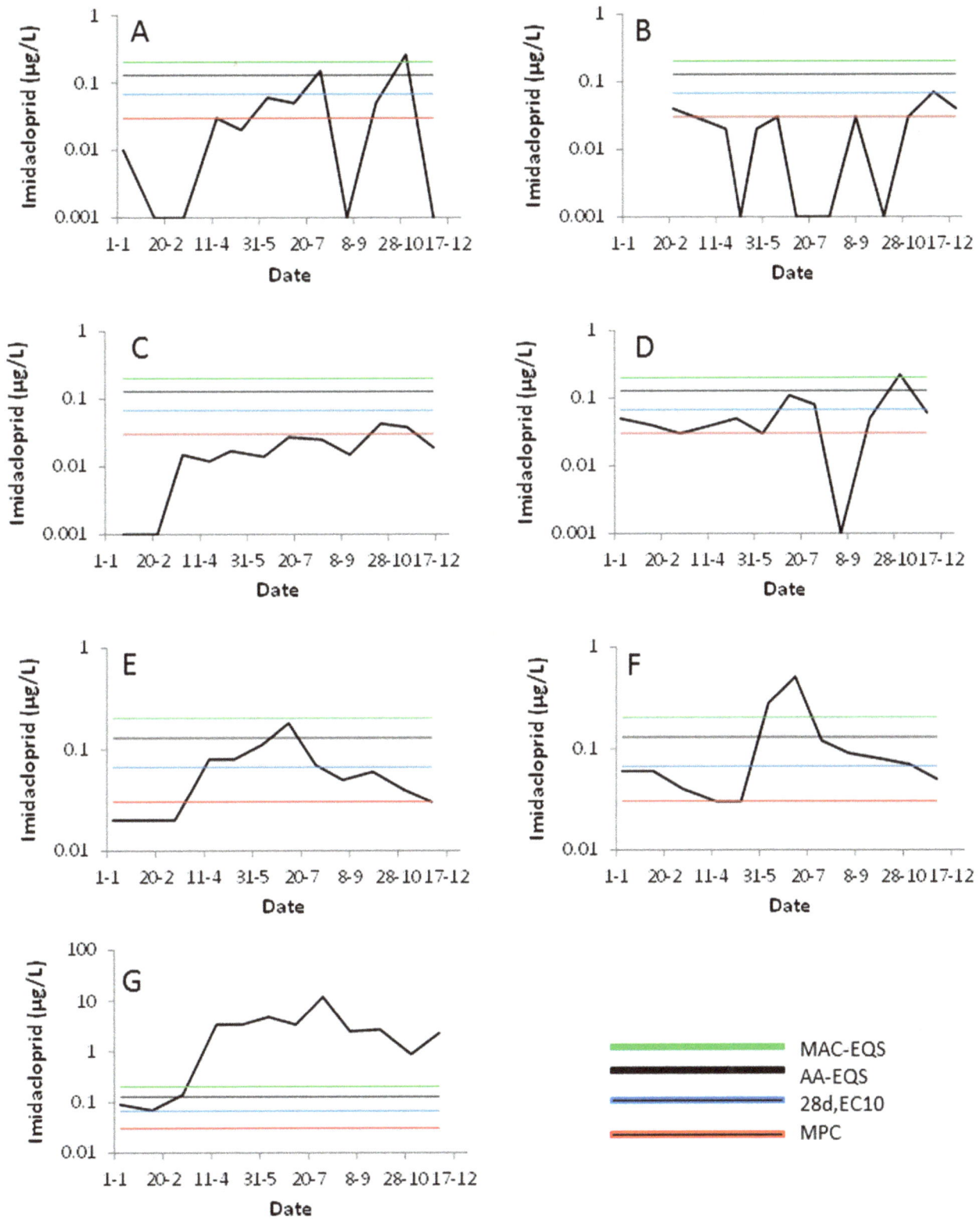

Figure 4. Concentration dynamics at the selected sampling sites (see text for procedure). The sampling sites 4A through 4G have X,Y coordinates of 108313,456412, 105888,455853, 103707,455196, 105927,453177, 170370,518957, 106781,503700 and 105079,453602, respectively. The horizontal lines denotes the MAC-EQS, the AA-EQS, the 28d, EC10 value for the mayflies C. horaria and C. dipterum (Roessink et al., 2013) and the MPC (top to bottom).

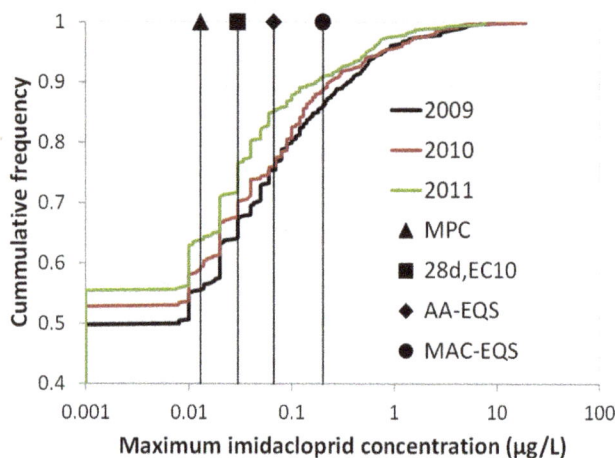

Figure 5. The cumulative frequency of the maximum imidacloprid concentrations of the sampling sites in 2009, 2010 and 2011, together with three standards and the 28d, EC10 of Cloeon dipterum and Caenis horaria.

MAC-EQS threshold values are exceeded by 36, 28, 15 and 9% of the maximum concentrations at the sampling sites, respectively. Since the Hazardous Concentration 5% based on 96h,EC10 values of 0.083 µg/L [16] corresponds more or less with the AA-EQS, acute effects of imidacloprid exposure cannot be excluded at a relatively large proportion of the sites (≈15%). The maximum concentration is of course not a good predictor for the time weighted average concentration of 28d which should ideally be compared with the chronic threshold value of 0.03 µg/L. Still, when combining the results of the time-series (Fig. 3) and the exceedance of this threshold value by the maximum concentrations (Fig. 4) chronic effects of imidacloprid on insects like mayflies may be expected at a vast proportion of sites, with 28% being the most conservative estimate and 5% being the best guess. This 5% is calculated by multiplying the 28% chance of exceeding the threshold value by the maximum concentration and 15% chance of having above LOR measurements at more than 50% of the samples taken at a particular site where imidacloprid is measured at least 10 times. The comparison of the standards with the ecotoxicological threshold value for mayflies also suggests that the

MAC-EQS and AA-EQS are not fully protective for acute and chronic effects on insect taxa, respectively.

Exceedances of environmental quality standards

As stated in the Van Dijk et al [1] paper, in 2009 imidacloprid frequently exceeds quality standards for surface waters: 111 and 62 times for the AA-EQS and the MAC-EQS respectively [8,18]. In addition to the probability of exceeding a standard, also the magnitude of exceedance is important since it is likely that at higher magnitudes the ecological effects are more severe and maybe even last longer. Table 3 shows the compounds that exceeded the MPC most frequently in 2009, ranked according to degree of exceedance.

Imidacloprid was predicted to have a relatively large impact on the ecosystems compared to other pesticides, and gained third place in the Top 10 pesticides violating the environmental quality standards in respect to frequency and magnitude of exceedance. The number of measurements is high, as is also the number of locations from which the samples are taken. This means that monitoring is quite intensive for this compound, and surely covers many different surface waters belonging to different water managers and covering the geographical distribution of the different water types in the Netherlands. Although less intensively measured – a factor 5 to 10 – Table 3 also shows that other pesticides exceed the MPC more often. Thus although imidacloprid poses a significant ecological risk to surface waters in the Netherlands, it is not the only potential cause of degradation in macroinvertebrate abundance, as many other pesticides mentioned in Table 3 also exceed the MPC frequently (and in cases by orders of magnitude) and thus undoubtedly contribute to overall stress regime. It is a common flaw in ecological studies to selectively interpret individual causal agents within stressor regimes as the sole cause of observed phenomena, leading to erroneous conclusions.

Conclusion

Imidacloprid is one of several pesticides that can be detected in surface waters draining agricultural areas at levels frequently exceeding environmental quality standards. Despite this, we show here that key assumptions made by Van Dijk et al. [1] specifically relating to imidacloprid toxicity are not supported by observational data and, therefore, their assessment is unsuitable to determine threshold levels of effects. Specifically, the validity of

Table 3. Top10 pesticides exceeding the MPC in the Netherlands in the year 2009.

Pesticides name	No. of monitoring sites	% Exceedance	No. of measurements	% Exceedance
Captan	38	47	194	13
desethyl-terbuthylazin	63	37	299	10
Imidacloprid	451	44	2133	28
Triflumuron	24	21	142	4
Dicofol	24	17	142	3
Omethoaat	31	16	169	3
Foraat	51	14	313	2
Captafol	15	27	29	14
Fipronil	69	12	230	7
Pyraclostrobin	66	17	341	7

No. = number. The ranking of pesticides is based on frequency and magnitude of exceedances.

two assumptions: 1) that imidacloprid levels are not correlated with toxic levels of other pesticides residues and 2) that chemical exposure data can be extrapolated over a 1 km distance and 160 day time window are here shown to be highly questionable. The ecological status of field sites can be attributed to a complex suite of stressors resulting from a range of anthropogenic practices in the highly managed landscape of the Netherlands, of which pesticides are just one factor, and imidacloprid only one of many pesticides being applied, albeit an important one in terms of ecological risks. We therefore propose that any risk assessment should base the ecological threshold values not solely on field observations but also largely rely on the results of controlled experiments, since these types of experiments allow a full control of separating the imidacloprid stress from other stressors.

Supporting Information

Table S1 Acute toxicity values of imidacloprid (source eTox database, EPA database downloaded Oct 23th

2013). Legend: Species selected for the toxicity test were given with their scientific name and with their species group. Toxicity data were given as log10 effect concentrations at which 50% of the organisms showed adverse effects. The scientific papers from which those data are collected are given.

Acknowledgments

The authors thank Donald Baird for his critical comments and language suggestions. We thank Dick de Zwart for providing the eTox database. All pesticides measurements compared to the different EU and MPC quality standards can be found and freely downloaded at www. bestrijdingsmiddelenatlas.nl [8].

Author Contributions

Conceived and designed the experiments: MGV PJB. Performed the experiments: MGV PJB. Analyzed the data: MGV PJB. Contributed reagents/materials/analysis tools: MGV PJB. Wrote the paper: MGV PJB.

References

1. Van Dijk TC, Van Staalduinen MA, Van der Sluijs JP (2013) Macro-invertebrate decline in surface water polluted with imidacloprid. PLOS ONE 8 (5) e62374.
2. Vijver MG, De Snoo GR (2012) Overview of the state-of-art of Dutch surface waters in the Netherlands considering pesticides. (chapter (9) In: The impact of pesticides, M. Jokanovic (ed.) AcedemyPublish.org, WY, USA. ISBN: 978-0-9835850-9-1.
3. Vijver MG, Van 't Zelfde M, Tamis WLM, Musters CJM, De Snoo GR (2008) Spatial and Temporal Analysis of Pesticides Concentrations in Surface Water: Pesticides Atlas. J Environ Sci Health Part B 43: 665–674.
4. Yamamoto I (1999) "Nicotine to Nicotinoids: 1962 to 1997". In Yamamoto, Izuru; Casida John. Nicotinoid Insecticides and the Nicotinic Acetylcholine Receptor. Tokyo: Springer-Verlag. pp. 3–27 ISBN: 443170213X.
5. McDonald-Gibson C (29 April 2013). *The Independent.* Retrieved 1 May 2013.
6. Van Vliet J, Vlaar LNC, Leendertse PC (2013) Toepassingen, gebruik en verbod van drie neonicotinoïden in de Nederlandse land en tuinbouw. CLM 825- 2013. Available: www.clm.nl. Accessed 2013 May 5.
7. Mineau P, Palmer C (2013) The impact of the nation's most widely used insecticides on birds. Neonicotinoid Insecticides and Birds. American Bird Conservancy. Available: http://www.abcbirds.org/abcprograms/policy/toxins/neonic_final.pdf.
8. Dutch pesticides atlas website. Available: http://www.bestrijdingsmiddelenatlas. nl, version 2.0. Institute of Environmental Sciences (CML) at Leiden University and Waterdienst of the Dutch Ministry of Infrastructure and Environment. Accessed 2013 Oct 23.
9. Jolliffe IT (2002) Principal Component Analysis, Series: Springer Series in Statistics, 2nd ed., Springer, NY. ISBN 978-0-387-95442-4.
10. Ter Braak CJF, Šmilauer P (2012) Canoco reference manual and user's guide: software for ordination, version 5.0. Microcomputer Power, Ithaca, USA, 496 pp.
11. Zafar MI, Belgers JDM, Van Wijngaarden RPA, Matser A, Van den Brink PJ (2012) Ecological impacts of time-variable exposure regimes to the fungicide azoxystrobin on freshwater communities in outdoor microcosms. Ecotoxicol 21:1024–1038.
12. Van Wijngaarden RPA, Van den Brink PJ, Oude Voshaar JH, Leeuwangh P (1995) Ordination techniques for analyzing response of biological communities to toxic stress in experimental ecosystems. Ecotoxicol 4: 61–77.
13. Posthuma L, Suter GW II, Traas TP (eds) (2002) Species Sensitivity Distributions in Ecotoxicology. Lewis Publishers, Boca Raton, FL, USA.
14. De Zwart D (2005) Ecological effects of pesticide use in the Netherlands: Modeled and observed effects in the field ditch. Integrated Environmental Assessment and Management 1:123–134.
15. Hewlett PS, Plackett RL (1959) A unified theory for quantal responses to mixtures of drugs: non-interactive action. Biometrics 15:591–610.
16. Roessink I, Merga LB, Zweers HJ, Van den Brink PJ (2013) The neonicotenoid imidacloprid shows high chronic toxicity to mayfly nymphs. Environ Toxicol Chem 32: 1096 – 1100.
17. Statistics Netherlands. Available: http://www.statline.cbs.nl. Accessed 2013 May 21.
18. De Snoo GR, Vijver MG (eds) (2012) Bestrijdingsmiddelen en waterkwaliteit. Universiteit Leiden, 180 pp., ISBN: 978-90-5191-170-1.

12

Neonicotinoid Binding, Toxicity and Expression of Nicotinic Acetylcholine Receptor Subunits in the Aphid *Acyrthosiphon pisum*

Emiliane Taillebois[1], Abdelhamid Beloula[1], Sophie Quinchard[1], Stéphanie Jaubert-Possamai[2], Antoine Daguin[3], Denis Servent[4], Denis Tagu[2], Steeve H. Thany[1*◑], Hélène Tricoire-Leignel[1◑]

1 Laboratoire Récepteurs et Canaux Ioniques Membranaires (RCIM), UPRES EA 2647 USC INRA 1330, SFR QUASAV 4207, Université d'Angers, Angers, France, 2 Institut de Génétique Environnement et Protection des Plantes (IGEPP), Institut National de la Recherche Agronomique (INRA), UMR 1349, Le Rheu, France, 3 Groupement Interprofessionnel de Recherche sur les Produits Agropharmaceutiques (GIRPA), Angers, France, 4 Institut de Biologie et Technologie (iBiTecS), Service d'Ingénierie Moléculaire des Protéines (SIMOPRO), Commisariat à l'Energie Atomique (CEA), Gif-sur-Yvette, France

Abstract

Neonicotinoid insecticides act on nicotinic acetylcholine receptor and are particularly effective against sucking pests. They are widely used in crops protection to fight against aphids, which cause severe damage. In the present study we evaluated the susceptibility of the pea aphid *Acyrthosiphon pisum* to the commonly used neonicotinoid insecticides imidacloprid (IMI), thiamethoxam (TMX) and clothianidin (CLT). Binding studies on aphid membrane preparations revealed the existence of high and low-affinity binding sites for [^{3}H]-IMI (Kd of 0.16 ± 0.04 nM and 41.7 ± 5.9 nM) and for the nicotinic antagonist [^{125}I]-α-bungarotoxin (Kd of 0.008 ± 0.002 nM and 1.135 ± 0.213 nM). Competitive binding experiments demonstrated that TMX displayed a higher affinity than IMI for [^{125}I]-α-bungarotoxin binding sites while CLT affinity was similar for both [^{125}I]-α-bungarotoxin and [^{3}H]-IMI binding sites. Interestingly, toxicological studies revealed that at 48 h, IMI (LC$_{50}$ = 0.038 μg/ml) and TMX (LC$_{50}$ = 0.034 μg/ml) were more toxic than CLT (LC$_{50}$ = 0.118 μg/ml). The effect of TMX could be associated to its metabolite CLT as demonstrated by HPLC/MS analysis. In addition, we found that aphid larvae treated either with IMI, TMX or CLT showed a strong variation of nAChR subunit expression. Using semi-quantitative PCR experiments, we detected for all insecticides an increase of Apisumα10 and Apisumβ1 expressions levels, whereas Apisumβ2 expression decreased. Moreover, some other receptor subunits seemed to be differently regulated according to the insecticide used. Finally, we also demonstrated that nAChR subunit expression differed during pea aphid development. Altogether these results highlight species specificity that should be taken into account in pest management strategies.

Editor: Guy Smagghe, Ghent University, Belgium

Funding: This work was supported by a specific grant from the French Institute for Agricultural Research (INRA), the Department 'Santé des Plantes et Environnement (SPE)': USC RCIM/UMR BIO3P. E. Taillebois acknowledges the French 'Ministère de la Recherche et de l'Enseignement Supérieur' for financial support in the framework of Ph.D grant. The funders had no role in study design, data collection and analysis, decision to publish, or preparation of the manuscript.

Competing Interests: The authors have declared that no competing interests exist.

* E-mail: steeve.thany@univ-angers.fr

◑ These authors contributed equally to this work.

Introduction

Neonicotinoid insecticides include several compounds such as imidacloprid (IMI), clothanidin (CLT) and thiametoxam (TMX). They are efficient agonists of insect neuronal nicotinic acetylcholine receptors (nAChRs) which are pentameric receptors formed by identical (homomeric) or different (heteromeric) subunits [1,2,3]. In the context of a stronger legislation on insecticide use to limit environmental and health concerns, approaches to describe and understand the cellular and molecular mechanisms involved in insecticide resistance are needed. Insects represent a very diverse group of animals and most Orders diverged approximately 300 million years ago [4]. As a consequence, adaptive mechanisms that confer insecticide resistance can vary from one order/species to another. Among insect pests that cause damage to agriculture, aphids (Hemiptera) have a particular biology: they feed from phloem sap (not by chewing plant tissues),

and thus transmit plant viruses very efficiently [5]. Their pest status is also attributable to their peculiar reproductive mode [5]. Asexual reproduction of aphids by parthenogenesis (during spring and summer) leads to extremely rapid population growth [6]. Several insecticides, such as neonicotinoids, are used as seed treatment to limit the impact of sucking-pest like aphids, *Sitobion avenae*, *Aphis craccivora* and *Myzus persicae* [7,8].

Bioassay studies have revealed that neonicotinicotinoid susceptibility varies between insect species [8,9,10]. In Hemiptera, IMI showed susceptibility differences with LC$_{95}$ values between 0.32 and 40 mg.L^{-1} [9]. Moreover, studies performed with the cotton aphid *Aphis gossipii* demonstrated that one IMI resistant-strain was still susceptible toward TMX and CLT suggesting no cross resistance [10] whereas a cross resistance was found in *Myzus persicae*, with resistance factors of 11, 18 and 100 for IMI, TMX and CLT, respectively [8]. This discrepancy suggested that some aphid species could carry particular resistance mechanisms. Similar differences could be found using competitive binding studies. In *Aphis craccivora* it was demonstrated that TMX was a non-competitive inhibitor for [3H]-IMI suggesting that it binds to a different site or in a different mode than IMI while in *Myzus persicae*, it was found that the resistant strain with the R81T mutation on the Mpβ1 subunit developed cross-resistance against IMI and TMX demonstrating that they interact with the same site [3]. In addition, saturation studies suggest that IMI binds to high- and low-affinity binding sites in the aphids *M. persicae* and *A. craccivora* [11,12]. Similar high and low affinity binding sites were also identified in *M. persicae* for the nicotinic antagonist, α-Bungarotoxin (α-Bgt) [13]. Altogether these data demonstrate that aphid species can carry different sensitivities against neonicotinoid insecticides.

In this study, we evaluated the binding properties and toxicological effects of IMI, TMX and CLT on the pea aphid and showed that IMI and TMX bind two different nAChR populations, both binding CLT, although the toxicity of CLT is lower than toxicities of IMI and TMX. In addition, using the full sequenced genome of *A. pisum* [14], we demonstrated that the eleven *A. pisum* nAChR subunits previously identified [15] are differentially regulated during aphid development and after neonicotinoid intoxication.

Materials and Methods

Insects

The pea aphid (*A.pisum*) sequenced strain LSR1 (corresponding to the reference genome) was generously provided by INRA-Rennes IGEPP. Unwinged parthenogenetic females were reared on faba bean (*Vicia fabae*) plants in a 16L: 8D photoperiod at constant temperature of 22°C in a climate chamber. Under these conditions, aphids reproduce by viviparous parthenogenesis, as clonal female: new born larvae become adults after four molts. Adults and larvae at each of the 4 stages were collected. Larval stages were determined by identifying the number of antennal segments.

Insecticides

IMI, TMX and CLT were purchased from Sigma-Aldrich (Saint-Louis, USA). Insecticides were dissolved in DMSO to give final concentrations of 50 mg.ml^{-1}. For binding experiments, insecticides were dissolved in the corresponding buffer (PBS buffer or Tris-HCl buffer for [125I]-α-Bgt and [3H]-IMI experiments respectively). For intoxication experiments, insecticides were dissolved in artificial diet at a final concentration of 0.2% DMSO for 100µg/ml and 2% DMSO for 1000 µg/ml. These concentrations of DMSO were used as controls in insecticidal assays.

Binding assays

Membrane preparations were isolated from frozen aphids according to the Wiesner and Kayser protocol [12]. Whole aphids were homogenized with a pestle motor in 4°C dissociation medium at pH 7.0. The dissociation medium contained: 20 mM sodium phosphate, 150 mM sodium chloride, 1 mM EDTA, 0.1 mM phenylmethanesulfonyl fluoride (dissolved in acetone), and 2 µg each of pepstatin, chymostatin, and leupeptin (dissolved in methanol, DMSO, and water, respectively). After homogenization, samples were centrifuged 10 min at 1000 g and supernatant was collected and ultracentrifuged 30 min at 4°C and 43000 g. The precipitate was washed with cold dissociation medium and then ultracentrifuged. The final pellet was resuspended in 3 ml of dissociation medium. Total protein was

quantified by spectrofluorometry at 750 nm according to the Lowry colorimetric method (DC protein assay, Biorad, France) with a range of BSA as a standard. Membrane preparations were conserved at −80°C until use.

α-Bgt binding experiments were performed using 40 µg of aphid total membrane protein in a total volume of 300 µl of PBS Buffer (Na2HPO4, NaH2PO4, NaCl, pH = 7.2) + 0.1% of bovine serum albumin and [125I]-α-Bgt (2200 Ci/mmol, PerkinElmer, USA) as radiolabeled ligand for total binding measurement. For non-specific binding determination, 1 µM of cobratoxin was added prior to membrane incubation. In saturation assays, the concentration of [125I]-α-Bgt varied from 18 nM to 0.9 pM to obtain a complete saturation binding curve. Competitive assays were performed with IMI, TMX and CLT. For this purpose, membranes were incubated with various concentrations of unlabeled competitor and [125I]-α-Bgt at 0.08 nM for CLT and TMX and 0.6 nM for IMI, respectively. Incubations were performed at room temperature during 4 hours and terminated by rapid vacuum filtration using GF/C Glass microfiber filters presoaked in polyethyleneiminine 0.5%. Filters were rapidly washed (< 20 s) twice with 5 ml of cold PBS Buffer at 0.01 M and transferred in tubes for immediate counting on a γ-counter.

IMI binding experiments were performed using 200 µg of aphid total membrane protein in a final volume of 300 µl of Tris-HCl Buffer (10 mM, pH = 7.4) and [3H]-IMI (ARC, 40 Ci/mmol) as radiotracer for total binding measurement. For non-specific binding determination, 0.3 mM of unlabeled IMI was added prior to membrane incubation. In saturation assays, the concentration of [3H]-IMI varied from 500 nM to 0.5 pM to obtain a complete binding curve. Competitive assays were achieved for CLT, TMX, IMI and α-Bgt using 25 nM of [3H]-IMI. Incubations were performed at room temperature during 4 hours and terminated by rapid vacuum filtration using GF/C Glass microfiber filters presoaked in 0.5% polyethyleneiminine. Filters were rapidly washed (< 20s) with cold Tris-HCl Buffer and dried for 1 hour before incubation in 5 ml of scintillation liquid (PerkinElmer, USA) and counting.

Insecticidal assays

The susceptibility of *A. pisum* to IMI, TMX and CLT was determined using an artificial diet bioassay according to Sadeghi *et al.*[16]. In brief, adults were put on a feeding apparatus (day D-1) containing 200 µl of artificial diet. Then, the first-instar nymphs were transferred (day D0) to freshly prepared diet with insecticide added (treatment series) or DMSO added (control series). For each insecticide, eight concentrations ranging from 0.001 to 1000 µg.ml^{-1} were tested. The mortality was scored after 24 h (day D+1) and 48 h (day D+2). Aphids that were unable to walk were considered dead [8] and were removed. Corrected mortality percentages were calculated using Henderson Tilton's formula after 24 h (day D+1) and 48 h (day D+2) of insecticide exposure.

HPLC-MS/MS analysis

For tissue extraction, 2.4 g of TMX (at LC50) treated aphid larvae were extracted with an acidified (0.2% acetic acid) water-methanol mixture (50/50) in a 50 ml centrifuge tube. Sample extracts were then filtered and purified on Oasis HLB Cartridges (200 mg) (Waters SAS, France). Elution was performed using 6 ml of acetonitrile. The obtained acetonitrile extract after the elution of the Oasis cartridges was reduced to dryness and the residue was re-dissolved in 1 ml methanol-water mixture (10/90). 40 µl of each sample was analyzed by high-performance liquid chromatography coupled to tandem mass spectrometry (HPLC-MS/MS).

HPLC–MS/MS was performed with an Ultimate 3000 rapid separation liquid chromatography system (Dionex, USA) coupled to an API 4000 Qtrap MS/MS from Applied Biosystems (Foster City, CA, USA). Separation was performed on a Phenomenex (Torrance, CA, USA) C18 column at 35°C with a gradient of water/methanol/acetic acid at a flow rate of 0.2 ml.min^{-1} and 5 mM ammonium acetate. MS/MS detection was performed in the multi-reaction-monitoring (MRM) mode using an ESI interface in the positive ion mode. The ionization voltage was 5500V, and the nebulizer and curtain gases were at 50 psi and 25 psi, respectively. The drying gas to assist the solvent evaporation in the source (600°C) was at 40 psi.

Optimisation of MRM transitions, collision energies and cone voltage were performed by direct injection of standard solutions. The optimized parameters for the detection of the two compounds (TMX and CLT) are listed in Table 1. With these parameters, calibration curves were linear over the concentration range of 0.9 to 20 µg.l^{-1} with a correlation coefficient (r) greater than 0.99.

Expression of nAChR subunits during developmental stages

Total RNAs were extracted from *A. pisum* adults, at different larval stages (Stage L1 to L4), using RNA Easy mini Plant Kit (Qiagen, Courtaboeuf France). To avoid genomic DNA (gDNA) contamination, total RNAs were treated using DNAse I kit (Invitrogen, Carlsbad, USA) according to manufacturer recommendations. RNAs were retro-transcribed using random hexamers with RevertAid kit (Thermoscientific, Waltham, USA), dissolved in RNAse-free water and conserved at −20°C. DNAse treatment was validated by PCR using primers set amplifying intron-containing sequence. Primer sets (Table 2) were designed using Primer3 software based on the *A. pisum* genome (http://www.ncbi. nlm.nih.gov/genbank/). Because Dale et al. identified potential alternatively spliced isoforms for Apisumα4 (exon 4) Apisumα6 (exon 6) and Apisumα7 (exon 6 and 7), primers were designed out of these exons, using genome information [14]. Amplification specificity of each primer set was also verified by cloning and sequencing the amplification products (data not shown). Amplification efficiencies were between 88 and 109%, allowing validation of each primer set for qPCR experiments. Because none of the endogenous reference genes had stable expression during developmental stages, external reference gene, *luciferase*, was used for normalization, as previously described [17,18]. Thus, 10 pg/1000 ng of luciferase RNA (Promega, Fitchburg, Wisconsin USA) were added after RNA extraction [17,18].

qPCR experiments were optimized according to MIQE Guideline recommendations [19] using ABI Prism 7700 instrument and 2X SYBR Green PCR Master Mix (Applied Biosystems, Courtaboeuf, France). Experiments were performed in triplicate using 100 ng of total RNA and 150 nM primers in a final volume of 25 µl. Product specificity was further assessed by electrophoresis on a 2% agarose gel with a 50 bp ladder and by dissociation curves giving rise to a single peak at the specific melting temperature [20]. Relative expression ratio (R) was calculated according to the Pfaffl formula [21], using primer efficiency (E) and CP value variation between control and sample (ΔCP) for each nAChR subunit. Ratio were normalized to reference genes and expressed in percentage: $R = (E_{subunit})^{\Delta CP subunit(control - sample)} / (E_{reference})^{\Delta CP reference (control - sample)}$. Luciferase was used as reference gene and quantification was relative to the first larval stage (L1).

Table 1. Multireaction monitoring conditions used for the HPLC-MS/MS analysis.

Compound	Ion	Transition	Declustering potential (DP)	Collision energy (CE)	Collision cell exit potential (CXP)	Dwell time (s)	HPLC retention time (tr)(min)
clothianidin	[M+H]$^+$	250>169	46	19	8	250	4,19
clothianidin	[M+H]$^+$	250>132	46	21	10	250	
thiamethoxam	[M+H]$^+$	292>211	56	17	10	250	3,84
thiamethoxam	[M+H]$^+$	292>181	56	33	8	250	

Table 2. Primers used to amplify nicotinic acetylcholine receptors subunits in quantitative PCR experiments.

gene	forward primer		reverse primer		Size (bp)
	name	nucleotidic sequence	name	nucleotidic sequence	
α 1	qpA1S1	CGGTCATTGTCGGTCAGTTG	qpA1R1	TGGCATCGGCACTTCCAT	60
α 2	qpA2S2	GGTCGTCACCATCATCATC	qpA2R2	CCACGACGGTATCTTGTGC	68
α 3	qpA3S1	GCGAGATTCACGGTCCAATAA	qpA3R1	GGCCATTTTGGTTTGTTTCG	60
α 4	qpA4S1	GAGTATGGTGGCGTGCAAATG	qpA4R1	GATATCCGGCCGCCAAAT	60
α 6	qpA6S1	TGGAGAGACCTGTATCCAACGA	qpA6R1	TGCTGTAGCGTGATGCCAAA	64
α 7	qpA7S1	CATGTATAATAGCGCTGACGAAGGT	qpA7R1	CTGTTGACCACCACGTTGGTT	63
α 8	qpA8S1	GAGGCACATCGACCAATCG	qpA8R1	CGCTTAGATCAATGCCAACATC	59
α 9	qpA9S1	GTGCAACCCGTGCAGTACAG	qpA9R1	TGCGTGTCATACGGCCAATA	65
α 10	qpA10S1	GCACATGGTTCATAGCGAACTG	qpA10R1	GGTGTTCATATTCGCTCGGATT	66
β1	qpB11S1	CGCCGTCCAAACACAAGAT	qpB11R1	CTTGCAGTTGGGATGATGCA	62
β2	qpB2S1	CCGTGAAGAGGAAAATACCG	qpB2R1	GAACACGACGACTATCGCTG	65
rpl7	qpRPL7F	GCGCGCCGAGGCTTAT	qpRPL7R	CCGGATTTCTTTGCATTTCTTG	81
actin	qpactinF	AGCTCTATTCCAACCTTCCTTCT	qpactinR	TGTATGTAGTCTCGTGGATACCG	62

Expression of nAChR subunits after neonicotinoid intoxication

To study subunit expression levels after insecticide exposure, relative qPCR was performed on first larval stage L1 exposed during 48 h with each neonicotinoid at LC_{50} or with DMSO (control condition). Total RNAs were extracted from intoxicated or control first larval stage (L1) using RNA Easy mini plant Kit and the same primers as described above. The results were normalized using the geometric mean of two reference genes, *actin* and *rpl7* and validated using Normfinder software [22,23,24]. Expression levels were relative to control condition.

Statistical analyses

Statistical analyses were performed using GraphPad Prism 5 (GraphPad Software Inc., La Jolla, CA). Data from binding experiments and insecticidal assays were analyzed by nonlinear regression analysis. A T-test ($P<0.05$; t-test with Welch's correction) was used for insecticidal assays and to compare Ki values. One-Way ANOVA ($p<0.05$) was used for qPCR experiments and binding assays.

Results

Binding properties of IMI, TMX and CLT on *A.pisum* native nAChRs

Saturation binding experiments were carried out with both $[^{125}I]$-α-Bgt and $[^{3}H]$-IMI on adult aphid membranes. Results are means of four experiments and the saturation binding parameters are summarized in Table 3. A saturation curve was first determined for $[^{125}I]$-α-Bgt (Figure 1A and 1B) and revealed the presence of two binding sites: a high affinity (Kd = 0.008±0.002 nM and Bmax = 12.86±5.92 fmol/mg protein) and a low-affinity binding site (Kd = 1.135±0.213 nM and Bmax = 135.9±6.0 fmol/mg protein). The saturation data obtained for $[^{3}H]$-IMI (Figure 2A and 2B) were also consistent with the presence of a high affinity (Kd = 0.16±0.04 nM and Bmax = 0.051±0.003 fmol/mg protein) and a low-affinity binding site (Kd = 41.7±5.9 nM and Bmax = 0.434±0.037 fmol/mg protein). For both $[^{3}H]$-IMI and $[^{125}I]$-α-

Bgt saturations curves, the presence of two binding sites was supported by the slope change in the Scatchard representation (Figure 1B and 2B). In addition, we noticed that high affinity binding sites only represented 8.6±3.8% and 10.4±5.2% of total $[^{125}I]$-α-Bgt and $[^{3}H]$-IMI binding sites, respectively (Table 3). The difference between Bmax values for high- and low-affinity binding sites, for both $[^{125}I]$-α-Bgt and $[^{3}H]$-IMI, is consistent with the presence of these two binding sites on different nAChR populations in the aphid membrane preparation. In addition, the comparison of Bmax values highlighted a larger proportion of $[^{125}I]$-α-Bgt binding sites compared to $[^{3}H]$-IMI (Table 3). Thus in the pea aphid *A. pisum*, α-Bgt-sensitive nAChRs seem to represent a large majority of nAChR populations.

In a second set of experiments, we studied the binding properties of IMI, TMX and CLT to the different nAChRs. Results are means of four experiments and the competition binding parameters are summarized in Table 3. For both radiotracers ($[^{3}H]$-IMI and $[^{125}I]$-α-Bgt), we studied the low affinity binding sites, which represent the majority of nAChR subtypes. Inhibition curves with IMI showed the presence of 20% of $[^{125}I]$-α-Bgt residual binding in excess of IMI, suggesting that some of the α-Bgt binding sites were insensitive to IMI (Figure 3A). On the contrary, inhibition was complete using TMX (Figure 3B) and CLT (Figure 3C). Indeed, the inhibition constant (Ki) for α-Bgt with a Kd value of 0.16 nM showed a better binding affinity for CLT and TMX (Ki = 0.18±0.05 μM and 1.53±0.65 μM, respectively) compared to IMI (Ki = 14.61±1.13 μM; One-Way ANOVA, $p<0.05$, table 3). With $[^{3}H]$-IMI, no specific binding inhibition was found using α-Bgt, indicating that low-affinity $[^{3}H]$-IMI binding sites were insensitive to α-Bgt (Figure 4A). Among the tested insecticides, homologous competition (Figure 4B) showed that IMI presents a Ki value of 38.14±6.88 nM which is consistent with the Kd value determined in the saturation experiment (41.7 nM, table 3). Interestingly, high concentrations of TMX and CLT were not able to completely displace $[^{3}H]$-IMI from its binding sites, with maximal inhibition of 35% for TMX (Figure 4C) and 75% for CLT (Figure 4D). The apparent Ki values calculated from these binding curves were 1.05±0.07 μM for TMX and 127±42.5 nM for CLT (Table 3). The residual

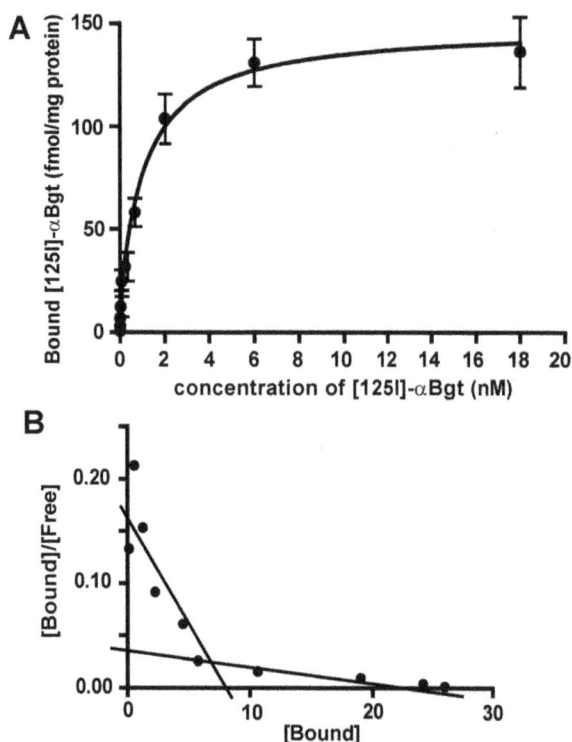

Figure 1. [^{125}I]-α-**Bungarotoxin specific binding on pea aphid.** Saturation curves (A) and Scatchard plots (B) for [^{125}I]-α-Bungarotoxin (α-Bgt) specific binding. Membranes were extracted from whole parthenogenetic adults of pea aphid *Acyrthosiphon pisum* LSR1. Results are means of four experiments. Error bars represent the SEM.

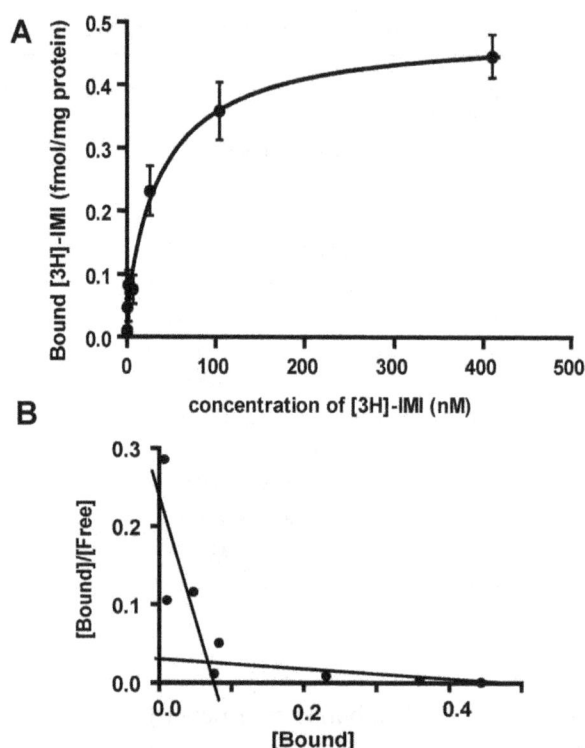

Figure 2. [^{3}H]-**imidacloprid specific binding on pea aphid.** Saturation curves (A) and Scatchard plots (B) for [^{3}H]-imidacloprid specific binding. Membranes were extracted from whole parthenogenetic adults of pea aphid *Acyrthosiphon pisum* LSR1. Results are means of four experiments. Error bars represent the SEM.

binding could be explained by the inability of TMX and CLT to interact with all the nAChR subtypes recognized by IMI, or by an interaction of these ligands that were not strickly competitive with IMI.

Toxicological effects of neonicotinoids on first-instar aphid larvae

The toxicological effects of neonicotinoids upon *A. pisum* larvae have been previously studied [16]. Using the same method, we found that the three neonicotinoids TMX, CLT, and IMI have

Table 3. [^{125}I]-α-Bungarotoxin ([^{125}I]-α-Bgt) and [^{3}H]-imidacloprid ([^{3}H]-IMI) binding parameters determined on aphid membranes.

Saturation binding

	[^{125}I]-α-Bgt			[^{3}H]-IMI		
	Kd (nM)	Bmax (fmol/mg)	% of high affinity	Kd (nM)	Bmax (fmol/mg)	% of high affinity
High affinity	0.008±0.002	12.86±5.92	8.6±3.8	0.16±0.04	0.051±0.003	10.4±5.2
Low affinity	1.135±0.213	135.9±6.0		41.7±5.9	0.434±0.037	

Competition binding

	[^{125}I]- α-Bgt			[^{3}H]-IMI		
	assay level (nM)	IC50 (µM)	Ki (nM)	assay level (nM)	IC50 (µM)	Ki (nM)
Imidacloprid	0.6	22.33±1.73	14.6±1.13[a]	25	0.061±0.011	38.14±6.88[c]
Thiamethoxam	0.08	1.636±0.701	1.53±0.65[a]	25	1.675±0.101	1047.2±63.1[d]
Clothianidin	0.08	0.194±0.052	0.18±0.05[b]	25	0.203±0.068	126.9±42.5[c]
α-Bgt	-	-	-	25	n.d.	n.d.

Ki values were calculated according to Cheng and Prusoff formula considering Kd of low affinity binding sites. n.d: not determined. IC50: half maximal inhibitory concentration. Results are mean of four experiments and are represented ± SD. Ki values that are significantly different using One-Way ANOVA ($p<0.05$) are noted with different letters.

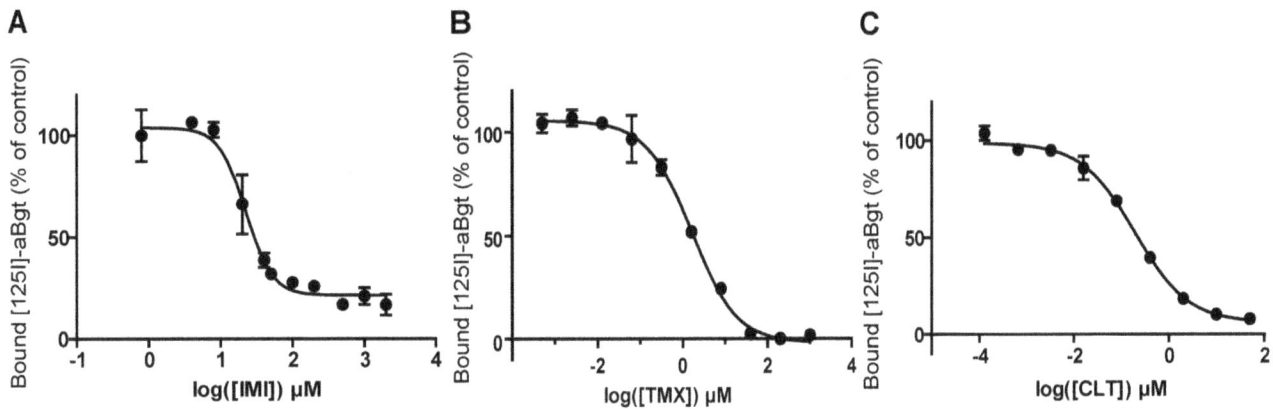

Figure 3. Neonicotinoids inhibition of [^{125}I]-α-Bungarotoxin specific binding. Inhibition curves were determined on membranes of whole parthenogenetic adults of pea aphid *Acyrthosiphon pisum* for three neonicotinoids: A) imidacloprid (IMI), B) thiamethoxam (TMX) and C) clothianidin (CLT). Results are means of four experiments. Error bars represent the SEM.

different toxicities against *A. pisum*. All results are presented in table 4 as means of 6 to 8 experiments. We found that TMX was the most toxic (LC$_{50}$ = 0.259 μg/ml) and CLT was the least toxic (LC$_{50}$ = 3.458 μg/ml) after 24 h of exposure. The toxicity of IMI was intermediate with an LC$_{50}$ of 0.913 μg/ml. The LC$_{50}$ values were significantly lower after 48 h of exposure. Interestingly IMI and TMX showed similar effect (LC$_{50}$ = 0.038 and 0.034 μg/ml, respectively) whereas CLT remained the least toxic (LC$_{50}$ = 0.118 μg/ml). We suggest that the potency of TMX could be associated to its double action: directly and after metabolization to CLT as previously proposed [25]. Indeed HPLC/MS analysis showed that TMX was metabolized to CLT (Figure 5). In TMX-treated aphids (using TMX at LC$_{50}$ = 0.034 μg/ml) we obtained

final concentrations of 1.34 μg/kg of TMX and 1.76 μg/kg of CLT, after 48 h of exposure.

Is the expression of aphid nAChR subunits influenced by developmental stage or by exposure to neonicotinoids?

Recently, using the full genome of *A. pisum*, Dale et al. highlighted the presence of 11 genes encoding putative nAChR subunits [15]. We confirmed the expression of these 11 genes in the pea aphid and studied the expression profile of these subunits according to the developmental stage and neonicotinoid exposure. First, qPCR experiments on the different developmental stages demonstrated that the expression of Apisumα1, Apisumα2, Apisumα6, Apisumα8 and Apisumβ2 was stable at the beginning

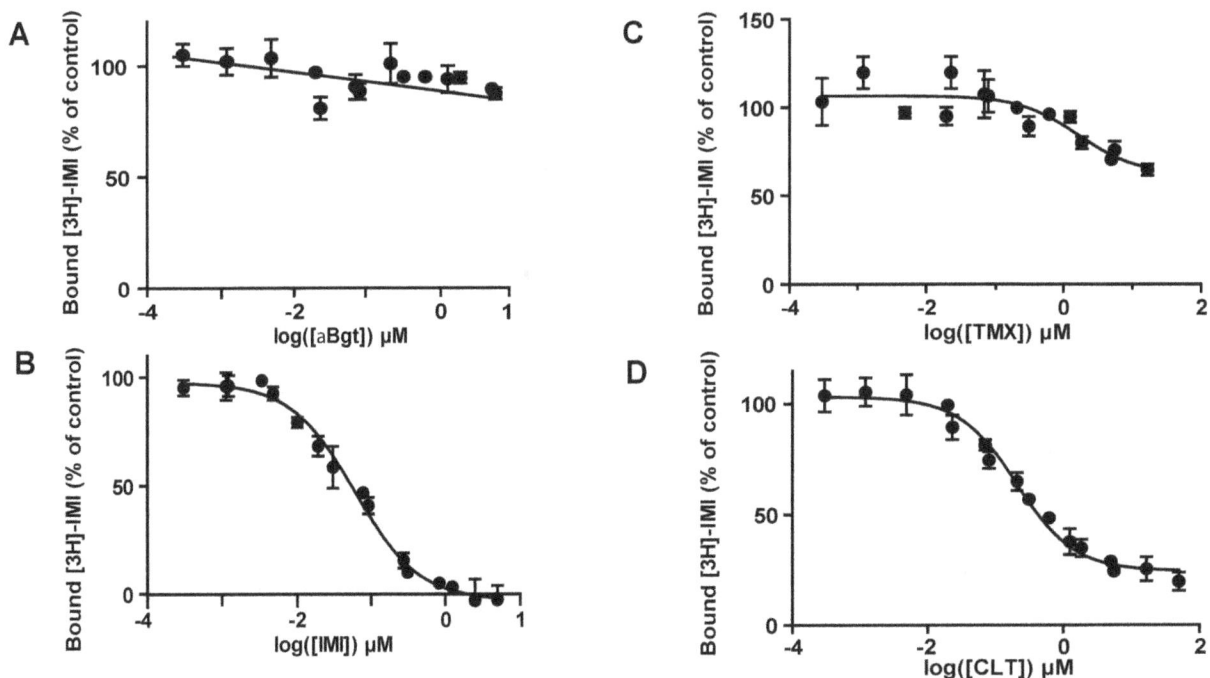

Figure 4. Neonicotinoid inhibition of [^{3}H]-imidacloprid specific binding. Inhibition curves were determined on membranes of whole parthenogenetic adults of pea aphid *Acyrthosiphon pisum* for (A) α-Bungarotoxin (α-Bgt) and three neonicotinoids: (B) imidacloprid (IMI), (C) thiamethoxam (TMX) and (D) clothianidin (CLT). Results are means of four experiments. Error bars represent the SEM.

A

B

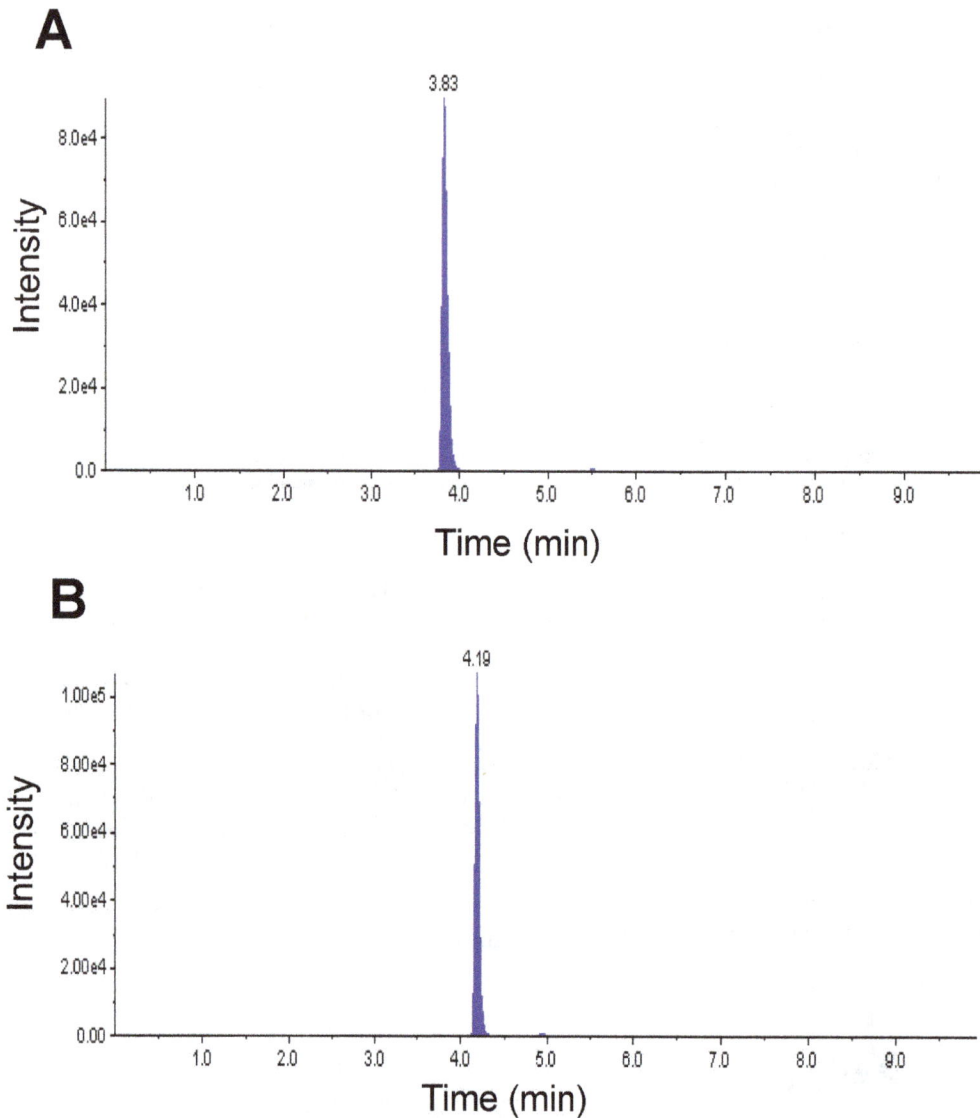

Figure 5. MRM chromatograms. Chromatograms of thiamethoxam (A) and its metabolite clothianidin (B) in 13,400 pea aphid larvae exposed to thiamethoxam at LC_{50} for 48 h. Intensity represents the peak area of the detected signal.

of aphid development and then was significantly reduced during adulthood (One Way Anova, $p<0.05$, $n=3$ experiments in triplicate, figure 6). On the contrary Apisumα3 expression increased with developmental stages although Apisumα7, Apisumα10 and Apisumβ1 transcript levels remained stable. The expression level of Apisumα4 and Apisumα9 subunits showed

Table 4. Neonicotinoid toxicity in the pea aphid on first-instar larvae for 24 h and 48 h of insecticide exposure using an artificial diet system.

	24 h		48 h		
	LC 50 (µg/ml)	CI 95%	LC 50 (µg/ml)	CI 95%	n
Imidacloprid	0.913 [a]	0.266 – 3.133	0.038 [d]	0.023 – 0.064	2381
Thiamethoxam	0.259 [b]	0.039 – 1.718	0.034 [d]	0.012 – 0.101	2613
Clothianidin	3.458 [c]	0.834 – 14.34	0.118 [e]	0.009 – 1.62	3016

n = number of insects tested; CI = confidence interval; LC_{50} = Lethal concentration leading to 50% mortality. Results were corrected using Henderson-Tilton's formula. Values followed by different letters are significantly different ($P<0.05$; t-test with Welch's correction). Toxicity curves were determined with 8 concentrations and 6 to 8 replicates were made for each concentration.

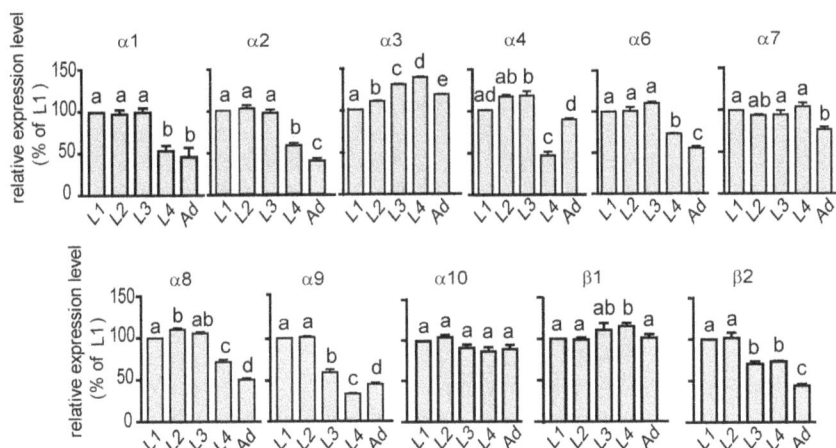

Figure 6. Expression level of nAChR mRNA subunits according to developmental stages of the pea aphid. Quantitative experiments were performed on whole individuals in triplicate. Results are mean of three independent experiments. Relative expression ratio were calculated relative to first-instar nymphs and normalized with external reference gene *luciferase*. Statistical analysis (One-Way ANOVA) was carried out using GraphPad Prism 5 software. For each subunit, expression ratio statistically different according to larval stage are designated by different letters.

greater variability, with a lower expression level at the fourth larval stage. Thus, *A. pisum* subunits expression was regulated during developmental stages, suggesting that different nAChR subtypes could be expressed.

Second, qPCR experiments were performed on surviving aphid larvae exposed to IMI, TMX and CLT after 48 h exposure. For this purpose, aphid larvae were intoxicated at the LC_{50} determined in insecticidal assays. We found that IMI induced a strong variation of nAChR subunits expression compared to control condition, with the exception of Apisumα4, Apisumα6, Apisumα7 and Apisumα9 (Figure 7, n = 4 to 7 experiments in triplicate). We also observed a significant increase of Apisumα10 (+218±40%), Apisumβ1 (+240±40%), Apisumα1 (+120±27%), Apisumα2 (+104±17%) and Apisumα3 (+61±10%), respectively. On the contrary, a decrease was found with Apisumα8 (−34±4%) and Apisumβ2 (−40±4%), respectively (Figure 7A). Aphid larvae treated with TMX showed a significant decrease of Apisumα2 (−23±5%), Apisumα7 (−46±13%), Apisumβ2 (−29±4%) and an increase for Apisumα10 (+90±31%) and Apisumβ1 (+39±13%; figure 7B). Exposure to CLT led to a significant diminution of Apisumα4 (−49±4%), Apisumα8 (−73±4%) and Apisumβ2 (−48±3%) whereas we found a significant increase of Apisumα10 (+56±12%; figure 7C). These data confirmed that the expression of aphid nAChR subunits was differentially modified after exposure to various neonicotinoids.

Discussion

The pea aphid presents several pharmacological binding sites with different affinity for neonicotinoids

In the present studies, saturation binding experiments demonstrated that $[^3H]$-IMI and $[^{125}I]$-α-Bgt labeled high- and low-affinity nAChR binding sites in *A. pisum*. Binding properties of α-Bgt were not well documented in insects but two binding sites have also been found in the aphid *M. persicae* [13]. Interestingly, two specific $[^3H]$-IMI binding sites were likewise reported in the aphids *M. persicae* and *A. craccivora* [11,12]. In the pea aphid, the large difference between Bmax values for high and low affinity binding sites for both $[^3H]$-IMI and $[^{125}I]$-α-Bgt was in accordance with the presence of these sites on distinct nAChR subtypes and not at various subunit interfaces on the same receptor. In addition,

competitive data showed that $[^3H]$-IMI low affinity binding site was insensitive to α-Bgt. We proposed that IMI could bind to α-Bgt-insensitive nAChR subtypes, which was consistent with data obtained in *M. persicae* and *A. craccivora* [12]. Moreover, in the pea aphid, it seemed that the majority of binding sites was sensitive to α-Bgt, as previously demonstrated in *D. melanogaster* and *M. persicae* [11,12,13]. Competitive experiments also revealed that CLT bound to both $[^{125}I]$-α-Bgt and $[^3H]$-IMI binding sites. CLT was known to interact well with IMI-binding sites in the aphids *M. persicae* and *A. craccivora* [3,26]. Interestingly, only one study referred to competitive experiments between CLT and labeled α-Bgt. Zhang et al. demonstrated a weak inhibitor potency of CLT to $[^3H]$-α-Bgt binding sites in *D. melanogaster* [27]. The apparent discrepancy between these results and ours could be attributed to species specificity. Furthermore, TMX, which was metabolized to CLT, showed a weak binding capacity for $[^3H]$-IMI binding sites and a better binding potency for $[^{125}I]$-α-Bgt binding sites in *A. pisum*. These results are consistent with previous studies describing a lack of TMX competition with $[^3H]$-IMI in other aphid species such as *M. persicae* and *A. craccivora* [3,12]. Unfortunately, there was no data on TMX competitive binding to α-Bgt sites, despite that $[^3H]$-TMX could bind directly in *M. persicae* and *A. craccivora* [28]. We propose that TMX binds to α-Bgt-binding sites in the pea aphid and that this mechanism could be present in other aphid species.

The neonicotinoids IMI, TMX and CLT have different toxicological effects on *A. pisum*

Acute toxicological assays demonstrated that TMX and IMI were more toxic than CLT. Similar data have been found with *A. gossypii* in which IMI was more toxic than both TMX and CLT [10]. Interestingly, in *M. persicae* CLT was found to be a more potent insecticide than IMI [8]. This discrepancy could be linked to variation in the intoxication method and/or susceptibility of aphid species. In other studies, the neonicotinoid susceptibility was evaluated using topical application and a dipping method for *M. persicae* and *A. gossypii*, respectively [8,10]. By contrast, we used an artificial diet protocol previously described by Sadeghi et al. [16]. The LC_{50} at 48 h for IMI corresponded to the LC_{50} at 72 h in Sadeghi's study which indicated that the pea aphid strains could be differentially sensitive to neonicotinoids. Moreover, because

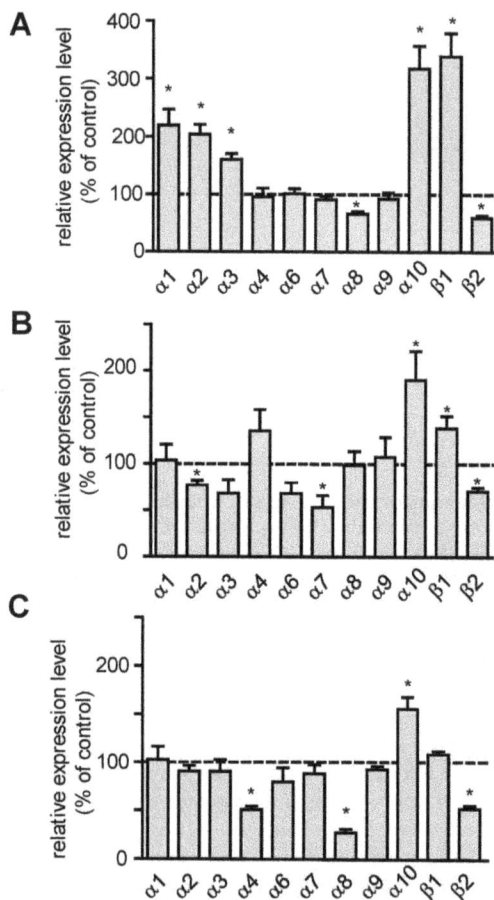

Figure 7. Expression levels of nAChR mRNA subunits after neonicotinoid exposure. Experiments were assessed on whole survivingl larvae exposed to neonicotinoids at LC_{50} for 48h. Aphids were intoxicated with imidacloprid (A) thiamethoxam (B) or clothianidin (C). Each qPCR experiment was performed in triplicate and results are represented as the mean of four to seven independent experiments after normalization with actin and ribosomal *rpl7* gene. Error bars represent the SEM. Results are expressed in % of the expression level in control conditions (no insecticide, corresponding to 100%). Statistical analysis (t-test, $\alpha = 0.05$) was carried out using Graphpad Prism 5 software.

uncommon nAChR subunit lacking one cysteine in the Cys-loop and could be involved in distinct functional properties [15]. Previous studies performed with electric ray *Torpedo* demonstrated that α subunit lacking one cysteine in the Cys-loop could co-assemble to form functional receptors that are expressed at the membrane [30]. In *Torpedo* the lack of one cysteine also led to the loss of α-Bgt binding sites [30]. As neonicotinoids bind to α-Bgt binding sites in the pea aphid, we propose that increased expression of Apisumα10 subunit could likewise lead to increased expression of nAChR subtypes that are less sensitive to neonicotinoids. After neonicotinoid exposure, we also observed that Apisumβ1 was over-expressed after TMX and IMI exposure and Apisumβ2 under-expressed after treatment with the three insecticides. Thus, in the pea aphid we proposed that both Apisumβ1 and Apisumβ2 could be differently involved in the regulation of neonicotinoid sensitivity. Indeed, studies performed on the brown planthopper *Nilaparvata lugens* and the aphid *M. persicae* showed that β1 was part of the IMI binding sites. Mutation of arginine to threonine at position 81 in this subunit induced an increase of resistance against neonicotinoids [31,32,33]. Moreover, recent studies demonstrated that nAChR subunit expression level was associated to neonicotinoid sensitivity [34,35]. For example, a decrease in Accβ1 and Accβ2 subunit expression in the Asiatic honey bee *Apis cerana* was described after IMI exposure [35]. Our results also demonstrated that neonicotinoid toxicity was associated with specific nAChR subunit regulation. For example, Apisumα10 and Apisumβ2 were always up- or down-regulated, following treatment with any of the three insecticides. On the contrary, we found that Apisumα7 and Apisumα4 were decreased only after exposure to TMX and CLT and the expression of Apisumα1 and Apisumα3 was increased after IMI exposure. These results suggest that some subunits could be involved in specific insecticide action. Thus, we suggest that high and low affinity binding sites could involved several nAChR subtypes.

Conclusions

Previous studies conclude that nAChR subunits influence the pharmacological properties of nicotinic receptors and thus could modify the neonicotinoid sensitivity [36]. Our results demonstrated that pea aphid nAChR subunits were differentially expressed, first between developmental stages, as previously demonstrated in *Drosophila* and *Apis cerana cerana* [35,37,38], and also according to the neonicotinoid exposure. Neonicotinoid sensitivity could then be dependent on either physiological status and/or environmental conditions in the pea aphid. Moreover, using toxicological and binding studies, we highlighted differences in neonicotinoid sensitivity in the pea aphid as compared to other aphid species and strains [8,10,16]. Thus, the insecticide strategies against aphid pests should be optimized for each particular species. In the pea aphid, the role of divergent subunits Apisumα9, Apisumα10 and Apisumβ2 could be of particular interest to further understand the neonicotinoid mode of action.

Acknowledgments

Sylvie Tanguy, Sylvie Hudaverdian and Joël Bonhomme (IGEPP INRA Rennes) are greatly acknowledged for their help in setting up aphid rearing. We thank Morgane Reynaud and Elodie Marcon (CEA Saclay) for their contribution in preliminary test for binding assays. We also acknowledge Benedicte Dubuc and Cyril Lecorre (UFR Sciences Angers) for horse bean cultivation. Finally we greatly thank Sylvia Anton and Kali Esancy for their critical comments.

part of TMX was metabolized to CLT, we propose that the unusually high toxicity of TMX in *A.pisum* compared to other aphid species, was associated to its metabolite CLT [8,10]. This hypothesis has been previously demonstrated using the moth *Spodoptera frugiperda* and the cockroach *Periplaneta americana*. Indeed, in *S. frugiperda* and *P. americana*, TMX was metabolized 24 h after treatment [25,29]. Nevertheless, in the present study, the proportion of metabolized TMX was different than previous studies [25,29]. Thus, the high insecticidal effect of TMX on the pea aphid could be due to its double action, by acting on nAChRs sensitive to α-Bgt and IMI.

Dale et al identified 11 putative genes encoding nAChR subunits in the pea aphid genome among which three were divergent (Apisumβ2, Apisumα9 and Apisumα10) and did not belong to conserved subunit groups between insects species [15]. Using qPCR experiments on surviving larvae, we demonstrated that IMI, TMX and CLT significantly influenced nAChR subunit expression. For all neonicotinoids tested, we found that Apisum α10 was highly expressed after treatment. This subunit is an

Author Contributions

Conceived and designed the experiments: ET HTL SHT DT DS SJP. Performed the experiments: ET AB AD SQ. Analyzed the data: ET HTL SHT DS. Wrote the paper: ET HTL SHT.

References

1. Tomizawa M, Casida JE (2003) Selective toxicity of neonicotinoids attributable to specificity of insect and mammalian nicotinic receptors. Annu Rev Entomol 48: 339–364.

2. Tomizawa M, Casida JE (2005) Neonicotinoid insecticide toxicology: Mechanisms of selective action. Annual Review of Pharmacology and Toxicology 45: 247 p-+.

3. Kayser H, Lee C, Decock A, Baur M, Haettenschwiler J, et al. (2004) Comparative analysis of neonicotinoid binding to insect membranes: I. A structure-activity study of the mode of [3H]imidacloprid displacement in Myzus persicae and Aphis craccivora. Pest Manag Sci 60: 945–958.

4. Engel MS, Grimaldi DA (2004) New light shed on the oldest insect. Nature 427: 627–630.

5. Dedryver CA, Le Ralec A, Fabre F (2010) The conflicting relationships between aphids and men: A review of aphid damage and control strategies. Comptes Rendus Biologies 333: 539–553.

6. Le Trionnaire G, Hardie J, Jaubert-Possamai S, Simon JC, Tagu D (2008) Shifting from clonal to sexual reproduction in aphids: physiological and developmental aspects. Biol Cell 100: 441–451.

7. Miao J, Du ZB, Wu YQ, Gong ZJ, Jiang YL, et al. (2013) Sub-lethal effects of four neonicotinoid seed treatments on the demography and feeding behaviour of the wheat aphid Sitobion avenae. Pest Manag Sci.

8. Foster SP, Cox D, Oliphant L, Mitchinson S, Denholm I (2008) Correlated responses to neonicotinoid insecticides in clones of the peach-potato aphid, Myzus persicae (Hemiptera: Aphididae). Pest Manag Sci 64: 1111–1114.

9. Jeschke P, Nauen R (2008) Neonicotinoids-from zero to hero in insecticide chemistry. Pest Manag Sci 64: 1084–1098.

10. Shi XB, Jiang LL, Wang HY, Qiao K, Wang D, et al. (2011) Toxicities and sublethal effects of seven neonicotinoid insecticides on survival, growth and reproduction of imidacloprid-resistant cotton aphid, Aphis gossypii. Pest Management Science 67: 1528–1533.

11. Lind RJ, Clough MS, Reynolds SE, Earley FGP (1998) [H-3]imidacloprid labels high- and low-affinity nicotinic acetylcholine receptor-like binding sites in the aphid Myzus persicae (Hemiptera: Aphididae). Pesticide Biochemistry and Physiology 62: 3–14.

12. Wiesner P, Kayser H (2000) Characterization of nicotinic acetylcholine receptors from the insects Aphis craccivora, Myzus persicae, and Locusta migratoria by radioligand binding assays: relation to thiamethoxam action. J Biochem Mol Toxicol 14: 221–230.

13. Lind RJ, Clough MS, Earley FGP, Wonnacott S, Reynolds SE (1999) Characterisation of multiple alpha-bungarotoxin binding sites in the aphid Myzus persicae (Hemiptera: Aphididae). Insect Biochemistry and Molecular Biology 29: 979–988.

14. Consortium TIAG (2010) Genome Sequence of the Pea Aphid Acyrthosiphon pisum. PLOS BIOLOGY 8.

15. Dale RP, Jones AK, Tamborindeguy C, Davies TG, Amey JS, et al. (2010) Identification of ion channel genes in the Acyrthosiphon pisum genome. Insect Mol Biol 19 Suppl 2: 141–153.

16. Sadeghi A, Van Damme EJ, Smagghe G (2009) Evaluation of the susceptibility of the pea aphid, Acyrthosiphon pisum, to a selection of novel biorational insecticides using an artificial diet. J Insect Sci 9: 1–8.

17. Smith RD, Brown B, Ikonomi P, Schechter AN (2003) Exogenous reference RNA for normalization of real-time quantitative PCR. Biotechniques 34: 88–91.

18. Johnson DR, Lee PK, Holmes VF, Alvarez-Cohen L (2005) An internal reference technique for accurately quantifying specific mRNAs by real-time PCR with application to the tceA reductive dehalogenase gene. Appl Environ Microbiol 71: 3866–3871.

19. Bustin SA, Benes V, Garson JA, Hellemans J, Huggett J, et al. (2009) The MIQE guidelines: minimum information for publication of quantitative real-time PCR experiments. Clin Chem 55: 611–622.

20. Ririe KM, Rasmussen RP, Wittwer CT (1997) Product differentiation by analysis of DNA melting curves during the polymerase chain reaction. Analytical Biochemistry 245: 154–160.

21. Pfaffl MW (2001) A new mathematical model for relative quantification in real-time RT-PCR. Nucleic Acids Research 29.

22. Andersen CL, Jensen JL, Orntoft TF (2004) Normalization of real-time quantitative reverse transcription-PCR data: A model-based variance estimation approach to identify genes suited for normalization, applied to bladder and colon cancer data sets. Cancer Research 64: 5245–5250.

23. McCulloch RS, Ashwell MS, O'Nan AT, Mente PL (2012) Identification of stable normalization genes for quantitative real-time PCR in porcine articular cartilage. J Anim Sci Biotechnol 3: 36.

24. Zhou ZJ, Zhang JF, Xia P, Wang JY, Chen S, et al. (2014) Selection of suitable reference genes for normalization of quantitative real-time polymerase chain reaction in human cartilage endplate of the lumbar spine. PLoS One 9: e88892.

25. Benzidane Y, Touinsi S, Motte E, Jadas-Hecart A, Communal PY, et al. (2010) Effect of thiamethoxam on cockroach locomotor activity is associated with its metabolite clothianidin. Pest Management Science 66: 1351–1359.

26. Zhang AG, Kayser H, Maienfisch P, Casida JE (2000) Insect nicotinic acetylcholine receptor: Conserved neonicotinoid specificity of [H-3]imidacloprid binding site. Journal of Neurochemistry 75: 1294–1303.

27. Zhang NJ, Tomizawa M, Casida JE (2004) Drosophila nicotinic receptors: evidence for imidacloprid insecticide and alpha-bungarotoxin binding to distinct sites. Neuroscience Letters 371: 56–59.

28. Wellmann H, Gomes M, Lee C, Kayser H (2004) Comparative analysis of neonicotinoid binding to insect membranes: II. An unusual high affinity site for [H-3]thiamethoxam in Myzus persicae and Aphis craccivora. Pest Management Science 60: 959–970.

29. Nauen R, Ebbinghaus-Kintscher U, Salgado VL, Kaussmann M (2003) Thiamethoxam is a neonicotinoid precursor converted to clothianidin in insects and plants. Pesticide Biochemistry and Physiology 76: 55–69.

30. Sumikawa K, Gehle VM (1992) Assembly of Mutant Subunits of the Nicotinic Acetylcholine-Receptor Lacking the Conserved Disulfide Loop Structure. Journal of Biological Chemistry 267: 6286–6290.

31. Li J, Shao Y, Ding ZP, Bao HB, Liu ZW, et al. (2010) Native subunit composition of two insect nicotinic receptor subtypes with differing affinities for the insecticide imidacloprid. Insect Biochemistry and Molecular Biology 40: 17–22.

32. Bass C, Puinean AM, Andrews M, Cutler P, Daniels M, et al. (2011) Mutation of a nicotinic acetylcholine receptor beta subunit is associated with resistance to neonicotinoid insecticides in the aphid Myzus persicae. Bmc Neuroscience 12.

33. Slater R, Paul VL, Andrews M, Garbay M, Camblin P (2012) Identifying the presence of neonicotinoidresistant peach-potato aphid (Myzus persicae) in the peach-growing regions of southern France and northern Spain. Pest Manag Sci 68: 634–638.

34. Markussen MDK, Kristensen M (2010) Low expression of nicotinic acetylcholine receptor subunit Md alpha 2 in neonicotinoid-resistant strains of Musca domestica L. Pest Management Science 66: 1257–1262.

35. Yu XL, Wang M, Kang MJ, Liu L, Guo XQ, et al. (2011) Molecular Cloning and Characterization of Two Nicotinic Acetylcholine Receptor Beta Subunit Genes from Apis cerana Cerana. Archives of Insect Biochemistry and Physiology 77: 163–178.

36. Lansdell SJ, Millar NS (2000) The influence of nicotinic receptor subunit composition upon agonist, alpha-bungarotoxin and insecticide (imidacloprid) binding affinity. Neuropharmacology 39: 671–679.

37. Grauso M, Reenan RA, Culetto E, Sattelle DB (2002) Novel putative nicotinic acetylcholine receptor subunit genes, D alpha 5, D alpha 6 and D alpha 7 in Drosophila melanogaster identify a new and highly conserved target of adenosine deaminase acting on RNA-mediated A-to-I pre-mRNA editing. Genetics 160: 1519–1533.

38. Gao JR, Deacutis JM, Scott JG (2007) The nicotinic acetylcholine receptor subunit Md alpha 6 from Musca domestica is diversified via post-transcriptional modification. Insect Molecular Biology 16: 325–334.

13

Tracking Contributions to Human Body Burden of Environmental Chemicals by Correlating Environmental Measurements with Biomarkers

Hyeong-Moo Shin[1]*, Thomas E. McKone[2,3], Michael D. Sohn[2], Deborah H. Bennett[1]

1 Department of Public Health Sciences, University of California Davis, Davis, California, United States of America, **2** Environmental Energy Technologies Division, Lawrence Berkeley National Laboratory, Berkeley, California, United States of America, **3** School of Public Health, University of California, Berkeley, California, United States of America

Abstract

The work addresses current knowledge gaps regarding causes for correlations between environmental and biomarker measurements and explores the underappreciated role of variability in disaggregating exposure attributes that contribute to biomarker levels. Our simulation-based study considers variability in environmental and food measurements, the relative contribution of various exposure sources (indoors and food), and the biological half-life of a compound, on the resulting correlations between biomarker and environmental measurements. For two hypothetical compounds whose half-lives are on the order of days for one and years for the other, we generate synthetic daily environmental concentrations and food exposures with different day-to-day and population variability as well as different amounts of home- and food-based exposure. Assuming that the total intake results only from home-based exposure and food ingestion, we estimate time-dependent biomarker concentrations using a one-compartment pharmacokinetic model. Box plots of modeled R^2 values indicate that although the R^2 correlation between wipe and biological (e.g., serum) measurements is within the same range for the two compounds, the relative contribution of the home exposure to the total exposure could differ by up to 20%, thus providing the relative indication of their contribution to body burden. The novel method introduced in this paper provides insights for evaluating scenarios or experiments where sample, exposure, and compound variability must be weighed in order to interpret associations between exposure data.

Editor: Alok Deoraj, Florida International University, United States of America

Funding: This research is funded by the American Chemistry Council (Grant#: 3-DBACC01, http://www.americanchemistry.com/). The funders had no role in study design, data collection and analysis, decision to publish, or preparation of the manuscript.

Competing Interests: The American Chemistry Council funded this study. There are no patents, products in development, or marketed products to declare.

* E-mail: hmshin@ucdavis.edu

Introduction

Correlation coefficients between environmental and biomarker measurements are widely used in environmental health assessments and epidemiology to explain the exposure associations between environmental media and human body burdens [1–4]. As a result considerable attention and effort have been given to interpretation of these coefficients [5–7]. However, there is limited information available on how the variance in environmental measurements, the relative contribution of exposure sources, and the elimination half-life affect the reliability of the resulting correlation coefficients. To address this information gap, we conducted a simulation study for various exposure scenarios of home-based exposure (e.g., inhalation, dermal uptake, non-dietary dust ingestion) to explore the impacts of pathway-specific scales of exposure variability on the resulting correlation coefficients between environmental and biomarker measurements.

Biomonitoring data, including those from blood, urine, hair, etc., have been used extensively to identify and quantify human exposures to environmental and occupational contaminants [8,9]. However, because the measured levels in biologic samples result from multiple sources, exposure routes, and environmental media,

the levels mostly fail to reveal how the exposures are linked to the source or route of exposure [10]. Thus, comparison of biologic samples with measurements from a single environmental medium (e.g., dust or air) results in weak correlations and lacks statistically significance. In addition, cross-sectional biological sample sets that track a single marker have large population variability and do not capture longitudinal (i.e. day-to-day) variability, especially for compounds with relatively short biologic half-lives, which can be on the order of days such as pesticides and phthalates. Therefore, in the case where the day-to-day variability of biological sample measurements is large, the use of biomarker samples with a low number of biological measurements in epidemiologic studies as a dependent variable can result in a misclassification of exposure as well as questions of reliability [11].

For chemicals frequently found at higher levels in indoor residential environments than in outdoor environments, it is common to assume that major contributions to cumulative intake are home-based exposure and/or food ingestion. This simplification can be further justified because people generally spend more than 70 percent of their time indoors [12,13]. Compounds with significant indoor sources and long half-lives in the human body– on the order of years for chemicals such as polybrominated diphenyl ethers (PBDEs)–have been found to have positive associations between indoor dust or air concentrations and serum concentrations in U.S. populations [4,14–17]. On the other hand, extant research has not reported significant associations between indoor samples and biomarkers for chemicals primarily associated with food-based exposures, for example, bisphenol-A [18] and perfluorinated compounds [19]. For chemicals with both home- and food-based exposure pathways and short body half-lives (on the order of days), as is the case for many pesticides, a significant association between indoor samples and biomarkers is found less frequently or relatively weak compared to PBDEs [1,20–23]. To better interpret these types of findings, we provide here a simulation study for various exposure scenarios to explore the role of the chemical properties and exposure conditions that are likely to give rise to a significant contribution from indoor exposures. We then assess for these situations the magnitude and variance of the associated correlation coefficients between biomarker and indoor levels.

The objectives of this study are (1) to generate simulated correlation coefficients between environmental measurements and biomarkers with different contributions of home-based exposure to total exposure and different day-to-day and population variability of intake from both residential (home) environments and food, (2) to interpret the contribution of home-based exposure to human body burden for two hypothetical compounds whose half-lives are on the order of days and years, and (3) to determine how the pattern of variability in exposure attributes impacts the resulting correlation coefficients linking biomarker levels to exposure media concentrations.

Materials and Methods

2. 1. Overview

In this study, our first step is to synthetically generate daily environmental concentrations and food exposure concentrations based on variations of day-to-day intake from residential environments and food as well as different relative contributions of home-based and food-based exposure. As different chemicals are likely to have different relative contributions from the home-based and food-based exposure pathways, we conducted our simulations across the full range of relative contributions between the two pathways to address all plausible scenarios for various compounds. We combine the simulated home-based exposures associated with indoor environmental concentrations and food concentrations, assuming that the total intake results only from home-based exposure and food ingestion. From these inputs we estimate time-dependent biomarker concentrations using a one-compartment pharmacokinetic model. We then computed correlation coefficients between simulated environmental and biomarker concentrations.

In order to facilitate numerous simulations, several simplifications are made regarding (1) a representative environmental medium for home exposure, (2) a distribution of environmental (inhalation/dermal) and food intake, and (3) sources of exposure. First, we select chemical concentrations from indoor wipe samples (C_{wipe}) as a way to represent home-based exposures that result from all potential exposure routes, including inhalation, non-dietary dust ingestion, and dermal uptake. From these wipe concentrations, resulting home-based exposure (E_{home}) can be assumed to be linearly related to C_{wipe} and E_{home} and C_{wipe} are assumed for simplicity to be equal. In addition, we assume that a contaminated food intake rate represents food exposures (E_{food}). Second, we select C_{wipe} and E_{food} from log-normal distributions of variability across both population and time [6]. Lastly, we assume

that the total intake accounting for biomonitoring data results from E_{home} and E_{food}, excluding any other exposure pathways.

Calculating the correlation coefficient between environmental and biomarker measurements requires a number of steps. First, we generate synthetic wipe concentrations for a subject's home i on a day 1 ($C_{wipe,i,1}$) and food exposure for a subject i on a day 1 ($E_{food,i,1}$). Second, we generate a wipe concentration for a subject's home i on a given day j ($C_{wipe,i,j}$) by correlating it with a wipe sample on the previous day ($C_{wipe,i,j-1}$). We then apply this approach for generating $C_{wipe,i,j}$ to generate synthetic food exposures for a subject i on a given day j ($E_{food,i,j}$). Third, we vary the contribution of home exposure to total exposure (X_1) to generate a different contribution of home and food exposures, based on the assumption that E_{home} is linearly related and equal to C_{wipe}. Fourth, we add $E_{home,i,j}$ and $E_{food,i,j}$ for a total daily intake rate for a subject i on a given day j. Fifth, time-dependent biological concentrations are estimated using a one-compartment pharmacokinetic model. Finally, we compute Pearson's correlation coefficients between wipe and biological (e.g., serum) concentrations for our simulated population of 500 on each of 30 days.

2. 2. Monte Carlo Simulations

2.2.1. Simulated home and food exposures. We assumed that wipe concentrations across the population are log-normally distributed with mean ($\mu_{wipe} = 1.0$ μg/g) and standard deviation expressed as a coefficient of variation ($CV_{wipe\_pop} = \mu_{wipe}/\sigma_{wipe}$). We used three different CVs–1.0, 2.0, and 4.0–in order to generate synthetic wipe concentrations for a subject's home i on a day 1 ($C_{wipe,i,1}$). We estimated parameters (a, mean and b, standard deviation) of the associated normal distribution, ln ($C_{wipe,i,1}$), by the following method of moments [24].

$$b_{wipe\_pop} = \sqrt{\log\left(CV_{wipe\_pop}^2 + 1\right)} \quad (1)$$

$$a_{wipe\_pop} = \log\left(\mu_{wipe}\right) + b_{wipe\_pop}^2 \quad (2)$$

where a_{wipe_pop} and b_{wipe_pop} are the mean and standard deviation of ln ($C_{wipe,i,1}$), respectively. We then used the following lognormal inverse cumulative distribution function (cdf) to generate wipe concentrations for 500 homes with a residential receptor population for the first day of exposure.

$$C_{wipe,i,1} = F^{-1}(p \,|\, a_{wipe\_pop}, b_{wipe\_pop}) \quad (3)$$

where $C_{wipe,i,1}$ is the wipe concentration selected with probability of p from the inverse lognormal cdf with parameters a_{wipe_pop} and b_{wipe_pop} for a subject's home i on a day 1. Since the wipe concentration for a subject's home i on a given day j ($C_{wipe,i,j}$) is likely to be correlated to that on a previous day ($C_{wipe,i,j-1}$), we used a log-Gaussian random walk to generate auto-correlated $C_{wipe,i,j}$. In other words, we first generated random numbers that are log-normally distributed using mean ($\mu = 0$) and standard deviation ($\sigma_{wipe\_day} = 1.0$, 2.0, and 4.0). Then, we randomly multiplied 1 or −1 by the randomly generated numbers and computed cumulative sums. This allows us to approximate the temporal autocorrelation expected for the same house from day to day. In addition, since wipe concentrations should be positive, they were scaled up to assure positive values, maintaining the distribution of concentrations from random walk.

The method to generate 'home' and 'food' exposures is the same, but the simulated numbers are different as we used a random number generator for each exposure source. Thus, for food exposures, Equations 1 through 3 are used to generate food exposures for each simulated subject i on a given day j ($E_{food,i,j}$) by replacing μ_{wipe}, σ_{wipe_day}, and CV_{wipe_pop} with μ_{food}, σ_{food_day}, and CV_{food_pop}. Auto-correlated wipe concentrations and food exposures are provided in Figure S1.

2. 2. 2. Biological concentrations. Because we assumed that the biological levels result from different combinations of average home exposure (E_{home}) and food exposure (E_{food}), we computed the relative E_{food} to E_{home} ratio using the following equation.

$$E_{food} = E_{home} \times (X_2/X_1) = E_{home} \times ((1 - X_1)/X_1) \quad (4)$$

where X_1 and X_2 are the percent contribution of exposure from home and food, respectively. Here, because we assumed that total exposure (E_{total}) is equal to the sum of E_{home} and E_{food}, the sum of X_1 and X_2 is 100%.

Using the different contributions to exposure from the home (X_1), we added $E_{home,i,j}$ and $E_{food,i,j}$ to obtain a total daily intake rate for a subject i on a given day j ($I_{i,j}$). Then, we used the one-compartment pharmacokinetic model described in Equation 5 to estimate time-dependent biological concentrations [3,25] using serum as the representative biological medium.

$$C_{serum,i,j} = C_{serum,i,j-1} \cdot e^{-k} + \left(1 - e^{-k}\right)\frac{f}{k \cdot V}I_{i,j} \quad (5)$$

where $C_{serum,i,j}$ is the serum concentration of the compound for a subject i at time j ($\mu g/L$), k is an excretion rate coefficient of the compound ($1/day$), f is the fraction of the ingested compound present in the blood after absorption across the gastrointestinal tract and distribution throughout the body (unitless), V is the volume of blood (L), and $I_{i,j}$ is the intake rate of the compound for a subject i at time j ($\mu g/day$), summed from $E_{home,i,j}$ and $E_{food,i,j}$.

In this model, the excretion rate coefficient k can be expressed as $\ln(2)/t_{1/2}$ where $t_{1/2}$ is the half-life of the compound in the human body. We assumed that the fraction f is assumed to be 1 for all compounds and the blood volume V is about 5 L for all subjects [26]. This approach can be applied for urine concentrations and can be adjusted as needed.

2. 3. Sensitivity Analysis

Identifying the most important sources of overall exposure variability allows researchers to concentrate resources on obtaining the most important exposure data [6]. Thus, we conducted a sensitivity analysis to determine which sources of variability have relatively more influence on the R^2 value for a given home-exposure contribution. Four types of exposure variability, σ_{wipe_day}, CV_{wipe_pop}, σ_{food_day}, and CV_{food_pop}, were considered in our study. We computed the mean R^2 for compounds with short and long half-lives by varying one exposure variability (e.g., CV_{wipe_day}) from 0.2 to 4.0, but fixing other exposure variability at 1.0 and then repeated this computation for other variability.

Results

3. 1. Correlation Coefficient and Home Exposure Contribution

In this study, we applied various exposure scenarios to investigate the relationship between R^2 and a relative contribution of home exposure to total exposure for compounds with different biological half-lives. Figure 1 shows that the R^2 between wipe and

serum concentrations increases with the increasing contribution of home exposure. Overall, as the home contribution increases, the gap between the median R^2 for a long half-life compound (empty box) and that for a short half-life compound (filled box) increases. In addition, the median R^2 is almost always larger for a compound with a short biological half-life compared to a compound with a long half-life when these compounds have the same average exposure contribution from the home environment. This is because biologic concentrations for the compound with a short half-life are more sensitive to home exposure with large variance, while concentrations for the compound with a long half-life remain relatively stable due to the longer body retention of the compound, which to a large extent buffers the variations. This result also indicates that for compounds primarily associated with food-based exposure, in other words, for those with little contribution from home exposure (e.g., BPA and outdoor use pesticides) [1,18,20–23], the R^2 value becomes very small, as expected. In addition, for compounds with a large fraction of exposure resulting from indoor residential environments, such as PBDEs, the median R^2 at 90–100% of home contribution is approximately 0.6 [4,14–17] as shown in Figure 1.

To look at the results in Figure 1 in a different point of view, we plot the percent of home exposure contribution with different R^2 values to reveal the relationship between the biological half-life of the compound and the relative contribution of home exposure in Figure 2. This figure illustrates that, although the R^2 between wipe and serum concentrations for two compounds with different half-lives is within the same range, the relative contribution of home exposure to total exposure differs by up to 20% between compounds. For example, when the R^2 values for two compounds with different half-life values is between 0.3 and 0.4, the resulting contribution from the home environment for the short half-life compound is 20% smaller than that for the long half-life compound.

In actual exposure situations, we expect the day-to-day variability of wipe concentrations for semivolatile organic compounds to be small, due to their strong persistence on surface materials and dust [27]. In this study, we did not include the day-to-day variability associated with the relationship between the concentration in the home environment and the resulting exposure. There are two basic models for relating home concentrations to exposure. First, there are models that assume that exposure is driven by direct surface contacts, which are likely to have high day-to-day variability [28]. Second, there are models that assume that air-to-skin trans-dermal uptake becomes more significant than dermal uptake from surface contacts, and air-to-skin transfer is likely to be less variable day to day [12,29]. These model choices are important because R^2 values will also be linked to whether intake is primarily associated with air-to-skin trans-dermal uptake or dermal uptake from surface contacts. Thus, under the same conditions used in Figures 1 and 2, but with equal contributions from home and food ($X_1 = X_2 = 0.5$), we also investigated relative changes of R^2 with different day-to-day variability of wipe concentrations (i.e., σ_{wipe_day}) with results shown in Figure 3. The gap between the median R^2 for a long half-life compound (empty box) and a short half-life compound (filled box) increases with increasing σ_{wipe_day}. For all values of σ_{wipe_day}, the median R^2 for a compound with a short half-life is larger than that with a long half-life. This result indicates that day-to-day variability of wipe concentrations determines not only the magnitude of R^2 for both compounds, but also the relative magnitude of R^2 between compounds.

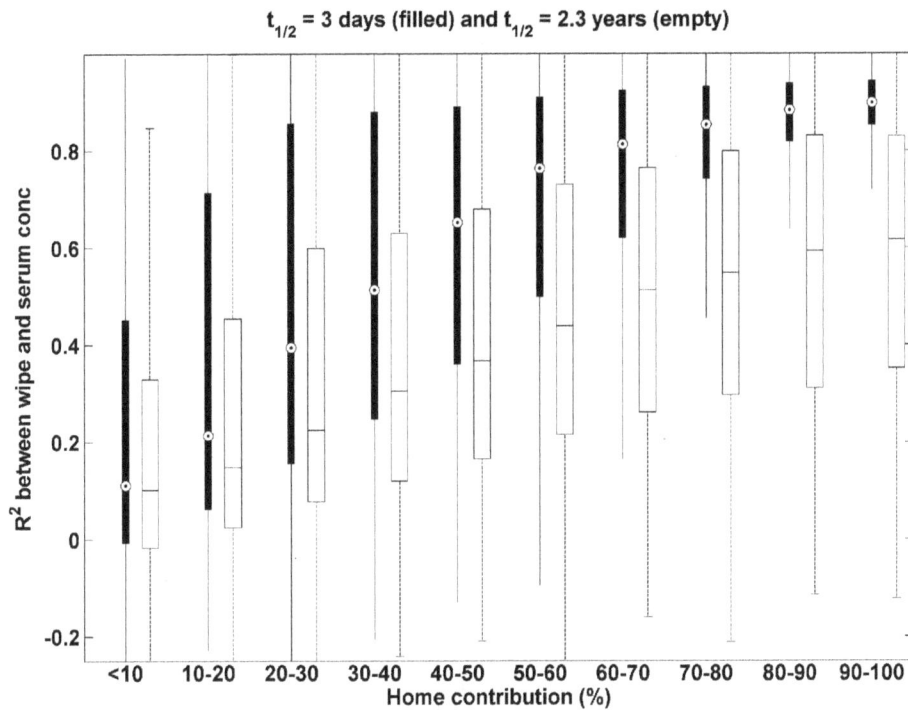

Figure 1. R² between wipe and serum concentrations with different contribution of home exposure for two compounds with 3 days of half-life (filled) and 2.3 years of half-life (empty).

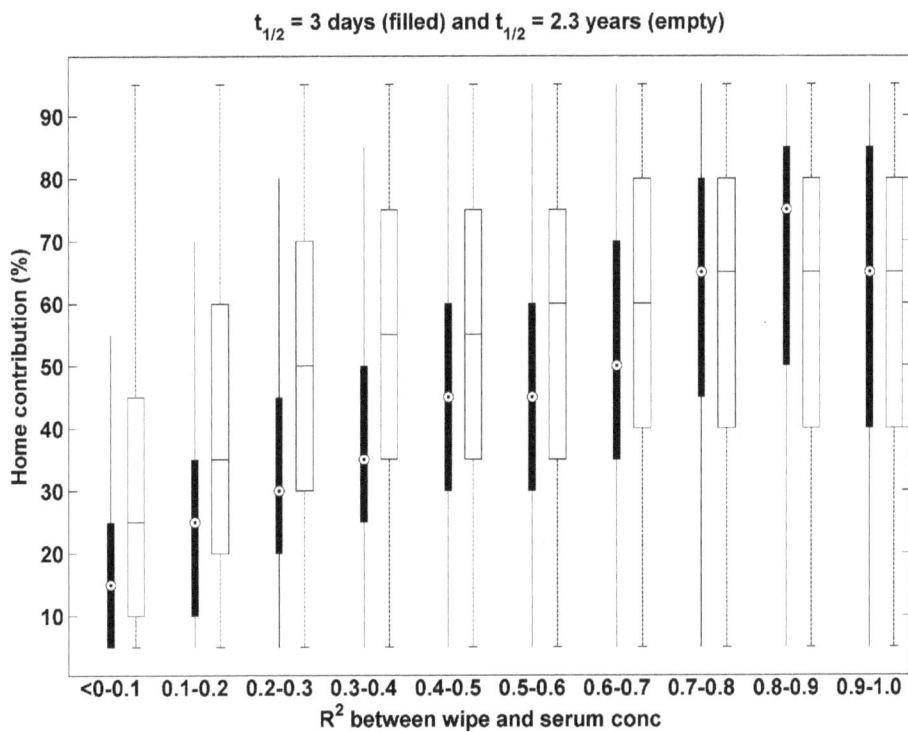

Figure 2. Contribution of home exposure (%) to total exposure with different R² for two compounds with 3 days of half-life (filled) and 2.3 years of half-life (empty).

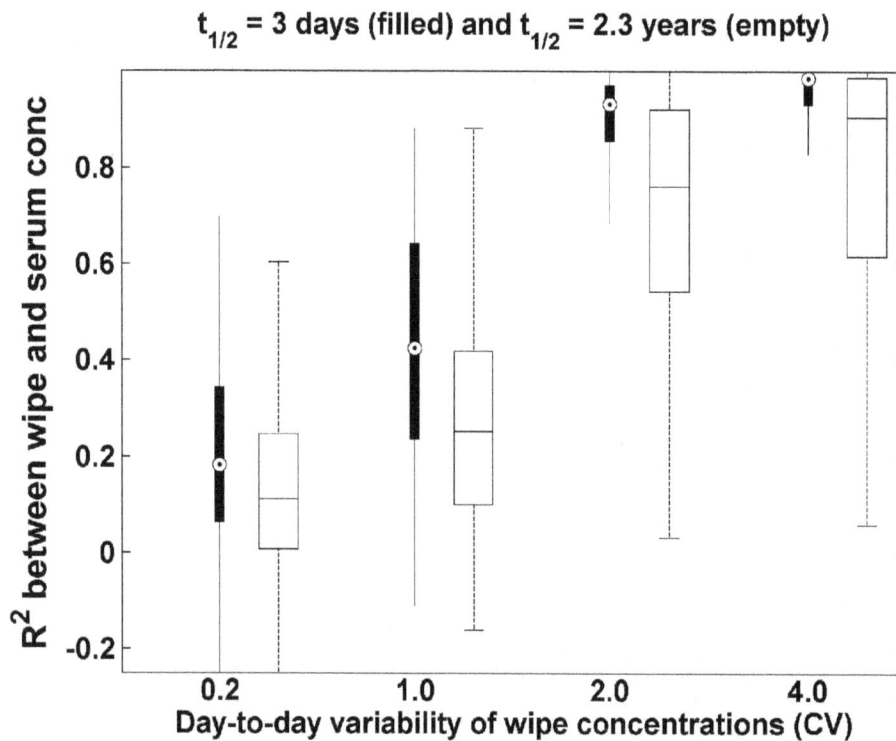

Figure 3. R^2 with different day-to-day variability of wipe concentrations for two compounds with 3 days of half-life (filled) and 2.3 years of half-life (empty).

3. 2. Influential Source of Variability on Correlation Coefficients

We determined the sensitivity of the mean R^2 value to each of the four different types of variability (i.e., σ_{wipe_day}, CV_{wipe_pop}, σ_{food_day}, and CV_{food_pop}) across a range of scales of variability (e.g., $CV = 0.2, 1.0, 2.0,$ and 4.0). Table 1 shows the mean R^2 at a specific variability for two compounds with different biological half-lives ($t_{1/2}$). In terms of changes in mean R^2 between compounds, a compound with a 2.3 year half-life is shown to be less sensitive to day-to-day variability of food concentrations (i.e., σ_{food_day}) than one with a 3 day half-life and both compounds have similar sensitivity to day-to-day variability of wipe concentrations (i.e., σ_{wipe_day}). In terms of changes in mean R^2 within compounds, R^2 is most sensitive to day-to-day variability of wipe concentrations for both compounds. In addition, the contributions of population variability of wipe concentrations and food exposures to changes in mean R^2 are minimal.

3. 3. Implications/Limitations

Because some indoor contaminants are considered potential threats to human health, many studies have applied significant resources to examine the relationship between exposure to indoor pollutants and adverse health effects. However, these studies are potentially limited by the use of a single or a few environmental and biological samples. The significant implications of this situation are reflected in our results. Multi-day, multi-person sample analyses are costly and labor-intensive. In addition, the resulting R^2 values from these studies are not interpreted or poorly interpreted in terms of variability and contribution of exposure sources and the biological half-life of a compound. In this regard, the simulation study in this paper provides an important step towards interpreting the relative contribution of home-based exposure to human body burden for two compounds whose biological half-lives are significantly different (days versus years). Although these two compounds do not cover the full range of chemical substances, bracketing half lives allows us to quantify the

Table 1. Mean R^2 at a specific variability for four types of variability (coefficient of variation (CV) or standard deviation (σ)) for two compounds with different half-lives ($t_{1/2}$).

σ or CV	$t_{1/2} = 3$ days					$t_{1/2} = 2.3$ years				
	0.2	1.0	2.0	4.0	range	0.2	1.0	2.0	4.0	range
σ_{wipe_day}	0.36	0.71	0.91	0.94	0.36–0.94	0.20	0.44	0.69	0.80	0.20–0.80
CV_{wipe_pop}	0.72	0.71	0.72	0.75	0.71–0.75	0.45	0.44	0.45	0.50	0.44–0.50
σ_{food_day}	0.84	0.71	0.44	0.13	0.13–0.84	0.52	0.44	0.30	0.07	0.07–0.52
CV_{food_pop}	0.71	0.71	0.73	0.68	0.68–0.73	0.43	0.44	0.47	0.42	0.42–0.47

significance of source, measurement, and exposure pattern variability for disaggregating body burden. In particular, it shows that exposure variability and different contributions of exposure sources are more interconnected than commonly considered in many experimental studies. The work also brings to attention the need to understand the impact of a chemical half-life on the relationship between environmental exposures and biomonitoring data. The sensitivity of day-to-day variability of wipe concentrations and food exposures on the resulting R^2 values also points to the importance of understanding variability and contribution of exposure sources. Finally, future work includes computing the relative number of samples needed for various levels of confidence to disaggregate body burden for various types of compounds (half lives), environments, and exposure pathways.

Despite the lack of experimental data, the simulated results provide key insights on the role of the variability and contribution of exposure sources and biological half-lives in quantifying a relationship between indoor exposure and human body burden. This approach will be useful for designing future exposure and epidemiologic studies that includes indoor environmental samples and biomonitoring data.

Author Contributions

Conceived and designed the experiments: HMS DHB. Performed the experiments: HMS MDS. Analyzed the data: HMS. Wrote the paper: HMS TEM MDS DHB.

References

1. Bradman A, Whitaker D, Quiros L, Castorina R, Henn BC, et al. (2007) Pesticides and their metabolites in the homes and urine of farmworker children living in the Salinas Valley, CA. J Expo Sci Env Epid 17: 331–349.
2. Fraser AJ, Webster TF, Watkins DJ, Nelson JW, Stapleton HM, et al. (2012) Polyfluorinated Compounds in Serum Linked to Indoor Air in Office Environments. Environ Sci Technol 46: 1209–1215.
3. Shin HM, Vieira VM, Ryan PB, Steenland K, Bartell SM (2011) Retrospective Exposure Estimation and Predicted versus Observed Serum Perfluorooctanoic Acid Concentrations for Participants in the C8 Health Project. Environ Health Perspect 119: 1760–1765.
4. Watkins DJ, McClean MD, Fraser AJ, Weinberg J, Stapleton HM, et al. (2011) Exposure to PBDEs in the Office Environment: Evaluating the Relationships Between Dust, Handwipes, and Serum. Environ Health Perspect 119: 1247–1252.
5. Chen CC, Wu KY, Chang MJW (2004) A statistical assessment on the stochastic relationship between biomarker concentrations and environmental exposures. Stoch Env Res Risk A 18: 377–385.
6. Loomis D, Kromhout H (2004) Exposure variability: Concepts and applications in occupational epidemiology. Am J Ind Med 5: 113–122.
7. Rappaport SM, Symanski E, Yager JW, Kupper LL (1995) The relationship between environmental monitoring and biological markers in exposure assessment. Environ Health Perspect 103: 49–53.
8. Centers for Disease Control and Prevention (CDC) (2005) Third National Report on Human Exposure to Environmental Chemicals. Atlanta, GA.
9. Centers for Disease Control and Prevention (CDC) (2009) Fourth National Report on Human Exposure to Environmental Chemicals. Atlanta, GA.
10. Shin HM, McKone TE, Bennett DH (2013) Evaluating environmental modeling and sampling data with biomarker data to identify sources and routes of exposure. Atmos Environ 69: 148–155.
11. Griffith W, Curl CL, Fenske RA, Lu CA, Vigoren EM, et al. (2011) Organophosphate pesticide metabolite levels in pre-school children in an agricultural community: Within- and between-child variability in a longitudinal study. Environ Res 111: 751–756.
12. Shin HM, McKone TE, Bennett DH (2012) Intake Fraction for the Indoor Environment: A Tool for Prioritizing Indoor Chemical Sources. Environ Sci Technol 46: 10063–10072.
13. Klepeis NE, Nelson WC, Ott WR, Robinson JP, Tsang AM, et al. (2001) The National Human Activity Pattern Survey (NHAPS): a resource for assessing exposure to environmental pollutants. Journal of Exposure Analysis and Environmental Epidemiology 11: 231–252.
14. Johnson PI, Stapleton HM, Slodin A, Meeker JD (2010) Relationships between Polybrominated Diphenyl Ether Concentrations in House Dust and Serum. Environ Sci Technol 44: 5627–5632.
15. Stapleton HM, Eagle S, Sjoedin A, Webster TF (2012) Serum PBDEs in a North Carolina Toddler Cohort: Associations with Handwipes, House Dust, and Socioeconomic Variables. Environ Health Perspect 120: 1049–1054.
16. Watkins DJ, McClean MD, Fraser AJ, Weinberg J, Stapleton HM, et al. (2012) Impact of Dust from Multiple Microenvironments and Diet on PentaBDE Body Burden. Environ Sci Technol 46: 1192–1200.
17. Wu N, Herrmann T, Paepke O, Tickner J, Hale R, et al. (2007) Human exposure to PBDEs: Associations of PBDE body burdens with food consumption and house dust concentrations. Environ Sci Technol 41: 1584–1589.
18. Geens T, Aerts D, Berthot C, Bourguignon JP, Goeyens L, et al. (2012) A review of dietary and non-dietary exposure to bisphenol-A. Food Chem Toxicol 50: 3725–3740.
19. Xu Z, Fiedler S, Pfister G, Henkelmann B, Mosch C, et al. (2013) Human exposure to fluorotelomer alcohols, perfluorooctane sulfonate and perfluorooctanoate via house dust in Bavaria, Germany. Sci Total Environ 443: 485–490.
20. Coronado GD, Vigoren EM, Thompson B, Griffith WC, Faustman EM (2006) Organophosphate pesticide exposure and work in pome fruit: Evidence for the take-home pesticide pathway. Environ Health Perspect 114: 999–1006.
21. Curwin BD, Hein MJ, Sanderson WT, Striley C, Heederik D, et al. (2007) Urinary pesticide concentrations among children, mothers and fathers living in farm and non-farm households in Iowa. Ann Occup Hyg 51: 53–65.
22. Quiros-Alcala L, Bradman A, Smith K, Weerasekera G, Odetokun M, et al. (2012) Organophosphorous pesticide breakdown products in house dust and children's urine. J Expo Sci Env Epid 22: 559–568.
23. Rothlein J, Rohlman D, Lasarev M, Phillips J, Muniz J, et al. (2006) Organophosphate pesticide exposure and neurobehavioral performance in agricultural and nonagricultural Hispanic workers. Environ Health Perspect 114: 691–696.
24. Ramaswami A, Milford JB, Small MJ (2005) Integrated Environmental Modeling-Pollutant Transport, Fate, and Risk in the Environment. Hoboken, New Jersey: John Wiley & Sons. Inc.
25. Bartell SM (2003) Statistical Methods for Non-Steady-State Exposure Estimation Using Biomarkers. Davis, CA: University of California, Davis.
26. Lorber M, Egeghy PP (2011) Simple Intake and Pharmacokinetic Modeling to Characterize Exposure of Americans to Perfluoroctanoic Acid, PFOA. Environ Sci Technol 45: 8006–8014.
27. Shin HM, McKone TE, Tulve NS, Clifton MS, Bennett DH (2013) Indoor Residence Times of Semivolatile Organic Compounds: Model Estimation and Field Evaluation. Environ Sci Technol 47: 859–867.
28. Cohen Hubal EA, Sheldon LS, Burke JM, McCurdy TR, Berry MR, et al. (2000) Children's exposure assessment: A review of factors influencing children's exposure, and the data available to characterize and assess that exposure. Environ Health Perspect 108: 475–486.
29. Weschler CJ, Nazaroff WW (2012) SVOC exposure indoors: fresh look at dermal pathways. Indoor Air 22: 356–377.

14

Quantification of the Pirimicarb Resistance Allele Frequency in Pooled Cotton Aphid (*Aphis gossypii* Glover) Samples by TaqMan SNP Genotyping Assay

Yizhou Chen*, Daniel R. Bogema, Idris M. Barchia, Grant A. Herron

Elizabeth Macarthur Agricultural Institute, NSW Department of Primary Industries, Menangle, New South Wales, Australia

Abstract

Background: Pesticide resistance monitoring is a crucial part to achieving sustainable integrated pest management (IPM) in agricultural production systems. Monitoring of resistance in arthropod populations is initially performed by bioassay, a method that detects a phenotypic response to pesticides. Molecular diagnostic assays, offering speed and cost improvements, can be developed when the causative mutation for resistance has been identified. However, improvements to throughput are limited as genotyping methods cannot be accurately applied to pooled DNA. Quantifying an allele frequency from pooled DNA would allow faster and cheaper monitoring of pesticide resistance.

Methodology/Principal Findings: We demonstrate a new method to quantify a resistance allele frequency (RAF) from pooled insects via TaqMan assay by using raw fluorescence data to calculate the transformed fluorescence ratio k' at the inflexion point based on a four parameter sigmoid curve. Our results show that k' is reproducible and highly correlated with RAF (r >0.99). We also demonstrate that k' has a non-linear relationship with RAF and that five standard points are sufficient to build a prediction model. Additionally, we identified a non-linear relationship between runs for k', allowing the combination of samples across multiple runs in a single analysis.

Conclusions/Significance: The transformed fluorescence ratio (k') method can be used to monitor pesticide resistance in IPM and to accurately quantify allele frequency from pooled samples. We have determined that five standards (0.0, 0.2, 0.5, 0.8, and 1.0) are sufficient for accurate prediction and are statistically-equivalent to the 13 standard points used experimentally

Editor: Shu-Biao Wu, University of New England, Australia

Funding: This study is funded by the CRDC (DAN1203). The funders had no role in study design, data collection and analysis, decision to publish, or preparation of the manuscript.

Competing Interests: The authors have declared that no competing interests exist.

* E-mail: yizhou.chen@dpi.nsw.gov.au

Introduction

Insecticide resistance has long been a problem of agriculture but has risen in prominence since the introduction of synthetic organic insecticides in the 1950's [1]. While the use of toxins remains a fundamental method of pest control, resistance will continue to threaten sustainable agriculture. The threat remains despite the introduction of new transgenic cotton varieties and Integrated Pest Management (IPM). This is because transgenics, such as *Bt*-cotton, also rely on toxins and so expose pests to high selection for resistance. Furthermore, IPM systems favour the use of more selective compounds, thereby narrowing the range of chemicals used. One such compound is pirimicarb (Pirimor), an insecticide that is highly effective at killing aphids but not the desirable and beneficial predatory species associated with aphids [2].

Traditional monitoring of pesticide resistance in arthropods is performed by bioassay in which insects are exposed to insecticide and mortality is recorded at specific post-exposure interval(s) [3]. Resistance levels are determined from dose-response mortality data and expressed as LC_{50} values, which are an estimate of the lethal concentration required to cause 50% mortality in the target population tested [3]. Additionally, resistance can be monitored via a single diagnostic or discriminating dose but these are difficult to accurately set [4] and require the generation of significant base line data which, due to tedious laboratory process, can take weeks or months to produce.

Insecticide resistance can be behaviourally- or physiologically-based with the latter involving three distinct mechanisms: target site insensitivity, enhanced detoxification and reduced pesticide penetration [5]. With the recent advance of genomics it has been possible to study many possible target resistance genes often associated with the insect nervous system.

Examples of this include the point mutation in the *GABA* receptor conferring insecticide resistance in *Drosophila melanogaster* [6] and a mutation in the acetylcholinesterase gene causing pesticide resistance in a variety of insect species [7]. In other cases resistance is caused by detoxification linked to a single nucleotide mutation [8] or even single gene duplication or deletions [9]. However, only when these molecular mechanisms are identified can rapid molecular methods be developed, allowing more effective monitoring of pesticide resistance.

The cotton or melon aphid, *Aphis gossypii* Glover is a serious pest of many crop species including cotton, pumpkin, citrus and melons [10]. This species has developed resistance to multiple insecticides including the carbamate pirimicarb (Pirimor) and some specific organophosphates that has led to chemical control failures in Australian cotton production regions [11]. The causal mechanism of pirimicarb resistance in *A. gossypii* has been identified as target site mutation in the acetylcholinesterase gene [12,13]. A double nucleotide substitution (TCA → TT[T/C]) in *ACE1* causes the replacement of a serine with a phenylalanine (S431F) and has been confirmed to be the cause of the pirimicarb resistance seen in Australian field collections of *A. gossypii* associated with control failure [14]. In Australian cotton IPM, a PCR-RFLP assay has been used to monitor pirimicarb resistance in the field by individually genotyping 20–50 individual aphids [2]. However, individual genotyping by PCR-RFLP limits the number of sites that can be monitored as it is labour intensive and offers limited benefits over the traditional bioassay. It is critical to have cost effective methods to monitor resistance allele frequencies (RAF) in field populations to maintain successful IPM strategies.

An alternative method to individual aphid genotyping is to estimate allele frequency from pooled DNA using real-time PCR technology with allele-specific probes or allele-specific primers [15–18]. However these pooled DNA approaches are often designed for specific assays and, due to the complexity of non-specific binding or amplification, are not widely used.

Currently, the most widely used qPCR platform for the estimation of allele frequency from pooled DNA is the 5′ nuclease assay. It utilizes TaqMan probes that possess a minor-groove binding (MGB) molecule and a fluorescent dye attached to the 3′ and 5′ ends, respectively. The 'gold standard' for this technique uses two probes with different reporter dyes, allowing the detection of both alleles. Quantification of allele frequency is achieved by using the threshold cycle (C_t) or crossing point (CP) to calculate allele ratios based on $2^{-\Delta Ct}$ [19,20]. However, significant variation can arise if the fluorescent probes differ significantly in their binding efficiency or if amplification efficiency varies between resistant and susceptible alleles. Yu *et al* [18] have used the normalized fluorescence ratio in the exponential phase of PCR with known premixed allele ratios and generated a linear regression from which an allele ratio can be estimated. However, this method suffers as the exponential phase of PCR is selected arbitrarily.

Here we have developed a simple method to estimate allele frequency using TaqMan assays. We show that by selecting a single, standard reference point RAF can be predicted from the ratio of the two fluorescence intensities. Additionally, we demonstrate that RAF is a function of the transformed fluorescence ratio (k') and that five standard-points are sufficient to develop the equation of prediction.

Materials and Methods

PCR Assay and Probe Design for S431F

The TaqMan SNP assay was designed based on the Genbank sequence (AF502802) using RealTimeDesign Software (Biosearch Technologies) with forward primer 5′-AACCAATATACT-CATGGGTAGTAACTC-3′ and the reverse primer 5′-AACC-GCCGCATCTGCATT-3′. A dual-labeled probe, 5′-Quasar 670-CGAAGAGGGTTACTATTCAA-3′- BHQ2 for the susceptible allele was designed based on a known susceptible *A. gossypii* sequence for a strain known as 'Sonya'. Two dual-labeled probes were designed for previously-identified resistance alleles, probe 5′-Fam- CGAAGAGGGTTACTATTTTA-3′-BHQ1 matching the

allele identified in pirimicarb-resistant strain Adam and probe 5′-Fam-CGAAGAGGGTTACTAYTTCA-3′-BHQ1 for the allele identified in pirimicarb-resistant strain Togo. All primers and probes were synthesized by Biosearch Technologies Inc (Biosearch Technologies Inc, Novato USA).

Predefined RAF with Plasmid DNA and Pooled Cotton Aphids

Fragments, 667 bp in size and containing the S431F mutation site, were amplified from the susceptible Sonya, and resistant Adam and Togo strains and cloned into the pCR4 vector (Invitrogen, USA) using RFLP genotyping primers. Plasmid DNA concentration was then measured by a Nanodrop 2000 (Nanodrop Technologies). To create a standard curve, a series of standards (T/S) with predefined RAF of 1.0, 0.95, 0.9, 0.8, 0.7, 0.6, 0.5, 0.4, 0.3, 0.2, 0.1, 0.05 and 0.0 were constructed by mixing plasmids containing the resistant Togo and susceptible Sonya alleles. A duplicate standard series (A/S) was made by mixing plasmids containing the Adam and Sonya alleles.

In addition to plasmid standards, a series of standards was prepared using susceptible and resistant aphids. Thirteen pools of 20 aphids were prepared with RAF of 1.0, 0.95, 0.9, 0.8, 0.7, 0.6, 0.5, 0.4, 0.3, 0.2, 0.1, 0.05 and 0.0. As an example, the pool for RAF 0.95 was constructed by extracting a tube containing 19 aphids from the resistant strain and 1 aphid from the susceptible strain.

2011/2012 Aphis Gossypii Field Collection

Methods for the collection, transport, culture and bioassay of *A. gossypii* samples have been described previously [11,14]. A total of 35 *A. gossypii* samples (or strains) collected from cotton producing farms across eastern Australia during the 2011/2012 season were genotyped individually by PCR-RFLP. Resistance allele frequencies were estimated by genotyping 20 individual aphids from each sample. Samples were further confirmed susceptible or resistant via bioassay using methods outlined in detail by Herron *et al* [11].

DNA Extraction

Aphis gossypii DNA was extracted from pooled or individual aphids using Chelex −100 resin (BioRad, USA) as described in [14]. Briefly, individual or 200 pooled aphids were placed inside a 1.5 mL microcentrifuge tube containing 80 µL of 5% Chelex − 100 resin. The sample was thoroughly homogenized with a sterile micropestle and incubated first at 56°C for 30 min, then at 100°C for 5 min. The crude DNA sample was then used for real-time PCR or PCR-RFLP or stored at −20°C for future use.

Individual Genotyping of S431F by PCR-RFLP

RFLP genotyping of the S431F mutation has been described previously [14]. Briefly, a 667 bp fragment containing the mutation was amplified with forward primer 5′- CAAGCCAT-CATGGAATCAGG-3′ and reverse primer 5′-TCATCAC-CATGCATCACACC-3′. The PCR product was digested by restriction endonuclease *Ssp*I by adding 5 units of enzyme and *Ssp*I buffer (1×) to a completed PCR for 3 hours at 37°C. The resultant PCR-RFLP profile was visualized by agarose gel electrophoresis. The pirimicarb-susceptible allele shows a single intense band at 336 bp (digested by *Ssp*I), whereas the pirimicarb-resistant allele shows a single intense band at 667 bp.

Real-time PCR with TaqMan Assay

PCRs contained 400 nmol forward primer and reverse primer, 200 nmol susceptible and resistant probe, in a 1×TaqMan

Universal PCR Master Mix (Applied Biosystems, USA) comprising a total 25 μl reaction volume. Each sample was set up in triplicate and one negative control sample was included in each run. Real-time PCR was performed in an ABI7500 Real-Time PCR System (Applied Biosystems, Foster City, CA, USA) with 10 min at 95°C followed by 47 cycles of 15 s at 95°C and 1 min at 60°C.

Data Analysis

Sigmoid 4 parameter curve fitting statistical analysis was carried out with GENSTAT release 10 software [21] using nonlinear regression and linear regression functions.

Principle of Quantification

Real-time PCR quantification is measured as the incremental change in signal (ΔRn) that is directly proportional to the amount of amplicons produced at any cycle [18,22] and is defined as follows:

$$[amplicon]_{synthesized} = \Delta R_n / \Delta \varphi \tag{1}$$

Where $\Delta \varphi$ represents the difference between the specific fluorescence of the free fluorophore and the specific fluorescence of the probe-bound fluorophore.

The synthesized amplicon is determined by the initial template number copy (N_0), the number of cycles (n) and the amplification efficiency (E).

$$[amplicon]_{synthesized} = N_0 * E^n \tag{2}$$

Combining equation 1 with 2 yields:

$$N_0 = \Delta R_n / \Delta \varphi / E^n \tag{3}$$

Quantification of the two alleles (susceptible and resistant) was achieved with TaqMan real-time SNP assays (Figure 1). Allele R (resistance allele) and allele S (susceptible) were detected by dual-labelled probes, 5′ FAM and 3′ BHQ and 5′ Quasar 670 and 3′ BHQ, respectively.

For the resistant allele R;

$$A_0 = \Delta R_n A / \Delta \varphi A / E_A^n \tag{4}$$

where A_0 is the initial copy number of allele R
$\Delta R_n A$ is the fluorescence intensity of Fam at cycle n.
$\Delta \varphi A$ is the parameter for fluorescence Fam.
E_A^n is the compound amplification efficiency of allele R.
For the susceptible allele S:

$$B_0 = \Delta R_n B / \Delta \varphi B / E_B^n \tag{5}$$

Where $\Delta R_n B$ is the fluorescence increment of Quasar 670 at cycle n.

The initial allele ratio:

$$\frac{A_0}{B_0} = \frac{\Delta R_n A}{\Delta R_n B} \times \frac{\Delta \varphi B}{\Delta \varphi A} \times \frac{E_B^n}{E_A^n} \tag{6}$$

While for any given assay, the ratio of parameter $\Delta \varphi B / \Delta \varphi A$ and E_B^n / E_A^n will be a relative constant, there will be constant relationship between $R = A_0 / B_0$ and $R' = \Delta R_n A / \Delta R_n B$.

Therefore R can be predicted by the ratio $R' = \Delta R_n A / \Delta R_n B$.

Estimation of ΔR_n

Real-time PCR is modelled via a four parametric sigmoid function [23,24]:

$$y(x) = \frac{a}{1 + \exp^{-(\frac{x-b}{c})}} + y_0 \tag{7}$$

Where:
x is cycle number,
$y(x)$ is raw fluorescence of cycle x,
y_0 is the background fluorescence,
a is the maximal height of the curve (the difference between the maximal fluorescence and background fluorescence).
b is the first derivative maximum of the function (the inflexion point of the curve) and c describes the slope of the curve.

If you subtract the background fluorescence the equation above can be rewritten as:

$$f(x) = \frac{a}{1 + \exp^{-(\frac{x-b}{c})}} \tag{8}$$

Where $f(x)$ is the fluorescence minus the background which is equivalent to ΔR_n at cycle n.

Selecting a Single Point in the Exponential Phase

For allele R, with fluorescence Fam,

$$f_{fam}(x) = \frac{a_{fam}}{1 + \exp^{-(\frac{x-b_{fam}}{c_{fam}})}} \tag{9}$$

Where:
a_{fam} is the maximal height of the curve for fluorescence Fam.
b_{fam} is the inflexion point of the curve of allele R,
c_{fam} is the slope of the curve of allele R.
For allele S with fluorescence Quasar.

$$f_{qua}(x) = \frac{a_{qua}}{1 + \exp^{-(\frac{x-b_{qua}}{c_{qua}})}} \tag{10}$$

Where:
a_{qua} is the maximal height of the curve for fluorescence Quasar.
b_{qua} is the inflexion point of the curve of allele S,
c_{qua} is the slope of the curve of allele S.

Use the ratio $f_{fam}(x)/f_{qua}(x)$ when one of the alleles is at its maximum speed, for example, if $b_{fam} < b_{qua}$ where the Fam reaches to its maximum speed first (Figure 2).

$$R' = f_{fam}(b_{fam})/f_{qua}(b_{fam})$$

$$R' = \frac{0.5 a_{fam}}{f(b_{fam})} = 0.5 \times \frac{a_{fam}}{\dfrac{a_{qua}}{1 + \exp\left(\frac{-(b_{fam}-b_{qua})}{c_{qua}}\right)}}$$

A.

B.

C.

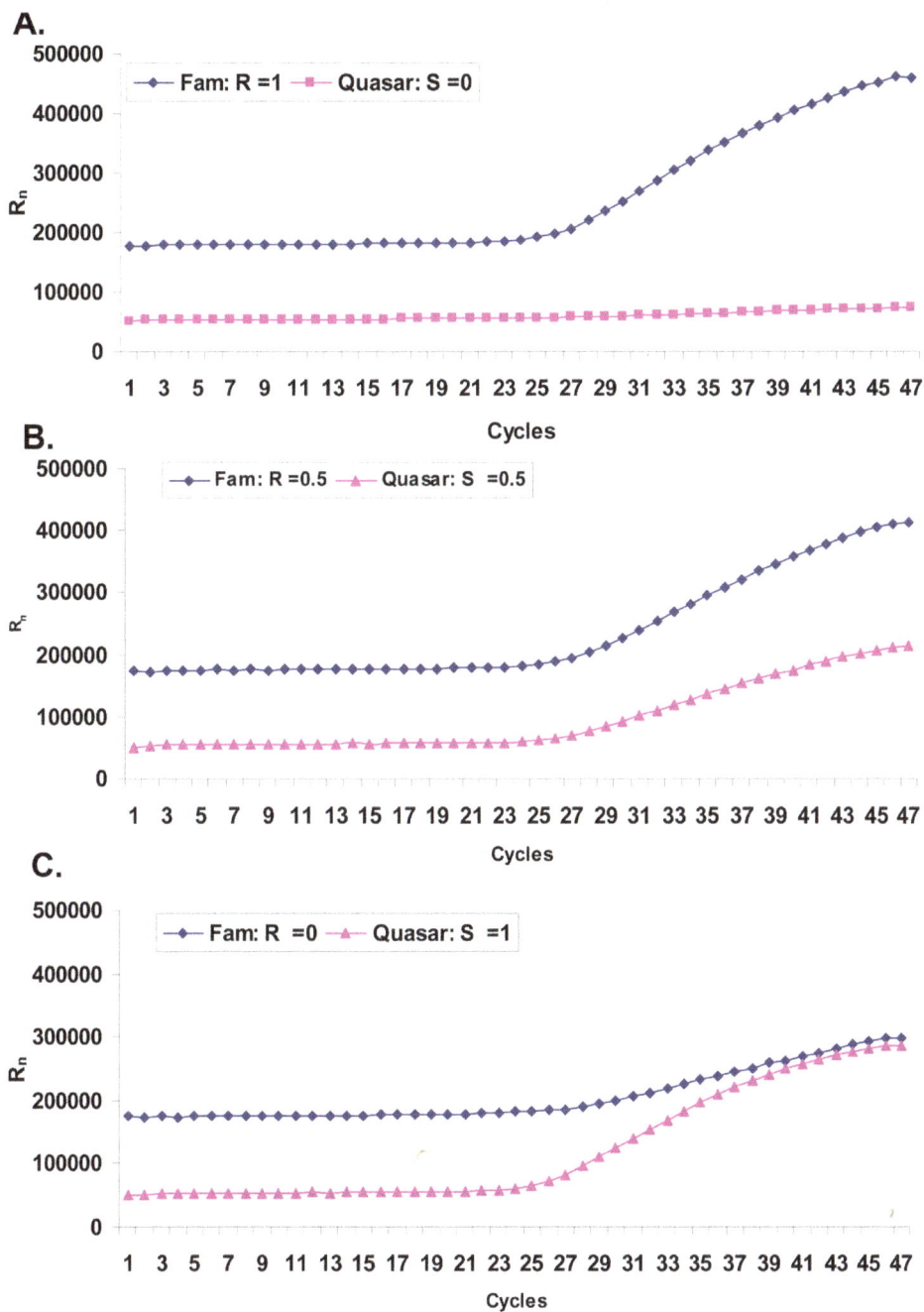

Figure 1. Raw fluorescence plot of TaqMan assay with two probes. Fam probe (blue) was from resistance allele and Quasar probe (red) was from susceptible allele.

$$= 0.5 \times \frac{a_{fam}}{a_{qua}} \times (1 + \exp\left(\frac{-(b_{fam} - b_{qua})}{c_{qua}}\right)) \qquad (11)$$

$$R' = \frac{f_{fam}(b_{qua})}{0.5 a_{qua}} = \frac{\dfrac{a_{fam}}{1 + \exp\left(\frac{-(b_{qua} - b_{fam})}{c_{fam}}\right)}}{0.5 a_{qua}}$$

However If $b_{fam} > b_{qua}$ where Quasar reaches to its maximum speed first.

$$= 2 \times \frac{a_{fam}}{a_{qua}} \times \frac{1}{1 + \exp\left(\frac{-(b_{qua} - b_{fam})}{c_{fam}}\right)} (12)$$

The frequency of allele can be expressed as k.

$$k = \frac{A_0}{A_0 + B_0} = \frac{A_0/B_0}{A_0/B_0 + 1} = \frac{R}{R+1} \qquad (13)$$

The ratio of fluorescence R' can be expressed as k'.

$$k' = \frac{R'}{R'+1} \qquad (14)$$

A Two-step Sigmoid Curve Fitting for Standardized Parameters

To reduce parameter estimation bias caused by variable number of cycles in PCR plateau phase, the raw fluorescence data was fitted to a sigmoid curve twice. Fitting was performed first with all data points (in our case 47 cycles), and second fitting used data points with one slope after the inflexion point (cycle b+c) (see Table S1). An example of this calculation of k' is demonstrated in Table S2.

Results

Transformed Fluorescence Ratio k'

Essentially the transformed fluorescence ratio k' is the transformation of the ratio of two fluorescence intensities when one of these intensities reaches its inflexion point (Figure 2). The transformed fluorescence ratio k' comprising 4 runs of plasmid mix and 3 runs of pooled aphids with predefined RAF is summarized in Table 1 with original data included in Table S3 and Table S9.

The transformed fluorescence ratio k' is highly consistent between triplicates from each standard. The average coefficient of variation within runs (intra-run) is between 1–2%. However the variation of k' between runs (inter-run) for the same standard mixes is more variable and range from 1–18%. The value of k' within runs for each predefined standard, follow the trends of the initial RAF. As RAF becomes higher, k' also becomes higher.

Testing the Relationship between RAF and k'

Using both plasmid and aphid standards, we first tested the relationship between RAF and k' by linear regression (as equation 6 predicts a linear relationship between these variables). In four runs using purified plasmids, a strong linear relationship was demonstrated with a high coefficient of determination ($R^2 > 0.99$). However, the linear model did not fit as well for standards made from Chelex-extracted aphids, where the three runs produced a coefficient of determination for aphid standards ranging from 0.93 to 0.95 (also see Table S4).

In addition to linear regression, we attempted a non-linear, 4 parameter sigmoid curve fitting model. Using this non-linear model, we found that the relationship between RAF and k' for the three runs that used extracted aphids produced a higher coefficient of determination ($R^2 > 0.98$), which was comparable to the purified

plasmid samples. Interestingly, the purified plasmid standards also show improved coefficient of determinations using this sigmoid curve fitting model (Table S4).

Inter-run Correlation of k'

Further analyses were performed to determine if there was correlation in the values of k' between runs. Table 2 summarizes the inter-run coefficient of determination for k' by linear and non-linear (sigmoid) regression. When compared like-for-like, the four plasmid and three aphid standard runs generally demonstrated a strong linear relationship ($R^2 > 0.98$). However, in contrast to above, the linear relationship was poorer when plasmid standard runs were compared to the standards made from Chelex-extracted aphids ($R^2 = 0.89–0.97$). When standard reactions were analyzed with the non-linear model a high correlation was observed between all runs. Coefficients of determination were higher ($R^2 > 0.99$) when plasmid and Chelex-extracted aphid runs were examined with non-linear regression and compared like-for-like. When plasmid standard runs were compared to the standards made from Chelex-extracted aphids using non-linear regression, coefficients of determination were also higher ($R^2 > 0.97$) than those generated from linear regression analysis (Table 2).

The Number of Standards Required for Accurate Prediction

The sigmoid relationship allows for a reduction in the number of data-points required to build the standard curve and hence increases the number of samples that can be examined per run. We have determined that five standards (0.0, 0.2, 0.5, 0.8, and 1.0) are sufficient for accurate prediction and are statistically-equivalent to the 13 standard points used experimentally (see Table S5, Table S6 and Table S7). We have examined RAFs using a full 13-standard model and a reduced 5-standard model for both the purified plasmid (Table S6) and Chelex-extracted aphid standard runs (Table S7). If the predefined RAF standards (0.05, 0.10, 0.30, 0.40, 0.60, 0.70, 0.90 and 0.95) were treated as unknowns, the RAFs predicted using the reduced model are highly accurate for all runs (see Table S6 and Table S7). The correlation between actual RAFs and those predicted using the reduced model standard curve is very high ($R^2 > 0.999$).

Combined Analysis of Multiple Runs

The sigmoid relationship between runs allows the analysis of multiple runs by normalizing all samples into a single run. The transformed fluorescence ratio k' for all runs was adjusted by sigmoid function using five standards shared between each run (Table S8). By normalizing to Run 1 T/S, allele frequency can now be predicted for all runs using the equation derived from this run (Table 3). The accuracy of prediction is statistically-equivalent to intra-run prediction with 13-standard-points (Figure 3).

Testing Pirimicarb Resistance Allele Frequency in Aphids Collected during the 2011/2012 Australian Cotton Season

To demonstrate the principle, we used qPCR to examine 35 *A. gossypii* samples collected from cotton producing farms across eastern Australia during the 2011/2012 season. Premixed DNA standards of known RAF were run simultaneously with DNAs extracted from a pool of 200 adult aphids. Table 4 lists the predicted resistance allele frequency based on k'. These results are consistent with resistance allele frequency obtained by individual genotyping.

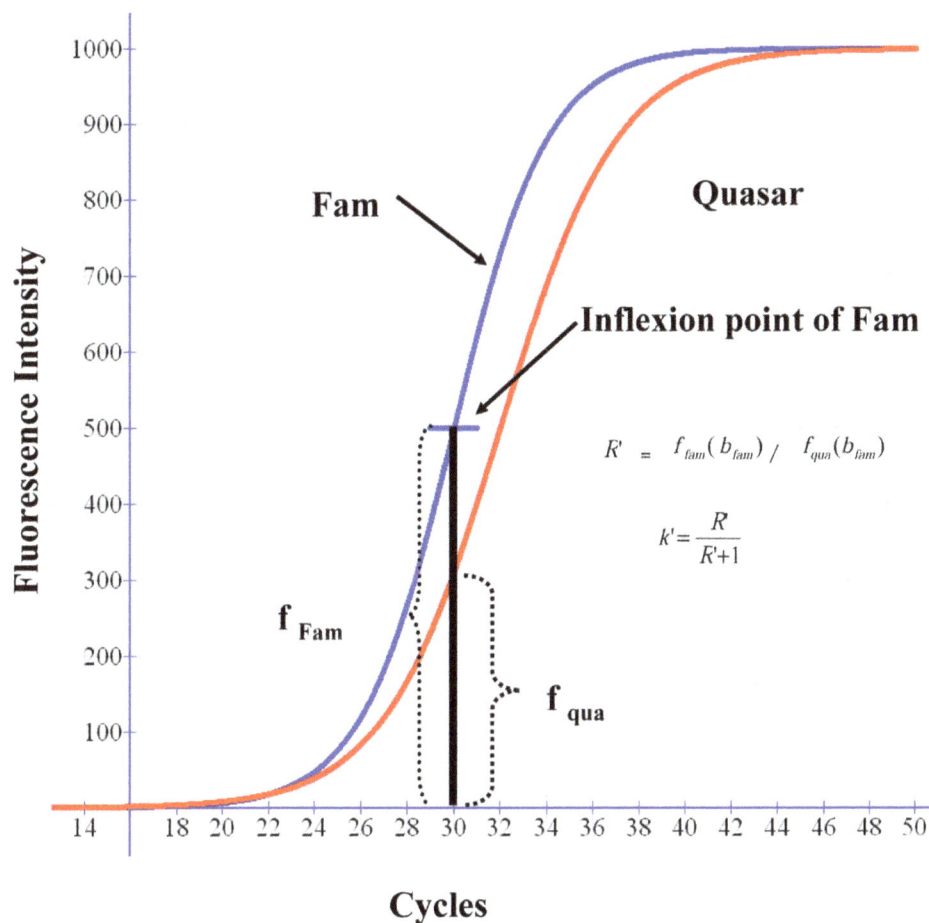

Figure 2. Schematic of the calculation of transformed fluorescence ratio k'.

Discussion

The principle of quantification (Equation 6) states that the ratio of fluorescence from the two allele specific reporter dyes is a function of the initial allele ratio. Our method using the transformed fluorescence ratio at a single, standard time point is able to accurately quantify the allele frequency from pooled DNA samples and fully complies with the principle of PCR quantification. The method is less affected by background variation and so has the potential to overcome the intra-run and inter-run variation.

Independence from Background Signal

A major contributor to observed variance in qPCR data outputs is baseline assignment and significant variation in the baseline fluorescence is often observed in replicate qPCR experiments [18,25]. Baseline variation affects the determination of the reaction threshold yet this parameter is often set automatically by the instrument software at 10 times the standard deviation of baseline. The fluorescence baseline commonly fluctuates between wells, runs and specific instrument being used [18]. Therefore, normalizing background fluorescence often reduces the well-to well variation [25].

The transformed fluorescence ratio k' uses raw fluorescence data points modeled by a four-parametric sigmoid function [23,24]. By using the transformation given in equation 8, the parameters; (a) the maximal height of the curve, (b) the first derivative maximum

of the function and (c) the slope of the curve are less dependent on background fluorescence and the estimation of ΔRn is standardized across different wells and runs.

Single Time Point (Inflexion Point) from Consistent Parameter Estimates

In the past decade, 'assumption free' quantification methods of PCR based on non-linear regression (NLR) have been developed to fit observed parameters and calculate the initial number of target molecules at cycle 0 [24,26–28]. Although these models are mathematically sound and have been reported to contain less well-to-well variation, independent studies show that quantification based on these NLR methods do not outperform the conventional cycle of quantification (C_t) method due to the increased random error of qPCR [29,30].

One factor often unnoticed when using these models is that parameter estimates are significantly influenced by the number of cycles in the plateau phase of PCR. Sigmoid fitting methods are often not reproducible when replicate samples reach the plateau phase at slightly different cycle numbers. Our two-step sigmoid curve fitting method enables a more consistent sigmoid parameter estimate. In undertaking this method, we first fitted a sigmoid curve with all data points to obtain the proximal inflexion point (b) and the slope of the curve (c). Next the sigmoid curve was refitted with only data points from the b+c cycles. By doing that we standardized the data points so that a similar data range exists after the inflexion point for all datasets. Having an equal number

Table 1. The transformed fluorescence ratio k′ comprising 4 runs of plasmid mix (run1-4) and 3 runs of pooled aphids (run5–7) with predefined RAF.

RAF	Run1 T/S		Run2 A/S		Run3 A/S		Run4 T/S		Run5 MP/S		Run6 MP/S		Run7 MP/S		Inter run CV (%)
	k'	CV (%)	k'	CV (%)	k'	CV (%)	k'	CV (%)	k'	CV (%)	k'	CV (%)	k'	CV (%)	
100	0.924	0.083	0.908	0.091	0.917	0.177	0.916	0.211	0.917	2.790	0.888	0.259	0.925	0.459	1.358
95	0.887	0.115	0.862	0.771	0.869	0.544	0.890	0.121	0.886	4.036	0.871	0.042	0.916	0.362	1.923
90	0.851	0.334	0.823	1.344	0.828	0.506	0.861	0.155	0.881	0.259	0.862	2.041	0.885	0.352	2.574
80	0.791	0.485	0.752	*	0.755	0.499	0.802	0.145	0.821	1.872	0.826	0.263	0.849	0.639	4.045
70	0.710	1.095	0.683	2.293	0.679	2.077	0.738	1.666	0.730	0.113	0.757	4.986	0.819	0.440	6.487
60	0.654	1.490	0.622	2.788	0.621	1.254	0.672	0.440	0.721	1.014	0.738	0.492	0.778	1.165	8.421
50	0.588	0.442	0.564	3.107	0.559	1.953	0.615	0.764	0.665	0.855	0.687	3.880	0.705	0.399	9.271
40	0.531	1.410	0.498	0.874	0.491	1.177	0.554	2.422	0.632	0.697	0.632	1.940	0.651	1.817	11.404
30	0.444	0.146	0.395	1.279	0.386	3.190	0.496	1.287	0.551	1.623	0.564	1.068	0.528	1.210	14.763
20	0.392	1.547	0.359	1.052	0.347	2.974	0.444	2.894	0.453	0.419	0.446	2.304	0.508	0.051	12.900
10	0.321	2.345	0.308	1.258	0.295	3.603	0.382	1.324	0.424	1.176	0.421	1.422	0.383	0.948	14.500
5	0.286	2.290	0.280	*	0.272	8.648	0.310	5.008	0.262	0.187	0.254	4.973	0.281	7.650	8.021
0	0.255	3.547	0.246	9.724	0.233	2.572	0.248	2.798	0.184	0.145	0.165	3.689	0.161	10.051	18.964
Intra run CV (%)		1.179		2.235		2.244		1.479		1.168		2.105		1.965	

The transformed fluorescence ratio k′ is the transformation of the ratio of two fluorescence intensities when one fluorescence reaches its inflexion point.

RAF: Predefined Resistance allele frequency (RAF) expressed as percentage.

K′: Average of transformed fluorescence ratio (equation 14) from 3 replicates.

Inter run CV (%): the average coefficient of variation for a standard among 7 runs.

Intra run CV (%): the average coefficient of variation for 13 standards within a run.

*: No replicate.

Table 2. Coefficient of determination R^2 of inter-run k' with linear and 4 parameter sigmoid curve fitting.

R^2	Run2 A/S	Run3 A/S	Run4 T/S	Run5 MP/S	Run6 MP/S	Run7 MP/S
Run1 T/S - linear	0.9975	0.997	0.9931	0.948	0.9262	0.9364
Run1 T/S - sigmoid	0.9989	0.9988	0.9953	0.9747	0.9855	0.9906
Run2 A/S – linear		0.9996	0.9859	0.9246	0.9057	0.9184
Run2 A/S - sigmoid		0.9999	0.9895	0.9675	0.982	0.9922
Run3 A/S - linear			0.9801	0.9229	0.8869	0.9091
Run3 A/S - sigmoid			0.9916	0.9743	0.9805	0.9884
Run4 T/S - linear				0.9722	0.9557	0.9582
Run4 T/S - sigmoid				0.9918	0.9946	0.9968
Run5 MP/S - linear					0.9948	0.9822
Run5 MP/S - sigmoid					0.9991	0.986
Run6 MP/S - linear						0.9869
Run6 MP/S - sigmoid						0.9874

Run1–4 are plasmid mix and run5–7 are pooled aphids.

of cycles after the infection point enables a more robust estimation of the parameters.

A Single Reference Point for Fluorescence Ratio Determination

Earlier work by Oliver *et al* [31] to quantify the initial allele ratio by examining the qPCR end point fluorescence ratio is not ideal for accurate quantification due to the dramatic decrease of amplification efficiency in late PCR cycles. To more accurately predict the initial allele ratio, Yu *et al* [18] used background-normalized fluorescence from both fluorophores in exponential phase. However, the selection of the exponential cycles in this method was arbitrary, particularly when one fluorescence signal reaches the exponential phase much earlier than the other which is often the case when one or the other allele frequency is quite low. In our two-step sigmoid curve fitting method, the fluorescence ratio is measured when a fluorescence signal first reaches the inflexion point (equation 11 and 12) and allows for a standard method of identifying the exponential phase. As the inflexion point is always in the middle of the exponential phase it shows very similar kinetics between replicate samples and so has the potential to be more accurate.

Transformed Fluorescence Ratio k'

The transformed fluorescence ratio k' (Equation 13) permits the development of a standard curve with allele frequency ranging from 0 to 1. The inclusion of a zero allele frequency is critical as a control to assess the sensitivity of the assay. In TaqMan assays,

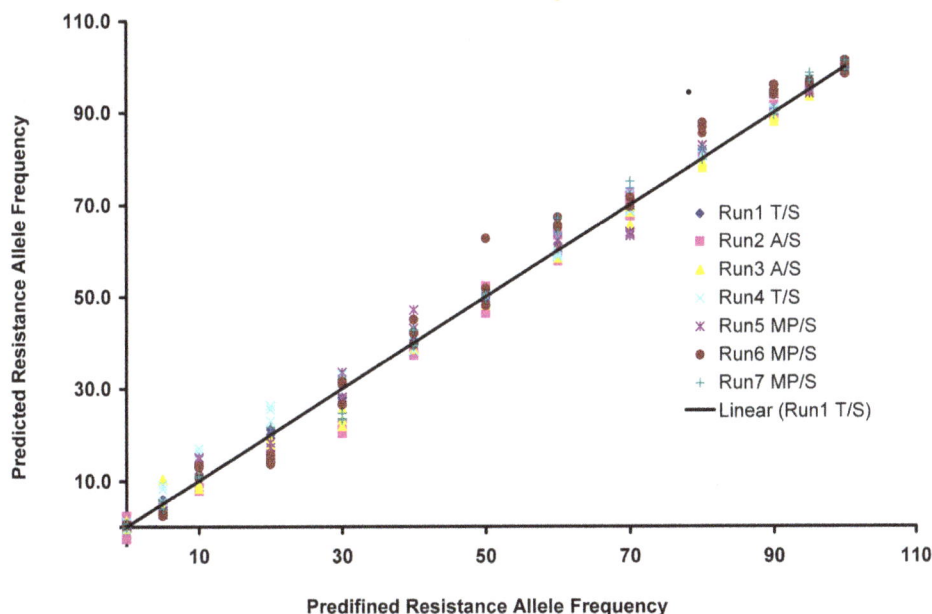

Figure 3. Predicted resistance allele frequency (RAF) for standards based on Run1 T/S. Prediction were based on five standards from Run1 T/S and all calculated transformed fluorescence ratio k' were adjusted to Run1 T/S by sigmoid function.

Table 3. Prediction of RAF by normalizing k' to Run1 T/S.

RAF	Run1 T/S		Run2 A/S		Run3 A/S		Run4 T/S		Run5 MP/S		Run6 MP/S		Run7 MP/S	
Standard	k'	RAF*	k' adj**	RAF*	k' adj	RAF*	k' adj	RAF*	k' adj	RAF*	k' adj	RAF*	k' adj	RAF*
100	0.924	99.9	0.927	100.3	0.926	100.2	0.921	99.5	0.925	100.0	0.918	99.0	0.923	99.7
95	0.887	94.5	0.885	94.2	0.885	94.2	0.895	95.6	0.881	93.5	0.883	93.9	0.907	97.4
90	0.851	89.2	0.849	88.9	0.849	88.8	0.864	91.1	0.873	92.5	0.865	91.2	0.854	89.6
80	0.791	80.2	0.783	79.0	0.784	79.1	0.798	81.3	0.789	80.0	0.798	81.2	0.794	80.7
70	0.71	68.1	0.717	69.1	0.713	68.6	0.724	70.2	0.668	61.8	0.686	64.5	0.746	73.5
60	0.654	59.7	0.657	60.1	0.659	60.4	0.646	58.4	0.657	60.1	0.659	60.4	0.684	64.2
50	0.588	49.7	0.599	51.3	0.599	51.3	0.579	48.3	0.590	50.0	0.593	50.4	0.585	49.2
40	0.531	41.1	0.531	41.1	0.531	41.2	0.510	38.0	0.553	44.4	0.531	41.2	0.522	39.7
30	0.444	28.0	0.422	24.8	0.424	25.0	0.449	28.8	0.471	32.1	0.467	31.5	0.409	22.7
20	0.392	20.2	0.384	19.0	0.383	18.8	0.399	21.3	0.391	20.1	0.381	18.6	0.394	20.5
10	0.321	9.7	0.328	10.6	0.327	10.5	0.345	13.3	0.371	17.0	0.366	16.3	0.322	9.8
5	0.286	4.5	0.297	6.0	0.302	6.8	0.292	5.3	0.284	4.2	0.289	4.9	0.284	4.2
0	0.255	−0.1	0.258	0.4	0.259	0.5	0.253	−0.4	0.255	0.0	0.260	0.7	0.254	−0.2
R^2		0.999		0.998		0.998		0.998		0.989		0.995		0.995

RAF*: Predicted resistance allele frequency from Run1 T/S.
k' adj**: normalized transformed fluorescence ratio k' based on Run1 T/S.

Table 4. Field isolates of *Aphis gossypii* collected during the 2011/2012 season showing pirimicarb resistance status determined by individual PCR-RFLP.

Strain	Region	RAF by qPCR with pooled DNA	RAF by RFLP genotyping of 20 individual aphids	Bioassay
Alch	Darling Downs, QLD	−2.4	0	S
And	Fitzroy, QLD	−1.6	0	S
Aral	Darling Downs, QLD	−0.4	0	S
Arra	Darling Downs, QLD	−1.0	0	S
Bal F3	S. West QLD	−2.3	0	S
Bal Vol	S. West QLD	−2.6	0	S
Boo Dry	Darling Downs, QLD	−1.7	0	S
Boo Irr	Darling Downs, QLD	0.4	0	S
Bor P	S. West QLD	0.6	0	S
Both	Kimberley, WA	0.9	0	S
Both B	Kimberley, WA	102.0	100	R
Bro Cle	S. West QLD	−1.1	0	S
Bro Tre	S. West QLD	1.0	0	S
Bud	Darling Downs, QLD	0.9	0	S
Bur Dry	S. West QLD	102.0	100	R
Car F3	N. Inland, NSW	−1.1	0	S
Car Vol	N. Inland, NSW	1.0	0	S
Carring	N. Inland, NSW	0.9	0	S
Cly	S. West QLD	−1.4	0	S
Doo 1	S. West QLD	−1.4	0	S
Doo 2	S. West QLD	−0.3	0	S
Eum	Darling Downs, QLD	−0.6	0	S
Fair	Darling Downs, QLD	0.0	0	S
Gra 148	Fitzroy, QLD	−0.1	0	S
Mon P	Northern QLD	104.6	100	R
Over	Darling Downs, QLD	−1.2	0	S
P Seed	Kimberley, WA	−0.7	0	S
Spri	N. Inland, NSW	−0.4	0	S
Terr	Darling Downs, QLD	−1.3	0	S
T Sand	Kimberley, WA	104.1	100	R
Walt	Darling Downs, QLD	−1.7	0	S
Wanh F	Kimberley, WA	103.9	100	R
Wise	N. Inland, NSW	−1.3	0	S
Wyad	N. Inland, NSW	−1.0	0	S
Zig	S. West QLD	−1.7	0	S

R: Resistant to Pirimicarb.
S: Susceptible to Pirimicar.

potential errors occur due to significant cross-binding of probes. Even when one allele is absent its corresponding florescence signal can still be observed due to cross binding. Including an allele frequency of 0 and 1 makes it possible to accurately estimate unknown samples with allele frequency <0.05 or <0.95.

Non-linear Relationship between Transformed Fluorescence Ratio k' and RAF

Our results demonstrate a sigmoid relationship between RAF and transformed fluorescence ratio k'. The predicted linear relationship between the initial allele frequency and transformed

fluorescence ratio k' can be achieved for PCR in optimal conditions. However, a non-linear model is more universal given most PCRs are performed in conditions that are not optimal, particularly when unknown PCR inhibitors are present.

The sigmoid function between RAF and transformed fluorescence ratio k' theoretically enables the construction of a standard curve using only 4 standard points and our results demonstrate that the prediction model obtained from 5 rather than 4 standard points was as accurate as the model base on 13 standard points. This reduction in the number of standards required for each run allows for a considerable increase in the number of wells that can be used for samples rather than standards.

Inter-run Correlation

Additionally, we have found that there is a sigmoid relationship between the transformed fluorescence ratios k' across multiple runs. This enables the normalization of samples from multiple experiments into a single run if at least 4 samples are shared in each. Therefore, a single analysis can be performed for all samples across all runs.

Practical Implementation of the Method

Our two-step sigmoid curve fitting method has the potential to be used broadly for high throughput/low cost genotyping. Although this method was developed using a TaqMan assay on an ABI 7500 real-time thermocycler, the principle can theoretically be applied to other fluorescent dye and instrument platforms (such as SYBR green). In some cases, when one allele is absent, there is no PCR amplification or irregular amplification, it is possible to manually estimate fluorescence height above the background at the approximate inflexion point of the other allele for the k' calculation. Alternatively, a predefined allele frequency 0.01 and 0.99 can be used as standard points.

Diagnostic Testing the Pirimicarb Resistance Allele Frequency in Aphids Collected during the 2011/2012 Cotton Season

To examine the effectiveness of our two-step sigmoid curve fitting method we used field samples to predict the pirimicarb-RAF in 35 field isolates of *A. gossypii* and compared those estimated allele frequencies with individual genotyping of 20 aphids from each isolate. A remarkable consistency was observed between the RAF predicted by qPCR and allele frequency predicted by individual genotyping. Unfortunately, the pirimicarb-RAF observed in the 2011/2012 season where either 0% or 100%, making it difficult to statistically assess the precision of the prediction. While this data limitation could not be overcome, the method demonstrated good sensitivity when RAF is low.

This method allows for a dramatic decrease in the amount of labor required for the high-throughput monitoring of RAF in insects of agricultural importance, so aiding sustainable IPM systems. Interestingly, we found that a similar amount of time was required for an experienced worker to extract DNA from 20 aphids individually or 200 aphids combined in one tube. However, there was a great difference in the amount of time required to genotype these samples. Genotyping of 35×20 aphids individually required almost three weeks of work while genotyping of 35×200 aphids using pooled DNA, a TaqMan assay and our two-step sigmoid curve fitting method could be performed in as little as three days.

Conclusion

We have developed a method using a TaqMan SNP assay to accurately estimate the allele frequencies from pooled DNA samples. The method uses the transformed fluorescence ratio based on a single reference point and has proven precise at predicting unknown allele frequencies. The prediction model can be built using five standard points and results can be normalized across multiple runs. The method can dramatically reduce time and labour required for insecticide resistance monitoring and has the potential for broad applications in high throughput genotyping such as genome –wide association studies, population studies even the quantitative assessment of post transplant chimeras in medicine.

Supporting Information

Table S1 An example of two-step four-parameter sigmoid curve fitting.

Table S2 Example of Calculation of k'.

Table S3 The transformed fluorescence ratio k' comprising 4 runs of plasmid mix (run1-4) and 3 runs of pooled aphids (run5-7) with predefined resistance allele frequency (RAF). The transformed fluorescence ratio k' is the transformation of the ratio of two fluorescence intensity when one fluorescence reaches its inflexion point.

Table S4 Test of the linear and non-linear (sigmoid) relationship between RAF and transformed fluorescence ratio k'. Run1-4 are plasmid mix and run5-7 are pooled aphids.

Table S5 Five standard points used for the reduced prediction model.

Table S6 Resistance allele frequencies (RAF) predicted from full and reduced prediction models in plasmid mix runs.

Table S7 Resistance allele frequencies (RAF) predicted from full and reduced prediction models in aphid mix runs.

Table S8 Transformed fluorescence ratio k' between runs can be normalized to Run1 T/S by using five standards based on sigmoid function.

Table S9 Raw fluorescence data for 7 runs.

Author Contributions

Conceived and designed the experiments: YC GH. Performed the experiments: DB. Analyzed the data: IB YC. Contributed reagents/materials/analysis tools: GH. Wrote the paper: YC GH.

References

1. Georguiou GP, Mellon RB (1983) Pesticide resistance in time and space. In: Georguiou GP, Saito T, editors. Pest resistance to pesticides. New York: Plenum Press. 1–46.
2. Mass S (2012) Cotton pest management guide 2012–13. Toowoomba: Greenmount Press.
3. Busvine JR (1971) A critical review of the techniques for testing insecticides. the University of Wisconsin - Madison: Commonwealth Agricultural Bureaux. 345 p.
4. Ffrench-Constant R, Roush R (1991) Resistance detection and documentation: The relative roles of pesticidal and biochemical assays. In: Roush R, Tabashnik B, editors. Pesticide resistance in arthropods: Springer US. 4–38.
5. Yu SJ (2008) The toxicology and biochemistry of insecticides. Boca Raton: CRC Press. 296 p.
6. Ffrenchconstant RH, Rocheleau TA, Steichen JC, Chalmers AE (1993) A point mutation in a drosophila gaba receptor confers insecticide resistance. Nature 363: 449–451.

7. Weill M, Fort P, Berthomieu A, Dubois MP, Pasteur N, et al. (2002) A novel acetylcholinesterase gene in mosquitoes codes for the insecticide target and is non–homologous to the ace gene Drosophila. Proc Biol Sci 269: 2007–2016.

8. Newcomb RD, Campbell PM, Ollis DL, Cheah E, Russell RJ, et al. (1997) A single amino acid substitution converts a carboxylesterase to an organophosphorus hydrolase and confers insecticide resistance on a blowfly. Proc Natl Acad Sci USA 94: 7464–7468.

9. Daborn PJ, Yen JL, Bogwitz MR, Le Goff G, Feil E, et al. (2002) A single p450 allele associated with insecticide resistance in drosophila. Science 297: 2253–2256.

10. Barbagallo S, Cravedi P, Pasqualini E, Patti I (1997) Aphids of the principal fruit bearing crops. Milan: Bayer. 123 p.

11. Herron GA, Powis K, Rophail J (2001) Insecticide resistance in *Aphis gossypii* Glover (Hemiptera : Aphididae), a serious threat to Australian cotton. Aust J Entomol 40: 85–89.

12. Andrews MC, Callaghan A, Field LM, Williamson MS, Moores GD (2004) Identification of mutations conferring insecticide-insensitive AChE in the cotton-melon aphid, Aphis gossypii Glover. Insect Mol Biol 13: 555–561.

13. Toda S, Komazaki S, Tomita T, Kono Y (2004) Two amino acid substitutions in acetylcholinesterase associated with pirimicarb and organophosphorous insecticide resistance in the cotton aphid, Aphis gossypii Glover (Homoptera: Aphididae). Insect Biochem Mol Biol 13: 549–553.

14. McLoon MO, Herron GA (2009) PCR detection of pirimicarb resistance in Australian field isolates of *Aphis gossypii* Glover (Aphididae: Hemiptera). Aust J Entomol 48: 65–72.

15. Billard A, Laval V, Fillinger S, Leroux P, Lachaise H, et al. (2012) The allele-specific probe and primer amplification assay, a new real-time pcr method for fine quantification of single-nucleotide polymorphisms in pooled DNA. Appl Environ Microbiol 78: 1063–1068.

16. Breen G, Harold D, Ralston S, Shaw D, St Clair D (2000) Determining SNP allele frequencies in DNA pools. Biotechniques 28: 464–466, 468, 470.

17. Psifidi A, Dovas C, Banos G (2011) Novel Quantitative Real-Time LCR for the Sensitive Detection of SNP Frequencies in Pooled DNA: Method Development, Evaluation and Application. PLoS One 6: e14560.

18. Yu A, Geng H, Zhou X (2006) Quantify single nucleotide polymorphism (SNP) ratio in pooled DNA based on normalized fluorescence real-time PCR. BMC Genomics 7: 143.

19. Chen J, Germer S, Higuchi R, Berkowitz G, Godbold J, et al. (2002) Kinetic polymerase chain reaction on pooled DNA: a high-throughput, high-efficiency alternative in genetic epidemiological studies. Cancer Epidemiol Biomarkers Prev 11: 131–136.

20. Germer S, Holland MJ, Higuchi R (2000) High-throughput snp allele-frequency determination in pooled DNA samples by kinetic pcr. Genome Res 10: 258–266.

21. Payne RW, Harding SA, Murray DA, Soutar DM, Baird DB, et al. (2007) Genstat release 10 reference manual. Hemel Hempstead: VSN International.

22. Swillens S, Goffard J, Marechal Y, de Kerchove d'Exaerde A, El Housni H (2004) Instant evaluation of the absolute initial number of cDNA copies from a single real-time PCR curve. Nucleic Acids Res 32: e56.

23. Johnson M, Haupt L, Griffiths L (2004) Locked nucleic acid (LNA) single nucleotide polymorphism (SNP) genotype analysis and validation using real-time PCR. Nucleic Acids Res 32: e55.

24. Liu W, Saint D (2002) Validation of a quantitative method for real time PCR kinetics. Biochem Biophys Res Commun 294: 347–353.

25. Carr AC, Moore SD (2012) Robust quantification of polymerase chain reactions using global fitting. PLoS One 7: e37640.

26. Ramakers C, Ruijter JM, Deprez RHL, Moorman AFM (2003) Assumption-free analysis of quantitative real-time polymerase chain reaction (PCR) data. Neurosci Lett 339: 62–66.

27. Rutledge RG (2004) Sigmoidal curve-fitting redefines quantitative real-time PCR with the prospective of developing automated high-throughput applications. Nucleic Acids Res 32: e178.

28. Spiess AN, Feig C, Ritz C (2008) Highly accurate sigmoidal fitting of real-time PCR data by introducing a parameter for asymmetry. BMC Bioinformatics 9: 221.

29. Karlen Y, McNair A, Perseguers S, Mazza C, Mermod N (2007) Statistical significance of quantitative PCR. BMC Bioinformatics 8: 131.

30. Bar T, Kubista M, Tichopad A (2012) Validation of kinetics similarity in qPCR. Nucleic Acids Res 40: 1395–1406.

31. Oliver DH, Thompson RE, Griffin CA, Eshleman JR (2000) Use of single nucleotide polymorphisms (snp) and real-time polymerase chain reaction for bone marrow engraftment analysis. J Mol Diagn 2: 202–208.

15

Mortality from and Incidence of Pesticide Poisoning in South Korea: Findings from National Death and Health Utilization Data between 2006 and 2010

Eun Shil Cha[1], Young-Ho Khang[2,3], Won Jin Lee[1]*

1 Department of Preventive Medicine, College of Medicine, Korea University, Seoul, South Korea, **2** Department of Preventive Medicine, University of Ulsan College of Medicine, Seoul, South Korea, **3** Institute of Health Policy and Management, Seoul National University College of Medicine, Seoul, South Korea

Abstract

Pesticide poisoning has been recognized as an important public health issue around the world. The objectives of this study were to report nationally representative figures on mortality from and the incidence of pesticide poisoning in South Korea and to describe their epidemiologic characteristics. We calculated the age-standardized rates of mortality from and the incidence of pesticide poisoning in South Korea by gender and region from 2006 through 2010 using registered death data obtained from Statistics Korea and national healthcare utilization data obtained from the National Health Insurance Review and Assessment Service of South Korea. During the study period of 2006 through 2010, a total of 16,161 deaths and 45,291 patients related to pesticide poisoning were identified, marking respective mortality and incidence rates of 5.35 and 15.37 per 100,000 population. Intentional self-poisoning was identified as the major cause of death due to pesticides (85.9%) and accounted for 20.8% of all recorded suicides. The rates of mortality due to and incidence of pesticide poisoning were higher in rural than in urban areas, and this rural-urban discrepancy was more pronounced for mortality than for incidence. Both the rate of mortality due to pesticide poisoning and its incidence rate increased with age and were higher among men than women. This study provides the magnitude and epidemiologic characteristics for mortality from and the incidence of pesticide poisoning at the national level, and strongly suggests the need for further efforts to prevent pesticide self-poisonings, especially in rural areas in South Korea.

Editor: Suryaprakash Sambhara, Centers for Disease Control and Prevention, United States of America

Funding: This work was supported by the National Research Foundation of Korea (NRF) Grant funded by the Korean government (MOE) (No. 2010-0021742), Republic of Korea. The funders had no role in study design, data collection and analysis, decision to publish, or preparation of the manuscript.

Competing Interests: The authors have declared that no competing interests exist.

* E-mail: leewj@korea.ac.kr

Introduction

Pesticide poisoning has been recognized as an important public health issue around the world [1]. Approximately 350,000–440,000 annual suicides by means of deliberate pesticide poisoning have been estimated to occur worldwide [2], and the numbers of victims of nonfatal pesticide poisoning are assumed to be much greater. However, few studies [3–5] have been conducted to evaluate nationally representative figures of acute pesticide poisoning. Identifying at the national level the magnitude of fatal and nonfatal pesticide poisoning and high risk individuals would well provide scientific evidence regarding the overall disease burden and serve as a fundamental step forward in developing national strategies to curb this grave issue.

Previously, we reported the total number of pesticide poisoning deaths from 1996 through 2005 to be 25,360 and that the age-standardized mortality rates due to pesticide poisoning significantly increased during this period from 4.42 to 6.42 per 100,000 population, accounting for the largest proportion of deaths among all forms of poisoning [6]. We also estimated an annual rate of pesticide-related hospitalization in South Korea of 17.8 per 100,000 population [7] and of the emergency department visit of 26.8 per 100,000 population [8]. However, these figures were derived from sampled hospitals accounting for about 30% of all

hospitals across South Korea. Moreover, relatively recent figures on the time trends in mortality from pesticide poisoning in South Korea have not been presented since 2005.

Therefore, we aimed to update the rate of mortality using registered death data and investigate the incidence and describe the epidemiological characteristics of pesticide poisoning using National Health Insurance claims data, which cover the entire population of South Korea.

Methods

Mortality Data

By law, all deaths of South Koreans must be reported to Statistics Korea within one month of their occurrence. Mortality data for South Korea between 2006 and 2010 were obtained from the registered death data provided by Statistics Korea [9]. The registered death data include information on age, gender, administrative district of residence, educational level, month of death, occupation, and marital status. Underlying causes of disease are coded in the data according to the *International Classification of Diseases and Related Health Problems, 10th Revision* (ICD-10) [10]. Pesticide poisoning deaths were defined as those featuring the code for toxic effect of pesticides (T60.0–T60.9). External cause of injury codes were used to classify the causes of pesticide poisoning deaths into intentional self-poisoning (X68) and unintentional poisoning [i.e., accidental poisoning (X48), assault (X87), and undetermined intent poisoning by and exposure to pesticides (Y18)]. Based on the 251 administrative residential districts used in the death data, urbanity levels were identified. Metropolises included Seoul and six other metropolitan cities in South Korea. Districts in the nine provinces other than these metropolises were classified into either cities or rural areas depending on governmental administrative divisions determined by population size and rural characteristics.

National Health Insurance Claims Data

The National Health Insurance of South Korea was established in 1989 and by statute covers the entire South Korean population and all medical institutions in the nation. Health care utilization data were extracted from the Korea National Health Insurance claims database retrieved from the National Health Insurance Review and Assessment Service between 2006 and 2010. All inpatient and outpatient medical utilization related to pesticide poisoning is expected to be included in the database. The National Health Insurance claims data contain primary and additional diagnostic codes per ICD-10. Identical ICD-10 codes (T60.0–T60.9) as in mortality data were used for this incidence data. The initial episode of emergency department visit or hospital admission during the study period of January 1, 2006 through December 31, 2010 was treated as one incident case of pesticide poisoning when a patient was examined at multiple hospitals or at a single hospital for several different conditions. In National Health Insurance claims data, only few records were obtained by external cause of injury code, and it was nearly impossible to differentiate intention of pesticide poisoning. The National Health Insurance claims data also contained information on types of care received, dates of hospitalization, address of medical institutions, and patients' genders and ages. Due to the lack of information regarding patients' addresses in our data, the locations of medical institutions were used as a proxy measure for patients' residential area and classified into metropolis, small/medium city, and rural area. The season in which pesticide poisoning occurred was extracted using the month of the occurrence of hospital visit.

Data Analysis

The rates of mortality and incidence were directly standardized to 10-year age groups, using the 2000 World Standard Population [11] as the standard population in this direct standardization. The population data and the calculation process of age-standardized rate for pesticide poisoning were presented as Appendix (Tables S1 and S2 in Appendix S1). Population registration data by gender, 10-year age groups, and the 251 administrative districts were obtained from Statistics Korea [9]. The annual rates of pesticide poisoning mortality and incidence were calculated as number of deaths and incident cases per 100,000 population, and were stratified by gender and urbanization level. Descriptive statistics by sociodemographic variables for pesticide poisoning were presented. All statistical tests were performed via STATA, version 12.0 (StataCorp, College Station, Texas).

Ethics Approval

We used publicly available mortality and healthcare utilization data without any personal identifiers, and thus ethical approval was unnecessary.

Results

As shown in Table 1, a total of 16,161 deaths caused by pesticide poisoning were identified during the study period between 2006 and 2010, with an average annual death rate due to pesticide poisoning of 5.35 per 100,000 population. Those rates declined slightly from 5.74 to 4.85 per 100,000 population over the course of the five years. Intentional pesticide poisoning was the major cause of death from pesticide poisoning, accounting for 85.9% of all pesticide poisoning deaths (13,890 deaths over five years). The number of patients with pesticide poisoning found in the National Health Insurance data was 45,291 during the study period, with an average annual incidence rate of 15.37 per 100,000 population. The rates remained generally stable over the five years, with a slight decline witnessed in 2010.

Table 2 illustrates that pesticide poisoning accounted for 66.9% of total poisoning deaths, whereas only 7.5% of all incident poisoning cases were due to pesticides. The proportion of death from pesticide poisoning among total poisoning deaths was similar between men and women (66.8% and 67.0%, respectively), but the proportion of incidence was higher among men than women (8.3% and 6.5%, respectively). The proportion of pesticide poisoning among total poisoning increased with age for both mortality and incidence data. Rural areas recorded the highest proportion (79.1%) of pesticide poisoning deaths among total poisoning deaths, but this was not the case with incidence. The proportion of pesticide poisoning among total suicide deaths was 20.8% overall and showed a wide variation between the 9.3% found in metropolises and 47.4% in rural areas.

Figure 1 shows urban-rural distinctions in and age patterns of pesticide poisoning mortality and incidence. Both mortality and incidence showed the highest rates in rural areas, with 14.67 and 33.75 per 100,000 population, respectively, while those in metropolises showed the lowest (2.17 per 100,000 population for mortality and 7.19 per 100,000 population for incidence) (Figure 1 (a)). The rural-metropolis ratio in the rate of pesticide poisoning appears to be more pronounced in mortality (14.67/2.17 = 6.76) than in incidence (33.75/7.19 = 4.69). The mortality and incidence rates of pesticide poisoning exponentially increased with age and were higher among men than women across all age groups (Figure 1 (b), (c)). When we calculated the age-specific mortality rate by intention of death, the mortality rate for intentional self-poisoning demonstrated a greater increase with age compared to unintentional pesticide poisoning deaths (data not shown here).

As shown in Table 3, the mortality rate of intentional pesticide poisoning was the highest (11.25 per 100,000 population) in rural areas, followed by the rates in cities (5.25 per 100,000 population) and metropolises (1.95 per 100,000 population). Regardless of urbanization level, the mortality rate of intentional pesticide poisoning was higher among men than women, and the age-specific rate increased with age. In rural areas, the majority of pesticide poisonings occurred among agricultural workers and the unemployed, and herbicides and fungicides were most-widely used by suicide victims, but the roles of these agents were more important in mortality than in incidence (data not shown here). Similar seasonal variations were observed for all areas.

Discussion

The results of this study indicate that pesticide poisoning is a prevalent public health problem, resulting in 5.35 deaths per 100,000 population and 15.37 incidence cases per 100,000 population annually in South Korea during the period of 2006–2010. Pesticide poisoning accounted for a large portion of suicide in South Korea. Pesticide poisoning rates were higher in rural

Table 1. Age-standardized mortality and incidence rate of pesticide poisoning per 100,000 population by year in South Korea, 2006–2010.

Year	Mortality									Incidence								
	Total (T60)			Intentional (X68)			Unintentional (X48, X87, Y18)			Total			Hospitalization			Outpatient		
	Cases	ASR[a]	(95% CI)	Cases	ASR	(95% CI)	Cases	ASR	(95% CI)	Cases	ASR	(95% CI)	Cases	ASR	(95% CI)	Cases	ASR	(95% CI)
2006	3,201	5.74	(5.54–5.94)	2,747	4.91	(4.72–5.09)	454	0.83	(0.75–0.91)	9,186	16.57	(16.23–16.92)	4,999	8.97	(8.72–9.22)	4,187	7.60	(7.37–7.83)
2007	3,288	5.68	(5.49–5.88)	2,881	4.96	(4.78–5.14)	407	0.72	(0.65–0.79)	9,660	16.87	(16.53–17.21)	5,185	8.98	(8.74–9.23)	4,475	7.89	(7.66–8.13)
2008	3,296	5.50	(5.31–5.68)	2,800	4.66	(4.48–4.83)	496	0.84	(0.76–0.91)	10,046	17.16	(16.82–17.50)	5,638	9.50	(9.25–9.75)	4,408	7.66	(7.43–7.89)
2009	3,170	5.07	(4.89–5.24)	2,743	4.38	(4.22–4.55)	427	0.68	(0.62–0.75)	9,921	16.45	(16.12–16.78)	5,732	9.35	(9.11–9.60)	4,189	7.09	(6.87–7.31)
2010	3,206	4.85	(4.68–5.02)	2,719	4.11	(3.96–4.27)	487	0.73	(0.67–0.80)	9,367	14.65	(14.35–14.95)	5,240	8.12	(7.90–8.35)	4,127	6.53	(6.32–6.73)
Total	16,161	5.35	(5.27–5.44)	13,890	4.59	(4.52–4.67)	2,271	0.76	(0.73–0.79)	45,291[b]	15.37	(15.22–15.51)	26,274	8.81	(8.70–8.91)	19,016	6.56	(6.47–6.66)

[a] Age-standardized rates per 100,000 population using the 2000 World Standard Population.
[b] Excluding duplicated patients during the study period, 2006–2010.

Table 2. The proportion of pesticide poisoning among total poisoning, and total suicides by sex, age, and area of residence in South Korea, 2006–2010.

	Death cases				Incident cases	
	Pesticide poisoning deaths	PPP[a]	Suicide by pesticides	PPS[b]	Pesticide poisoning cases	PPP
	Number (%)	%	Number (%)	%	Number (%)	%
Total	16,161 (100)	66.9	13,890 (100)	20.8	45,291 (100)	7.5
Sex						
Men	10,844 (67.1)	66.8	9,361 (67.4)	21.5	26,855 (59.3)	8.3
Women	5,317 (32.9)	67.0	4,529 (32.6)	19.5	18,436 (40.7)	6.5
Age group						
0–9	7 (0.1)	11.9	0 (0.0)	0.0	1,256 (2.8)	2.0
10–19	40 (0.2)	21.4	35 (0.3)	2.1	510 (1.1)	1.4
20–29	282 (1.7)	26.3	265 (1.9)	3.4	1,702 (3.8)	3.0
30–39	1,033 (6.4)	44.4	935 (6.7)	8.8	4,318 (9.5)	5.4
40–49	2,211 (13.7)	57.7	2,015 (14.5)	16.0	8,254 (18.2)	7.3
50–59	2,564 (15.9)	66.9	2,290 (16.5)	21.0	8,449 (18.7)	8.0
60–69	3,570 (22.1)	75.3	3,076 (22.2)	31.9	9,500 (21.0)	11.6
70–79	4,235 (26.2)	80.5	3,536 (25.4)	40.0	8,077 (17.8)	15.6
≥80	2,218 (13.7)	77.6	1,737 (12.5)	36.7	3,224 (7.1)	21.4
Area						
Rural	5,488 (34.0)	79.1	4,560 (32.8)	47.4	11,137 (24.6)	7.1
City	7,808 (48.3)	68.6	6,812 (49.0)	22.8	24,915 (55.0)	8.6
Metropolis	2,865 (17.7)	49.3	2,518 (18.2)	9.3	9,239 (20.4)	5.8

[a]Proportion of pesticide poisoning among total poisoning.
[b]Proportion of pesticide poisoning among total suicide deaths.

areas and among the elderly, agricultural workers, and unemployed individuals. This study presented pesticide poisoning mortality and incidence simultaneously at the national level and emphasized the urgency of constructing preventive national strategies to reduce pesticide poisoning in South Korea.

Although the mortality and incidence of pesticide poisoning declined during the study period, our estimates are much greater than the rates derived in a number of other developed countries such as Taiwan [3], Japan [4], and US [5,12]. The magnitudes of pesticide poisoning mortality (5.35 per 100,000 population) shown in our study surpass those of a number of major causes of deaths in South Korea in 2010, such as breast cancer (mortality rate of 4.0 per 100,000 population), and Alzheimer's disease (mortality rate of 4.8 per 100,000 population) [9].

The most important reason underlying this high mortality from pesticide poisoning is the elevated proportion of suicide by means of ingestion of pesticides, which accounted for 85.9% of total pesticide poisoning deaths. South Korea has the highest suicide rate (33.3 out of 100,000 population, based on 2011 data) among all OECD (Organization for Economic Co-operation and Development) countries and this suicide rate has been sharply increasing [13]. Based on South Korea's country profile from the recent Global Burden of Disease Study 2010 [14], suicide is now the second most important cause of premature deaths as measured by years of life lost in South Korea. Suicide through pesticides is the second-most frequently used method, which accounted for 20.8% of total suicides, followed by hanging (50.9%) in South Korea during the study period of 2006 through 2010. South Korea has a relatively large proportion of suicide cases from

pesticide poisoning compared with other developed countries such as Japan and Taiwan [15,16].

The high rate of pesticide ingestion in suicide in South Korea may be explained by their easy accessibility in South Korea. Easy access to pesticides is believed make pesticide self-poisoning a preferred means of self-harm [17,18]. Although South Korea does maintain regulations covering the buying or selling of restricted pesticides, in the absence of licensing requirements individuals experience little difficulty in bypassing the regulations and purchasing pesticides. The regional and seasonal variations in pesticide poisoning presented in this study may also corroborate this explanation of accessibility to pesticides. Considering that widespread access to pesticides may easily convert a number of impulsive acts into suicide by means of pesticide ingestion, restrictions on pesticides should be a priority for suicide prevention efforts in South Korea.

The study results showed that pesticide self-poisoning is a largely rural phenomenon and is the most common method of self-harm resulting in death, accounting for 47.4% of total rural suicides in South Korea. Rural suicides involving pesticides have been documented in other Asian countries such as Sri Lanka [19], Taiwan [20], and China [21]. Numerous aspects of rural life, such as socioeconomic disadvantages, limited availability of and access to emergency medical services, an aging population, and geographic and interpersonal isolation may all contribute to the elevated rates of suicide in rural areas [22,23]. Such explanations may also be applied in a Korean context, compounded by easy access to pesticides. The number of people aged 65 years and older has been rapidly increasing in South Korea and the proportion of the elderly was 33.7% in rural areas in 2011 compared to the

(a)

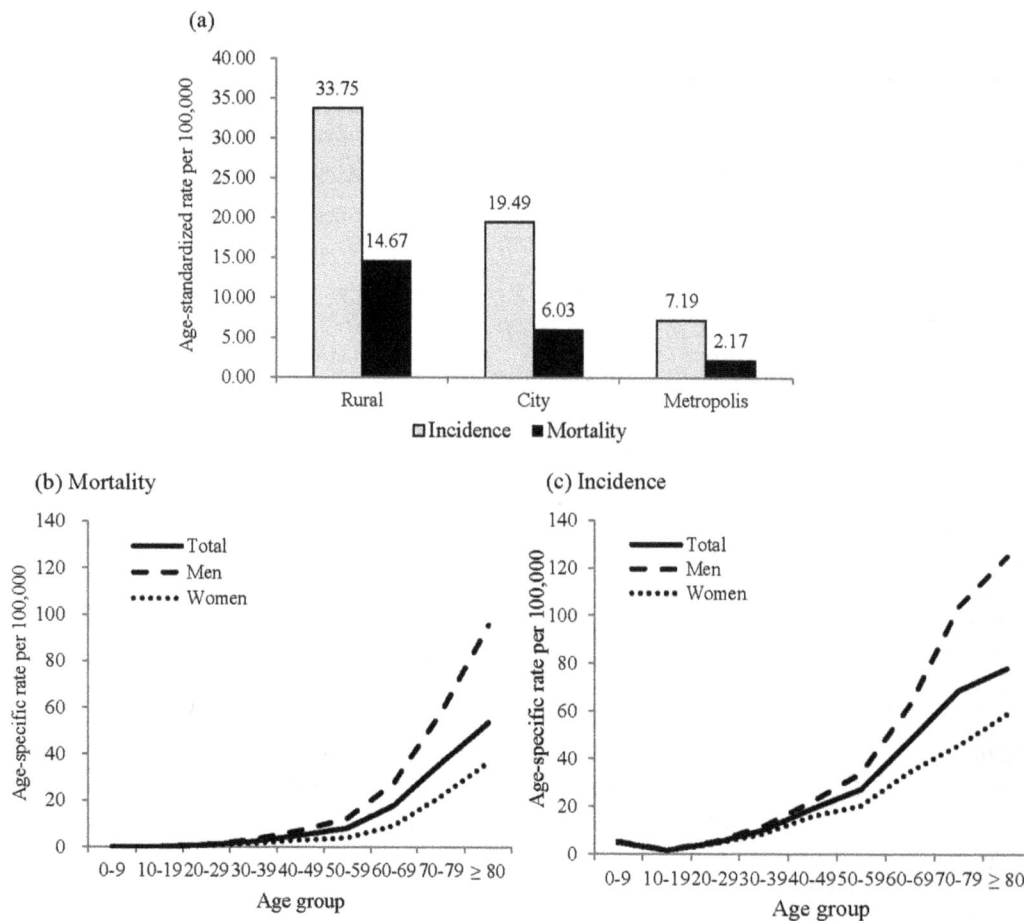

(b) Mortality

(c) Incidence

Figure 1. (a) Age-standardized incidence and mortality rate of pesticide poisoning per 100,000 population by area, (b) Age-specific mortality and (c) incidence rate of pesticide poisoning per 100,000 population by sex in South Korea, 2006–2010.

national average of 11.4% [24]. The elderly are the most vulnerable group in terms of self-destructive behavior and disproportionately represent the socially disadvantaged in contemporary Korean society [25]. The elderly make more serious suicide attempts using more-lethal methods [26] and access to lethal means such as pesticide ingestion seems to play an important role in suicides by the elderly in rural settings.

The results of the study revealed that men in rural areas were 2.4 times (16.58 vs. 6.85 per 100,000 population) more likely to die as a result of intentional pesticide poisoning than were women in rural areas. Corresponding figures for cities and metropolises were 2.6 times (8.04 vs. 3.09 per 100,000 population) and 2.8 times (3.05 vs. 1.09 per 100,000 population), respectively. The gender ratio in deaths by suicide stands at around 2.4 to 1 in South Korea [25] while the suicide rates in many high-income countries are typically more than three times higher among men than women [27]. Men have traditionally been more commonly engaged in occupations involving pesticides and have greater access to pesticides than do women. Thus, it can be suggested that the access to pesticides has allowed a relatively lower gender ratio in intentional pesticide poisoning in rural areas than that found in urban areas, as well as high suicide rates due to pesticide poisoning both in rural areas and across the country. In addition, this accessibility might have in part contributed to the higher suicide mortality rates and relatively lower gender ratio in suicide mortality compared with Western high-income countries. More-

over, our results that the rate of pesticide self-poisoning was the highest among men aged above 80 in rural areas (85.5 per 100,000 population) reflects a combined effect of risk factors for pesticide-related suicide among rural elderly men.

The poor quality of medical service in rural areas can be an important risk factor for elevated overall mortality, including mortality due to pesticide poisoning. One recent study reported that in South Korea rural areas showed higher mortality rates from major diseases, including poisoning, than did urban areas, although the prevalence of the diseases were similar in both rural and urban areas across the country [28]. Our results that the rural-metropolis ratio in the rate of pesticide poisoning appears to be more pronounced in mortality than in incidence support these results.

The large proportion of farmers in rural areas may also contribute to the high mortality and incidence of pesticide poisoning in such areas. In South Korea, an elevated occupational pesticide poisoning incidence of 24.7 per 100 population among male agricultural workers has been reported using nationwide survey data [29], and the incidence was significantly related with work-related factors such as poor personal hygiene practices and failure to make use of personal protective equipment [30]. Average pesticide application by unit of agricultural land in South Korea is also much higher than in other developed countries [31]. Furthermore, it has been suggested that the exposure to high levels of pesticides, including poisoning, experienced by agriculture

Table 3. Geographic difference of intentional pesticide poisoning in South Korea, 2006–2010.

Characteristic	Total	Rural	City	Metropolis
Total	13,890 (100)	4,560 (100)	6,812 (100)	2,518 (100)
ASR (95% CI)[a]				
Total	4.59	11.25	5.25	1.95
	(4.52–4.67)	(10.90–11.61)	(5.12–5.37)	(1.88–2.03)
Men	7.07	16.58	8.04	3.05
	(6.93–7.22)	(15.97–17.19)	(7.80–8.28)	(2.90–3.20)
Women	2.69	6.85	3.09	1.09
	(2.61–2.77)	(6.45–7.25)	(2.96–3.22)	(1.01–1.17)
Age-specific rate[b]				
0–19	0.06	0.4	0.05	0.01
20–29	0.7	3.1	0.9	0.3
30–39	2.2	9.9	2.5	0.9
40–49	4.7	18.9	5.0	2.1
50–59	7.4	21.4	8.9	3.3
60–69	15.5	37.1	18.1	6.5
70–79	30.2	54.9	35.2	12.1
≥80	42.1	85.5	45.3	15.4
Occupation (case, %)				
Managers	58 (0.4)	11 (0.2)	30 (0.4)	17 (0.7)
Professionals (including technicians)	210 (1.5)	40 (0.9)	113 (1.7)	57 (2.3)
Clerical support workers	325 (2.3)	59 (1.3)	180 (2.6)	86 (3.4)
Service and sales workers	668 (4.8)	166 (3.6)	318 (4.7)	184 (7.3)
Skilled agricultural, forestry and fishery workers	3,923 (28.2)	2,052 (45.0)	1,736 (25.5)	135 (5.4)
Craft and related trades workers	229 (1.7)	45 (1.0)	130 (1.9)	54 (2.1)
Plant and machine operators, and assemblers	106 (0.8)	13 (0.3)	68 (1.0)	25 (1.0)
Elementary occupations	519 (3.7)	153 (3.4)	267 (3.9)	99 (3.9)
Unemployed (including students and homemakers)	7,411 (53.4)	1,864 (40.9)	3,772 (55.4)	1,775 (70.5)
Unknown, Military	441 (3.2)	157 (3.4)	198 (2.9)	86 (3.4)
Causative agents (case, %)				
Organophosphate and carbamate insecticides (T60.0)	1,449 (10.4)	345 (7.6)	602 (8.8)	502 (19.9)
Halogenated insecticides (T60.1)	22 (0.1)	4 (0.1)	7 (0.1)	11 (0.5)
Other insecticides (T60.2)	112 (0.8)	31 (0.7)	57 (0.8)	24 (1.0)
Herbicides and fungicides (T60.3)	9,329 (67.2)	3,348 (73.4)	4,704 (69.1)	1,277 (50.7)
Rodenticides (T60.4)	25 (0.2)	9 (0.2)	10 (0.2)	6 (0.2)
Other pesticides (T60.8)	206 (1.5)	47 (1.0)	96 (1.4)	63 (2.5)
Pesticides, unspecified (T60.9)	2,747 (19.8)	776 (17.0)	1,336 (19.6)	635 (25.2)
Season (case, %)				
Spring	4,026 (29.0)	1,344 (29.5)	1,954 (28.7)	728 (28.9)
Summer	4,215 (30.4)	1,403 (30.8)	2,073 (30.4)	739 (29.4)
Fall	3,255 (23.4)	1,045 (22.9)	1,623 (23.8)	587 (23.3)
Winter	2,394 (17.2)	768 (16.8)	1,162 (17.1)	464 (18.4)

[a]Age-standardized rate per 100,000 population using the 2000 World Standard Population.
[b]Age-specific rate per 100,000 population.

workers and rural residents may result in an elevated risk of neuropsychiatric sequelae such as mood disorders, depression, and suicide attempts [32,33]. Recently, we found that depressive symptoms [34] and suicidal ideation [35] were significantly associated with a history of acute occupational pesticide poisoning and that this relationship was further associated with the severity of the symptoms of poisoning among Korean male farmers.

We also found that highly toxic herbicides as a suicide method were more commonly used in rural areas than in metropolises, which may well explain higher ratio in mortality due to pesticide poisoning in rural areas. Paraquat is a non-selective herbicide that has been ranked as one of the most commonly used pesticides, and has been widely used as the main causative agent for suicide attempts in South Korea [36]. A survey based on nationwide 38 hospitals reported that the fatality of paraquat was 78%, whereas the average fatality of pesticides poisoning was as 22% in South Korea [37]. Although South Korea implemented the Act on Paraquat Regulations in 1999 and revised it in 2005, mortality due to paraquat was still seen to be high thereafter. Recently, the South Korean government banned the selling of paraquat from the end of 2012 but the paraquat sold prior to the ban continues to exist in South Korea due to the lack of further progressive policies such as recalling paraquat from the market and from farmers. Although paraquat poisoning cases can be anticipated to decrease with the new government policy, it is important to monitor the change in suicide by pesticide ingestion for the identification of alternative pesticides in the future.

This study features an important strength in terms of the simultaneous report of nationally representative figures on mortality from and the incidence of pesticide poisoning. However, because the two databases used in this study were not individually linked, certain limitations exist in the identification of the fatality of pesticide poisoning or related risk factors. In addition, the National Health Insurance claims data may not include deaths occurring outside hospitals. Patients who had been referred to other hospitals or voluntarily discharged from hospitals in the National Health Insurance claims data were not determined the status of death in our analysis.

Another limitation of our study is that the death certificates and National Health Insurance data we used were mainly for administrative purposes, therefore, they do not include detailed information for research such as the detailed place of death, time to death from ingestions and other co-morbid factors. In particular, the National Health Insurance claims data had no information regarding intent of poisoning and suicide, which hinders further analyses. In addition, the magnitude of pesticide poisoning based on hospital records or mortality data in this study may underestimate the full impact of pesticide poisoning.

Suboptimal levels of validity for the coding of pesticide poisoning cases may also be a potential weakness of our study.

However, in South Korea, the death registration rate has reached near completeness and 97.5% of all deaths in the national registered death data in 2011 are confirmed by a physician's diagnosis [38]. According to a validation study on the diagnostic codes found in the National Health Insurance claims database in South Korea for the years 1999 and 2001, the consistency of diagnostic codes with medical records was determined to be approximately 70% and tended to be higher for hospital admissions and severe conditions than for outpatients or mild cases [39].

Conclusions

We have demonstrated through the national death and healthcare utilization data that the magnitude of acute pesticide poisoning in South Korea is higher than that in other developed countries. The majority of pesticide poisoning deaths were the result of intentional poisoning; in particular, elderly suicide by pesticide ingestion in rural areas was shown to be a serious social problem. Easy access to pesticides and the lack of management of suicide by pesticide ingestion in rural area are suggested as major factors related with the high rate of pesticide poisoning in South Korea. Therefore, intensive intervention efforts, such as the strict regulation of toxic pesticides and prevention efforts directed at controlling suicide are critically needed to reduce the burden of pesticide poisoning in South Korea.

Supporting Information

Appendix S1 Contains the files: Table S1. Number of population and age-specific death rate for pesticide poisoning in 2010. Table S2. Age-standardized death rates of pesticide poisoning in 2010 by WHO World Standard Population.

Author Contributions

Conceived and designed the experiments: WJL. Performed the experiments: ESC. Analyzed the data: ESC. Wrote the paper: ESC YHK WJL.

References

1. Bertolote JM, Fleischmann A, Eddleston M, Gunnell D (2006) Deaths from pesticide poisoning: a global response. Br J Psychiatry 189: 201–203.
2. Gunnell D, Eddleston M, Phillips MR, Konradsen F (2007) The global distribution of fatal pesticide self-poisoning: systematic review. BMC Public Health 7: 357.
3. Chien WC, Lin JD, Lai CH, Chung CH, Hung YC (2011) Trends in poisoning hospitalization and mortality in Taiwan, 1999–2008: a retrospective analysis. BMC Public Health 11: 703.
4. Ito T, Nakamura Y (2008) Deaths from Pesticide Poisoning in Japan, 1968–2005: Data from Vital Statistics. J Rural Med 3: 5–9.
5. Langley RL, Mort SA (2012) Human exposures to pesticides in the United States. J Agromedicine 17: 300–315.
6. Lee WJ, Cha ES, Park ES, Kong KA, Yi JH, et al. (2009) Deaths from pesticide poisoning in South Korea: trends over 10 years. Int Arch Occup Environ Health 82: 365–371.
7. Kim HJ, Cha ES, Ko Y, Kim J, Kim SD, et al. (2012) Pesticide poisonings in South Korea: findings from the National Hospital Discharge Survey 2004–2006. Hum Exp Toxicol 31: 751–758.
8. Ko Y, Kim HJ, Cha ES, Kim J, Lee WJ (2012) Emergency department visits due to pesticide poisoning in South Korea, 2006–2009. Clin Toxicol (Phila) 50: 114–119.
9. Statistics Korea. Available: http://kostat.go.kr. Accessed 2 December 2013.
10. World Health Organization (1992) International statistical classification of diseases and related health problems-tenth revision. Geneva: World Health Organization.
11. Ahmad OB, Boschi-Pinto C, Lopez AD, Murray CJL, Lozano R, et al. (2001) Age standardization of rates: a new WHO Standard. No. 31. Geneva: World Health Organization. Available: http://www.who.int/healthinfo/paper31.pdf. Accessed 2 December 2013.
12. Badakhsh R, Lackovic M, Ratard R (2010) Characteristics of pesticide-related hospitalizations, Louisiana, 1998–2007. Public Health Rep 125: 457–467.
13. OECD. OECD Health Data 2013. Available: http://www.oecd.org/health/health-systems. Accessed 2 December 2013.
14. Institute for Health Metrics and Evaluation. GBD profile: South Korea 2010. Available: http://www.healthmetricsandevaluation.org/gbd. Accessed 2 December 2013.
15. Wu KC, Chen YY, Yip PS (2012) Suicide methods in Asia: implications in suicide prevention. Int J Environ Res Public Health 9: 1135–1158.
16. Kim SY, Kim MH, Kawachi I, Cho Y (2011) Comparative epidemiology of suicide in South Korea and Japan: effects of age, gender and suicide methods. Crisis 32: 5–14.
17. Gunnell D, Eddleston M (2003) Suicide by intentional ingestion of pesticides: a continuing tragedy in developing countries. Int J Epidemiol 32: 902–909.
18. Mohamed F, Manuweera G, Gunnell D, Azher S, Eddleston M, et al. (2009) Pattern of pesticide storage before pesticide self-poisoning in rural Sri Lanka. BMC Public Health 9: 405.
19. Senarathna L, Jayamanna SF, Kelly PJ, Buckley NA, Dibley MJ, et al. (2012) Changing epidemiologic patterns of deliberate self poisoning in a rural district of Sri Lanka. BMC Public Health 12: 593.
20. Chang SS, Lu TH, Sterne JA, Eddleston M, Lin JJ, et al. (2012) The impact of pesticide suicide on the geographic distribution of suicide in Taiwan: a spatial analysis. BMC Public Health 12: 260.
21. Kong Y, Zhang J (2010) Access to farming pesticides and risk for suicide in Chinese rural young people. Psychiatry Res 179: 217–221.
22. Hirsch JK (2006) A review of the literature on rural suicide: risk and protective factors, incidence and prevention. Crisis 27: 189–199.
23. Judd F, Cooper AM, Fraser C, Davis J (2006) Rural suicide-people or place effects? Aust N Z J Psychiatry 40: 208–216.

24. Statistics Korea (2012) The 2011 survey of agriculture, fishery, and forestry. Daejeon: Statistics Korea. (In Korean).

25. Kim MH, Jung-Choi K, Jun HJ, Kawachi I (2010) Socioeconomic inequalities in suicidal ideation, parasuicides, and completed suicides in South Korea. Soc Sci Med 70: 1254–1261.

26. Conwell Y, Thompson C (2008) Suicidal behavior in elders. Psychiatr Clin North Am 31: 333–356.

27. Phillips MR, Cheng HG (2012) The changing global face of suicide. Lancet 379: 2318–2319.

28. Ki M, Na BJ, Lee WJ, Jeong BK, Lee HY, et al. (2012) Evaluation of the project for the improvements of rural health service and suggestions for the direction of future investment. Seoul: Ministry of Health & Welfare, Korea Health Promotion Foundation, and Eulji University. (In Korean).

29. Lee WJ, Cha ES, Park J, Ko Y, Kim HJ, et al. (2012) Incidence of acute occupational pesticide poisoning among male farmers in South Korea. Am J Ind Med 55: 799–807.

30. Kim JH, Kim J, Cha ES, Ko Y, Kim DH, et al. (2013) Work-related risk factors for acute pesticide poisoning by severity among male farmers in South Korea. Int J Environ Res Public Health 10: 1100–1112.

31. OECD. OECD Environmental Data Compendium 2008. Agriculture. Available: http://www.oecd.org/statistics. Accessed 2 December 2013.

32. Freire C, Koifman S (2013) Pesticides, depression and suicide: A systematic review of the epidemiological evidence. Int J Hyg Environ Health 216: 445–460.

33. London L, Beseler C, Bouchard MF, Bellinger DC, Colosio C, et al. (2012) Neurobehavioral and neurodevelopmental effects of pesticide exposures. Neurotoxicology 33: 887–896.

34. Kim J, Ko Y, Lee WJ (2013) Depressive symptoms and severity of acute occupational pesticide poisoning among male farmers. Occup Environ Med 70: 303–309.

35. Kim J, Shin DH, Lee WJ (2014) Suicidal ideation and occupational pesticide exposure among male farmers. Environ Res 128: 52–56.

36. Lee WJ (2011) Pesticide exposure and health. J Environ Health Sci 37: 81–93. (In Korean).

37. Rural Development Administration (2008) Researches on the actual state of pesticide poisoning in Korea and guidelines for diagnosis and treatment of paraquat and organophosphate poisoning. Suwon: Rural Development Administration. (In Korean).

38. Statistics Korea (2012) The 2011 annual report on the cause of death statistics. Daejeon: Statistics Korea. (In Korean).

39. Park BJ, Sung JH, Park KD, Seo SW, Kim SW (2003). Strategies to improve the validity of diagnostic codes of National Health Insurance claims data. Seoul: Technical report by Health Insurance Review and Assessment Services. (In Korean).

16

Xenobiotic Effects on Intestinal Stem Cell Proliferation in Adult Honey Bee (*Apis mellifera* L) Workers

Cordelia Forkpah[1], Luke R. Dixon[1], Susan E. Fahrbach[2], Olav Rueppell[1]*

1 Department of Biology, University of North Carolina, Greensboro, North Carolina, United States of America, **2** Department of Biology, Wake Forest University, Winston-Salem, North Carolina, United States of America

Abstract

The causes of the current global decline in honey bee health are unknown. One major group of hypotheses invokes the pesticides and other xenobiotics to which this important pollinator species is often exposed. Most studies have focused on mortality or behavioral deficiencies in exposed honey bees while neglecting other biological functions and target organs. The midgut epithelium of honey bees presents an important interface between the insect and its environment. It is maintained by proliferation of intestinal stem cells throughout the adult life of honey bees. We used caged honey bees to test multiple xenobiotics for effects on the replicative activity of the intestinal stem cells under laboratory conditions. Most of the tested compounds did not alter the replicative activity of intestinal stem cells. However, colchicine, methoxyfenozide, tetracycline, and a combination of coumaphos and tau-fluvalinate significantly affected proliferation rate. All substances except methoxyfenozide decreased proliferation rate. Thus, the results indicate that some xenobiotics frequently used in apiculture and known to accumulate in honey bee hives may have hitherto unknown physiological effects. The nutritional status and the susceptibility to pathogens of honey bees could be compromised by the impacts of xenobiotics on the maintenance of the midgut epithelium. This study contributes to a growing body of evidence that more comprehensive testing of xenobiotics may be required before novel or existing compounds can be considered safe for honey bees and other non-target species.

Editor: Nicolas Desneux, French National Institute for Agricultural Research (INRA), France

Funding: This project was supported by the North Carolina Biotechnology Center and the Agriculture and Food Research Initiative Competitive Grant no. 2010-65104-20533 from the USDA National Institute of Food and Agriculture. The funders had no role in study design, data collection and analysis, decision to publish, or preparation of the manuscript.

Competing Interests: The authors have declared that no competing interests exist.

* E-mail: olav_rueppell@uncg.edu

Introduction

The western honey bee, *Apis mellifera* (L), is the most important managed pollinator worldwide and provides economically important pollination services in natural and agricultural ecosystems [1,2]. Despite their significance to agriculture, the number of managed honey bee colonies in the United States has declined over the past decades [3]. Since 2006, severe annual losses have been reported by beekeepers in conjunction with declining honey bee health and a syndrome of collapsing colonies that accounts for some of these losses [4,5]. This colony collapse syndrome is characterized by the rapid disappearance of adult worker honey bees, arguing for research on adult honey bee health.

The causes of the observed decline in honey bee health are poorly understood [5,6]. Presumably, these causes are complex and heterogeneous with multiple, potentially interacting contributors [7,8,9]. Novel pathogens such as Israeli acute paralysis virus and combinations of parasites and pathogens have been associated with declining honey bee health in laboratory studies [10] and large-scale surveys [11,12]. General management stress reflecting changes in beekeeping practices and inadequate nutrition may also play important roles [13,14]. Additionally, pesticides and other xenobiotics have been associated with mass killings of honey bees [15], and novel compounds, formulations, and applications may contribute to recent declines in honey bee health [16,17,18].

Honey bees are exposed to a large number of xenobiotics, some of which accumulate in their hives [16,18]. Over 120 pesticides and metabolites have been identified to enter the hive with returning foragers or as a result of direct application by beekeepers [19,20]. This large number is concerning because substances can harm honey bee health via synergistic interactions [16,21,22]. Modern systemic insecticides are incorporated into all plant parts, including the pollen and nectar that honey bees collect [23]. Through food-storage and -sharing these substances are distributed throughout the hive although substances that are directly applied to the hive, such as the miticides coumaphos and fluvalinate, are typically found in higher concentrations [18,21,24].

Field-relevant concentrations of some pesticides not only kill honey bees but also produce sublethal effects detectable as behavioral deficiencies [25,26,27,28], shortened lifespan [29,30], or increased susceptibility to diseases [8,31,32]. Because many pesticides target the nervous system, tests of sublethal effects on honey bees have concentrated primarily on behavior and direct measures of neuronal activities [33,34]. Sublethal effects on other functions and organs have been rarely studied, although pesticides and other xenobiotics are known to affect several physiological functions. For example, compromised hypopharyngeal gland development caused by exposure of nurse bees to four different pesticides [35] can be linked to decreased brood production at the colony level [36]. Exposure of the midgut epithelium of honey bee larvae to sublethal concentrations of a broad range of pesticides resulted in increased apoptosis [37]. Both of these observations predict smaller colony sizes that eventually translate into reduced colony survival [38]. Neither of these studies, however, directly addresses the topic of sublethal physiological effects in adult workers. This gap in the literature is significant given a context of colony collapse without reduced brood production [9].

The digestive system is a critical organ for honey bee health because it is the site of contact with many pathogens and xenobiotics [39,40]. The midgut epithelium is for many pathogens the principal barrier to invasion of the honey bee host, and it is the main site for establishment of other pathogens, such as *Nosema* sp. [41]. Additionally, the midgut epithelium is responsible for detoxification of ingested xenobiotics [42], and some insecticides specifically target the midgut epithelium [40,43]. Damage to the midgut epithelium of honey bees has also been reported as a consequence of acute exposure to the insecticides malathion, deltamethrin, and thiamethoxam [44]. This spatial overlap between immunity and detoxification may facilitate synergistic interactions between pesticides and pathogens to the detriment of honey bee health [7,39].

The midgut epithelium is the only tissue of adult honey bees that exhibits widespread cell proliferation [45]. Proliferation also occurs in the midgut of stingless adult bees, although at a lower rate than reported for honey bees. [46]. Proliferative cells (Figure 1) continuously replace the columnar and goblet cells that form the functional epithelium [47,48,49]. The proliferative activity of the intestinal stem cells (ISCs) varies with age and social function and responds dynamically to high digestive activity [45,50]. The proliferation rate of the ISCs could therefore be a sensitive indicator of sublethal effects of ingested xenobiotics in the honey bee. On the one hand, toxic effects may increase the rate of proliferation by increasing the demand for cellular replacement. If replicative capacity of the ISCs is unable to compensate, epithelial function may be compromised and lifespan may be shortened. On the other hand, toxins may directly damage the ISCs, directly resulting in a decreased proliferation rate which may also compromise epithelial function and shorten lifespan.

We have examined the impact of a number of pesticides and other xenobiotics on ISC proliferation in honey bees. To investigate this potential mode of xenobiotic action, we used relatively high doses in a controlled cage environment. We concomitantly monitored survival but our focus was on the question of whether ISC proliferation is altered by sublethal exposure to common xenobiotics.

Materials and Methods

Experiment 1

Ten xenobiotics were studied along with solvent controls (Table 1). We used colchicine, an inhibitor of mitosis [51], as a control to demonstrate that our method was sensitive enough to detect the inhibition of ISC proliferation by a xenobiotic [45]. The insect steroid 20-hydroxyecdysone was selected as a positive control because of previous reports of a positive effect of this hormone on ISC proliferation in other insect species [52]. The trials involved monitoring survival during continuous exposure to one concentration of each xenobiotic over seven days, followed by a standardized assessment of intestinal stem cell proliferation. The chosen concentrations either represented the maximum concentrations reported from bee hives in the literature or, in the case of compounds typically applied to colonies by beekeepers, the maximum allowable dose per manufacturer instructions.

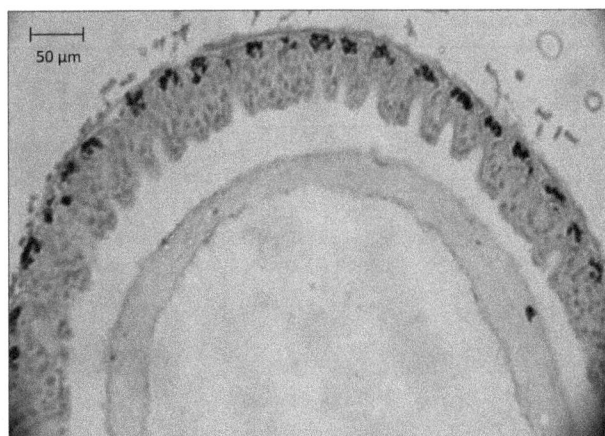

Figure 1. Cross section of the honey bee midgut, showing the midgut epithelium consisting of discrete crypts. The peritrophic membrane is visible in the midgut lumen. In the midgut epithelium, BrdU-labeled nuclei are brown, indicating that DNA replication occurred during the 24 h exposure to the marker. An index of proliferative activity has been developed based on counting the number of labeled nuclei in 10 μm thick cross sections relative to the number of active crypts. This index can be used to rank proliferative activity in different samples and assess possible sublethal effects of ingested xenobiotics on the midgut epithelium. Sections are counterstained with hematoxylin (in blue) to facilitate detection of crypts and other tissue features.

Workers (*Apis mellifera* L) from 4–10 hives maintained at the University of North Carolina at Greensboro bee yard were used. Colonies were maintained following standard practices without chemical disease control or artificial diets. Combs with ready-to-emerge workers were transferred to an incubator (complete dark cycle, 35°C, 60% rel. hum.) and collected from the combs upon emergence. Newly emerged bees were randomly assigned to treatment or control groups. Four groups of 25 bees per treatment were kept in separate Plexiglas feeding cages (10 cm×7.5 cm ×10 cm) in an incubator (complete dark cycle, 33°C, 60% rel. hum.), fed *ad libitum* queen candy (9:3:1, powdered sugar: water: honey), and provided with water. Dead bees were removed and counted from the cages daily. Although cage studies are widely used in honey bee research [53], they can be problematic [54] and have been reported to compromise the natural colonization of the gut by bacteria [55]. We preferred the controlled cage environment for these initial studies because our goal was to link a known xenobiotic exposure to quantitative effects on ISC proliferation.

All substances except tau-fluvalinate were mixed with the queen candy food for direct delivery to the midgut epithelium. Tau-fluvalinate was delivered via Apistan strips (Zoëcon, USA), the form typically used by beekeepers. Two of the four tau-fluvalinate cages were terminated after three days instead of the planned seven day exposure to ensure that a sufficient number of living honey bees could be obtained for our studies of ISC proliferation. For all other treatments, living honey bees were collected from the cages after seven days for ISC proliferation assays.

Experiment 2

On the basis of the results of the first experiment only methoxyfenozide, tetracycline, and tau-fluvalinate were tested further in large scale studies, using three different dosages (Table 2). Because of the potential for synergistic effects [22], a combination of tau-fluvalinate and coumaphos was also tested. The experiment comprised eleven trials, each with its own water

Table 1. Xenobiotics tested in the first experiment for effects on the intestinal stem cell proliferation.

Xenobiotic	Supplier	Dosage	Source for selecting concentrations
Fumagillin	Mann Lake Ltd	2 mg/g	Highest dose allowed per manufacturer guidelines
Tau-fluvalinate*	Mann Lake Ltd	Permanent exposure	Practical dosage under experimental conditions exceeds manufacturer guidelines
Tetracycline	Sigma- Aldrich	3 mg/g	[60]
Imidacloprid	Sigma-Aldrich	500 ppb	[16]
Coumaphos	Sigma-Aldrich	5000 ppb	[16]
Chlorothalonil	Fluka	1000 ppb	[16]
Methoxyfenozide	ChemSevice Inc.	400 ppb	[16]
Colchicine	Sigma-Aldrich	5 mg/g	[51]
20-Hydroxyecdysone	MP Chemicals	200 ppb	[68]
DMSO	Acros Chemical	0.01 mg/g	Control for imidacloprid
Acetone	Mallinckrodt Chemicals	0.1 mg/g	Control for methoxyfenozide, coumaphos, and chlorothalonil
Isopropanol	Fisher	200 ppb	Control for 20-hydroxyecdysone
Water	N/A	N/A	Control for tetracycline, colchicine, tau-fluvalinate, and fumagilin

* Tau-fluvalinate was exposed to bees using the commercial Apistan strip. A half of one standard strip was placed in each cage for the indicated exposure time per day for 7 days.

or acetone control groups to account for seasonal effects and the use of honey bees from different sources. As described, newly emerged bees were caged, housed in an incubator, and fed xenobiotics in food provided *ad libitum* for seven days. In this study, food was provided as a 30% sucrose solution in liquid feeders. Two-four replicate cages were used per treatment, with 120–155 bees housed per cage (same dimensions as in Experiment 1). Survival was monitored daily, and a subset of the surviving bees assayed for ISC proliferation after 7 days (see below). Fresh and dry weights of the head and thorax of random samples of additional bees from all treatments were determined to test for differences in food uptake.

After the collection on day 7, the remaining honey bees were continued to be daily monitored for survival in their cages without xenobiotic exposure: They were provisioned with distilled water and sucrose solution. A second sample of honey bees of each treatment group was assayed for delayed treatment effects on ISC proliferation between ages 19–22 days, or earlier if mortality of the experimental cohort exceeded 90% before that age.

ISC Proliferation Assay

Following our previous methods [45,50], assessment of proliferation rate of intestinal stem cells relied on immunohistochemical

labeling of the thymidine analog 5-bromo-2-deoxyuridine (BrdU) incorporated into newly synthesized DNA. Briefly, workers without signs of morbidity such as reduced mobility or responsiveness to stimuli were selected for this assay. These individuals were fed 5 mg/ml BrdU (Life Technologies, CA) in queen candy *ad libitum* for a 24-hour period. Shorter feeding periods were evaluated in a pilot study with newly emerged workers (Figure 2), and a 24-hour period was selected for the actual experiments because this survival reliably produced a substantial number of labeled nuclei.

Only individuals that appeared healthy after this feeding period were selected for analysis. Dissected midguts rinsed with saline were fixed in Carnoy's fixative for 24 hours and embedded in Paraplast (Thermo Fisher Scientific, MA) for sectioning (10 μm) using a HM315-Microm microtome (Thermo Fisher Scientific, MA). Sections were mounted on Superfrost Fisher plus microscope slides (Thermo Fisher Scientific, MA), dewaxed in xylene, rehydrated via a graded alcohol series, and permeabilized in phosphate-buffered saline containing 0.01% Triton X-100 detergent (PBS-T; Sigma-Aldrich, MO). Samples were denatured with 2N hydrochloric acid, washed in phosphate-buffered saline (PBS), blocked with normal goat serum (Thermo Fisher Scientific, MA), and incubated with anti-BrdU antibody (Phoenix Flow Systems,

Table 2. Summary of the second experiment testing different dosages of select xenobiotics for effects on intestinal stem cell proliferation.

Xenobiotic	Supplier	Dosage		
		Low	Mid	High
Methoxyfenozide	ChemSevice Inc.	40 ppb	400 ppb	2000 ppb
Tau-fluvalinate*	Mann Lake Ltd	3 minutes randomized	3 minutes sequential	15 minutes
Tetracycline	Sigma-Aldrich	1.2 μg/g	30 μg/g	60 μg/g
Tau-fluvalinate* and 500 ppb Coumaphos	Mann Lake Ltd Sigma-Aldrich	3 minutes randomized	3 minutes sequential	15 minutes

* Tau-fluvalinate was exposed to bees using the commercial Apistan strip. A full strip was placed in each cage for the indicated exposure time per day for 7 days.

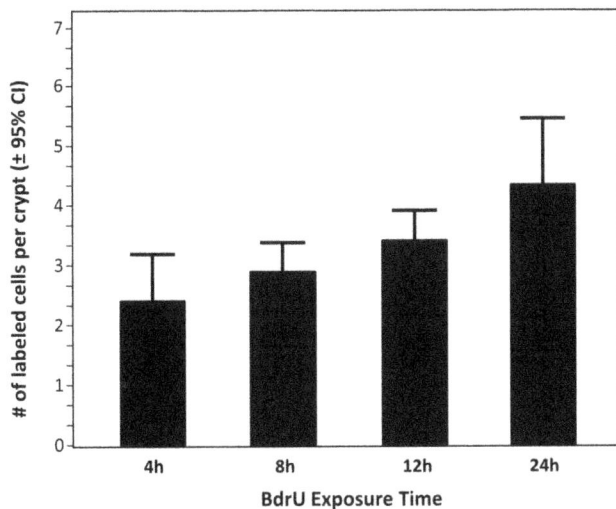

Figure 2. The number of labeled nuclei in a single midgut cross-section is a function of duration of exposure to BrdU. This presumably reflects the number of cell cycle events that occur during the exposure. To control for this effect, a standardized 24 h duration of exposure to BrdU via feeding was used in the main experiments that assessed the effects of xenobiotics on intestinal stem cell proliferation. The longer exposure time is also expected to increase the accuracy of ISC proliferation estimate, although the inter-individual variation among samples in this experiment was lower after 8 and 12 h.

PRB1U) for 24 h at 4°C. After several washes in PBS-T and PBS, sections were incubated at room temperature for two hours with a peroxidase-conjugated anti-mouse secondary antibody (Jackson ImmunoResearch Laboratories, PA), washed again, and incubated with the chromogen diaminobenzidine (Sigma-Aldrich, MO). All nuclei containing DNA synthesized after ingestion of BrdU were labeled with a dark brown reaction product. Slides were counterstained for approximately five minutes using Gill hematoxylin (Thermo Fisher Scientific, MA) followed by 0.1% sodium-bicarbonate solution for one minute. After dehydration in ethanol, the tissue was cleared with CitriSolv (Thermo Fisher Scientific, MA). Slides were coverslipped using Permount and viewed under a Nikon Eclipse E200 microscope.

In the first experiment, all BrdU-labeled nuclei and active centers of proliferation (crypts) were counted in one randomly selected intact section per individual (Figure 1). An active crypt was defined as any containing one or more cells with a labeled nucleus. The average number of labeled nuclei per crypt visible in the selected section was calculated. In the second experiment, the labeled nuclei of 10–22 random crypts from multiple, arbitrarily selected intact cross sections were counted, and the average number of labeled nuclei per crypt was determined to reduce bias associated with analysis of a single section. Observers evaluated slides without knowledge of treatment group identity.

Analyses

In the initial screening experiment, differences in survival between xenobiotic exposed-groups and control groups were assessed by simple contingency analyses with Yates' correction because standard survival estimates and statistical comparisons could not be computed in groups with 100% survival until the end of the experiment. In the follow-up experiments, survival was compared among the treatment and vehicle control groups by pairwise Kaplan-Meier analysis (log-rank tests), censoring any individuals that were sampled for quantification of their intestinal

stem cell proliferation or weight determination. We separately assessed acute mortality (during xenobiotic exposure) and legacy mortality (after xenobiotic exposure was terminated). Cages were treated as separate replicates in the overall evaluation of each experimental treatment.

In the first experiment, the effects of each xenobiotic on the number of labeled nuclei, active crypts, and number of labeled nuclei per crypt were assessed by simple ANOVAs. In the second experiment, the effect of each xenobiotic on the number of labeled nuclei per crypt was analyzed by ANOVA using age group (acute versus legacy effects) as one independent fixed factor and treatment as the second factor. The treatment factor divided the samples into honey bees that were exposed to the three different concentrations of each xenobiotic and the appropriate vehicle control. The overall analyses were followed by separate analyses of the two age groups in which interactions between treatment and age were indicated. Because of unequal variances among groups, *post hoc* comparisons among the different doses of a specific xenobiotic treatment were performed with Dunnett's T3 test.

Results

Experiment 1

ISC proliferation was significantly affected by feeding on colchicine, tetracycline, and methoxyfenozide, but not by feeding on fumagilin, imidacloprid, coumaphos, chlorothalonil, or by fluvalinate treatment (Table 3). Compared with untreated controls, colchicine significantly reduced the number of labeled nuclei per active crypt (2.8 versus 4.7). Workers that fed on tetracycline had fewer active crypts per section (32.8 versus 55.3), fewer labeled nuclei per section (55.9 versus 254.9), and fewer labeled nuclei per active crypt (1.7 versus 4.7). In contrast, methoxyfenozide significantly increased labeled nuclei per section (324.9 versus 251.5) and per crypt (5.3 versus 4.7), relative to controls (Figure 3). The survival of individuals across experimental groups was positively correlated with the average number of labeled nuclei per crypt (Pearson's $R_P = 0.79$, n = 13, p = 0.001). The proportion of surviving individuals varied among experimental groups from 6.4% to 100%. None of the solvent controls affected honey bee survival but survival was significantly reduced by the high experimental exposure to colchicine ($\chi^2 = 154.7$, p<0.001), tetracycline ($\chi^2 = 169.2$, p<0.001), fluvalinate ($\chi^2 = 119.2$, p<0.001), fumagillin ($\chi^2 = 76.4$, p<0.001), imidacloprid ($\chi^2 = 20.9$, p<0.001), and coumaphos ($\chi^2 = 6.4$, p = 0.012) relative to their respective controls.

Experiment 2

The experimental groups did not differ significantly in fresh or dry weights of the head ($F^{fresh}_{(9,379)} = 1.4$, p = 0.167, $F^{dry}_{(9,379)} = 1.2$, p = 0.267) or dry weight of the thorax ($F^{dry}_{(9,379)} = 1.3$, p = 0.218). In contrast, thorax fresh weight was significantly affected ($F^{fresh}_{(9,379)} = 2.5$, p = 0.010). *Post hoc* comparisons revealed that a significantly lower thorax weight was found in the acetone control group than in the water control and in the highest tetracycline dosage group.

Xenobiotic feeding effects on the number of labeled nuclei per active crypt were variable (Figure 4). Tetracycline showed significant concentration ($F_{(3,99)} = 2.8$, p = 0.042), age group ($F_{(1,99)} = 18.5$, p<0.001), and interaction ($F_{(3,99)} = 2.9$, p = 0.040) effects. Overall, sections contained more labeled nuclei directly after termination of treatment than two weeks later. Differences among treatments were not significant in the young age group directly after xenobiotic exposure ($F_{(3,57)} = 2.4$, p = 0.081), but the highest dosage of tetracycline was associated with a significant

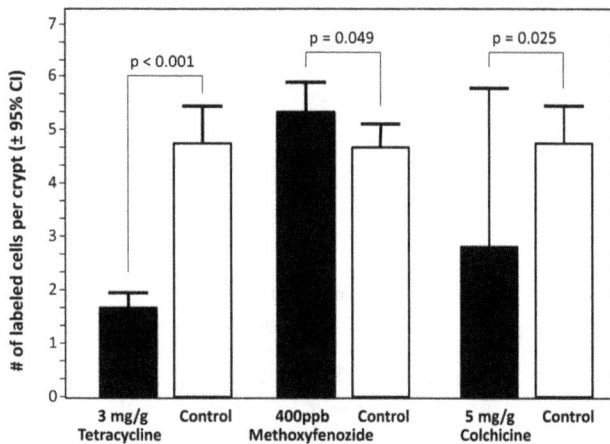

Figure 3. In the first experiment, 3 of the 12 xenobiotics tested had significant effects on the number of labeled nuclei per active crypt. While colchicine and tetracycline decreased the proliferation of ISCs, methoxyfenozide increased activity. Means are shown with 95% confidence intervals.

decline in the number of labeled nuclei per crypt compared with the control group in the older age group ($F_{(3,42)} = 3.5$, $p = 0.025$). Analysis of the results of the methoxyfenozide experiment revealed a smaller number of labeled nuclei in older bees ($F_{(1,71)} = 29.0$, $p < 0.001$), no overall effect of treatment ($F_{(3,71)} = 1.1$, $p = 0.349$), but a significant interaction between the two factors ($F_{(3,71)} = 3.5$, $p = 0.019$). Analyzed separately, no significant effect of treatment was apparent in either age group (young: $F_{(3,34)} = 2.5$, $p = 0.074$; old: $F_{(3,37)} = 1.4$, $p = 0.250$). Fluvalinate alone exhibited no overall age group ($F_{(1,98)} = 2.3$, $p = 0.129$) or treatment ($F_{(3,98)} = 1.1$, $p = 0.371$) effects but a significant interaction effect ($F_{(2,98)} = 3.1$, $p = 0.048$). Separate analyses did not reveal specific treatment effects in either age group (young: $F_{(3,57)} = 2.3$, $p = 0.087$; old: $F_{(2,41)} = 0.5$, $p = 0.616$). Coumaphos and fluvalinate in combination showed a significant treatment effect ($F_{(3,58)} = 6.8$, $p = 0.001$), no age group effect ($F_{(1,58)} = 3.1$, $p = 0.084$), and a significant interaction between the two factors ($F_{(2,58)} = 4.0$, $p = 0.023$). Treatment significantly affected the labeled nuclei per active crypt in the younger group ($F_{(3,30)} = 7.8$, $p = 0.001$), with the highest dosage significantly reducing the counts relative to the control and lowest dosage. In the older group, no significant treatment effect was found ($F_{(2,28)} = 1.0$, $p = 0.385$).

Across all treatment groups, there was no significant relation between ISC proliferation directly after xenobiotic exposure and its measure at older ages ($Rs = 0.02$, $n = 13$, $p = 0.943$). The average ISC proliferation in the groups sampled at the older age was positively associated with survival after treatment ($R_S = 0.67$, $n = 13$, $p = 0.013$), while no association between ISC proliferation and survival at the younger age (during xenobiotic exposure) was found ($R_S = -0.19$, $n = 15$, $p = 0.499$).

Mortality in the caged experimental cohorts was generally higher than in the first experiment with seven-day survival ranging from 44–68% and significant variation among cages of the same treatment groups, including control groups (see File S1). Overall, the acute mortality was different among treatment groups for tetracycline ($\chi^2 = 10.1$, $p = 0.018$; Figure 5a), methoxyfenozide ($\chi^2 = 8.8$, $p = 0.032$; Figure 5b), fluvalinate ($\chi^2 = 18.7$, $p < 0.001$; Figure 5c), and the combination of fluvalinate and coumaphos ($\chi^2 = 38.9$, $p < 0.001$, Figure 5d). After Bonferroni correction, only the 3-minute fluvalinate treatment ($\chi^2 = 11.8$, $p_{corr} = 0.002$) and the 3-minute fluvalinate exposure combined with coumaphos ($\chi^2 = 31.0$, $p_{corr} < 0.001$) increased mortality compared with the respective solvent controls.

Overall legacy mortality after the treatment was different among experimental groups for tetracycline ($\chi^2 = 113.6$, $p < 0.001$; Figure 5a), fluvalinate ($\chi^2 = 27.3$, $p < 0.001$; Figure 5c), and the combination of fluvalinate and coumaphos ($\chi^2 = 114.4$, $p < 0.001$, Figure 5d). Treatments that significantly increased legacy mortality relative to their respective controls were the medium ($\chi^2 = 32.5$, $p_{corr} < 0.001$) and high ($\chi^2 = 33.7$, $p_{corr} < 0.001$) dose of tetracycline and the 3×1 min exposure of fluvalinate ($\chi^2 = 11.5$, $p_{corr} = 0.002$).

Discussion

This study demonstrated that select xenobiotics can decrease the proliferative rate of ISCs of adult worker honey bees. Reduced ISC proliferation represents a novel, possibly important effect of xenobiotics because the midgut epithelium provides the first line of defense against many pathogens, is responsible for nutrient uptake, and detoxifies many ingested toxins [39].

Most of the tested substances did not significantly affect ISC proliferation although they were directly ingested and therefore must have come into close contact with the midgut epithelium of the studied honey bees. This finding contrasts with widespread pesticide effects on apoptosis in the midgut of honey bee larvae [37], suggesting that juvenile stages might be more susceptible to pesticides than adults. Only colchicine (included as a technical

Table 3. Xenobiotic feeding effects* on ISC proliferation pooled across replicate cages.

Xenobiotic	Effect on # of labeled cells per cross-section	Effect on # of crypts per cross-section	Effect on # of labeled cells per crypt
Fumagillin	$F_{(1,17)} = 0.0$, $p = 0.936$	$F_{(1,17)} = 0.0$, $p = 0.900$	$F_{(1,17)} = 0.4$, $p = 0.536$
Tau-fluvalinate	$F_{(1,31)} = 0.0$, $p = 0.946$	$F_{(1,20)} = 0.2$, $p = 0.626$	$F_{(1,20)} = 1.8$, $p = 0.195$
Tetracycline	**$F_{(1,14)} = 56.3$, $p < 0.001$**	**$F_{(1,13)} = 17.7$, $p = 0.001$**	**$F_{(1,13)} = 56.3$, $p < 0.001$**
Imidacloprid	$F_{(1,16)} = 0.0$, $p = 0.983$	$F_{(1,16)} = 0.0$, $p = 0.871$	$F_{(1,16)} = 0.0$, $p = 0.967$
Coumaphos	$F_{(1,19)} = 0.6$, $p = 0.440$	$F_{(1,17)} = 0.2$, $p = 0.678$	$F_{(1,17)} = 3.2$, $p = 0.091$
Chlorothalonil	$F_{(1,23)} = 0.4$, $p = 0.544$	$F_{(1,20)} = 0.1$, $p = 0.725$	$F_{(1,20)} = 0.8$, $p = 0.384$
Methoxyfenozide	**$F_{(1,21)} = 6.9$, $p = 0.016$**	$F_{(1,20)} = 3.4$, $p = 0.079$	**$F_{(1,20)} = 4.4$, $p = 0.049$**
Colchicine	$F_{(1,11)} = 4.2$, $p = 0.065$	$F_{(1,9)} = 0.4$, $p = 0.534$	**$F_{(1,10)} = 7.2$, $p = 0.025$**
20-Hydroxyecdysone	$F_{(1,20)} = 0.3$, $p = 0.592$	$F_{(1,18)} = 1.3$, $p = 0.273$	$F_{(1,18)} = 0.3$, $p = 0.596$

*Significant effects in bold.

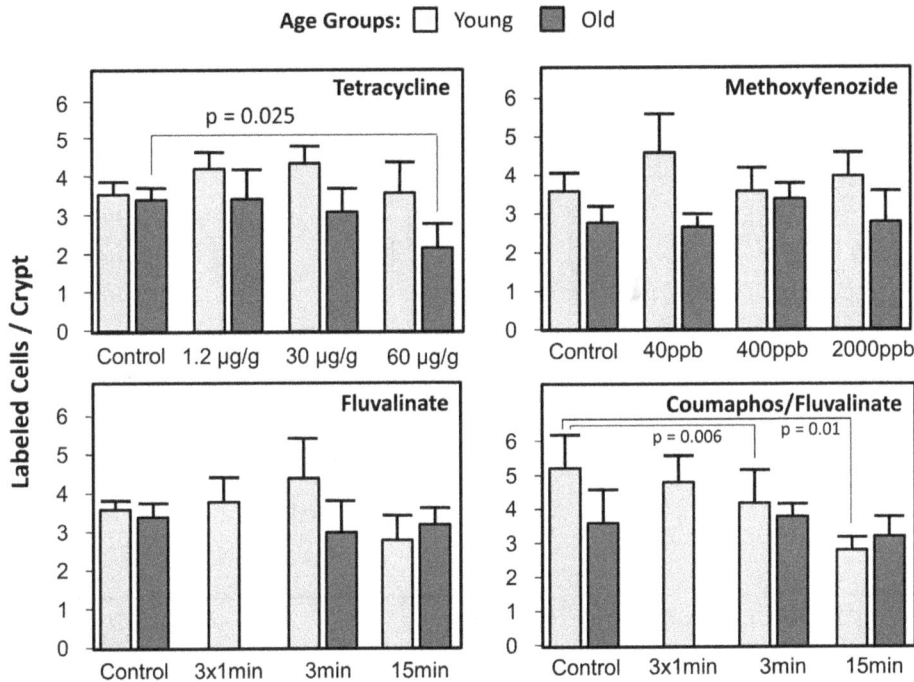

Figure 4. ISC proliferation, indicated by the number of BrdU-labeled nuclei per active crypt, was significantly decreased in older bees. The combination of coumaphos and fluvalinate reduced proliferation measured immediately after treatment, while tetracycline decreased proliferation only at older ages, over ten days after exposure to the xenobiotic had ended.

control), tetracycline, methoxyfenozide, and a combination of fluvalinate and coumaphos showed effects on ISC proliferation. The effects were moderate, dose-dependent, and inconsistent between experiments for methoxyfenozide. Our overall results indicate that the replication rate of ISCs is quite robust after ingestion of most xenobiotics under cage conditions. In contrast, in-hive studies have shown that ISC replication rates decrease in worker honey bees with age and reduced digestive activity [45,50]. Future experiments that better mimic hive conditions and field-relevant exposure levels will be necessary to assess the threat of xenobiotics for intestinal health of honey bees.

Xenobiotic-induced alteration of ISC proliferation may directly harm the affected worker honey bees, causing an increase in immediate or delayed mortality. Overall, our results suggest that reduced ISC proliferation is associated with mortality. Specifically in the second experiment, one of the coumaphos and fluvalinate combination treatments decreased ISC proliferation and survival during exposure; the high tetracycline dosage exhibited delayed effects on both ISC proliferation and mortality. Under field conditions, these effects would result in smaller and/or collapsed colonies due to increased mortality of adult workers. However, we cannot rule out that xenobiotic-induced alteration of ISC proliferation also occurs independently of increased mortality. Under field conditions such effects may increase individual disease susceptibility, for example to *Nosema* [7,56], and compromise the physiological capacity of nurse bees to produce sufficient brood food. Additional studies are needed to address these questions because exposures to sublethal levels of xenobiotics are likely to be more common than exposures to lethal levels [32,38] and sublethal effects are important but difficult to integrate into pesticide regulation [33,57].

In the first experiment, mortality in the cages was increased by several xenobiotics presented at high dosages. Therefore, we employed lower dosages in the second experiment and extended

our mortality and ISC proliferation measurements to include potentially subtle long term effects on ISC proliferation. The sample sizes required for the additional long term analyses resulted in crowded cages and higher mortality, even in the untreated control cages. The increased mortality likely reflects a variety of factors, including poorer hygiene and competition for access to the feeder [53]. However, the determination of worker body weight at the end of the second experiment did not indicate significant differences in food intake between xenobiotic and control groups. We excluded all moribund individuals when assessing ISC proliferation but the concomitant assessment of potential mortality effects of the administered treatment is problematic, particularly because significant variation among replicate cages existed and effects on mortality were inconsistent when cages were analyzed separately (see File S1). Thus, we are reluctant to label any of the measured effects as lethal or sublethal, although mortality was increased by treatments that reduced ISC proliferation. Similarly, the insecticides thiamethoxam, deltamethrin, and malathion have been shown to disrupt the integrity of the honey bee midgut at concentrations that increase mortality [44].

Tetracycline is widely used by beekeepers to combat *Paenibacillus larvae* and *Melissococcus plutonius*, the bacterial agents of American and European foulbrood, respectively [58], but it is a general antibiotic with a wide range of target microorganisms [59]. Compared with controls, caged honey bees exposed to tetracycline exhibited lower ISC proliferation in both experiments. In the first experiment, a dosage that was 1000-fold higher than that typically found in hives [60] significantly reduced ISC proliferation directly after the seven days of treatment. In the second experiment a 50-fold reduced dosage, but not a 100-fold or 2500-fold reduced dosage, also reduced ISC proliferation in the long term. No short-term effects were observed for the lower dosages in the second experiment. Thus, exposure of honey bees to very high doses of tetracycline may result in acute deterioration of the gut physiology

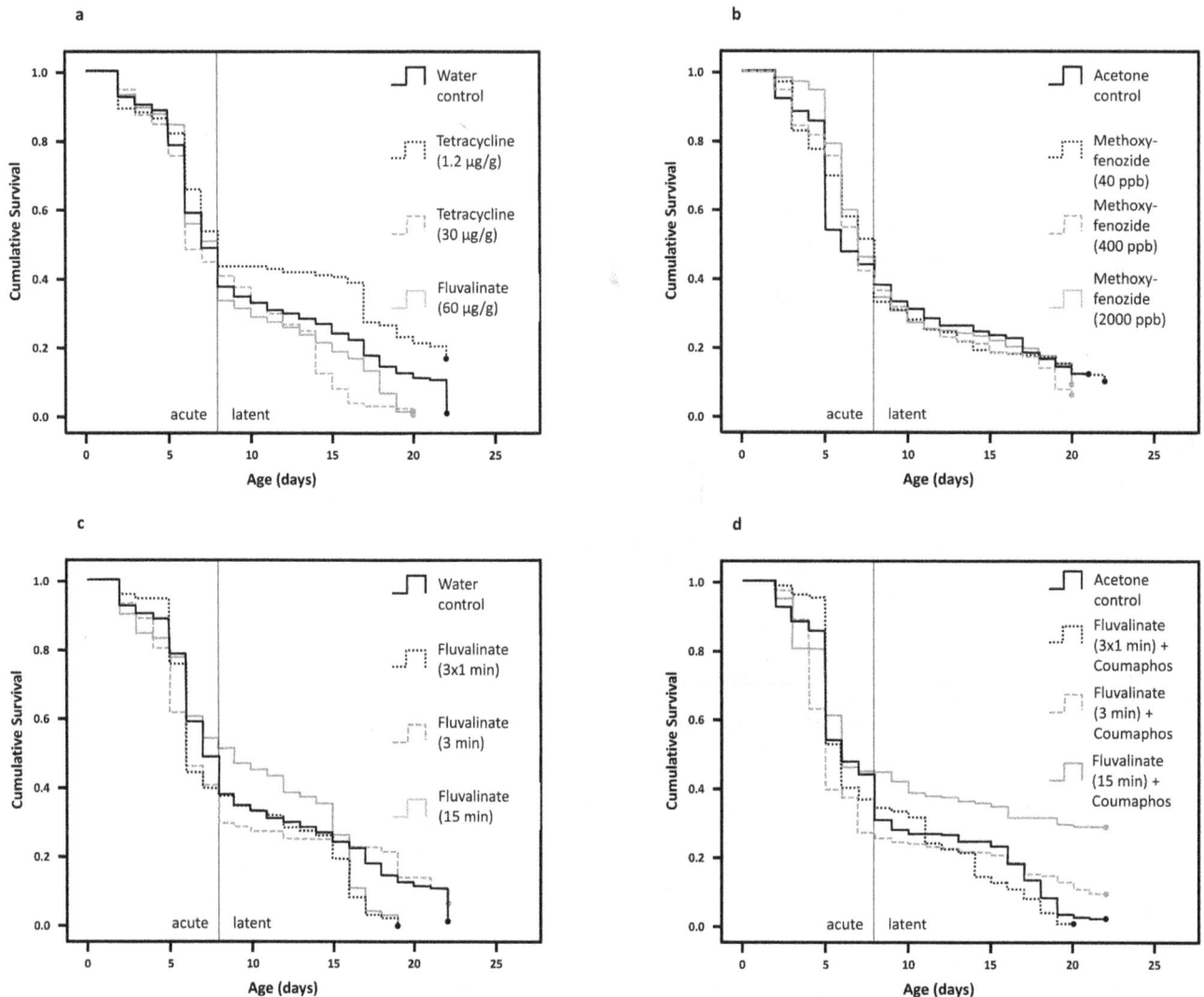

Figure 5. Survival of worker cohorts under high-density cage conditions was lower than in the initial screening experiment and varied inconsistently among treatments. Different panels summarize cumulative survival of honey bee workers grouped from different cage replicates according to the tested xenobiotic: (a) tetracycline, (b) methoxyfenozide, (c) tau-fluvalinate, and (d) the combination of tau-fluvalinate and coumaphos. Acute mortality effects were measured during the first 7 days of exposure, while legacy mortality effects were measured on the days after treatment had ended.

or compromise ISCs directly, while lower concentrations of tetracycline appear to produce a delayed effect. The delayed effect could be due to changes in the intestinal microbial community that can disrupt honey bee health [61,62]. Although we did not monitor the intestinal microbiome, our results could be explained by an interaction between the intestinal microbiome and the physiology of its honey bee host, similar to findings reported in Drosophila that linked the intestinal microbiome to stem cell proliferation [63].

The insect growth regulator methoxyfenozide has not been demonstrated to be harmful to adult honey bees [64], but this compound also accumulates in honey bee hives at significant concentrations [18]. The results of our first experiment suggested that methoxyfenozide may have physiological effects in honey bees by stimulating ISC proliferation. This observation is consistent with the role of methoxyfenozide as an ecdysteroid agonist in the insect midgut [52,65]. In contrast, direct feeding of 20-hydro-xyecdysone, did not affect ISC proliferation, which may reflect the

efficient metabolic conversion of the natural hormone by the gut [66]. Under the crowded conditions of the second experiment, the increases in acute ISC proliferation produced by methoxyfenozide exposure were not significant, and at the older age the low exposure group actually showed a slightly lower number of labeled nuclei than the respective control bees. Thus, the effect of methoxyfenozide on the ISCs is subtle and might not have any health consequences, particularly when considering that the concentrations found in honey bee hives are typically lower than the tested concentrations [18].

Fluvalinate is used by beekeepers to control *Varroa* and tracheal mites. The high dosage of fluvalinate in the first experiment proved so toxic that we quantified ISC proliferation after three days, a time at which most of the exposed workers had already died. At this time no significant effect on ISC proliferation was apparent. In the second experiment we reduced the daily exposure to the fluvalinate strip by over 80-fold, resulting in lower mortality. No effect on ISC proliferation was found after 7 days of exposure and 2 weeks after exposure was terminated. The second commonly used miticide, coumaphos, also did not show significant effects on ISC proliferation in the first experiment despite a very high dosage only found in rare cases under natural conditions [18,30]. However, the combination of coumaphos and fluvalinate significantly decreased short-term ISC proliferation. Concomitantly, this combination treatment decreased survivorship at all dosage levels relative to the corresponding fluvalinate-only treatments (log rank tests: 3×1 min: $\chi^2 = 15.0$, p<0.001; 3 min: $\chi^2 = 3.9$, p = 0.049; 15 min: $\chi^2 = 5.9$, p = 0.015). Thus, the mortality data and ISC proliferation rates indicate synergism between fluvalinate and coumaphos, which has been reported in other contexts [21,22]. The dosages used in the second experiment may be higher than average field exposure but they fall within the limits of concentrations measured in honey bee hives [18,67] and the findings may therefore be relevant for honey bee health. Coumaphos and fluvalinate both target primarily the nervous system: Coumaphos, when converted to its metabolite coumaphos oxon, inhibits the acetylcholinesterase enzyme and fluvalinate serves as an agonist of the voltage-gated sodium channel [22]. Our results may therefore be explained by effects on the neural control of the digestive system or changes in behavior that may have indirectly decreased ISC proliferation. However, we cannot rule out other, non-neural effects. The synergism between the two miticides may be due to inhibition of the detoxification mechanism [22].

We did not find support for the hypothesis that ISCs increase proliferation to compensate for xenobiotic damage to the midgut epithelium [37]. Instead, tetracycline and the combination of fluvalinate and coumaphos decreased ISC proliferation, suggesting direct or indirect effects that decrease ISC activity. The number of labeled nuclei per active crypt also declined with age in all control and treatment groups, except for the groups with the highest

exposure to fluvalinate. This finding confirms our earlier results that ISC proliferation declines with age in honey bees [45]. The age-related decline under natural conditions may reflect the fact that digestive demand is higher in young workers, which are typically nurse bees [45]. In our cage experiments, however, workers did not transition from nursing to foraging behavior. Thus, the age-related decline of ISC proliferation occurred independently of diet or behavioral changes, suggesting the possibility of intrinsic aging of the replicative capacity of ISCs.

Supporting Information

File S1 Significant variability in mortality among separate cages was observed within each treatment of the second experiment. This file details the mortality results with respect to the separate cages in each treatment. Due to unexplainable variation and the focus of our study on ISC proliferation, we omitted these details from the main text.

Acknowledgments

We thank Laura Willard, Jeffery Jackson, and Candice Harrison for their practical help with the project. In addition, we thank all members of the UNCG Social Insect lab and the North Carolina Honey Bee Research Consortium for the suggestions and encouragement. The comments of three anonymous reviewers and our editor helped to improve the quality of the manuscript.

Author Contributions

Conceived and designed the experiments: OR. Performed the experiments: CF LRD. Analyzed the data: CF SEF OR. Contributed reagents/materials/analysis tools: SEF OR. Wrote the paper: CF SEF OR.

References

1. Gallai N, Salles JM, Settele J, Vaissiere BE (2009) Economic valuation of the vulnerability of world agriculture confronted with pollinator decline. Ecological Economics 68: 810–821.
2. Morse RA, Calderone NW (2000) The value of honey bees as pollinators of U.S. crops in 2000. Bee Culture 128: 1–15.
3. vanEngelsdorp D, Meixner MD (2010) A historical review of managed honey bee populations in Europe and the United States and the factors that may affect them. Journal of Invertebrate Pathology 103: S80–S95.
4. Ellis JD, Evans JD, Pettis JS (2010) Colony losses, managed colony population decline and colony collapse disorder in the United States. Journal of Apicultural Research 49: 134–136.
5. vanEngelsdorp D, Speybroeck N, Evans JD, Nguyen BK, Mullin C, et al. (2010) Weighing risk factors associated with bee colony collapse disorder by classification and regression tree analysis. Journal of Economic Entomology 103: 1517–1523.
6. Neumann P, Carreck NL (2010) Honey bee colony losses. Journal of Apicultural Research 49: 1–6.
7. Pettis JS, vanEngelsdorp D, Johnson J, Dively G (2012) Pesticide exposure in honey bees results in increased levels of the gut pathogen *Nosema*. Naturwissenschaften 99: 153–158.
8. Boncristiani H, Underwood R, Schwarz R, Evans JD, Pettis J, et al. (2012) Direct effect of acaricides on pathogen loads and gene expression levels in honey bees *Apis mellifera*. Journal of Insect Physiology 58: 613–620.
9. vanEngelsdorp D, Evans JD, Saegerman C, Mullin C, Haubruge E, et al. (2009) Colony Collapse Disorder: A descriptive study. PLoS ONE 4: e6481.
10. Nazzi F, Brown SP, Annoscia D, Del Piccolo F, Di Prisco G, et al. (2012) Synergistic parasite-pathogen interactions mediated by host immunity can drive the collapse of honeybee colonies. PLoS Pathog 8: e1002735.
11. Cox-Foster DL, Conlan S, Holmes EC, Palacios G, Evans JD, et al. (2007) A metagenomic survey of microbes in honey bee colony collapse disorder. Science 318: 283–287.
12. Genersch E, von der Ohe W, Kaatz H, Schroeder A, Otten C, et al. (2010) The German bee monitoring project: a long term study to understand periodically high winter losses of honey bee colonies. Apidologie 41: 332–352.
13. Alaux C, Ducloz F, Crauser D, Le Conte Y (2010) Diet effects on honeybee immunocompetence. Biology Letters 6: 562–565.
14. Mattila HR, Otis GW (2006) Influence of pollen diet in spring on development of honey bee (Hymenoptera: Apidae) colonies. Journal of Economic Entomology 99: 604–613.
15. Wahl O, Ulm K (1983) Influence of pollen feeding and physiological condition on pesticide sensitivity of the honey bee *Apis mellifera* carnica. Oecologia 59: 106–128.
16. Johnson RM, Ellis MD, Mullin CA, Frazier M (2010) Pesticides and honey bee toxicity - USA. Apidologie 41: 312–331.
17. Gill RJ, Ramos-Rodriguez O, Raine NE (2012) Combined pesticide exposure severely affects individual- and colony-level traits in bees. Nature 491: 105–U119.
18. Mullin CA, Frazier M, Frazier JL, Ashcraft S, Simonds R, et al. (2010) High levels of miticides and agrochemicals in North American apiaries: Implications for honey bee health. PLoS One 5: e9754.
19. Krupke CH, Hunt GJ, Eitzer BD, Andino G, Given K (2012) Multiple routes of pesticide exposure for honey bees living near agricultural fields. PLoS One 7: e29268.
20. Hawthorne DJ, Dively GP (2011) Killing them with kindness? In-hive medications may inhibit xenobiotic efflux transporters and endanger honey bees. PLoS One 6.
21. Johnson RM, Dahlgren L, Siegfried BD, Ellis MD (2013) Acaricide, fungicide and drug interactions in honey bees (*Apis mellifera*). PLoS One 8: e54092.
22. Johnson RM, Pollock HS, Berenbaum MR (2009) Synergistic interactions between in-hive miticides in *Apis mellifera*. Journal of Economic Entomology 102: 474–479.
23. Blacquiere T, Smagghe G, van Gestel CAM, Mommaerts V (2012) Neonicotinoids in bees: a review on concentrations, side-effects and risk assessment. Ecotoxicology 21: 973–992.
24. Chauzat MP, Faucon JP, Martel AC, Lachaize J, Cougoule N, et al. (2006) A survey of pesticide residues in pollen loads collected by honey bees in France. Journal of Economic Entomology 99: 253–262.
25. Yang EC, Chang HC, Wu WY, Chen YW (2012) Impaired olfactory associative behavior of honeybee workers due to contamination of imidacloprid in the larval stage. PLoS One 7: e49472.
26. Yang EC, Chuang YC, Chen YL, Chang LH (2008) Abnormal Foraging Behavior Induced by Sublethal Dosage of Imidacloprid in the Honey Bee (Hymenoptera: Apidae). Journal of Economic Entomology 101: 1743–1748.

27. Henry M, Beguin M, Requier F, Rollin O, Odoux JF, et al. (2012) A common pesticide decreases foraging success and survival in honey bees. Science 336: 348–350.

28. Williamson SM, Wright GA (2013) Exposure to multiple cholinergic pesticides impairs olfactory learning and memory in honeybees. Journal of Experimental Biology 216: 1799–1807.

29. Smirle MJ, Winston ML, Woodward KL (1984) Development of a sensitive bioassay for evaluating sublethal pesticide effects on the honey bee (Hymenoptera, Apidae). Journal of Economic Entomology 77: 63–67.

30. Wu JY, Anelli CM, Sheppard WS (2011) Sub-lethal effects of pesticide residues in brood comb on worker honey bee (Apis mellifera) development and longevity. PLoS One 6: e14720.

31. James RR, Xu J (2012) Mechanisms by which pesticides affect insect immunity. Journal of Invertebrate Pathology 109: 175–182.

32. Pettis JS, Lichtenberg EM, Andree M, Stitzinger J, Rose R, et al. (2013) Crop pollination exposes honey bees to pesticides which alters their susceptibility to the gut pathogen Nosema ceranae. PLoS One 8: e70182.

33. Desneux N, Decourtye A, Delpuech JM (2007) The sublethal effects of pesticides on beneficial arthropods. Annual Review of Entomology 52: 81–106.

34. Palmer MJ, Moffat C, Saranzewa N, Harvey J, Wright GA, et al. (2013) Cholinergic pesticides cause mushroom body neuronal inactivation in honeybees. Nat Commun 4: 1634.

35. Heylen K, Gobin B, Arckens L, Huybrechts R, Billen J (2011) The effects of four crop protection products on the morphology and ultrastructure of the hypopharyngeal gland of the European honeybee, Apis mellifera. Apidologie 42: 103–116.

36. Bendahou N, Fleche C, Bounias M (1999) Biological and biochemical effects of chronic exposure to very low levels of dietary cypermethrin (Cymbush) on honeybee colonies (Hymenoptera: Apidae). Ecotoxicology and Environmental Safety 44: 147–153.

37. Gregorc A, Ellis JD (2011) Cell death localization in situ in laboratory reared honey bee (Apis mellifera L.) larvae treated with pesticides. Pesticide Biochemistry and Physiology 99: 200–207.

38. Bryden J, Gill RJ, Mitton RAA, Raine NE, Jansen VAA (2013) Chronic sublethal stress causes bee colony failure. Ecology Letters 16: 1463–1469.

39. Johnson RM, Evans JD, Robinson GE, Berenbaum MR (2009) Changes in transcript abundance relating to colony collapse disorder in honey bees (Apis mellifera). Proc Natl Acad Sci U S A 106: 14790–14795.

40. Han P, Niu CY, Biondi A, Desneux N (2012) Does transgenic Cry1Ac+CpTI cotton pollen affect hypopharyngeal gland development and midgut proteolytic enzyme activity in the honey bee Apis mellifera L. (Hymenoptera, Apidae)? Ecotoxicology 21: 2214–2221.

41. Higes M, Meana A, Bartolome C, Botias C, Martin-Hernandez R (2013) Nosema ceranae (Microsporidia), a controversial 21st century honey bee pathogen. Environmental Microbiology Reports 5: 17–29.

42. Mao WF, Schuler MA, Berenbaum MR (2011) CYP9Q-mediated detoxification of acaricides in the honey bee (Apis mellifera). Proceedings of the National Academy of Sciences of the United States of America 108: 12657–12662.

43. Vachon V, Laprade R, Schwartz JL (2012) Current models of the mode of action of Bacillus thuringiensis insecticidal crystal proteins: A critical review. Journal of Invertebrate Pathology 111: 1–12.

44. Kakamand FAK, Mahmoud TT, Amin ABM (2008) The role of three insecticides in disturbance the midgut tissue in honeybee Apis mellifera L. workers. Journal of Dohuk University 11: 144–151.

45. Ward KN, Coleman J, Clittin K, Fahrbach SE, Rueppell O (2008) Age, caste, and behavior determine the replicative activity of intestinal stem cells in honeybees (Apis mellifera L.). Exp Gerontol 43: 430–437.

46. Fernandes KM, Araujo VA, Serrao JE, Martins GF, Campos LAO, et al. (2010) Quantitative analysis of the digestive and regenerative cells of the midgut of Melipona quadrifasciata anthidioides (Hymenoptera: Apidae). Sociobiology 56: 489–505.

47. Ohlstein B, Spradling A (2006) The adult Drosophila posterior midgut is maintained by pluripotent stem cells. Nature 439: 470–474.

48. Hakim RS, Baldwin K, Smagghe G (2010) Regulation of midgut growth, development, and metamorphosis. Annual Review of Entomology 55: 593–608.

49. Snodgrass RE (1956) Anatomy of the Honey Bee. Ithaca, NY: Comstock Publishing Associates. 334 p.

50. Willard LE, Hayes AM, Wallrichs MA, Rueppell O (2011) Food manipulation in honeybees induces physiological responses at the individual and colony level. Apidologie 42: 508–518.

51. Sullivan JT, Castro L (2005) Mitotic arrest and toxicity in Biomphalaria glabrata (Mollusca: Pulmonata) exposed to colchicine. Journal of Invertebrate Pathology 90: 32–38.

52. Smagghe G, Vanhassel W, Moeremans C, De Wilde D, Goto S, et al. (2005) Stimulation of midgut stem cell proliferation and differentiation by insect hormones and peptides. Trends in Comparative Endocrinology and Neurobiology 1040: 472–475.

53. Williams GR, Alaux C, Costa C, Csaki T, Doublet V, et al. (2013) Standard methods for maintaining adult Apis mellifera in cages under in vitro laboratory conditions. Journal of Apicultural Research 52.

54. Rinderer TE, Danka RG, Stelzer JA (2012) Seasonal inconsistencies in the relationship between honey bee longevity in field colonies and laboratory cages. Journal of Apicultural Research 51: 218–219.

55. Martinson VG, Moy J, Moran NA (2012) Establishment of characteristic gut bacteria during development of the honeybee worker. Applied and Environmental Microbiology 78: 2830–2840.

56. Wu JY, Smart MD, Anelli CM, Sheppard WS (2012) Honey bees (Apis mellifera) reared in brood combs containing high levels of pesticide residues exhibit increased susceptibility to Nosema (Microsporidia) infection. Journal of Invertebrate Pathology 109: 326–329.

57. Decourtye A, Henry M, Desneux N (2013) Environment: Overhaul pesticide testing on bees. Nature 497: 188.

58. Martel AC, Zeggane S, Drajnudel P, Faucon JP, Aubert M (2006) Tetracycline residues in honey after hive treatment. Food Additives and Contaminants 23: 265–273.

59. Chopra I, Roberts M (2001) Tetracycline antibiotics: Mode of action, applications, molecular biology, and epidemiology of bacterial resistance. Microbiology and Molecular Biology Reviews 65: 232–260.

60. Thompson HM, Waite RJ, Wilkins S, Brown MA, Bigwood T, et al. (2005) Effects of European foulbrood treatment regime on oxytetracycline levels in honey extracted from treated honeybee (Apis mellifera) colonies and toxicity to brood. Food Additives and Contaminants 22: 573–578.

61. Rada V, Machova M, Huk J, Marounek M, Duskova D (1997) Microflora in the honeybee digestive tract: counts, characteristics and sensitivity to veterinary drugs. Apidologie 28: 357–365.

62. Gilliam M (1997) Identification and roles of non-pathogenic microflora associated with honey bees. Fems Microbiology Letters 155: 1–10.

63. Buchon N, Broderick NA, Chakrabarti S, Lemaitre B (2009) Invasive and indigenous microbiota impact intestinal stem cell activity through multiple pathways in Drosophila. Genes Dev 23: 2333–2344.

64. Carlson GR, Dhadialla TS, Hunter R, Jansson RK, Jany CS, et al. (2001) The chemical and biological properties of methoxyfenozide, a new insecticidal ecdysteroid agonist. Pest Management Science 57: 115–119.

65. Ninov N, Manjon C, Martin-Blanco E (2009) Dynamic control of cell cycle and growth coupling by ecdysone, EGFR, and PI3K signaling in Drosophila histoblasts. PLoS Biol 7: e1000079.

66. Feyereisen R, Lagueux M, Hoffmann JA (1976) Dynamics of ecdysone metabolism after ingestion and injection in Locusta migratoria. General and Comparative Endocrinology 29: 319–327.

67. Berry JA, Hood WM, Pietravalle S, Delaplane KS (2013) Field-level sublethal effects of approved bee hive chemicals on honey bees (Apis mellifera L). PLoS One 8: e76536.

68. Rharrabe K, Bouayad N, Sayah F (2009) Effects of ingested 20-hydroxyecdysone on development and midgut epithelial cells of Plodia interpunctella (Lepidoptera, Pyralidae). Pesticide Biochemistry and Physiology 93: 112–119.

Clustering of Vector Control Interventions Has Important Consequences for Their Effectiveness: A Modelling Study

Angelina Mageni Lutambi[1,2,3]*, **Nakul Chitnis**[1,2,4], **Olivier J. T. Briët**[1,2], **Thomas A. Smith**[1,2], **Melissa A. Penny**[1,2]

1 Epidemiology and Public Health, Swiss Tropical and Public Health Institute, Basel, Switzerland, **2** University of Basel, Basel, Switzerland, **3** Ifakara Health Institute, Dar es Salaam, Tanzania, **4** Fogarty International Center, National Institutes of Health, Bethesda, Maryland, United States of America

Abstract

Vector control interventions have resulted in considerable reductions in malaria morbidity and mortality. When universal coverage cannot be achieved for financial or logistical reasons, the spatial arrangement of vector control is potentially important for optimizing benefits. This study investigated the effect of spatial clustering of vector control interventions on reducing the population of biting mosquitoes. A discrete-space continuous-time mathematical model of mosquito population dynamics and dispersal was extended to incorporate vector control interventions of insecticide treated bednets (ITNs), Indoor residual Spraying (IRS), and larviciding. Simulations were run at varying levels of coverage and degree of spatial clustering. At medium to high coverage levels of each of the interventions or in combination was more effective to spatially spread these interventions than to cluster them. Suggesting that when financial resources are limited, unclustered distribution of these interventions is more effective. Although it is often stated that locally high coverage is needed to achieve a community effect of ITNs or IRS, our results suggest that if the coverage of ITNs or IRS are insufficient to achieve universal coverage, and there is no targeting of high risk areas, the overall effects on mosquito densities are much greater if they are distributed in an unclustered way, rather than clustered in specific localities. Also, given that interventions are often delivered preferentially to accessible areas, and are therefore clustered, our model results show this may be inefficient. This study provides evidence that the effectiveness of an intervention can be highly dependent on its spatial distribution. Vector control plans should consider the spatial arrangement of any intervention package to ensure effectiveness is maximized.

Editor: Joshua Yukich, Tulane University School of Public Health and Tropical Medicine, United States of America

Funding: Funding support was from the Bill and Melinda Gates Foundation provided through Swiss TPH. The funders had no role in study design, data collection and analysis, decision to publish, or preparation of the manuscript.

* E-mail: angelina-m.lutambi@unibas.ch

Introduction

Efforts to reduce malaria transmission have lead to the development of efficient vector control interventions, particularly insecticide treated nets (ITNs)(which includes conventional nets treated with a WHO recommended insecticide and long-lasting insecticidal nets). [1], indoor residual spraying (IRS), and larviciding [2–6]. These interventions are currently widely used in malaria endemic countries especially those in sub-Saharan Africa [7] and have lead to a substantial reduction in malaria morbidity and mortality. Nevertheless, malaria continues to claim hundreds of thousands of lives every year [7], thus necessitating a continued control effort to fight the disease. While over $2 billion is invested each year in procuring and distributing vector control interventions [8] for malaria control, this funding is insufficient to achieve universal coverage [8] and it is not clear if this will be sustained given current economic constraints.

Mosquito flight from one place to another [9–12] is affected by several factors including wind, odour, blood and nectar sources, availability of breeding sites, mating, and other ecological and environmental factors [13,14]. The probability that a mosquito will encounter areas that are in receipt of a particular vector control intervention while flying is dependent on the spatial arrangement of the intervention. This probability is also dependent on the complexity of how this interacts with patterns of mosquito movement. This means that it is not obvious how this dependence affects the effectiveness of interventions in controlling malaria. An understanding of how spatial clustering of interventions modifies effectiveness is particularly relevant when financial resources are insufficient, or when logistic constraints make it difficult to achieve universal coverage. It has been unclear how to prioritise the spatial allocation of interventions in such situations.

While the World Health Organization (WHO) strategy on vector management provides information on improving the efficacy, cost-effectiveness, ecological soundness and sustainability of vector control [6], there is limited relevant information on the influence of spatial distribution of these interventions on effectiveness. Approaches coupling both theory and empirical evidence are needed to evaluate and measure effectiveness of interventions at different degrees of spatial distribution for each level of intervention coverage. Despite the importance of these approaches, their development and integration in vector control programmes has been receiving inadequate attention.

Mathematical models play an important role in assessing interventions [15]. Many studies evaluate intervention effectiveness [16–25], depending on intervention coverage [16,22–24] and the significance of distribution of hosts and breeding sites for malaria transmission [20,25]. Some studies consider spatial and network models [19,20,25,26] while others consider spatial distributions of mosquito populations [27,28]. These models allow the evaluation of interventions by coverage or by any combination of intervention packages [19].

In contrast to these studies, this paper focuses on the spatial distribution of interventions rather than on heterogeneity in distribution of hosts and breeding sites. Using insights from a recent study on mosquito movements [29], a spatial model of vector population dynamics and interventions is used to assess the impact of spatial distribution of vector control interventions on reducing the population of biting mosquitoes. The effects are explored at different coverage levels to provide theoretical evidence on the existence of variability in intervention effectiveness, depending on their spatial distribution in small areas like villages.

Methods

A discrete-space continuous-time mathematical model of mosquito population dynamics and dispersal [29] was extended to incorporate ITN, IRS, and larviciding interventions. The model includes six stages of the mosquito life and feeding cycle: three juvenile stages (egg (E), larval (L), pupal (P)) and three adult stages (host seeking (A_h), resting (A_r), and oviposition site searching (A_o)). The population dynamics of mosquitoes in each stage are described by ordinary differential equations. The discrete space used in the model is a grid made up of hexagons called patches that allows any representation of spatial distribution of hosts and breeding sites and mosquito movement (dispersal) between patches. Dispersal of adult mosquitoes searching for hosts or breeding sites is restricted to the nearest six neighbouring patches.

Model Equations with Interventions

As described in more detail in [29], the population dynamics of mosquitoes are governed by the recruitment of new mosquitoes through the average number of eggs laid per oviposition, b, the development/progression rate from one stage to the next, ρ, the stage specific mortality, μ, the movement rates of host seeking, β^H, and oviposition site searching mosquitoes, β^B. The dynamics of each stage of the life cycle in patch (i,j) with interventions and movement are described using ordinary differential equations:

$$\frac{dE_{(i,j)}}{dt} = b_{(i,j)}\rho_{A_{o(i,j)}}A_{o(i,j)} - \left(\mu_{E_{(i,j)}} + \rho_{E_{(i,j)}}\right)E_{(i,j)},$$

$$\frac{dL_{(i,j)}}{dt} = \rho_{E_{(i,j)}}E_{(i,j)} - \left(\mu_{L_1(i,j)} + \rho_{L(i,j)}\right)L_{(i,j)} - \mu_{L_2(i,j)}L_{(i,j)}^2,$$

$$\frac{dP_{(i,j)}}{dt} = (1-\varepsilon_{\text{LV}})\rho_{L(i,j)}L_{(i,j)} - \left(\mu_{P(i,j)} + \rho_{P(i,j)}\right)P_{(i,j)},$$

$$\frac{dA_{h(i,j)}}{dt} = \rho_{P(i,j)}P_{(i,j)} + \rho_{A_{o(i,j)}}A_{o(i,j)} - \left(\mu_{A_{h(i,j)}} + \rho_{A_{h(i,j)}}\right)A_{h(i,j)} - \gamma_{\text{ITN}(i,j)}\mu_{A_{h(i,j)}}A_{h(i,j)} - \Psi_{\text{out}}^H A_{h(i,j)} + \Psi_{\text{in}}^H A_{h\xi'},$$

$$\frac{dA_{r(i,j)}}{dt} = \rho_{A_{h(i,j)}}A_{h(i,j)} - \left(\mu_{A_{r(i,j)}} + \rho_{A_{r(i,j)}}\right)A_{r(i,j)} - \gamma_{\text{IRS}(i,j)}\mu_{A_{r(i,j)}}A_{r(i,j)},$$

$$\frac{dA_{o(i,j)}}{dt} = \rho_{A_{r(i,j)}}A_{r(i,j)} - \left(\mu_{A_{o(i,j)}} + \rho_{A_{o(i,j)}}\right)A_{o(i,j)} - \Psi_{\text{out}}^B A_{o(i,j)} + \Psi_{\text{in}}^B A_{o\xi'}.$$

The terms $\gamma_{\text{ITN}(i,j)}\mu_{A_{h(i,j)}}A_{h(i,j)}$ and $\gamma_{\text{IRS}(i,j)}\mu_{A_{r(i,j)}}A_{r(i,j)}$ are additional mortality terms due to ITNs and IRS respectively. The term $(1-\varepsilon_{\text{LV}})\rho_{L(i,j)}L_{(i,j)}$ represents the reduced number of larvae developing to pupae from untreated breeding sites, where ε_{LV} represents the proportion of breeding sites in a given patch covered by larviciding. Parameters $\Psi_{out}^H = \sum_{\xi' \in N(i,j)} \beta_{(i,j)/\xi'}^H$ and $\Psi_{in}^H = \sum_{\xi' \in N(i,j)} \beta_{\xi'/(i,j)}^H$ represent dispersal out and into patch i,j for host seeking adults respectively, and $N(i,j)$ is a set of six nearest neighbours to patch (i,j) and $\xi' \in N(i,j)$ [29]. Similarly, $\Psi_{out}^B = \sum_{\xi' \in N(i,j)} \beta_{(i,j)/\xi'}^B$ and $\Psi_{in}^B = \sum_{\xi' \in N(i,j)} \beta_{\xi'/(i,j)}^B$ represent dispersal out and into patch i,j for oviposition site searching adults. Details of calculation of β are provided in [29]. H and B represent hosts and breeding sites respectively. The remaining parameter definitions and their corresponding values are given in Table 1.

Modelling of the Killing Effects of ITNs and IRS

ITNs kill and prevent access to people for host seeking malaria vectors, thus providing personal protection against malaria to the individuals using them [1,30]. ITNs also provide community protection to non-users [31] due to their killing effects which reduce mosquito longevity. Here, ITNs deployed in a patch are assumed to kill mosquitoes directly, hence affecting the density of host seeking adults in that patch. The killing effect of ITNs in the host seeking stage is modelled as additional mortality to normal mortality associated with host seeking process in the absence of ITNs.

IRS is the application of insecticides on the indoor walls and roofs of houses primarily to kill resting adult mosquitoes. IRS reduces malaria transmission by reducing the vector's life span and population density of vectors [32], but provides little direct personal protection against bites. Although some ingredients used in IRS may repel mosquitoes, this study considers only those

Table 1. Parameter definitions and values used in model simulations [29].

Parameter	Description	Units	Baseline	Source
b	number of eggs laid per oviposition	–	100	[53]
ρ_E	egg hatching rate	day^{-1}	0.50	[53], [54],[55]
ρ_L	rate at which larvae develop into pupae	day^{-1}	0.14	[56], [57], [58]
ρ_P	rate at which pupae develop into adults	day^{-1}	0.50	[53],[54]
μ_E	egg mortality rate	day^{-1}	0.56	[59]
μ_{L_1}	density-independent larval mortality rate	day^{-1}	0.44	[59]
μ_{L_2}	density-dependent larval mortality rate	day^{-1} mosq.$^{-1}$	0.05	
μ_P	pupal mortality rate	day^{-1}	0.37	[59]
ρ_{A_h}	rate at which host seeking mosquitoes enter the resting state	day^{-1}	0.46	[29,60]
ρ_{A_r}	rate at which resting mosquitoes enter oviposition site searching state	day^{-1}	0.43	[60]
ρ_{A_o}	oviposition rate	day^{-1}	3.0	[60]
μ_{A_h}	mortality rate of mosquitoes searching for hosts	day^{-1}	0.18	[29,60]
μ_{A_r}	mortality rate of resting mosquitoes	day^{-1}	0.0043	[60]
μ_{A_o}	mortality rate of mosquitoes searching for oviposition sites	day^{-1}	0.41	[60]

without repellency. Therefore, only the direct killing effect to resting adult mosquitoes is considered.

For ITNs, we let γ_{ITN} be the model parameter for additional mortality of host seeking adults and for IRS, we let γ_{IRS} be the model parameter for additional mortality of resting adults. To compare interventions, γ_{ITN} and γ_{IRS} are expressed as functions of intervention efficacy where efficacy is defined as the ability of an intervention to reduce mosquito survival proportionally. For ITNs or IRS, efficacy, ε_I, (where I represents ITNs or IRS) is given by

$$\varepsilon_I = \frac{S_0 - S_I}{S_0}. \qquad (1)$$

Here S_0 represents the survival probability of mosquitoes in the absence of an intervention in a given mosquito stage given by

$$S_0 = \frac{\rho_s}{\mu_s + \rho_s}, \qquad (2)$$

and S_I represents the survival probability of mosquitoes in the presence of interventions in a given stage given by

$$S_I = \frac{\rho_s}{\mu_T + \rho_s}. \qquad (3)$$

In equations (2) and (3), ρ_s is the development rate of a mosquito from stage s to the next stage, and μ_s (per unit time) is the natural mortality rate of a mosquito in stage s in the absence of an intervention. μ_T (per unit time) is the total mortality rates of mosquitoes in stage s in the presence of interventions expressed by:

$$\mu_T = \mu_s + \mu_s\gamma_I. \qquad (4)$$

Here, γ_I (unitless) is a multiplicative factor associated with the effect of intervention I (ITN or IRS). The term $\mu_s\gamma_I$ represents additional mortality of intervention, I. In order to obtain the expression for γ_I, we substitute equations (2), (3), and (4) into (1) to obtain

$$\gamma_I = \frac{\varepsilon_I(\rho_s + \mu_s)}{\mu_s(1 - \varepsilon_I)}. \qquad (5)$$

Using the stage specific parameter values for ρ_s, and μ_s [29], with $\varepsilon_I \in [0,1]$, the relationship between γ_I and ε_I is shown in Figure 1. As would be expected model intervention parameters γ_I

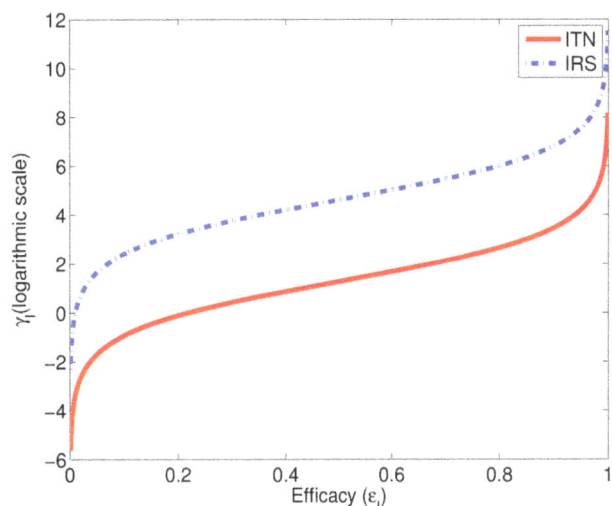

Figure 1. Relationship between ITN and IRS intervention parameters to efficacy (Equation 5 of main text).

increase with increasing efficacy of ITNs or IRS, with IRS showing higher values of γ_I compared to ITNs.

Modelling the Effect of Larviciding

Larviciding is the application of insecticides to mosquito breeding sites targeting the larval stages of the mosquitoes. Studies show that larviciding kills all larvae in treated breeding sites [33–35] and has proved to be important in suppressing the number of malaria transmitting mosquitoes in certain areas [3,33–36]. However, where breeding sites are scattered, field studies show that it is difficult to find and treat the majority of productive breeding sites [37]. The effect of larviciding in the model is to reduce the development of larvae into pupae and thus include a parameter representing the proportion of breeding sites identified and treated within patch (i,j), as $\epsilon_{LV}(i,j)$. The proportion $(1 - \epsilon_{LV}(i,j))$ represents the untreated breeding sites, where larvae develop into pupae.

Modelling ITN Repellency

In addition to the killing effect of ITNs that directly affects the density of host seeking adults, the pyrethroid insecticide used to treat nets has a repellent effect acting as a chemical barrier that irritates host seeking mosquitoes as they come close to the nets. Repellency of nets reduces the availability of blood to mosquitoes, increases host searching time, and subsequently prolongs the mosquito gonotrophic cycle duration which in turn impacts mosquito population size. We model the repellent effect of ITNs as follows:

Let Pc be the proportion of hosts within a patch who are covered by ITNs (patch coverage), and Z be the repellent effect of ITNs. If $H_{(i,j)}$ is the number of hosts in patch (i,j), and $I_{(i,j)} = H_{(i,j)}Pc_{(i,j)}Z$ is the number of protected hosts in patch (i,j), then the number of unprotected hosts $(U_{(i,j)})$ in that particular patch is given by

$$U_{(i,j)} = H_{(i,j)} - I_{(i,j)} = H_{(i,j)}\left(1 - Pc_{(i,j)}Z\right). \quad (6)$$

If the patch does not have ITNs ($Pc_{(i,j)} = 0$), then $U_{(i,j)} = H_{(i,j)}$.

Since the repellent effect of ITNs affects host seeking mosquitoes, their dispersal rate into patches containing ITNs is affected. This effect is included by assuming that ITN repellency reduces hosts availability to mosquitoes in a given patch so that attractiveness of the patch to hosts seeking mosquitoes is reduced. Hosts covered by ITNs are therefore protected as some mosquitoes are repelled during the host seeking process. The dispersal rate, $\beta^H_{\xi'/(i,j)}$, detailed in [29] was modified by replacing the number of hosts present in a patch by those who are not protected by ITNs in the particular patch as:

$$\beta^H_{\xi'/(i,j)} = De^{-\lambda\left(\bar{U}^{ij}_{\xi'} - \bar{U}^{ij}_{\xi}\right)} \quad (7)$$

where $\bar{U}^{ij}_{\xi'}$ is the proportion of unprotected hosts available in patch ξ' contained in $c_{i,j}$ given by $\bar{U}^{ij}_{\xi'} = U^{\xi'}/H^{ij}_u$, and H_u is the total number of unprotected hosts in $c_{i,j}$. Here, $c_{i,j}$ is a set of seven patches sharing boundaries (patch (i,j) and its 6 neighbours). Simulations of the repellent effect are performed by considering that only unprotected hosts are attracting mosquitoes in each of the patches in the neighbourhood.

Spatial Clustering

Ecological models have been developed and used to study effects of landscape spatial heterogeneity on population dynamics [38–40] with increasing interest in the field of epidemiology [41]. Some models have been used to investigate spatial clustering effects in ecology [41–46]. To our knowledge, such methods have not been used by the malaria community to investigate clustering of vector control interventions. The degree of clustering (in the context of this study) is defined as a measure of the degree to which patches/hexagons on the hexagonal grid tend to spatially cluster together. In the context of vector control interventions, we define spatial clustering as a measure of the extent to which areas under interventions on a landscape are aggregated together. This degree varies from 0 (if the spatial distribution of interventions is random) to 1 (if the spatial distribution of interventions is highly concentrated on a certain portion of the landscape, or highly grouped together).

To evaluate the effect of spatial clustering of interventions using the model, we distributed interventions on the spatial grid [29]. The spatial distribution of interventions was varied according to the degree of spatial clustering chosen. These spatial clusters used for distributing interventions were created using the pair approximation method [38,39]. Two pair states were used: intervention and non-intervention states. These two states were assigned after defining a coverage area (that is proportion of patches assumed to be under interventions). Following Hiebeler [39], the degree of clustering, q_{00} was defined as the probability that a randomly chosen neighbour to a patch with intervention also contains the intervention. Spatial clusters of varying degrees on the model grid were created in Matlab using the steps detailed by Hiebeler [39]. Several configurations of spatial clusters were created from different initial random distributions of the intervention states to account for stochasticity of the method. Figure 2 illustrates one such cluster configuration produced at different degrees of clustering, q_{00}, when intervention coverage is 50% over the entire grid.

For the vector control investigations, cluster configurations were created at 10%, 30%, 50%, and 70% coverage levels, with the degree of spatial clustering, q_{00} ranging from 0 to 1 at an interval of 0.1. However, it is only possible to create spatial clusters when $q_{00} \geq 2 - (1/p_0)$ [39] (where p_0 represent intervention coverage). This was due to the fact that when an intervention coverage is high, it is likely that neighbours of patches under intervention, are also under intervention. This implies a lower bound on q_{00} for high coverage. For example, at $p_0 = 70\%$, the lower bound for q_00 is 0.57. This means that, it is not possible to create clusters at a degree of spatial clustering less than 0.57.

Model Parameterizations and Assumptions

Parameter values on stage specific mortality, and development rates used to simulate the model are given in Table 1. Various experimental studies show that ITN killing efficacy is variable [47,48] as it depends on local entomological and epidemiological conditions [49]. For the parameter values of interventions, we make the assumption that ITNs and IRS are 80% efficacious so that ε_{ITN} and ε_{IRS} were fixed at 0.8.

When a larvicide is applied to a breeding site, all larvae experience an increased mortality. Field studies show that larviciding is likely to kill all larvae when applied to a breeding site [33–35]. However, not all breeding sites can be identified for larvicidal treatment. Here, 80% ($\epsilon_{LV} = 0.8$) of the breeding sites inside a patch are assumed to be identified and treated with larvicide. Thus, leaving 20% of breeding sites within a patch without larvicide, allowing larvae develop into pupae. We also make the assumption that larvae are distributed uniformly across breeding sites.

Field studies on mark release recapture experiments of *Anopheles gambiae* also show that daily flight range from 200 to 400 m [50] or 800 m a day [9]. Others show that about 90% of mosquitoes reach a distance of 1.5 km. These experimental results indicate that mosquito flight distance is variable. Due to these variations, the total area modelled in this study was limited to one square kilometre. The patch size, with patch centroids 50 m apart and used in this work, was based on flight distances of mosquitoes chosen and numerical ease.

A 25 by 21 hexagonal grid was used as a hypothetical representation of a landscape. At the edges of the grid, periodic boundary conditions were used. This assumes the area being modelled is comparable to its neighbourhood. For simplicity,

simulations were performed with all hexagons (patches) on the grid containing breeding sites and hosts. The dispersal related parameters for host seeking (β^H) and oviposition site searching (β^B) mosquitoes depend on the availability of hosts and breeding sites respectively and the diffusion rate, $D = 0.2$ per time was used in all simulations. The diffusion coefficient of dispersal ($D^* = D/A$, where A is the area of each patch contained in the hexagonal grid) scales with patch size and as a result, the equilibrium results presented in this study scale with increasing patch size or increasing number of patches (and total area modelled).

Measuring Intervention Effectiveness

We define intervention effectiveness as the reduction in the total equilibrium population of host seeking mosquitoes, over all

Figure 2. An example of spatial clusters generated at different degrees of clustering (q_{00}). An example of spatial clusters generated at different degrees of clustering (q_{00}) with a coverage of $p_0 = 0.5$ for the covered states (white) for intervention deployment and uncovered states (black). Clustering increases with increasing q_{00}.

patches on the grid. In malaria transmission control, the number of potentially infective mosquitoes should be reduced. Thus, only host seeking adults, which transmit malaria, are considered. From the model, the equilibrium total number of host seeking mosquitoes is calculated over the entire grid as

$$A_h^* = \sum_{\xi \in \Xi} A_{h\xi}^*, \tag{8}$$

where $A_{h\xi}$ is the equilibrium number of adult host seeking mosquitoes in patch ξ and Ξ is the set of all patches on the entire grid. In this context, we calculate intervention effectiveness, ϵ_{int}, as the proportionate reduction of an equilibrium population of host seeking mosquitoes, namely

$$\epsilon_{int} = 1 - \frac{A_h^{*(int)}}{A_h^*}, \tag{9}$$

where A_h^* is the equilibrium population of host seeking mosquitoes in the absence of interventions, and $A_h^{*(int)}$ is the equilibrium population of host seeking mosquitoes in the presence of an intervention.

Simulations

Simulations were carried out in Matlab 7.10.0 (R2010a). The adaptive step size Runge-Kutta method of fourth and fifth order (*ode45*) was used to solve the system of ordinary differential equations (Eqn. (1)). Simulations were performed at intervention coverage levels of 0% coverage (no intervention), 10%, 30%, 50%, and 70%. The 0% level scenario was included to compute intervention effectiveness (Equation 9).

Several simulations were performed in this study. The first set of simulations involved creations of cluster configurations at each value of q_{00} as described in the spatial clustering subsection. A total of four cluster configurations were generated for each q_{00}. After clusters were generated, each cluster (a matrix of zeros and ones) for each q_{00} at each coverage level was used as an input matrix for placing interventions. Interventions were placed in entries with ones and entries with zeros represented non-intervention areas. One simulation was performed for each cluster configuration for each intervention package. Simulations were run until the system (1) was at equilibrium. The resulting equilibrium values were recorded and used to evaluate intervention effectiveness. For each cluster configuration at each coverage, one simulation was performed to obtain the equilibrium value which was used as a baseline for computing effectiveness as described above.

For each scenario a representative total population of 2700 eggs, 1900 larvae, 2000 pupae, 2400 host seeking mosquitoes, 1800 resting, and 1200 oviposition site searching mosquitoes were initially distributed across the grid. Parameter values used to simulate the model are given in Table 1. We numerically tested that there exists only one equilibrium point given different initial conditions for both the non-intervention and intervention scenarios.

Statistical Analysis of the Relationship between Intervention Spatial Clustering and Effectiveness

Simulation results for each coverage level were further analysed using statistical methods. The aim was to quantify the relationships between effectiveness and the degree of spatial clustering of an intervention. Since the effectiveness is measured as the propor-tionate reduction in host seeking mosquitoes, its range lies within 0 and 1. Thus, robust generalized linear models with a logit link [51] were used. The outcome variable in each model was the simulation results of effectiveness of an intervention package with the explanatory variable being the degree of spatial clustering at a given coverage level of that particular intervention package.

Results

The effectiveness of ITNs, IRS, and larviciding is related to the degree of spatial clustering of interventions and coverage levels (Figure 3). When the coverage of larviciding and IRS is 10% (Figure 3A), simulation results indicate that these interventions tend to be more effective when highly clustered compared to low clustering. However, the benefits of highly clustering IRS are not statistically significant (Table 2). At 30% coverage, high clustering of IRS appears to be no longer more effective than low clustering. For larviciding, at 30% spatial coverage level, larviciding is more effective when highly clustered compared to when lowly clustered. For ITNs distributed at low coverages of 10% to 30% (Figure 3A–B), the intervention is more effective with a low degree of spatial clustering compared to with a high degree of spatial clustering (ITN effectiveness is negatively correlated to the degree of spatial clustering).

At a moderate intervention coverage level of 50% (Figure 3C), effectiveness of IRS and larviciding decreases with increasing clustering and distributing ITNs randomly in a non-clustered way is more beneficial than in a clustered way. At an intervention coverage level of 70% (Figure 3D), distributing interventions widely and randomly in a non-clustered manner is more effective than clustering for any of the interventions.

When interventions are combined (Figure 4), effectiveness decreases with increasing degree of spatial clustering, implying more benefits when widely distributed in space. However, the combination of IRS and larviciding was not associated with the degree of spatial clustering when coverage was less then 30%.

Effectiveness of an intervention at zero clustering is highest for ITNs and lowest for larviciding (given our parameter values) when interventions are singly deployed (Table 2). Effectiveness at zero clustering is highest when all interventions are combined together, but the additional effect over ITNs alone is small. The combination of IRS and larviciding had the lowest effectiveness at zero clustering, irrespective of the coverage level.

At lower spatial coverage levels of single interventions, the difference in effectiveness between one intervention and another decreases with increasing value of the degree of spatial clustering. This gap (difference) remains almost constant at high coverage levels (Figure 3). For combined interventions and at all coverage levels, there is almost no difference in effectiveness for all combinations of interventions that included ITNs (Figure 4). The effectiveness of a combination of IRS and larviciding is consistently lower across all coverage levels. In addition, the difference in effectiveness between a combination of IRS and larviciding and other combinations is always high. However, at lower coverage levels, this difference decreased with increasing degree of spatial clustering (Figure 4A and B).

The scatter plots also show that there is variability in effectiveness. These variations increase with increasing clustering (Figures 3 and 4), especially at low to moderate coverage levels.

Discussion

In this study, an existing mathematical model of mosquito dispersal [29] was extended to include vector control interventions. In order to distribute interventions heterogeneously across the

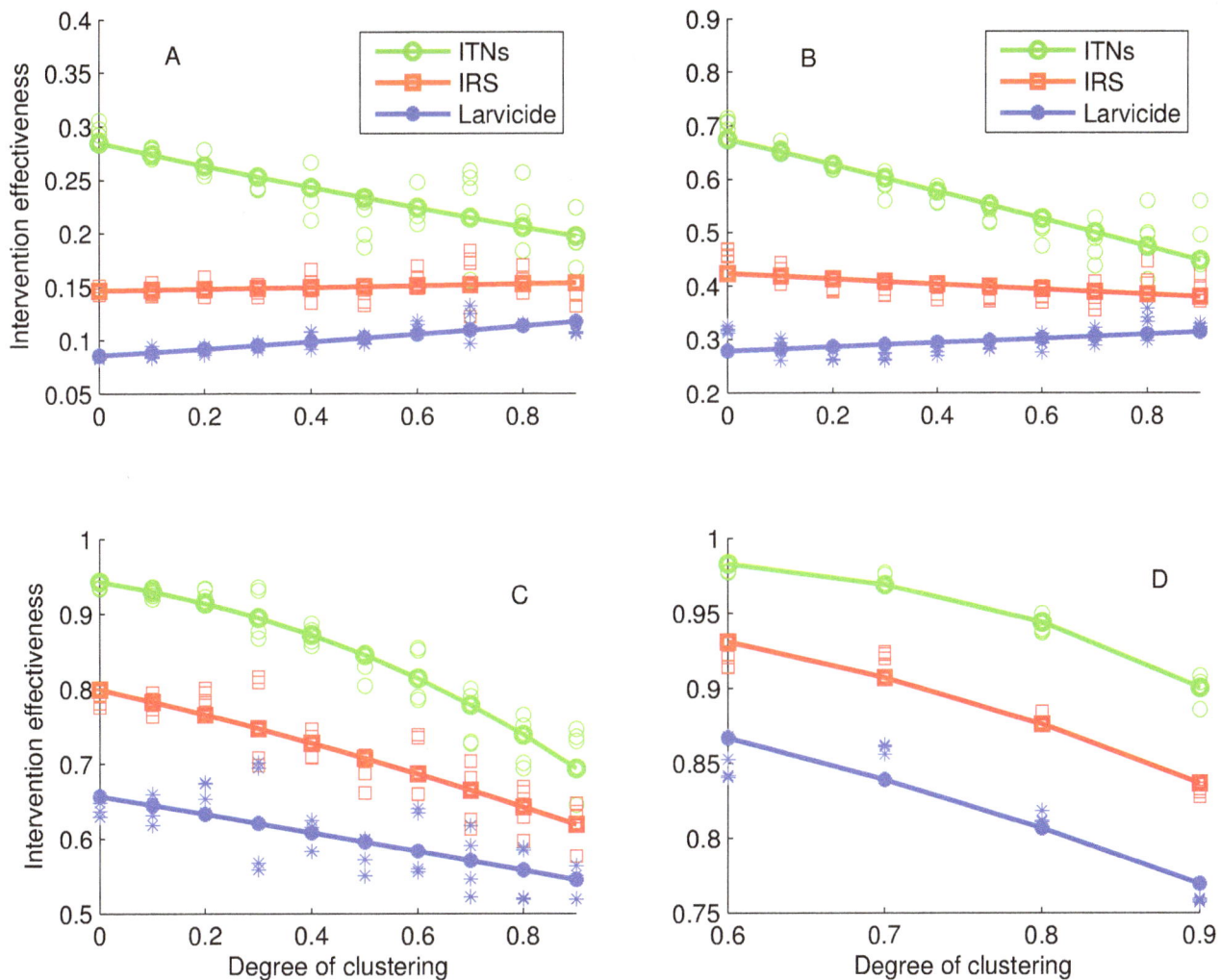

Figure 3. Intervention effectiveness by degree of spatial clustering of ITNs, IRS, and larviciding at different coverage levels. The symbols (scatter plots) represent simulated intervention effectiveness data from different configurations of intervention distribution to account for stochastic variations and the lines are the result of a linear fit on a logarithmic scale ($p = 1/(1 + \exp^{(-\beta_0 - \beta_1 C)})$). Effectiveness is measured as the proportionate reduction of the equilibrium population of host seeking mosquitoes. Hosts and breeding sites were homogeneously distributed across the grid. Coverage levels A: 10%, B: 30%, C: 50%, and D: 70%.

landscape, according to the degree of clustering chosen, this model was combined with an approach for modelling spatially hetero-geneous landscapes [39] to assess the effects of spatial clustering of vector control interventions on their effectiveness, at various levels of spatial coverage and intervention combinations. As in another study [22], the reduction in the overall vector population density was used as an indicator of the population-wide effect of interventions. The results have important implications for deployment strategies in situations where universal coverage is not achievable.

Our model indicates that, with a single intervention of either IRS or larviciding in an environment where breeding sites and hosts are homogeneously distributed and spatial coverage of the intervention is low (i.e. few patches are covered), there is a small increase in effectiveness when deployment is highly spatially clustered compared to widely distributed in space. However, with high spatial coverage, it is more effective to distribute these interventions randomly in an unclustered manner. ITNs were less effective at a higher degree of clustering than at a lower degree of clustering for any spatial coverage level.

At a spatial coverage of less than 50%, if larviciding is highly clustered, then treated areas become almost mosquito free. However, if larviciding is not clustered, mosquitoes that breed in neighbouring patches can still feed in areas that have been larvicided. If coverage is moderate to high (50% or larger), larviciding is more effective when randomly distributed and unclustered, because a greater proportion of the remaining adult mosquitoes is likely to encounter the intervention when ovipositing. When larviciding is clustered, most of the ovipositing occurs in non-larvicided areas because adult mosquitoes are rare in larvicided areas. When larviciding is widespread and unclustered, a proportion of adult mosquitoes emerging in non-larvicided patches will migrate to, feed and oviposit in larvicided breeding sites.

With adulticidal interventions, especially ITNs, the benefits of distributing the intervention widely and unclustered are greater, because the mosquitoes need to avoid intervention patches each gonotrophic cycle if they are to survive. Where adulticidal interventions are clustered, mosquitoes emerging in locations remote from the intervention area are unlikely to be killed,

Table 2. Association between intervention effectiveness and the degree of spatial clustering of interventions by coverage level.

Coverage	10%	30%	50%	70%
Effectiveness at zero clustering (β_0) (logit transformed)[a]				
ITNs	−0.92 (0.02)	0.73 (0.04)	2.80 (0.06)	6.56 (0.41)
IRS	−1.77(0.02)	−0.31 (0.03)	1.38 (0.04)	3.49 (0.28)
Larvicide	−2.37 (0.02)	−0.95 (0.04)	0.65 (0.04)	2.29 (0.24)
All	−0.82 (0.02)	0.88 (0.04)	3.14 (0.07)	7.78 (0.46)
ITNs and IRS	−0.86 (0.02)	0.82 (0.04)	3.01 (0.07)	7.41 (0.45)
ITNs and larviciding	−0.87 (0.02)	0.82 (0.04)	3.03 (0.07)	7.51 (0.45)
IRS and larviciding	−1.55 (0.02)	0.00 (0.04)	1.93 (0.05)	6.21 (0.38)
Effect of clustering (β_1) on the effectiveness (logit scale)				
ITNs	−0.54 (0.07)	−1.04 (0.10)	−2.20 (0.12)	−4.75 (0.50)
IRS	0.06 (0.05)[b,c]	−0.20 (0.07)	−0.99 (0.07)	−1.95 (0.36)
Larviciding	0.39 (0.04)[b]	0.19 (0.07)[b]	−0.52 (0.06)	−1.09 (0.32)
All	−0.61 (0.07)	−1.17 (0.10)	−2.5(0.14)	−6.02 (0.55)
ITNs and IRS	−0.59 (0.07)	−1.11 (0.10)	−2.42 (0.13)	−5.64 (0.54)
ITNs and larviciding	−0.57 (0.07)	−1.11 (0.10)	−2.44(0.13)	−5.75 (0.54)
IRS and larviciding	−0.05 (0.05)[c]	−0.43 (0.08)[c]	−1.46 (0.09)	−4.73 (0.49)

Association between intervention effectiveness and the degree of spatial clustering of interventions by coverage levels. β_1 is an estimate (gradient) of the effect of the degree of spatial clustering of an intervention and β_0 is an intercept measuring the effectiveness of the intervention at zero clustering. The higher β_0, the higher the effectiveness at zero clustering. Figures in parenthesis are standard errors.
[a]$\beta_0 = \ln\left(\frac{p_0}{1-p_0}\right)$, where p_0 is the actual effectiveness.
[b]Positive relationship, implying a benefit of clustering the intervention.
[c]Not statistically significant (i.e. p-value >0.05).

whereas when interventions are non-clustered, a mosquito will encounter them sooner or later. Consequently, at any spatial coverage level, average biting densities are reduced more by deploying ITNs in an unclustered manner than by clustering them. It also follows that widespread distribution of adulticidal interventions will reduce the number of old (potentially disease-transmitting) mosquitoes even more than it will reduce average densities. This finding, that the overall effect in the reduction of mosquito numbers is much greater if the intervention is spatially non-clustered and widely distributed, especially when coverage is moderate and insufficient to achieve universal coverage, contradicts the notion that a locally high coverage is needed to achieve a mass effect of ITNs or IRS for reduction in disease.

Highly clustered scenarios had lower ITN effectiveness. This is likely due to the fact that when intervention coverage is high, then the likelihood that any patch and its six neighbours are under intervention is high. In this aspect, patch attractiveness to biting mosquitoes is reduced. When this occurs, then all neighbouring patches produce the same repellency effect which results into fewer mosquitoes leaving the centre patch (because they are also repelled for each of their neighbours). In so doing, the repellency effect decreases and the killing effect becomes the main factor, rather than the combination of both repellency and killing.

While non-clustered deployment of most intervention packages is generally most effective, this may be expensive to achieve since it requires delivery even to remote locations. Interventions are often delivered preferentially to more accessible areas, and such clustered (and sometimes inequitable) distributions are likely to be the cheapest. To investigate how delivery costs affect cost-effectiveness, there is a need for modelling of different distribution schemes (for example for ITNs or IRS) of interventions given a fixed budget in various settings with different degrees of clustering, coverage levels and accessibility.

Efficacy, defined as the effect on the target stage of the vector as a proportion of the theoretical maximum effect, translates differently into effectiveness defined on some common metric of levels of transmission, disease control, or, in this paper, densities of host seeking mosquitoes. We have assumed 80% efficacious interventions throughout, and our results are consistent with other modelling work suggesting that at constant efficacy, ITNs have the highest impact on biting densities of mosquitoes [16,22] and in our simulations any combination of interventions which includes ITNs is also highly effective at all levels of coverage and across all spatial clustering. This may be accounted by the repellency effect of ITNs included in the model. The assumed 80% efficacy of ITNs in this work is representative of both the killing and repellency action of ITNs and of indoor biting coverage of individuals within a patch. Even with small patch sizes assuming an 80% efficacy for ITNs is likely too high. A further extension of the models would be to vary the level of intervention within each patch, and thus efficacy.

Comparing of Figures 3 and 4 indicates that although ITNs provide better protection alone compared to other interventions, results show that there are additional benefits if ITNs are combined with other interventions. Our study also shows that although larviciding is less effective compared to ITNs and IRS, treating a similar or higher level of coverage would result in a higher reduction of biting mosquitoes.

The current results are indicative of the effect of applying interventions within a small village, with a small number of dwellings or breeding sites per patch, but should also be broadly applicable to smaller patches corresponding to single individuals or breeding sites. We would not necessarily expect the same results to hold with very large patches, e.g. corresponding to whole villages where patch size might be comparable to the flight range of the mosquitoes and where other factors such as spatial variation within patches might be relevant.

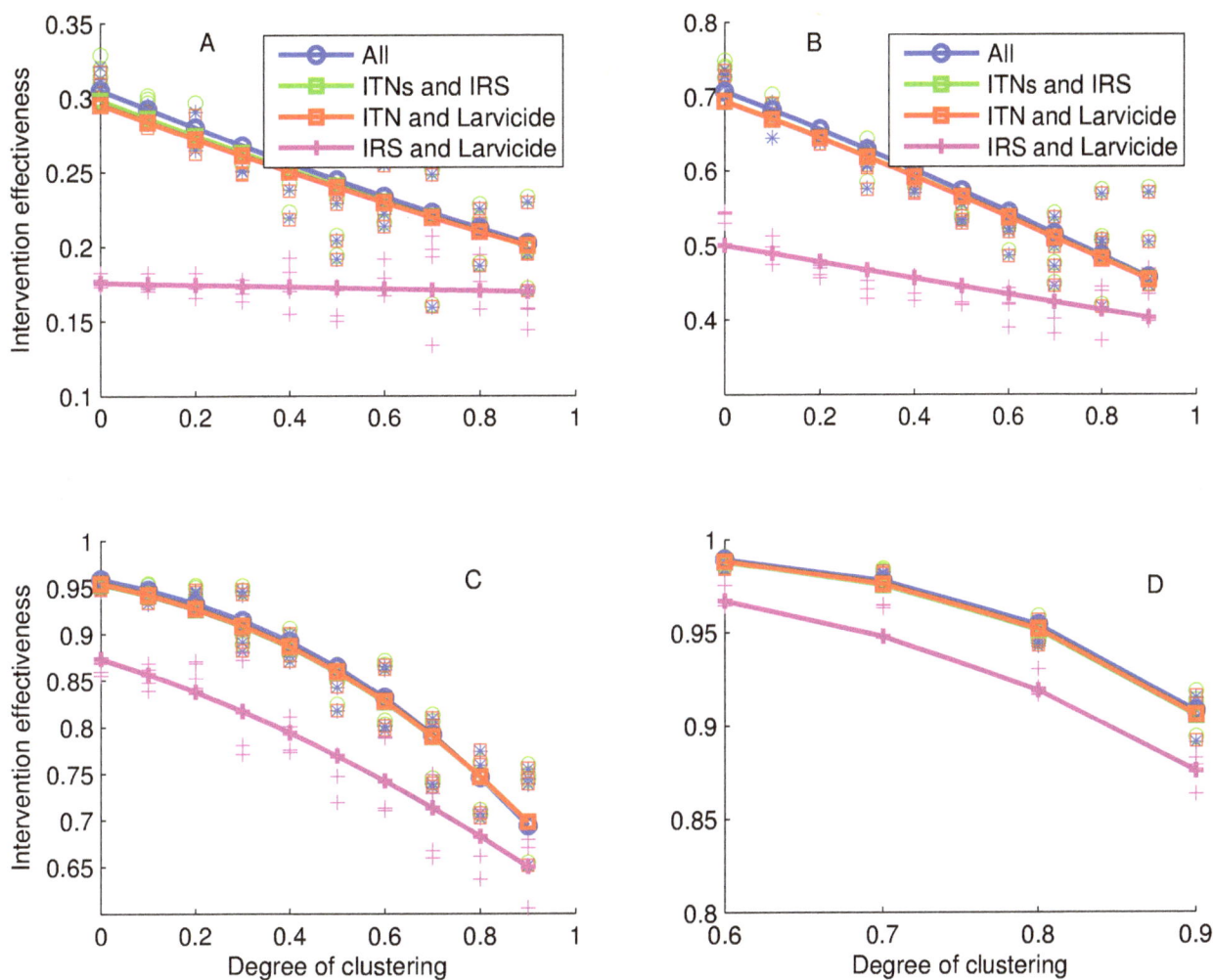

Figure 4. Effect of spatial clustering of interventions by coverage level for combined interventions. The symbols (scatter plots) represent simulated intervention effectiveness from different realizations depicting initial distribution of interventions before the process of clustering was undertaken to account for stochastic variations and the lines are the result of a linear fit on a logarithmic scale ($p = 1/(1 + \exp^{(-\beta_0 - \beta_1 C)})$). Hosts and breeding sites were homogeneously distributed over the grid. Coverage levels A: 10%, B: 30%, C: 50%, and D: 70%.

Modelling and simulation provides a much easier approach to investigate these issues than field studies do, but inevitably require making simplifying assumptions. To assess the effect of clustering, we simulated a homogeneous distributions of both human hosts and breeding sites. The cues that these human hosts and breeding sites provide that influence movement of mosquitoes cancel each other out, therefore movement was not influenced by the availability of these hosts or breeding sites [52]. Further investigations need to incorporate scenarios in which breeding sites and hosts are heterogeneously distributed. In such scenarios, knowledge about hotspots will allow targeted (and therefore likely spatially clustered) deployment of interventions and this may well be more cost-effective than non-clustered deployment. In other words, in scenarios with spatially heterogeneous hosts and/or breeding sites, the cost of knowledge about where these are may well compensate for potential gains in effectiveness. However, in the absence of knowledge about spatial location of hosts and breeding sites for mosquitoes (even for scenarios when they are heterogeneously distributed) non-clustered distribution may be most cost-effective.

Results from this study provide evidence that the effectiveness of an intervention can be highly dependent on its spatial distribution. Given logistical and financial constraints, vector control plans should consider the spatial arrangement of any intervention package to ensure effectiveness is maximized. In the case of high achievable coverage, and in the absence of information that allows targeting, it is of great help to ensure that the distribution is as equitable and as evenly spatially spread as possible for maximizing benefits.

Acknowledgments

We thank our colleagues in the malaria modelling group at the Swiss TPH for their contributions and critical comments they provided. We also thank the reviewers for their critical review of this manuscript. The author thank the Swiss Tropical and Public Health Institute and the Ifakara Health Institute for their support.

Author Contributions

Conceived and designed the experiments: AML. Performed the experiments: AML. Analyzed the data: AML. Contributed reagents/materials/analysis tools: AML MAP TAS NC OB. Wrote the paper: AML. Provided critical input on designing the experiments: MAP TAS NC OB. Provided critical input on the experiments: MAP. Provided critical input on data analysis: MAP TAS NC OB. Provided critical input on writing the manuscript: TAS MAP NC OB.

References

1. WHO (2007) Insecticide treated mosquito nets: a WHO position statement. Technical report, World Health Organization.

2. Bayoh MN, Mathias DK, Odiere MR, Mutuku FM, Kamau L, et al. (2010) *Anopheles gambiae*: historical population decline associated with regional distribution of insecticide-treated bed nets in western Nyanza province, Kenya. Malaria Journal 9.

3. Kroeger A, Horstick O, Riedl C, Kaiser A, Becker N (1995) The potential for malaria control with the biological larvicide Bacillus thuringiensis israelensis (Bti) in Peru and Ecuador. Acta Tropica 60.

4. Lengeler C (2004) Insecticide-treated bed nets and curtains for preventing malaria. Cochrane Database of Systematic Reviews.

5. Muturi EJ, Burgess P, Novak RJ (2008) Malaria vector management: where have we come from and where are we headed? The American Journal of Tropical Medicine and Hygiene 78.

6. WHO (2004) Global strategic framework for integrated vector management. Technical report, World Health Organization, Geneva.

7. WHO (2011) World Malaria Report. Technical report, World Health Organization, Geneva.

8. RBM (2013) Minutes of roll back malaria vector control working group 8th annual meeting. Technical report, Roll Back Malaria. Available: http://www.rollbackmalaria.org/mechanisms/vcwg.html.

9. Gillies MT (1961) Studies on the dispersion and survival of *Anopheles Gambiae Giles* in East Africa, by means of marking and release experiments. Bulletin of Entomological Research 52: 99–127.

10. Gillies MT, Wilkes TJ (1978) The effect of high fences on the dispersal of some West African mosquitoes (Diptera: Culicidae). Bulletin of Entomological Research 68: 401–408.

11. Gillies MT, Wilkes TJ (1981) Field experiments with a wind tunnel on the flight speed of some West African mosquitoes (Diptera: Culicidae). Bulletin of Entomological Research 71: 65–70.

12. Service MW (1997) Mosquito dispersal - The long and the short of it. Journal of medical entomology 34.

13. Edman J, Scott T, Costero A, Morrison A, Harrington L, et al. (1998) *Aedes aegypti* (diptera culicidae) movement influenced by availability of oviposition sites. Journal of Medical Entomology 35: 578–583.

14. Cummins B, Cortez R, Foppa IM, Walbeck J, Hyman JM (2012) A Spatial Model of Mosquito Host-Seeking Behavior. PLoS Computational Biology 8.

15. McKenzie FE, Samba EM (2004) The role of mathematical modeling in evidence-based malaria control. The American Journal of Tropical Medicine and Hygiene 71: 94–96.

16. Chitnis N, Schapira A, Smith T, Steketee R (2010) Comparing the effectiveness of malaria vectorcontrol interventions through a mathematical model. The American Journal of Tropical Medicine and Hygiene 83.

17. Eckhoff PA (2011) A malaria transmission-directed model of mosquito life cycle and ecology. Malaria Journal 303: 10.

18. Griffin JT, Hollingsworth TD, Okell LC, Churcher TS, White M, et al. (2010) Reducing *Plasmodium falciparum* malaria transmission in Africa: A model-based evaluation of intervention strategies. PLoS Medicine 7.

19. Gu W, Novak RJ (2009) Agent-based modelling of mosquito foraging behaviour for malaria control. Transactions of the Royal Society of Tropical Medicine and Hygiene 103: 1105–1112.

20. Gu W, Regens JL, Beier JC, Novak RJ (2006) Source reduction of mosquito larval habitats has unexpected consequences on malaria transmission. PNAS 103.

21. Menach LA, Takala S, McKenzie FE, Andre P (2007) An elaborated feeding cycle model for reductions in vectorial capacity of night-biting mosquitoes by insecticide-treated nets. Malaria Journal 6: 10.

22. White MT, Griffin JT, Churcher TS, Ferguson NM, Basanez MG, et al. (2011) Modelling the impact of vector control interventions on *Anopheles gambiae* population dynamics. Parasites and Vectors 4.

23. Worrall E, Connor SJ, Thomson MC (2007) A model to simulate the impact of timing, coverage and transmission intensity on the effectiveness of indoor residual spraying (IRS) for malaria control. Tropical Medicine and International Health 12.

24. Yakob L, Yan G (2009) Modeling the effects of integrating larval habitat source reduction and insecticide treated nets for malaria control. PLoS ONE 4.

25. Yakob L, Yan G (2010) A network population model of the dynamics and control of African malaria vectors. Transactions of the Royal Society of Tropical Medicine and Hygiene 104: 669–675.

26. Gu W, Novak RJ (2009) Predicting the impact of insecticide-treated bed nets on malaria transmission: the devil is in the detail. Malaria Journal 256: 8.

27. Nourridine S, Teboh-Ewungkem MI, Ngwa GA (2011) A mathematical model of the population dynamics of disease - transmitting vectors with spatial consideration. Journal of Biological dynamics 5: 335–365.

28. Otero M, Schweigmann N, Solari HG (2008) A Stochastic Spatial Dynamical Model for *Aedes Aegypti*. Bulletin of Mathematical Biology 70: 1297–325.

29. Lutambi AM, Penny MA, Smith T, Chitnis N (2013) Mathematical modelling of mosquito dispersal in a heterogeneous environment. Mathematical Biosciences 241: 198–216.

30. Takken W (2002) Do insecticide-treated bednets have an effect on malaria vectors? Tropical Medicine and International Health 7.

31. Hawley WA, Phillips-Howard PA, Ter Kuile FO, Terlouw DJ, Vulule JM, et al. (2003) Communitywide effects of permethrin-treated bed nets on child mortality and malaria morbidity in western Kenya. The American Journal of Tropical Medicine and Hygiene 68: 121–7.

32. WHO (2006) Indoor residual spraying: Use of indoor residual spraying for scaling up global malaria control and elimination. Technical report, World Health Organization.

33. Fillinger U, Lindsay SW (2006) Suppression of exposure to malaria vectors by an order of magnitude using microbial larvicides in rural Kenya. Tropical Medicine and International Health 11.

34. Majambere S, Lindsay SW, Green C, Kandeh B, Fillinger U (2007) Microbial larvicides for malaria control in The Gambia. Malaria Journal 6: 76.

35. Mwangangi JM, Kahindi SC, Kibe LW, Nzovu JG, Luethy P, et al. (2011) Wide-scale application of Bti/Bs biolarvicide in different aquatic habitat types in urban and peri-urban Malindi, Kenya. Parasitology Research 108: 1355–1363.

36. Majambere S, Pinder M, Fillinger U, Ameh D, Conway DJ, et al. (2010) Is mosquito larval source management appropriate for reducing malaria in areas of extensive flooding in The Gambia? A cross-over intervention trial. The American Journal of Tropical Medicine and Hygiene 82.

37. Killeen GF, Tanner M, Mukabana WR, Kalongolela MS, Kannady K, et al. (2006) Habitat targeting for controlling aquatic stages of malaria vectors in Africa. The American Journal of Tropical Medicine and Hygiene 74.

38. Hiebeler D (1997) Stochastic spatial models: From simulations to mean field and local structure approximations. Journal of Theoretical Biology 187.

39. Hiebeler D (2000) Populations on fragmented landscapes with spatially structured heterogeneities: Landscape generation and local dispersal. Ecology 81.

40. Okuyama T (2008) Intraguild predation with spatially structured interactions. Basic and Applied Ecology 9: 135–144.

41. Hiebeler D (2005) A cellular automaton SIS epidemiological model with spatially clustered recoveries. In: Sunderam VS, van Albada GD, Sloot PMA, Dongarra J, editors, International Conference on Computational Science (2). Springer, number 3515 in Lecture Notes in Computer Science, 360–367.

42. Lee SH, Su NY, Bardunias P (2007) Exploring landscape structure effect on termite territory size using a model approach. BioSystems 90.

43. Su M, Li W, Li Z, Zhang F, Hui C (2009) The effect of landscape heterogeneity on host-parasite dynamics. Ecological Research 24: 889–896.

44. Thomson NA, Ellner SP (2003) Pair-edge approximation for heterogeneous lattice population models. Theoretical Population Biology 64: 271–280.

45. Tsonis AA, Swanson KL, Wang G (2008) Estimating the clustering coefficient in scale-free networks on lattices with local spatial correlation structure. Physica A 387.

46. Westerberg L, Ostman O, Wennergren U (2005) Movement effects on equilibrium distributions of habitat generalists in heterogeneous landscapes. Ecological Modelling 188: 432–447.

47. Chouaibou M, Simard F, Chandre F, Etang J, Darriet F, et al. (2006) Efficacy of bifenthrinimpregnated bednets against *Anopheles funestus* and pyrethroid-resistant *Anopheles gambiae* in North Cameroon. Malaria Journal 5: 77.

48. Oxborough RM, Mosha FW, Matowo J, Mndeme R, Feston E, et al. (2008) Mosquitoes and bednets: testing the spatial positioning of insecticide on nets and the rationale behind combination insecticide treatments. Annals of Tropical Medicine and Parasitology 102.

49. Smith DL, Hay SI, Noor AM, Snow RW (2009) Predicting changing malaria risk after expanded insecticide-treated net coverage in Africa. Trends in Parasitology 25.

50. Midega JT, Mbogo CM, Mwambi H, Wilson MD, Ojwang G, et al. (2007) Estimating dispersal and survival of *Anopheles gambiae* and *Anopheles funestus* along the kenyan coast by using markrelease-recapture methods. Journal of Medical Entomology 44: 923–929.

51. Papke LE, Wooldridge MJ (1996) Econometric methods for fractional response variables with an application to 401(K) plan participation rates. Journal of applied econometrics 11: 619–632.

52. Killeen GF, McKenzie FE, Foy BD, Bogh C, Beier JC (2001) The availability of potential hosts as a determinant of feeding behaviours and malaria transmission by African mosquito populations. Transactions of the Royal Society of Tropical Medicine and Hygiene 95.

53. Service MW (2004) Medical Entomology for Students. Cambridge University Press, Third edition.

54. Holsetein MH (1954) Biology of *Anopheles Gambiae*: research in French West Africa. Monograph series number 9, World Health Organization, Palais des Nations, Geneva.

55. Yaro AS, Dao A, Adamou A, Crawford JE, Ribeiro JMC, et al. (2006) The distribution of hatching time in *Anopheles gambiae*. Malaria Journal 5: 19.

56. Bayoh MN, Lindsay SW (2003) Effect of Temperature on the development of the aquatic stages of *Anopheles gambiae sensu stricto* (Diptera: Culicidae). Bulletin of Entomological Research 93: 375–381.

57. Kirby MJ, Lindsay SW (2009) Effect of temperature and inter-specific competition on the development and survival of *Anopheles gambiae sensu stricto* and *An. arabiensis* larvae. Acta Tropica 109: 118–123.

58. Gething PW, Van Boeckel TP, Smith DL, Guerra CA, Patil AP, et al. (2011) Modelling the global constraints of temperature on transmission of *Plasmodium falciparum* and *P. vivax*. Parasites and Vectors 92: 2.

59. Okogun G (2005) Life-table analysis of *Anopheles* malaria vectors: generational mortality as tool in mosquito vector abundance and control studies. Journal of Vector borne Diseases 42: 45–53.

60. Chitnis N, Smith T, Steketee R (2008) A mathematical model for the dynamics of malaria in mosquitoes feeding on a heterogeneous host population. Journal of Biological Dynamics 2: 259–285.

18

Oral Delivery Mediated RNA Interference of a Carboxylesterase Gene Results in Reduced Resistance to Organophosphorus Insecticides in the Cotton Aphid, *Aphis gossypii* Glover

You-Hui Gong[1], **Xin-Rui Yu**[1], **Qing-Li Shang**[2], **Xue-yan Shi**[1]*, **Xi-Wu Gao**[1]*

1 Department of Entomology, China Agricultural University, Beijing, China, **2** College of Plant Science and Technology, Jilin University, Changchun, China

Abstract

Background: RNA interference (RNAi) is an effective tool to examine the function of individual genes. Carboxylesterases (CarE, EC 3.1.1.1) are known to play significant roles in the metabolism of xenobiotic compounds in many insect species. Previous studies in our laboratory found that *CarE* expression was up-regulated in *Aphis gossypii* (Glover) (Hemiptera: Aphididae) adults of both omethoate and malathion resistant strains, indicating the potential involvement of *CarE* in organophosphorus (OP) insecticide resistance. Functional analysis (RNAi) is therefore warranted to investigate the role of *CarE* in *A. gossypii* to OPs resistance.

Result: *CarE* expression in omethoate resistant individuals of *Aphis gossypii* was dramatically suppressed following ingestion of dsRNA-*CarE*. The highest knockdown efficiency (33%) was observed at 72 h after feeding when dsRNA-*CarE* concentration was 100 ng/μL. The CarE activities from the *CarE* knockdown aphids were consistent with the correspondingly significant reduction in *CarE* expression. The CarE activity in the individuals of control aphids was concentrated in the range of 650–900 mOD/per/min, while in the individuals of dsRNA-*CarE*-fed aphids, the CarE activity was concentrated in the range of 500–800 mOD/per/min. In vitro inhibition experiments also demonstrated that total CarE activity in the CarE knockdown aphids decreased significantly as compared to control aphids. Bioassay results of aphids fed dsRNA-*CarE* indicated that suppression of *CarE* expression increased susceptibility to omethoate in individuals of the resistant aphid strains.

Conclusion: The results of this study not only suggest that ingestion of dsRNA through artificial diet could be exploited for functional genomic studies in cotton aphids, but also indicate that *CarE* can be considered as a major target of organophosphorus insecticide (OPs) resistance in *A. gossypii*. Further, our results suggest that the *CarE* would be a propitious target for OPs resistant aphid control, and insect-resistant transgenic plants may be obtained through plant RNAi-mediated silencing of insect *CarE* expression.

Editor: Youjun Zhang, Institute of Vegetables and Flowers, Chinese Academy of Agricultural Science, China

Funding: This research was supported by National Basic Research and Development Program of China (Contract No. 2012CB114103) (http://program.most.gov.cn/) and the National Natural Science Foundation of China (31330064 and 30871661) (http://isisn.nsfc.gov.cn/). The funders had no role in study design, data collection and analysis, decision to publish, or preparation of the manuscript.

Competing Interests: The authors have declared that no competing interests exist.

* Email: shixueyan@cau.edu.cn (XYS); gaoxiwu@263.net.cn (XWG)

Introduction

Carboxylesterases (CarE, EC 3.1.1.1), or carboxyl/cholinesterases, are known to play significant roles in the metabolism of xenobiotic compounds in many insect species [1–2]. Many studies have reported that the elevation of esterase activity through gene amplification or up-regulated transcription accounts for some degree of resistance to insecticides in some insects [3]. This phenomenon is so common that, overexpression of esterases has become a dominant criterion in identifying the development of resistance to organophosphorus insecticides (OPs). This has been well documented in numerous insect species including *Myzus persicae* [4–6], *Aphis gossypii* [7–9], *Bemisia tabaci* [10], *Culex*

pipiens quinquefasciatus [11–15], other mosquitoes in *Culicine* [16–20], *Nilaparvata lugens* [21], and *Locusta migratoria manilensis* [22]. Metabolic resistance to OPs was also associated with mutations in esterase gene sequences in several insect species. Some mutations (G137D and W251L/S) in esterase genes that confer resistance to OPs have been reported in *Musca domestica* [2,23–25], *Lucilia cuprina* [26–27], *Cochliomyia hominivorax* [28] and *Culex pipiens* [29]. Cui et al. [30] reported that G/A151D or W271L mutations could be common mechanisms in the development of OP resistance in Dipteran species.

The cotton aphid, *Aphis gossypii* (Glover) (Hemiptera: Aphididae), is an important pest of a number of agriculturally important crops; not only because of its destructive damage to crops through feeding and virus transmission [31], but also due to its extreme ability to develop resistance to many classes of insecticides, including OPs, pyrethroids, and carbamates [7–8]. Elevation of esterase activity through up-regulated esterase transcription, as well as point mutations within esterase genes, which change substrate specificities, are two known mechanisms of esterase-mediated insecticide resistance [9,32–39]. Previous studies in our laboratory found that *CarE* was more highly expressed in the apterous adults of both omethoate- and malathion- resistant *Aphis gossypii* strains as compared to apterous adults from susceptible strains. These findings strongly suggested the potential involvement of *CarE* in OPs resistance in these strains [7–9]. As such, more detailed functional analysis investigating the role of *CarE* in OPs resistance in resistant *A. gossypii* strains is warranted.

Post-transcriptional gene silencing by RNA interference (RNAi) is a very useful tool to examine the functions of individual genes. RNAi is mediated by double-stranded RNA (dsRNA) that is cleaved into 21–23 nucleotide small interfering RNAs (siRNAs) by an RNase III-type enzyme known as Dicer [40,41]. RNAi has been successfully used to investigate gene function in the pea aphid *Acyrthosiphon pisum* and the green peach aphid *Myzus persicae* [42–47]. Aphids can be fed artificial diets which are sandwiched between thin parafilm membranes [47]. In *A. pisum*, both microinjection and dsRNA-feeding through artificial diets have been reported to be valuable methods for achieving RNAi [42–46]. In *M. persicae*, a plant-mediated RNAi approach was documented to knockdown gene expression by up to 60% in transgenic *Nicotiana benthamiana* and *Arabidopsis thaliana* [47]. RNAi has not yet been employed for gene functional studies in *A. gossypii*. It is difficult to perform microinjections without affecting aphids' survival rates, as cotton aphids, like *M. persicae*, are smaller than *A. pisum*. As such, dsRNA delivery through feeding may be an efficacious method for RNAi-based functional studies in *A. gossypii*.

In this study, we used artificial diet feeding and RNAi methods to functionally analyze the role of *CarE* in the omethoate resistance of an *A. gossypii* strain known to be resistant to omethoate. We measured aphid susceptibility to omethoate, CarE activity in individual aphids, and the *in vitro* inhibitory effects of S,S,S-tributyl phosphorotrithioate (DEF) on CarE activity in the omethoate resistant aphids 72 h after feeding on ds-*CarE*.

Materials and Methods

Insects

The omethoate-resistant aphid strain used in this study is the same strain in which overexpression of *CarE* was previously identified [7]. The resistant strain was initially established from a field population originally collected in 1999 from cotton fields in Xinjiang Uygur Autonomous Region, China. Many generations of this strain were subjected to omethoate selection pressure in our laboratory, using the leaf-dipping method described by Moores et al. [48]. The susceptible strain was supplied by Dr. Donghai Zhang (Shihezi University, Xinjiang Uygur Autonomous Region, China) in 1999, and was maintained over many generations without exposure to insecticides.

Total RNA isolation, synthesis of cDNA and dsRNA

Apterous adult aphids were homogenized in TRIzol reagent (Invitrogen, USA). The extracted RNA samples were treated with DNase (RNase free) (NEB, USA) to exclude DNA contamination. The total RNA was analyzed with gel electrophoresis and quantified using a spectrophotometer (UV-2550; SHIMADZU, Japan). The first-strand cDNA was synthesized according to the manual of the SuperScript III first-strand synthesis system for RT-PCR kit (Invitrogen, USA). All of the available nucleotide sequences of the *A. gossypii CarE* gene (Genbank No. AY485218 and Genbank No. AY485216) were retrieved from the NCBI GenBank database, and a homology search to define the conserved regions was carried out using Megalign (DNASTAR) software. The forward primer 5′-taatacgactcactataggg TAACCCTTGGGCGTTTACTG-3′and the reverse primer 5′-taatacgactcactataggg GGTCTCGTCGCAAAAATCAT-3′ were used to amplify a *CarE* gene fragment. A 686 bp fragment was amplified and confirmed by sequencing. The fragment was used as a template to generate the corresponding dsRNA using the MEG Ascript RNAi kit (Ambion, USA). dsRNA-*CarE* was dissolved in 50 μL diethypyrocarbonate (DEPC)-treated water, analyzed with gel electrophoresis (1% agarose), and quantified using a spectrophotometer (UV-2550; SHIMADZU, Japan).

Rearing on artificial diet and dsRNA feeding

The artificial diet recipe and the rearing device used for this study were developed based on the methods of Mittler [49] with some modifications. The diet was prepared in DEPC-treated water to ensure the absence of RNase activity. For the dsRNA feeding experiments, dsRNA-*CarE* was added into the artificial diet at 50, 100, and 500 ng/μL concentrations. Artificial diet lacking dsRNA-*CarE* was used as a control. Third instar (L3) stage aphids grown on cotton leaves were transferred onto the artificial diet device for rearing. The artificial diet was sealed between two layers of Parafilm in a 2 cm diameter feeding arena; twenty apterous adult aphids were placed in each arena. The arena was covered with a fine mesh to prevent their escape. The insects were reared under controlled growth conditions: $27 \pm 1°C$, $65 \pm 5\%$ relative humidity, and 16:8 h light:dark photoperiod. After feeding for 24 h, the aphids were collected or transferred onto cotton leaves for the subsequent experiments. In order to determine the optimal dsRNA-*CarE* concentration and the optimal silencing time (different intervals after feeding) to ascertain the maximum silencing efficacy of *CarE*, five silencing times (12 h, 24 h, 36 h, 48 h, and 72 h post-feeding) of each dsRNA-*CarE* concentration (50, 100, and 500 ng/μL) were sampled and evaluated by real time PCR (protocols described below). Based on the results from these optimization studies, aphids fed with dsRNA-*CarE* at the concentration of 100 ng/μL and with the interval of 72 h post-feeding were used for the omethoate toxicity, CarE activity, and *In vitro* CarE inhibition assays. We did not sample at the 500 ng/μL concentration in the formal assays; as the 100 ng/μL dsRNA concentration was found to have the same silencing efficacy as the 500 ng/μL dsRNA concentration at the 72 h post-feeding interval (Figure 1).

Quantitative real-time PCR (qRT–PCR)

Cotton aphid cDNA was prepared as described in section 2.2, above. The primer pair of CS2 (5′-CATACCCTACGCTCAAC-CAC-3′) and CA2 (5′-GCAATCTTCACTTCCAACGA-3′) was designed for detecting the transcript levels of *CarE*. The primer pair of R1 (5′-ATTGACGGAAGGGCACC-3′) and R2 (5′-CGCTCCACCAACTAAGAACG-3′) was designed based on the 18S rRNA gene (Genbank No. AF487716), and was used as an internal reference for the relative expression analysis. The qRT-PCR assays were conducted on an ABI 7300 Real time PCR system (ABI) following the manufacturer's recommendations. The reactions were performed in a 10 μL reaction mixture, which contained 4 μL SYBR, 0.2 μL ROX I, 2.6 μL ddH₂O, 0.2 μL primers, and 2 μL cDNA (equivalent to 0. 08 μg of total RNA).

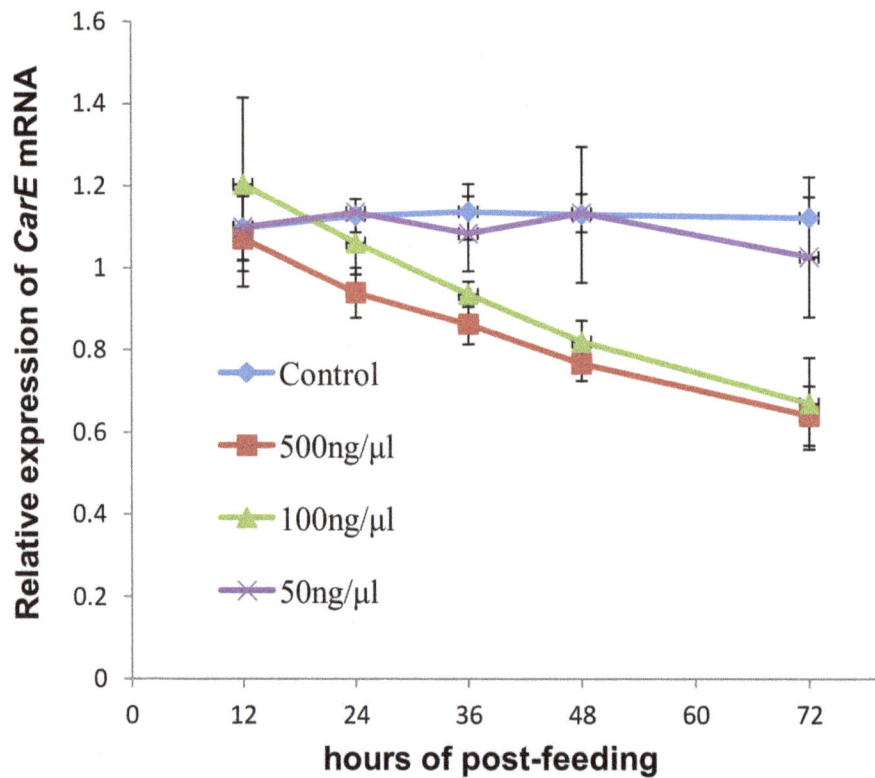

Figure 1. The effect of time course and dsRNA-*CarE* dose on RNAi efficiency of the *Aphis gossypii*. The relative expression of *CarE* mRNA with time course in Aphis gossypii fed on artificial diets with or without *CarE*-dsRNA were recorded. The final concentration of *CarE*-dsRNA in the artificial diet was 50, 100, 500 ng/μL respectively, the artificial diet without *CarE*-dsRNA used as the control. The transcript abundances were determined relative to the normalized calibrator, i.e. cDNA from non-RNAi apterous adults (only fed on cotton leaves), which was set to 1.0. Each treatment had three biological replicates, and 20 insects were used per pooled RNA sample. The results are shown as means ± S.D.

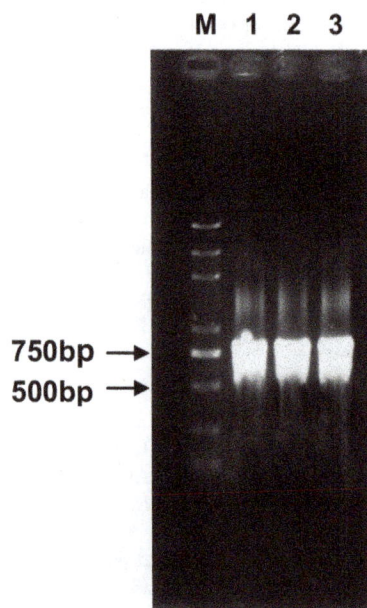

Figure 2. Electrophorosis of dsRNA-*CarE*. M: Molecular weight marker 5000; 1, 2, 3: dsRNA-*CarE*. Three groups of dsRNA-*CarE* were dissolved in 50 μL diethypyrocarbonate (DEPC)-treated water, analyzed with gel electrophoresis (1% agarose).

The cycling parameters were 95°C for 30 s; followed by 40 cycles of 95°C for 5 s; 60°C for 31 s. After the cycling protocol, the final step was applied to all reactions by continuously monitoring fluorescence through the dissociation temperature of the PCR product at a temperature transition rate of 0.1°C/s, to generate a melting curve. Quantification was conducted according to the $2^{-\Delta Ct}$ method [50]. The transcript abundances were determined relative to the normalized calibrator, i.e. cDNA from non-RNAi Apterous adults (fed with cotton leaves), which was set to 1.0. The experiment was conducted three times independently, with different RNA preparations. The qRT-PCR results are presented as means with standard errors (SE) of transcript levels, on a logarithmic scale. The statistical significance of changes in gene expression was calculated using a Student's t-test for all 2-sample comparisons. A value of $P \leq 0.05$ was considered to be statistically significant (* indicates $P \leq 0.05$; ** indicates $P \leq 0.01$; *** indicates $P \leq 0.001$).

Susceptibility of aphids to omethoate after RNAi of *CarE*

Omethoate toxicity in the resistant aphids was determined by the leaf-dipping method described by Moores et al. [48] and Cao et al. [7]. Briefly, for the bioassays, a stock of insecticide was prepared in acetone and diluted to a series of six concentrations with distilled water containing 0.05% (v/v) Triton X-100 and 1% acetone. Cotton leaf discs (15-mm diameter) were dipped in omethoate solutions for 5 s, placed in the shade to air dry, and then placed upside down on an agar bed (25 mm in depth) in the wells of 12-well tissue-culture plates. Bioassays were carried out by exposing 45 apterous adults (15 per well) to omethoate-treated

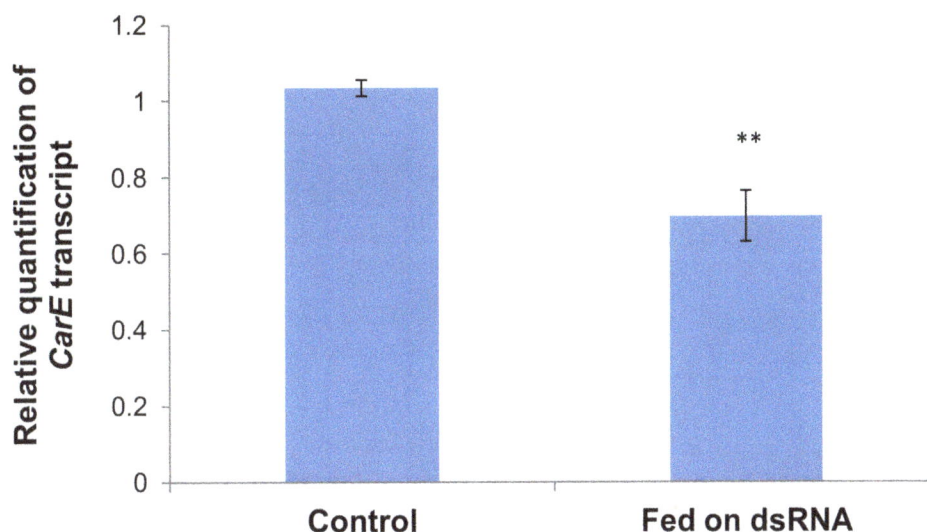

Figure 3. The dsRNA-mediated suppression of *CarE* transcript expression in aphids fed on the artificial diet with dsRNA. The final concentration of dsRNA in the artificial diet was 100 ng/μL. The transcript abundances were determined relative to the normalized calibrator, i.e. cDNA from non-RNAi apterous adults (fed on cotton leaves), which was set to 1.0. The values are means with standard errors of three biological replicates from RNA samples of 20 aphids. "**" indicates significant differences as determined by Student's t-test (P<0.01).

leaves. The aphids were confined by applying a ring of fluon to the exposed lip of each well. Assays of each concentration were replicated at least three times, and mortality was assessed at 25°C, 24 h after commencing the treatment. LC_{50} values were calculated with POLO software (LeOra Software Inc., Berkeley CA).

The LC_{50} value of the aphids to omethoate was used as the diagnostic dose for assessment of the sensitivity of cotton aphids to omethoate at 72 h post-feeding of dsRNA-*CarE*. The mortality was assessed at 25°C, 24 h after exposure to omethoate. The control was conducted by using the aphids fed with artificial diet only. This experiment was repeated three times. There were 60 individuals aphids in each replicate. The statistical significance of mortality rate was calculated using a Student's t-test for all 2-sample comparisons. A value of $P \leq 0.05$ was considered to be statistically significant (* indicates $P \leq 0.05$; ** indicates $P \leq 0.01$; *** indicates $P \leq 0.001$).

CarE activity in individual aphids

For these assays, each adult aphid was homogenized in 100 μL ice-cold phosphate buffer (0.04 M, pH 6.5). The homogenate was centrifuged at 4°C, 10,000 g for 15 min, and the supernatant was used as the enzyme source for measuring the activity of CarE. CarE activity was measured by the method of van Aspern [51] modified for use of a microplate reader (ACT-AMPR-750, ACTGene). 50 μL of homogenate (equivalent to half of one cotton aphid) and 50 μL phosphate buffer (0.04 M, pH 6.5) were added to each well in a microplate; freshly prepared 100 μL mixed solution (6 mg Fast Blue RR in 10 ml 100 μM α-NA) was then

added to each well. The absorbance (405 nm) was read 30 times over the course of 5 min. The slope, OD increase value per min per aphid, was taken to represent the CarE activity of a single aphid. 100 cotton aphids were analyzed for both the dsRNA-*CarE* fed treatment group and the control group.

In vitro CarE inhibition by DEF

S,S,S-tributyl phosphorotrithioate (DEF) is an inhibitor of CarE. It can inhibit the activity of CarE *in vitro*. The IC_{50} of DEF was determined according to the method of Young et al [52] with modifications. For these assays, one hundred apterous adults with similar color and size from 72 h post-feeding of dsRNA-*CarE* group and the control group were homogenized in 1 mL of ice-cold phosphate buffer (0.04 M, pH 7.0). The controls were aphids fed with artificial diet only. The homogenates were centrifuged at 4°C, 10,000 g for 15 min, The supernatant was used as an enzyme source for measuring the activity of CarE. Stock solutions (30 mM) of DEF were prepared in acetone, and serial dilutions of DEF solutions from 0.156 to 10 mM were prepared in ice-cold phosphate buffer (0.04 M, pH 7.0). Enzyme solutions in buffer, and buffer only served as positive and negative treatments, respectively. 50 μL of insect homogenate was incubated for 30 min with 5 μL DEF solution. 450 μL of phosphate buffer (pH 7.0, 0.04 M) and 1.8 mL 0.3 mM substrate solution (α-NA) were then added. The reaction was stopped by the addition of 0.9 mL of stop solution (two parts of 1% Fast Blue BB and five parts of 5% sodium dodecyl sulfate) after incubation at 30°C for 15 min. The color was allowed to develop for 15 min at room

Table 1. Susceptibility of cotton aphids to omethoate.

Strain	LC_{50} (mg/L) (95% CL[a])	Slope±SE[b]	χ^2
Resistant strain	5874 (4027.83–8912.50)	2.51±0.41	2.67

[a]CL: Confidence limited.
[b]SE: Standard error.

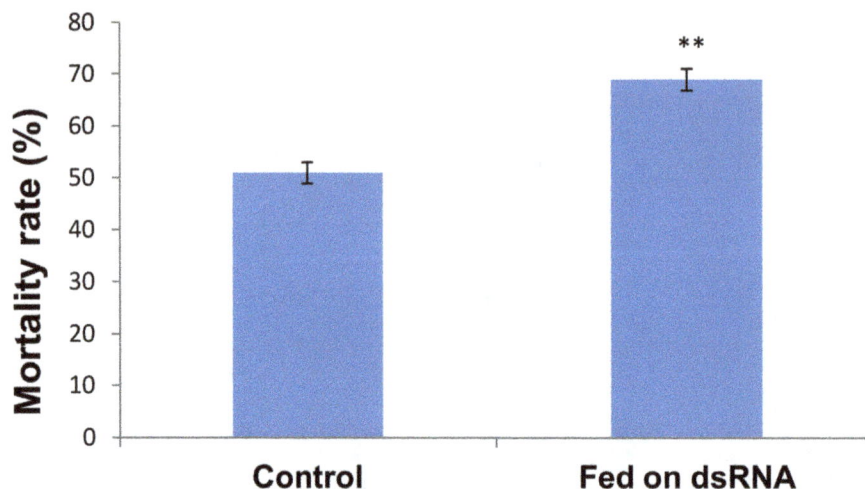

Figure 4. *CarE* knockdown increases resistant aphids' sensitivity to omethoate. The final concentration of dsRNA in the artificial diet was 100 ng/µL; the artificial diet lacking dsRNA was used as control. The mortality was recorded after 24 h exposure to omethoate (3 replicates, 60 individuals for each replicate). The values are presented as the means with standard errors. "**" indicates significant difference as determined by Student's t-test (P<0.01).

temperature, and the absorbance was measured at 600 nm for α-NA with a UV/VIS Spectrometer Lambda Bio40 (Perkin-Elmer, USA). Mean levels of residual esterase activity after inhibition with DEF were based on protein content and α-NA standard curves. Protein content was determined by the method of Bradford [53], using bovine serum albumin as the standard. The IC_{50} was calculated based on the concentration of DEF and the value of the residual CarE activity.

Results

CarE -dsRNA synthesis and dose effect and time course on RNA interference

In this study, A 686 bp *CarE* gene fragment was amplified and confirmed by sequencing. Figure 2 shows a 686 bp single band of dsRNA-*CarE* was amplified using a 686 bp *CarE* gene fragment as a template. The final concentration of the three groups of dsRNA-*CarE* (Figure 2) was 780 ng/µl, 831 ng/µl and 867 ng/µl, respectively, after dissolution in 50 µL of diethypyrocarbonate (DEPC)-treated water.

As mentioned in section 2.3, in order to determine the optimal dsRNA-*CarE* concentration and the optimal silencing time (different intervals after feeding) as well as to ascertain the maximum silencing efficacy of *CarE*, five silencing times (12 h, 24 h, 36 h, 48 h, and 72 h post-feeding) of each dsRNA-*CarE* concentration (50, 100, and 500 ng/µL) were sampled and evaluated by real time PCR. Figure 1 shows that the *CarE*-mRNA expression decreased as ds-RNA-*CarE* concentration increased. The effect at the 500 ng/µl level showed no significant difference from the 100 ng/µl level. At each increase of post feeding times, the *CarE* transcripts expression decreased, with the lowest expression level at post feeding 72 h. Finally, based on the results from these optimization studies, aphids fed with dsRNA-*CarE* at a concentration of 100 ng/µL 72 h post-feeding were examined again for *CarE* transcripts expression. Figure 3 shows that the expression level of *CarE* was significantly lower at 72 h post-feeding in the omethoate- resistant aphids which were fed 100 ng/µL dsRNA-*CarE*, as compared to the control aphids (Student's t test, t = 4.8, 4 degrees of freedom, P = 0.009). The average reduction of *CarE* expression observed was about 33%

(Figure 3), indicating that the dsRNA-mediated knock down of *CarE* transcripts was successful and these silencing conditions can be used for following experiments.

CarE knockdown increases sensitivity to omethoate in the aphids of the resistant strain

A Probit analysis of the susceptibility of omethoate resistant aphids exposed to omethoate is summarized in Table 1. The LC_{50} value of the *A. gossypii* resistant strain was 5874 mg/L, and this value was used as the dosage to evaluate the effect of RNAi of *CarE* on the susceptibility of cotton aphids to omethoate. The results demonstrated that mortality increased significantly (Student's t test, t = −6.182, 4 degrees of freedom, P = 0.003), from 50.78% in the control aphids to 68.44% in the dsRNA-CarE-fed aphids (Figure 4).

CarE knockdown decreases CarE activity in individual aphids

The CarE activities of individual aphids were grouped into different activity intervals; each interval level increased by 50 mOD/per/min (Figure 5). Frequency distributions of individuals on each interval level were then calculated for both the control and the dsRNA-fed aphids based on their CarE activities. The CarE activity in the control aphids was concentrated in the range of 650–900 mOD/per/min, with an average CarE activity of 742 mOD/per/min (Figure 5A), while the CarE activity in the dsRNA-fed aphids was concentrated in the range of 500–800 mOD/per/min with an average CarE activity of 677 mOD/per/min (Figure 5B).

CarE knockdown decreases the IC_{50} value of DEF inhibited CarE activity in resistant aphids

The IC_{50} value of DEF for inhibiting CarE activity was 1.2-fold higher in the control aphids than in the dsRNA-CarE-fed aphids (Table 2) and there was a significant difference (Student's t test, t = 3.771, 4 degrees of freedom, P = 0.020). This result illustrated that CarE activity decreased due to the suppression of *CarE* transcript expression in the dsRNA-*CarE* treated aphids.

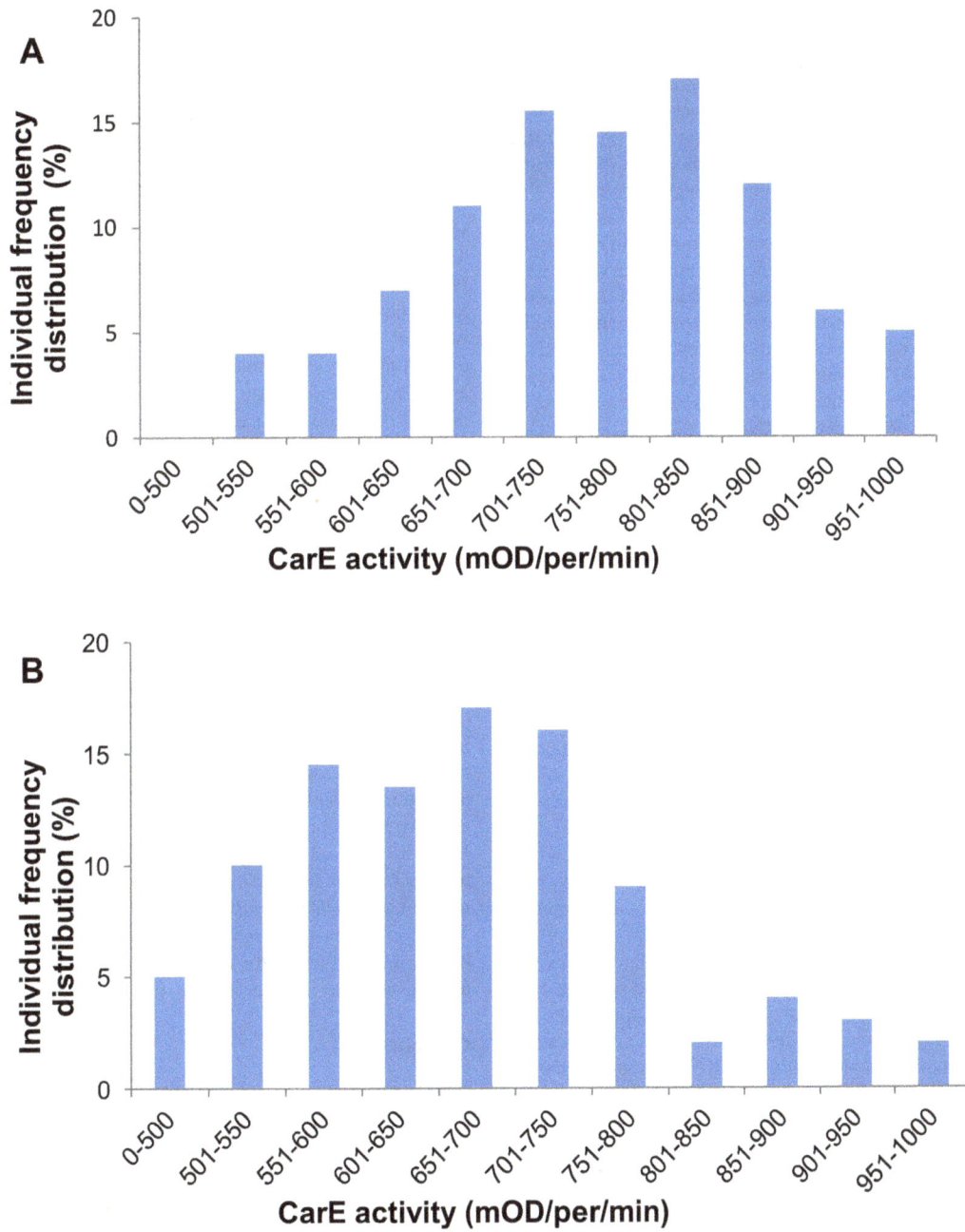

Figure 5. Frequency distribution of individual CarE activity in aphids fed on artificial diet lacking dsRNA-*CarE* or with dsRNA-*CarE*. Total 100 survival aphids were tested at 72 h after feeding on the artificial diet lacking dsRNA of *CarE* (A) or with dsRNA of *CarE* (B). The final concentration of dsRNA in the artificial diet was 100 ng/μL.

Table 2. CarE knockdown decreases the IC$_{50}$ value of DEF for inhibiting CarE activity in resistant aphids.

Inhibitor	IC$_{50}$ (mol/L)	
	fed on artificial diet	**fed on dsRNA**
DEF	0.094 ± 0.003	0.078 ± 0.003*

The values are presented as the means with standard errors. The asterisk indicates significant differences as determined by Student's t-test ($P \leq 0.05$). The final concentration of dsRNA in the artificial diet was 100 ng/μL.

Discussion

Both microinjection and oral delivery of dsRNA through artificial diets have been reported to be valuable methods for achieving RNAi in aphids [42–44,46]. However, there is evidence that micro-injection of dsRNA can cause negative impacts on the survival of the aphids [43]. These impacts may result from mechanical damage or from the sudden higher levels of the injected dsRNA in the hemolymph [47]. Cotton aphids are smaller than *A. pisum*, and it is difficult to do microinjections on cotton aphids without deleteriously affecting aphid survival rates. Therefore, in our study, we used an artificial diet containing dsRNA-*CarE* to knock down the expression of the cotton aphid *CarE* gene. The survival rate of the cotton aphids fed on dsRNA for 24 h was above 90% in our experiment (data not shown), so the delivery of dsRNA via artificial diet can now be considered to be a suitable method for achieving RNAi in this pest. qRT-PCR analyses showed that *CarE* transcript levels in aphids significantly decreased by up to 33% at 72 h after feeding the dsRNA-*CarE* at a 100 ng/μL concentration, as compared to the control group (Figure 3). This is the first example of RNAi in a cotton aphid. The *CarE* transcript level dropped gradually, over time after feeding (Figure 1), as was observed in the light brown apple moth by Turner et al. [54]. This may indicate that quite time is required for the dsRNA to enter midgut cells and/or for the induction of the RNAi process [54]. Although the silencing observed through RNAi in the present study is significant, it was not complete. This result is similar to the RNAi effect observed for the pea aphid [46]. This could be due to the degradation of dsRNA in the artificial diet prior to ingestion, or in the body of aphids. Recently, Allen and Walker [55] reported that the dsRNA present in the saliva of *Lygus lineolaris* is capable of degrading dsRNA. Yu et al. [56] also found that aphids feeding on an artificial diet containing dsGFP caused degradation of that dsRNA in the diet. In summary, our results demonstrate that RNAi by dsRNA feeding is possible in the *A. gossypii*, as it is in nematodes [57], ticks [58], *Epiphyas postvittana* [54] and *A. pisum* [45–46]. Oral delivery of dsRNA offers many advantages compared with injection and soaking. It is labor-saving, cost-effective, easy to perform, and applicable for high-throughput gene screening [59]. It is also a less invasive method than microinjection forinducing RNAi in small insects such as aphids and in first and second-instar larvae or nymphs [60].

In insects, CarEs are key components of defense against xenobiotic compounds, including insecticides [1]. CarE-based metabolic resistance to OPs has been observed in many insects, including cotton aphids [7–9]. The molecular mechanism of this resistance originates either from mutations in esterase-encoding sequences or from increased transcription of esterase genes [3,30]. RNAi is a useful tool to probe the functions of genes. To date, there are no reports analyzing the functions of *CarE* genes in *A. gossypii* using RNAi methods. In our study, the LC$_{50}$ dosage of the resistant cotton aphid strain was used to detect the susceptibility of dsRNA-*CarE*-fed cotton aphids to omethoate (Table 1). The mortality of dsRNA-*CarE*-fed aphids was 68.44%, while that of control aphids was 50.78% (Figure 4), indicating that RNAi of *CarE* expression reduced the detoxification metabolic effect of CarE to omethoate. This effect may be caused by reduced total CarE activity. This is consistent with a previous report that overexpression of a *CarE* gene was involved in the resistance to

OPs in a omethoate-resistant strain of *A. gossypii* [7]. Our results from the individual aphid CarE activity assays confirm to this as well. We found that the CarE activity in the control aphids was concentrated in the range of 650–900 mOD/per/min with an average CarE activity of 742 mOD/per/min (Figure 5A), while the CarE activity in the dsRNA-*CarE*-fed aphids was concentrated in the range of 500–800 mOD/per/min with an average CarE activity of 677 mOD/per/min (Figure 5B). The results showed significant reduction of CarE activity by dsRNA-*CarE* feeding. This reduction in activity likely reduces the insecticide detoxification ability of the aphids. This conclusion was further supported by the results of *in vitro* CarE inhibition assays, in which the IC$_{50}$ value of DEF was 1.2-fold higher in control aphids than in dsRNA-fed aphids (Table 2). In our study, the susceptibility of aphids to omethoate and enzyme assays after *CarE* knock down implied that the CarE plays an important role in OP resistance.

Feeding of dsRNA targeting vATPase transcripts from an artificial diet achieved a 30% decrease in transcripts of *A. pisum* and caused a significant increase in aphid mortality [47]. Since our results indicated that the suppression of *CarE* transcript levels increased the resistant cotton aphids' susceptibility to omethoate (Figure 4), it will be appealing to use dsRNA targeting of the *CarE* gene for controlling the OP resistant aphids, and possibly even other pest insects that have similar resistance mechanisms. The plant-mediated RNAi method has been used to effectively silence genes of Lepidopteran, Coleopteran, and Hemipteran insect species [61–64]. Researchers have noted that phloem sap-sucking insects, such as aphids, whiteflies, planthoppers and plant bugs, have evolved from minor pests to major pests. It would be a revolution in plant protection if plant-mediated RNAi can be used extensively to protect plants from sucking insect pests [65]. Fortunately, plant-mediated RNAi approaches were documented to knock down *M. persicae* gene expression by up to 60% on transgenic *Nicotiana benthamiana* and *Arabidopsis thaliana* [48]. Therefore, plant-mediated RNAi method targeting of *CarE* may be possible and useful for the control of OP resistant aphid pests, and this warrants further investigation in the future.

Conclusion

In conclusion, our findings provide insights about the role of *CarE* in the xenobiotic metabolism of an OP resistant *A. gossypii* strain. These results suggest that feeding of dsRNA through artificial diet can be exploited for functional studies in cotton aphids. Further, our results suggest that the *CarE* would be a promising potential target for OPs resistance management and aphid control.

Acknowledgments

This research was supported by National Basic Research Program of China (Contract No. 2012CB114103) and the National Natural Science Foundation of China (31330064 and 30871661).

Author Contributions

Conceived and designed the experiments: XWG XYS. Performed the experiments: XRY YHG. Analyzed the data: YHG XRY. Contributed reagents/materials/analysis tools: XWG. Contributed to the writing of the manuscript: YHG QLS XWG.

References

1. Oakeshott JG, Claudianos C, Russell RJ, Robin GC (1999) Carboxyl/cholinesterases: a case study of the evolution of a successful multigene family. Bioassays 21(12): 1031–1042.

2. Oakeshott JG, Claudianos C, Campbell PM, Newcomb RD, Russell RJ (2005) Biochemical genetics and genomics of insect esterases. In Gilbert LI, Latrou K, Gill SS, Editors. Comprehensive Molecular Insect Science Pharmacology:Else-:Elsevier, Oxford, 309–381.

3. Hemingway J (2000) The molecular basis of two contrasting metabolic mechanisms of insecticide resistance. Insect Biochem Mol Biol 30: 1009–1015.

4. Field LM, Devonshire AL (1997) Structure and organization of amplicons containing the E4 esterase genes responsible for insecticide resistance in the aphid Myzus persicae (Sulzer). Biochem J 322: 867–871.

5. Field LM, Blackman RL, Tyler-Smith C, Devonshire AL (1999) Relationship between amount of esterase and gene copy number in insecticide-resistant Myzus persicae (Sulzer). Biochem J 399: 737–742.

6. Bizzaro D, Mazzoni E, Barbolini E, Giannini S, Cassanelli S, et al. (2005) Relationship among expression, amplification, and methylation of FE4 esterase genes in Italian populations of Myzus persicae (Sulzer) (Homoptera:Aphididae). Pestic Biochem Physiol 81: 51–58.

7. Cao CW, Zhang J, Gao XW, Liang P, Guo HL (2008a) Overexpression of carboxylesterase gene associated with organophosphorous insecticide resistance in cotton aphids, Aphis gossypii (Glover). Pestic Biochem Physiol 90: 175–180.

8. Cao CW, Zhang J, Gao XW, Liang P, Guo HL (2008b) Differential mRNA expression levels and gene sequences of carboxylesterase in both deltamethrin resistant and susceptible strains of the cotton aphid, Aphis gossypii. Insect Sci 15: 209–216.

9. Pan YO, Guo HL, Gao XW (2009) Carboxylesterase activity, cDNA sequence, and gene expression in malathion susceptible and resistant strains of the cotton aphid, Aphis gossypii. Comp Biochem Physiol Part B 152: 266–270.

10. Alon M, Alon F, Nauen R, Morin S (2008) Organophosphates'resistance in the B-biotype of Bemisia tabaci (Hemiptera: Aleyrodidae) is associated with a point mutation in an ace1-type acetylcholinesterase and overexpression of carboxylesterase. Insect BiochemMol Biol 38(10): 940–949.

11. Karunaratne SHPP (1994) Characterisation of multiple variants of carboxylesterases which are involved in insecticide resistance in the mosquito Culex quinquefasciatus. PhD Thesis, University of London.

12. Karunaratne SHPP, Hemingway J, Jayawardena KGI, Dassanayaka V, Vaughan A (1995) Kinetic and molecular differences in the amplified and non-amplified esterases from insecticide resistant and susceptible Culex quinquefasciatus mosquitoes. J Biol Chem 270: 31124–31128.

13. Vaughan AM, Hawkes NJ, Hemingway J (1997) Co-amplification explains linkage disequilibrium of two mosquito esterase genes in insecticide resistant Culex quinquefasciatus. Biochem J 325: 359–365.

14. Paton MG, Karunaratne SHPP, Giakoumaki E, Roberts N, Hemingway J (2000) Quantitative analysis of gene amplification in insecticide resistant Culex mosquitoes. Biochem J 346: 17–24.

15. Liu Y, Zhang H, Qiao C, Lu X, Cui F (2011) Correlation between carboxylesterase alleles and insecticide resistance in Culex pipiens complex from China. Parasit Vectors 4: 236.

16. Devonshire AL, Field LM (1991) Gene amplification and insecticide resistance. Annu Rev Entomol 36: 1–23.

17. Vaughan A, Hemingway J (1995) Cloning and sequence of the full-length cDNA for a major insecticide resistance gene worldwide in the mosquito Culex quinquefasciatus. J Biol Chem 270: 17044–17049.

18. Hemingway J, Hawkes N, Prapanthadara L, Indrananda Jayawardenal KG, Ranson H (1998) The role of gene splicing, gene amplification and regulation in mosquito insecticide resistance. Phil Tran Roy Soc B 353: 1695–1699.

19. Hemingway J, Hawkes NJ, McCarroll L, Ranson H (2004) The molecular basis of insecticide resistance in mosquitoes. Insect Biochem Mol Biol 34: 653–665.

20. Mouches C, Pasteur N, Berge JB, Hyrien O, Raymond M, et al. (1986) Amplification of an esterase gene is responsible for insecticide resistance in a Californian Culex mosquito. Science 233: 778–780.

21. Vontas JG, Small GJ, Hemingway J (2000) Comparison of esterase gene amplification, gene expression and esterase activity in insecticide susceptible and resistant strains of the brown planthopper, Nilaparvata lugens (Stål). Insect Mol Bio 9(6): 647–653.

22. Zhang J, Zhang J, Yang M, Jia Q, Guo Y, et al. (2011) Genomics-based approaches to screening carboxylesterase-like genes potentially involved in malathion resistance in oriental migratory locust (Locusta migratoria manilensis). Pest Manag Sci 67: 183–190.

23. Claudianos C, Russell RJ, Oakeshott JG (1999) The same amino acid substitution in orthologous esterases confers organophosphate resistance on the house fly and a blowfly. Insect Biochem Mol Biol 29: 675–686.

24. Taskin V, Kence M (2004) The genetic basis of malathion resistance in house fly (Musca domestica L.) strains from Turkey. Russ J Genet 40: 1215–1222.

25. Taskin V, Kence M, Göçmen B (2004) Determination of malathion and diazinon resistance by sequencing the MdaE7 gene from Guatemala, Columbia, Manhattan and Thailand house fly (Musca domestica L.) strains. Russ J Genet 40: 377–380.

26. Newcomb JRD, Campbell PM, Ollis DL, Cheah E, Russell RJ, et al. (1997) A single amino acid substitution converts a carboxylesterase to an organophosphorus hydrolase and confers insecticide resistance on a blowfly. Proc Natl Acad Sci USA 94: 7464–7468.

27. Heidari R, Devonshire AL, Campbell BE, Bell KL, Dorrian SJ, et al. (2004) Hydrolysis of organophosphorus insecticides by in vitro modified carboxylesterase E3 from Lucilia cuprina. Insect Biochem Mol Biol 34: 353–363.

28. de Carvalho RA, Torres TT, de Azeredo-Espin AML (2006) A survey of mutations in the Cochliomyia hominivorax (Diptera: Calliphoridae) esterase E3 gene associated with organophosphate resistance and the molecular identification of mutant alleles. Vet Parasitol 140: 344–351.

29. Cui F, Qu H, Cong J, Liu X, Qiao C (2007) Do mosquitoes acquire organophosphate resistance by functional changes in carboxylesterases? FASEB J 21: 3584–3591.

30. Cui F, Lin Z, Wang HS, Liu SL, Chang HJ, et al. (2011) Two single mutations commonly cause qualitative change of nonspecific carboxylesterases in insects. Insect Biochem Mol Biol 41: 1–8.

31. Blackman RL, Eastop VF (1984) Aphids on the World's Crops. An Identification Guide. John, Wiley and Sons, NY.

32. Sun YQ, Feng GL, Yuan JG, Zhu P, Gong KY (1987) Biochemical mechanism of resistance of cotton aphids to organophosphorus insecticides. Acta Entomol Sin 30: 13–20.

33. O'Brien PJ, Abdel-Aal YA, Ottea JA, Graves JB (1992) Relationship of insecticide resistance to carboxylesterase in Aphis gossypii from midsouth cotton. J Econ Entomol 85: 651–657.

34. Saito T (1993) Insecticide resistance of the cotton aphid, Aphis gossypii Glover. VI. Qualitative variations of aliesterase activity. Appl Entomol Zool 28: 263–265.

35. Suzuki K, Hama H, Konno Y (1993) Carboxylesterase of the cotton aphid, Aphis gossypii Glover (Homoptera: Aphididae), responsible for fenitrothion resistance as a sequestering protein. Appl Entomol Zool 28: 439–450.

36. Owusu EO, Horiike M, Hirano C (1996) Polyacrylamide gel electrophoretic assessments of esterases in cotton aphid (Homoptera: Aphididae) resistance to dichlorvos. J Econ Entomol 89: 302–306.

37. Suzuki K, Hama H (1998) Carboxylesterase of the cotton aphid, Aphis gossypii Glover, Isoelectric point variants in an organophosphorus resistance clone. Appl Entomol Zool 33: 11–20.

38. Takada H, Murakami Y (1988) Esterase variation and insecticides resistance in Janpanese Aphis gossypii. Entomol Exp Appl 48: 37–41.

39. Sun LJ, Zhou XG, Zhang J, Gao XW (2005) Polymorphisms in a carboxylesterase gene between organophosphate-resistant and -susceptible Aphis gossypii (Homoptera: Aphididae). J Econ Entomo l 98: 1325–1332.

40. Fire A, Xu S, Montgomery MK, Kostas SA, Driver SE, et al. (1998) Potent and specific genetic interference by double-stranded RNA in Caenorhabditis elegans. Nature 391: 806–811.

41. Hannon GJ (2002) RNA interference. Nature 418: 244–251.

42. Mutti NS, Park Y, Reese JC, Reeck GR (2006) RNAi knockdown of a salivary transcript leading to lethality in the pea aphid, Acyrthosiphon pisum. J Insect Sci 6: 1–7.

43. Jaubert-Possamai S, Le Trionnaire G, Bonhomme J, Christophides GK, Rispe C, et al. (2007) Gene knockdown by RNAi in the pea aphid Acyrthosiphon pisum. BMC Biotechnol 7: 63.

44. Mutti NS, Louis J, Pappan LK, Pappan K, Begum K, et al. (2008) A protein from the salivary glands of the pea aphid, Acyrthosiphon pisum, is essential in feeding on a host plant, Proceedings of the National Academy of Sciences of the United States of America 105: 9965–9969.

45. Shakesby AJ, Wallace IS, Isaacs HV, Pritchard J, Roberts DM, et al. (2009) A water-specific aquaporin involved in aphid osmoregulation. Insect Biochem Mol Biol 39: 1–10.

46. Whyard S, Singh AD, Wong S (2009) Ingested double-stranded RNAs can act as species-specific insecticides. Insect Biochem Mol Biol 39: 824–832.

47. Pitino M, Coleman AD, Maffei ME, Ridout CJ, Hogenhout SA (2011) Silencing of Aphid Genes by dsRNA feeding from Plants. Plos One 6(10): e25709.

48. Moores GD, Gao XW, Denholm I, Devonshire AL (1996) Characterization of insensitive acetylcholinesterase in the insecticide-resistant cotton aphid, Aphis gossypii Glover (Homoperta: Aphidedae). Pestic Biochem Physiol 56: 102–110.

49. Mittler TE, Dadd RH (1964) An improved method for feeding aphids on artificial diets. Ann Entomol Soc Am 57: 139.

50. Pfaffl MW (2001) A new mathematical model for relative quantification in real-time RT-PCR. Nucl Acids Res 29: 2002–2007.

51. van Asperen K (1962) A study of housefly esterase by means of a sensitive colorimetric method. Insect Physiol 8: 401–406.

52. Young SJ, Gunning RV, Moores GD (2005) The effect of piperonyl butoxide on pyrethroid-resistance-associated esterases in Helicoverpa armigera (Hübner) (Lepido-petra, Noctuidae). Pest Manag Sci 61: 397–401.

53. Bradford MM (1976) A rapid and sensitive method for the quantitation of microgram quantities of protein, utilizing the principle of protein-dye binding. Anal Biochem 72: 248–254.

54. Turner CT, Davy MW, MacDiarmid RM, Plummer KM, Birch NP, et al. (2006) RNA interference in the light brown apple moth, Epiphyas postvittana (Walker) induced by double-stranded RNA feeding. Insect Mol Biol 15: 383–391.

55. Allen ML, Walker WB (2012) Saliva of Lygus lineolaris digests double stranded ribonucleic acids. J insect physiol 58: 391–396.

56. Yu N, Christiaens O, Liu JS, Niu JZ, Cappelle K, et al. (2013) Delivery of dsRNA for RNAi in insects: an over view and future directions. Insect Science 20: 4–14.

57. Kamath RS, Fraser AG, Dong Y, Poulin G, Durbin R, et al. (2003) Systematic functional analysis of the Caenorhabditis elegans genome using RNAi. Nature 421: 231–236.

58. Soares CAG, Lima CMR, Dolan MC, Piesman J, Beard CB, et al. (2005) Capillary feeding of specific dsRNA induces silencing of the *isac* gene in nymphal *Ixodes scapularis* ticks. Insect Mol Biol 14: 443–452.

59. Kamath RS, Martinez-Campos M, Zipperlen P, Frasher AG, Ahringer J (2000) Effectiveness of specific RNA-mediated interference through ingested double-stranded RNA in *Caenorhabditis elegans*. Genome Biology 2: research/0002.

60. Tian H, Peng H, Yao Q, Chen H, Xie Q, et al. (2009) Developmental control of a Lepidopteran pest *Spodoptera exigua* by ingestion of bacterial expressing dsRNA of a non-midgut gene. PLoS ONE 4: e6225.

61. Mao YB, Cai WJ, Wang JW, Hong GJ, Tao XY, et al. (2007) Silencing a cotton bollworm P450 monooxygenase gene by plant-mediated RNAi impairs larval tolerance of gossypol. Nat Biotechnol 25: 1307–1313.

62. Mao YB, Tao XY, Xue XY, Wang LJ, Chen XY (2011) Cotton plants expressing *CYP6AE14* double-stranded RNA show enhanced resistance to bollworms. Transgenic Res 20: 665–673.

63. Baum JA, Bogaert T, Clinton W, Heck GR, Feldmann P, et al. (2007) Control of coleopteran insect pests through RNA interference. Nat Biotechnol 25: 1322–1326.

64. Zha W, Peng X, Chen R, Du B, Zhu LL, et al.(2011) Knockdown of Midgut Genes by dsRNA-Transgenic Plant-Mediated RNA Interference in the Hemipteran Insect *Nilaparvata lugens*. PloS ONE 6(5): e20504.

65. Zhang H, Li HC, Miao XX (2013) Feasibility, limitation and possible solutions of RNAi-based technology for insect pest control. Insect Science 20: 15–30.

19

Environmental Fate of Soil Applied Neonicotinoid Insecticides in an Irrigated Potato Agroecosystem

Anders S. Huseth[1], Russell L. Groves[2]*

1 Department of Entomology, Cornell University, New York State Agricultural Experiment Station, Geneva, New York, United States of America, 2 Department of Entomology, University of Wisconsin-Madison, Madison, Wisconsin, United States of America

Abstract

Since 1995, neonicotinoid insecticides have been a critical component of arthropod management in potato, *Solanum tuberosum* L. Recent detections of neonicotinoids in groundwater have generated questions about the sources of these contaminants and the relative contribution from commodities in U.S. agriculture. Delivery of neonicotinoids to crops typically occurs as a seed or in-furrow treatment to manage early season insect herbivores. Applied in this way, these insecticides become systemically mobile in the plant and provide control of key pest species. An outcome of this project links these soil insecticide application strategies in crop plants with neonicotinoid contamination of water leaching from the application zone. In 2011 and 2012, our objectives were to document the temporal patterns of neonicotinoid leachate below the planting furrow following common insecticide delivery methods in potato. Leaching loss of thiamethoxam from potato was measured using pan lysimeters from three at-plant treatments and one foliar application treatment. Insecticide concentration in leachate was assessed for six consecutive months using liquid chromatography-tandem mass spectrometry. Findings from this study suggest leaching of neonicotinoids from potato may be greater following crop harvest in comparison to other times during the growing season. Furthermore, this study documented recycling of neonicotinoid insecticides from contaminated groundwater back onto the crop via high capacity irrigation wells. These results document interactions between cultivated potato, different neonicotinoid delivery methods, and the potential for subsurface water contamination via leaching.

Editor: Christopher J. Salice, Texas Tech University, United States of America

Funding: This research was supported by the Wisconsin Potato Industry Board and the National Potato Council's State Cooperative Research Program FY11-13. The funders had no role in study design, data collection and analysis, decision to publish, or preparation of the manuscript.

Competing Interests: RLG has received research funding, not related to this project, from Bayer CropScience, DuPont, Syngenta, and Valent U.S.A.

* E-mail: groves@entomology.wisc.edu

Introduction

The neonicotinoid group of insecticides is among the most broadly adopted, conventional management tools for insect pests of annual and perennial cropping systems [1]. Benefits of the neonicotinoid group of compounds include flexibility of application, diversity of active ingredients, and broad spectrum activity [2]. Moreover, growers have readily adopted neonicotinoids for two specific reasons: first, these compounds are fully systemic in plants after soil application and second, several new generic formulations have recently become available which have incentivized their continued use in many crops [1–3]. Since 2001, the United States Environmental Protection Agency (EPA) has classified several neonicotinoids as either conventional, reduced-risk pesticides, or as organophosphate alternatives [4],[5]. EPA certification often requires replacement of older, broad-spectrum pesticides with newer, more specific products for management of key economic pests. Critical attributes of replacement insecticides include documented reductions in human and environmental risk when compared to older, broad-spectrum pesticides [5]. Despite acceptance of neonicotinoid insecticides as reduced-risk by growers and regulatory agencies, nearly two decades of widespread, repetitive use has resulted in several insecticide resistance

issues, impacts on native and domestic pollinators, and unanticipated environmental impacts [6–9].

The environmental fate of several neonicotinoid active ingredients have been previously assessed. Previous studies focused on degradation and movement processes in soil, leachate, and runoff [10–15]. The leaching potential of the neonicotinoids into groundwater, as well as persistence in the plant canopy, is related to properties of the chemicals and delivery method of the compound to the crop (Fig. S1)[12],[15],[16]. Soil application (e.g., seed treatment or in-furrow) has been adopted as the principal form of insecticide delivery in potato production as it provides the longest interval of pest control, while also reducing non-target impacts, and limits exposure to workers when compared to foliar application methods. Since 1995, soil-applied neonicotinoids (i.e., clothianidin, imidacloprid, thiamethoxam) have been the most common pest management strategy used to control infestation of Colorado potato beetle, *Leptinotarsa decemlineata* Say; potato leafhopper, *Empoasca fabae* Harris; green peach aphid, *Myzus persicae* Sulzer; and potato aphid, *Macrosiphium euphorbiae* Thomas. The now widespread and extensive use of these systemic neonicotinoid insecticides, coupled with the recent detection of thiamethoxam in groundwater [17],[18], supports the hypothesis that potato pest management may contribute a portion of the documented neonicotinoid contaminants reported in

Wisconsin, USA. Furthermore, we hypothesized that neonicotinoid insecticides applied to potato are most vulnerable to leaching in the spring season when the root system of the plant has yet to fully exploit all of the active ingredient applied directly in the seed furrow. Large rain events at this time could drive insecticide leaching from potato and subsequent groundwater contamination at large scales. In this study, we examined how neonicotinoid concentrations in leachate were altered in response to different insecticide delivery methods using potatoes grown under commercial production practices. We also report the patterns of historic neonicotinoid insecticide detections in groundwater using water quality surveys collected by the Wisconsin Department of Agriculture, Trade and Consumer Protection-Environmental Quality Section (WI DATCP-EQ). Second, using potato as a model system, we analyzed leachate captured below different seed treatments, soil-applications, and foliar delivery treatments for thiamethoxam using liquid chromatography-tandem mass spectrometry (LC/MS/MS) over two consecutive field seasons. In this experiment, thiamethoxam was chosen as one representative insecticide in a broader group of water-soluble neonicotinoids. Moreover, this active ingredient represented the majority of positive neonicotinoid detections in groundwater monitoring surveys conducted by the WI DATCP-EQ [17], [18]. Third, using identical quantitative methods, we measured thiamethoxam concentration in irrigation water collected from operating, high-capacity irrigation wells at two time points in each sampling year. And finally, we characterize irrigation use and production trends of crops that may contribute to neonicotinoid detection in groundwater. Results of this study increase our understanding about the influence of insecticide delivery method on the neonicotinoid insecticides leaching from potato into the surrounding environment.

Materials and Methods

Ethics Statement

No specific permits were required for the field study described here. Access to field sites was granted by the private landholder to conduct leaching experiments. No specific permissions were needed to present publically available records provided by Wisconsin Department of Agriculture, Trade and Consumer Protection or Wisconsin Department of Natural Resources. Field studies did not involve any endangered or protected species.

Groundwater Contamination

Permanent groundwater monitoring wells, maintained by the WI DATCP-EQ, were used to measure neonicotinoid contamination of subsurface water resources as one component of an ongoing study documenting agrochemical (e.g., insecticides, herbicides, nutrients) impact on groundwater quality. Beginning in 2006, analytical water quality assessments for neonicotinoid contamination were conducted by the Wisconsin Department of Agriculture Trade and Consumer Protection-Bureau of Laboratory Services. Concentrations of acetamiprid, clothianidin, dinotefuran, imidacloprid, and thiamethoxam were monitored in 20–30 different monitoring well locations from 2006–2012. Presented are positive detections of those insecticides in different monitoring wells from 2006–2012 [17],[18]. Data provided by WI DATCP-EQ characterize the temporal and spatial profile of thiamethoxam and other neonicotinoid detections that occurred between 2008–2012. These data are presented in summary as a foundation for following objectives (Table 1).

Experimental Site and Design

In 2011 and 2012, leaching experiments were conducted 6 km west of Coloma, Wisconsin. Experiments were planted in two different fields approximately 0.5 km apart on 20 May 2011 and 11 May 2012. The soil at both sites consisted of Richford loamy sand (sandy, mixed, mesic, Typic Udipsamments) [19]. Soil composition was 7% clay, 82% sand, and 11% silt. Organic matter was 0.53 percent by weight. Study sites soils had a high infiltration rate (Hydrological Soil Group A), a high saturated hydraulic conductivity (K_{sat}) at 28 micrometers per second, and an available water capacity rating of 0.1 cm per cm [19]. No restrictive layer that would impede water movement through the soil has been documented [19]. Study site soil was formed in the bed of glacial Lake Wisconsin from parent material of glacial till overlain by glacial outwash [20]. Upper soil horizons (A and B) are sand with minimal structure. Subsurface soil (C horizon) had no structure. Irrigation pivots in sample fields withdrew water at a depth of 37 m and the water table depth (static water level) was approximately 6 m for both sites [21].

A randomized complete block design with four insecticide delivery treatments and an untreated control was established using the potato cultivar, 'Russet Burbank'. Plots were 0.067 ha in size and planted at a rate of one seed piece per 0.3 m with 0.76 m spacing between rows. Each year, experiments were nested within a different ~32 ha commercial potato field, and maintained under commercial management practices by the producer (e.g., nutrient application timing, chemical usage, tillage practices, etc.), with the exception of insecticide inputs. The decision to locate these experiments in commercial fields was, in part, based upon access to a center pivot irrigation system to best duplicate water inputs used to produce commercial potato in Wisconsin. All other inputs and production strategies (e.g. tillage, fumigation, fertility, and disease management) were conducted by the producer with equipment and products in a manner consistent with the best management practices for potato production in Wisconsin. Prior to planting in each season, a tension plate lysimeter (25.4×25.4×25.4 cm) was buried at a depth of 75 cm below the soil surface. Lysimeters were constructed of stainless steel with a porous stainless steel plate affixed to the top to allow water to flow into the collection basin over each sampling interval. Experimental blocks were connected with 9.5 mm copper tubing to a primary manifold and equipped with a vacuum gauge. A predefined, fixed suction was maintained under regulated vacuum at 107 ± 17 kPa (15.5 ± 2.5 lb per in^2) with a twin diaphragm vacuum pump (model UN035.3 TTP, KnF, Trenton, NJ) connected to a 76 L portable air tank. Each treatment block was equipped with a data-logging rain gauge (Spectrum Technologies, Inc. model # 3554WD1) recording daily water inputs at a five minute interval. Data was offloaded with Specware 9 Basic software (Spectrum Technologies, Inc., Plainfield, IL, USA) and aggregated into daily irrigation or rain event totals using the *aggregate* and *dcast* function in R (package: reshape2, [22]). Irrigation event records were obtained from the grower to identify days and estimated inputs of water application throughout the growing season.

Insecticides and Application

Thiamethoxam treatments (Platinum 75SG, 75% thiamethoxam per formulated unit, Syngenta, Greensboro, NC) were selected to represent a common, soil-applied insecticide in potato. A second formulation of thiamethoxam was selected to represent a common pre-plant insecticide seed treatment in potato (Cruiser 5FS, 47.6% thiamethoxam per formulated unit, Syngenta, Greensboro, NC). Each insecticide formulation is used to manage early season infestations of Colorado potato beetle, potato

Table 1. Positive (means±SD) neonicotinoid detections in groundwater from 2008–2012, State of Wisconsin Department of Agriculture Trade and Consumer Protection.

Year	County	Area potato (ha)[a]	Row crops (ha)[b]	Percent potato[c]	Well ID	N positive samples	Insecticide concentration (µg/L)[d]		
							clothianidin	imidacloprid	thiamethoxam
2008	Adams	2,617	21,385	10.9	6	2	-	-	4.34 (4.97)
	Grant	0	47,827	0.0	10	1	-	-	1.25
	Iowa	18	25,795	0.1	11,12,13	9	-	-	1.50 (0.67)
	Richland	29	9,582	0.3	16	1	-	-	0.69
	Sauk	30	31,931	0.1	17	2	-	-	2.41 (1.32)
	Waushara	2,630	29,447	8.2	20	2	-	-	0.67 (0.05)
2009	Adams	3,989	24,894	13.8	6	2	-	-	5.31 (5.12)
	Dane	22	101,527	0.0	9	1	-	-	1.61
	Iowa	343	33,375	1.0	11,12	3	-	-	1.31 (0.68)
	Richland	87	14,402	0.6	16	1	-	-	1.26
	Sauk	328	40,571	0.8	17	2	-	-	3.00 (0.94)
2010	Adams	4,188	24,871	14.4	6	4	3.43	-	2.97 (2.04)
	Brown	1	39,322	0.0	7	1	-	-	0.52
	Dane	34	110,979	0.0	8,9	4	0.54 (0.24)	0.54	1.08
	Grant	49	74,566	0.1	10	1	0.73	-	-
	Iowa	356	38,840	0.9	11,12,13	7	-	-	1.25 (1.02)
	Sauk	188	45,309	0.4	17	5	0.41	-	1.81 (0.88)
	Waushara	4,184	33,576	11.1	19,20	2	-	2.77 (0.81)	-
2011	Adams	4,066	27,693	12.8	2,5,6	9	0.63 (0.36)	0.33	0.63 (0.26)
	Brown	7	38,309	0.0	7	1	-	-	0.21
	Dane	33	107,214	0.0	8	2	0.62 (0.19)	-	-
	Grant	13	75,436	0.0	10	1	0.30	-	-
	Iowa	47	40,138	0.1	12	4	-	0.34 (0.09)	0.88 (0.23)
	Portage	7,364	45,324	14.0	15	1	-	-	0.32
	Sauk	213	46,686	0.5	17,18	5	0.54 (0.10)	-	1.92 (0.43)
	Waushara	4,536	36,676	11.0	19,20,21,23	23	0.25 (0.03)	0.78 (0.69)	1.40 (0.56)
2012	Adams	4,263	27,037	13.6	1,3,4,6	6	0.52 (0.30)	0.51 (0.26)	0.27
	Dane	11	115,501	0.0	8	1	0.67	-	-
	Grant	4	72,920	0.0	10	1	0.26	-	-
	Iowa	369	40,764	0.9	12	2	0.24	0.28	0.44
	Juneau	907	28,542	3.1	14	2	0.42 (0.18)	-	0.20
	Portage	7,622	46,337	14.1	15	2	-	0.47	0.47
	Waushara	5,904	38,999	13.1	21,22,23	13	-	0.68 (0.88)	1.51 (0.72)
			summary	N=23		67	25	30	68

Table 1. Cont.

Year	County	Area potato (ha)[a]	Row crops (ha)[b]	Percent potato[c]	Well ID	N positive samples	Insecticide concentration (µg/L)[d]		
							clothianidin	imidacloprid	thiamethoxam
						Average	0.62 (0.63)	0.79 (0.83)	1.59 (1.51)
						Range	0.21–3.34	0.26–3.34	0.20–8.93

[a]Acreage estimates generated from USDA National Agricultural Statistics Service - Cropland Data Layer, 2008–2012 [26].
[b]Row crops class is the sum of the following crop areas (ha): maize, soy, small grains, wheat, peas, sweet corn, and miscellaneous vegetables and fruits.
[c]Percent potato calculated as the potato area grown annually divided by total arable row crop acreage (other row crops + potato).
[d]Positive neonicotinoid detections extracted from long-term, groundwater wells maintained by the WI-DATCP-EQ Program.

leafhopper, and colonizing aphid in Wisconsin potato crops. Commercially formulated insecticides were applied at maximum labeled rates for in-furrow (140 g thiamethoxam ha^{-1}) and seed treatment (112 g thiamethoxam ha^{-1} at planting density of 1,793 kg seed ha^{-1}) for potato [23]. A calibrated CO_2 pressurized, backpack sprayer with a single nozzle boom was used to deliver an application volume of 94 liters per hectare at 207 kPa through a single, extended range, flat-fan nozzle (TeeJet XR80015VS, Spraying Systems, Wheaton, IL) for in-furrow applications. Spray applications were directed onto seed pieces in the furrow at a speed of one meter per second and furrows were immediately closed following application. Seed treatments were applied using a calibrated CO_2 pressurized backpack sprayer with a single nozzle boom delivering an application volume of 102.2 L per hectare at 207 kPa through a single, extended range, flat-fan nozzle (TeeJet XR80015VS, Spraying Systems, Wheaton, IL) was used for delivery of thiamethoxam in water (130 mL) directly to suberized, cut seed pieces (23 kg) 24 hours prior to planting. Seed treatments were allowed to dry in the absence of light at 20°C during that pre-plant period. A novel soil application method, impregnated copolymer granules, was included as another treatment in an attempt to stabilize applied insecticide in the soil. Polyacrylamide horticultural copolymer granules (JCD-024SM, JRM Chemical, Cleveland, OH) were impregnated at an application rate of 16 kg per hectare. The polyacrylamide treatment was included as a novel delivery method to stabilize insecticide in the rooting zone and possibly reduce leaching in the early season. Thiamethoxam (0.834 g, Platinum 75SG) was initially diluted in 250 mL of deionized water and 100 µL of blue food coloring was incorporated into solution to ensure uniform mixing (brilliant blue FCF). Insecticide solutions were mixed with 75 g polyacrylamide then stirred until the liquid was absorbed and a uniform color was observed. Impregnated granules were vacuum dried in the absence of light for 24 hours at 20°C. Treated granules were divided into even quantities per row and evenly distributed into the four treatment rows for each polyacrylamide plot. A single untreated flanking row was planted between plots. All soil-applied insecticides were applied on 20 May 2011 and 11 May 2012 at the time of planting.

Two foliar applications of thiamethoxam (Actara 25WG, 25% thiamethoxam per formulated unit, Syngenta, Greensboro, NC) sprayed on the same plot were included as a fourth delivery treatment. Two successive neonicotinoid applications are recommended for foliar control of pests in potato [23]. Foliar thiamethoxam was applied using a calibrated CO_2 pressurized backpack sprayer delivering an application volume of 187.1 liters per hectare at 207 kPa through four, extended range flat-fan nozzles (TeeJet XR80015VS, Spraying Systems, Wheaton, IL) spaced at 45.2 cm. The first foliar application was followed approximately seven days later with a second equivalent rate of thiamethoxam to total the season-long maximum labeled rate (105 g thiamethoxam ha^{-1}) [23] and were timed to coincide with the appearance of 1st and 2nd instar larvae of native populations of *L. decemlineata*. Foliar applications of thiamethoxam were applied on 28 June and 5 July in 2011 and 15 and 22 June in 2012. Although total amounts of active ingredient differ by formulation, these rates are identical to registered label recommendations [23] and reflect the maximum amount of active ingredient used on an average hectare of cultivated potato. Specific chemical properties of formulated thiamethoxam that affect solubility and leaching potential in soil can be found in Gupta et al. [15] and the references therein (Fig. S1).

Chemical Extraction and Quantification

Lysimeter leachate was sampled twice monthly beginning on June 1 of each year and concluding in October of 2011 and November of 2012. Total leachate volume was recorded for each plot. A 500 mL subsample was taken from each plot into a 0.5 L glass vessel and immediately placed on ice and refrigerated at 4–6°C in the laboratory prior to analysis. Samples were homogenized into a 400 mL monthly (i.e., two samples per month) sample as percent volume per volume dependent on total catch measured in the field. Neonicotinoid residues from monthly water samples were extracted using automated solid phase extraction (AutoTrace SPE workstation, Zymark, Hopkinton, MA) with LiChrolut EN SPE columns (Merk KGaA, Darmstadt, Germany). If visual inspection of sample found excessive sediment contamination, samples were filtered through a 0.45 μm filter prior to extraction. Columns were conditioned prior to extraction with 3 mL of methanol (MeOH) and 3 mL of water. 210 mL of sample were loaded onto columns and rinsed with 10 mL of water then dried under flowing nitrogen for 15 minutes (N-evap, Organomation, Berlin, MA). Samples were eluted using a 50% ethyl acetate (EtOAc) and 50% methanol solution to collect a 2 mL sample fraction. Sample extract fractions were analyzed using a Waters 2690 HPLC/Micromass Quattro LC/MS/MS (Waters Corporation, Milford, MA). All thiamethoxam residues were identified, quantified, and confirmed using LC/MS/MS by the Wisconsin Department of Agriculture Trade and Consumer Protection-Bureau of Laboratory Services. The method detection limit (MDL) of the extraction procedure was $0.2 \ \mu g \ L^{-1}$. Specific conditions for all quantitative procedures follow WI-DATCP Standard Operating Procedure #1009 developed from Seccia et al. [24] and references therein.

Irrigation Use and Crop Area

To determine the extent of irrigated agriculture present within the watershed, we utilized current high capacity well pumping data and irrigated agriculture estimates derived from digital imagery. Publically available operator reporting data for high capacity agricultural pivots were obtained from the Wisconsin Department of Natural Resources Bureau of Drinking Water and Groundwater. Records included location information and pumping volume for the year 2012. High capacity wells service several irrigated fields and often these fields are further divided into individual crop management units each with unique irrigation requirements. We digitized the area watered by all identifiable center pivot, linear move, and traveling gun irrigation systems using digital aerial photography to measure the total number of management units present within the greater Central Wisconsin Water Management Unit watershed [25] (ArcGIS version 10.1, Redlands, CA). Fields were subdivided into management units using the consistent divisions in crop types with a sequence of National Agricultural Statistics Service Cropland Data Layer (NASS-CDL) [26] thematic data and aerial photography images [25] from 2010–2012.

To determine agronomic trends in the Central Sands vegetable production region of Wisconsin, we used a combination of publically available land use data and current neonicotinoid registration information. A geospatial watershed management boundary layer delineated by the Wisconsin Department of Natural Resources [27] was used to generally define the spatial extent where agriculture could be contributing to the detection of neonicotinoid insecticides in subsurface water. The Central Wisconsin Water Management Unit extent was used to estimate annual crop composition using the NASS-CDL [26] from 2006–2012 using ArcGIS. From these data, we selected major crops that

frequently receive either seed or in-furrow soil-applied neonicotinoid insecticide treatments. Application rates were identical for several similar crops (e.g. soybean and green bean), and so, we chose to aggregate crops based on insecticide rate and crop type into three primary groups: maize, beans, and potato [23],[28–30]. These crop groups comprise the majority of production area in the Central Wisconsin Water Management Unit extent. To our knowledge, limited information exists documenting the proportion of different soil-applied neonicotinoid active ingredients that are used on a per crop basis in the Central Wisconsin Water Management Unit. Based on this level of uncertainty, we chose not to extend tabulated crop areas to a direct calculation or estimate of neonicotinoid active ingredients applied.

Data Analysis

To determine the impact of different insecticide delivery treatments on thiamethoxam leachate detected over time, we reported the mean concentration over a period of several months. All lysimeter analyses included samples where neonicotinoid insecticides were not detected (i.e., zero detections). All data manipulation and statistical analyses of leachate concentrations were performed in R, version 2.15.2 [31] using the base distribution package. Functions used in the analysis are available in the base package of R unless otherwise noted. Observed concentration for time points in each year were subjected to a repeated-measures analysis of variance (ANOVA) using a linear mixed-effects model to determine significant delivery (i.e. treatment), date, and delivery×date effects ($P<0.05$). Because the agronomic conditions differed between years and given that our comparison of interest was at the insecticide delivery treatment level, insecticide concentrations were analyzed separately for each year. Mixed-effects models (i.e., repeated-measures analysis of variance) were fit using the *lme* function (package nlme, [32]). Empirical autocorrelation plots from unstructured correlation model residuals were examined using the *ACF* function (package nlme, [32]). Correlation among within-group error terms were structured and examined in three ways: first, unstructured correlation, second, with compound symmetry using the function *corCompSymm* and third, with autoregressive order one covariance using the function *corAR1* (package nlme, [32])[33]. Since models were not nested, fits of unstructured, compound symmetry, and autoregression order one covariance were compared using Akaike's information criterion statistic with the function *anova* (test = "F"). Data were transformed with natural logarithms before analysis to satisfy assumptions of normality, however untransformed means are graphically presented. In 2012, a single lysimeter in the polyacrylamide treatment of the leachate study malfunctioned and these observations were dropped from subsequent analyses leading to an unbalanced replicate number for that treatment (N = 3) in 2012. Water input data collected from tipping bucket samplers were averaged across block by day and aggregated as cumulative water inputs using the *cumsum* function. All summary statistics and model estimates were extracted using *aggregate*, *summary*, and *anova* functions.

Results and Discussion

Groundwater Detections

Neonicotinoid insecticides were detected at 23 different well monitoring well locations by WI-DATCP-EQ surveys between the years 2008 and 2012 (Table 1). These annual surveys, administered by WI-DATCP-EQ, occur at sensitive geologic or hydrogeologic locations that are at high risk of non-point source agrochemical leaching. Specifically, two agriculturally intensive

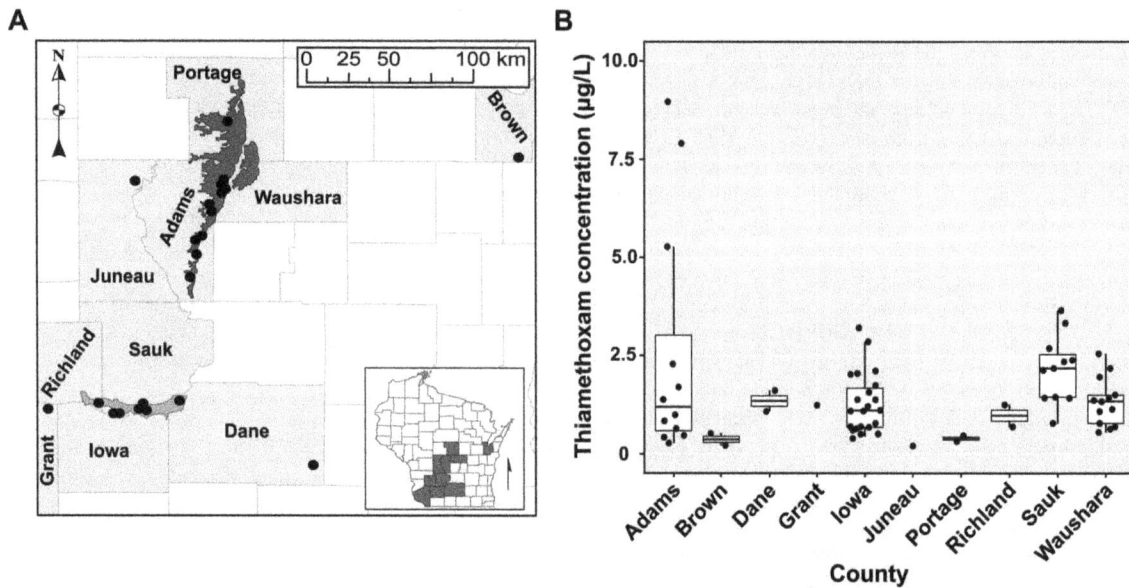

Figure 1. Positive thiamethoxam residue detections in groundwater 2008–2012. Points in the map (A) correspond to positive detection locations. Dark grey shaded region indicates the Central Sands potato production region. Light grey delimits the Lower Wisconsin River potato production region. Positive detections were obtained from established agrochemical monitoring wells collected by the Wisconsin Department of Agriculture, Trade and Consumer Protection (DATCP)-Environmental Quality division in collaboration with the Wisconsin DATCP Bureau of Laboratory Services. Boxplots (B) indicate average concentration detected from 2008–2012. Points show individual measured concentrations.

production regions of the state, the Central Sands and Lower Wisconsin River valley, are classified as high-risk areas for groundwater contamination and are frequently monitored for the presence of common agrochemicals (Fig. 1A). These regions have well-drained, sandy soils and easily accessible groundwater for irrigation that has driven agricultural intensification focused on vegetable production. Commercial potato is a key component in the agricultural production sequence, but is also rotated with many other specialty crops such as: carrots, onions, peas, pepper, processing cucumber, sweet corn, and snap beans. Unfortunately, the unique soil and water characteristics supporting a profitable

specialty crop production system are also particularly vulnerable to groundwater contamination with water-soluble agricultural products [34–36]. Regulatory exceedences of nitrates and herbicide products (e.g. triazines, triazinones, and chloroacetamide) have been commonplace for several years [34–37], but recent detections of neonicotinoid contaminants have created new groundwater quality concerns. Beginning in the spring of 2008, two wells had detections of 1.25 and 1.47 $\mu g\ L^{-1}$ thiamethoxam in Grant and Sauk Counties, WI (Fig. 1B, Table 1). Subsequent sampling later that season identified six additional locations for a total of 17 independent positive thiamethoxam detections that year. Since

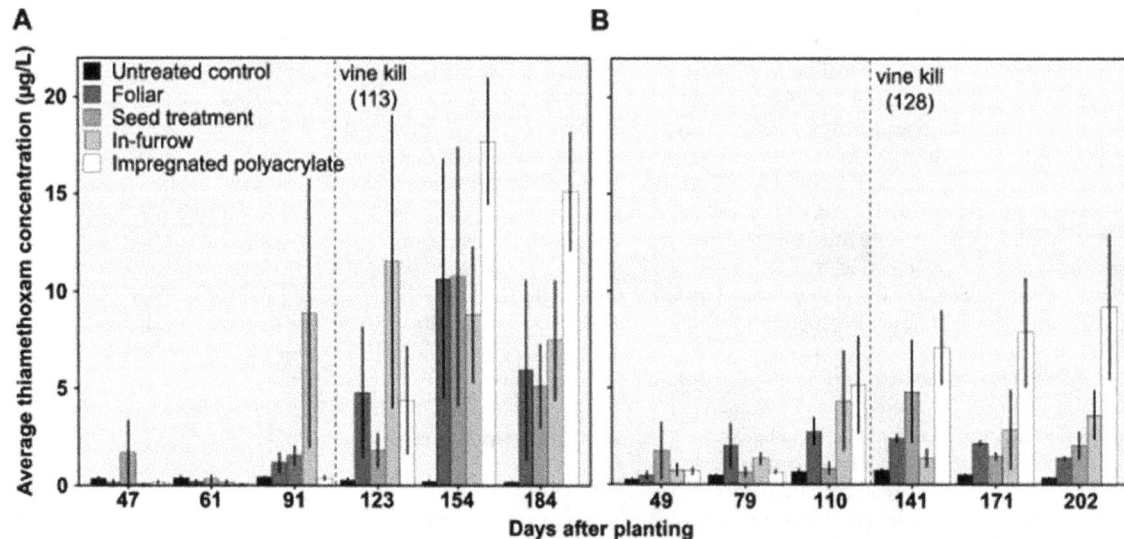

Figure 2. Thiamethoxam concentration in leachate from potato. Average thiamethoxam ($\pm$SD) recovered from in-furrow and foliar treatments in (A) 2011 an (B) 2012. Dotted lines indicate the date that the producer applied vine desiccant prior to harvest. Lysimeter studies continued in undisturbed soil following vine kill.

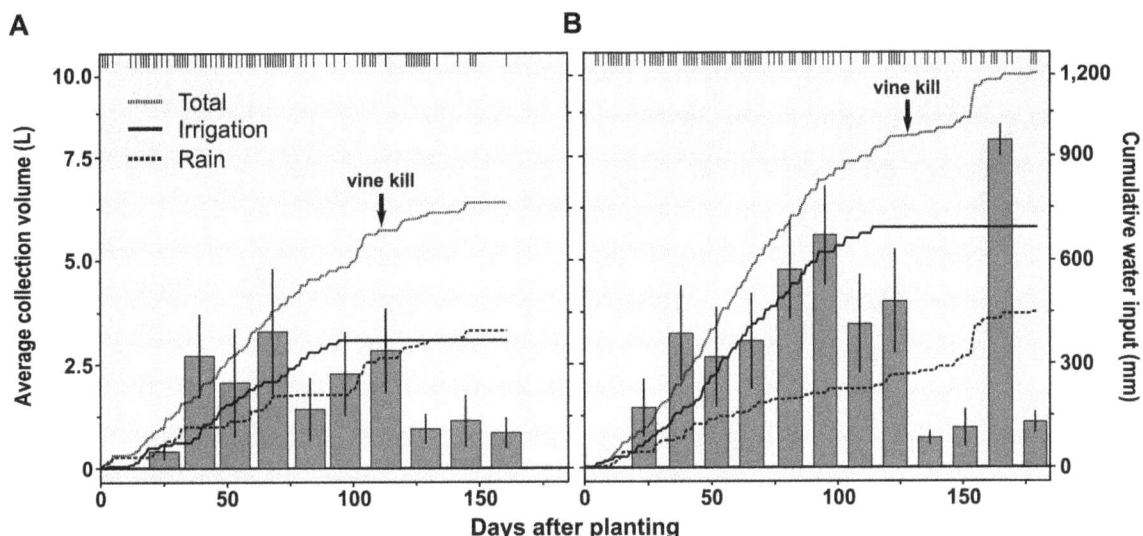

Figure 3. Water input volumes, 2011 and 2012. Water inputs and leachate volume collected in lysimeter studies in (A) 2011 and (B) 2012. Lines indicate cumulative water measured in tipping bucket rain gauges installed in plots each season. Bar plots indicate average leachate volume (±SD) collected in lysimeters on a bi-monthly sampling frequency. Hash marks at the top of each figure indicate days that overhead irrigation or rainfall occurred in each season.

these early detections, the WI-DATCP-EQ [17],[18] has repeatedly detected thiamethoxam, imidacloprid, and clothianadin residues at 23 different monitoring well locations over a five-year period (Table 1). Although the sampling effort was not uniformly distributed within the state, neonicotinoid detections often correspond to areas where intensive irrigated agricultural production occurs (Fig. 1A). As an indication of specialty crop production intensity, we used county-level potato abundance to better describe trends in historical neonicotinoid detections. Observed frequency and magnitude of neonicotinoid detections did not consistently correspond to potato abundance (Table 1). Although the contribution of potato production to the observed detections was not clear, regulatory agencies have continued to pursue this interaction by sampling where potato occurs at a high density, specifically the Central Sands and Wisconsin River Valley. Groundwater sampling strategies have provided a useful timeline of non-point source agrochemical pollution events in subsurface water resources. Identifying the origin of pollutants in the state is complicated by the diversity of neonicotinoid registrations, application methods and formulations; currently Wisconsin has 164 different registrations for field, forage, tree fruit, vegetable, turf, and ornamentals crops (6 acetamiprid, 18 clothianadin, 4 dinotefuran, 108 imidacloprid, 1 thiacloprid, 26 thiamethoxam) [38].

Neonicotinoid Losses and Concentrations in Leachate

The neonicotinoid insecticide thiamethoxam was included in field experiments to investigate the potential for leaching losses associated with different types of pesticide delivery. Specifically, formulations of thiamethoxam were applied as foliar and as at plant systemic treatments in commercial potato over two years and at two different irrigated fields. We hypothesized that thiamethoxam would be most vulnerable to leaching early in the season when plants were small and episodic heavy rains can be common. Interestingly, we observed the greatest insecticide losses following vine-killing operations which occurred more than 100 days after planting (Fig. 2). Detections of thiamethoxam in lysimeters varied between insecticide delivery treatments through time in 2011 (delivery×date interaction, $F=2.1$; d.f. $=20,88$; $P=0.0131$) and again in 2012 (delivery×date interaction, $F=1.8$; d.f. $=20,87$; $P=0.0384$). Moreover, the impregnated polyacrylamide delivery produced the greatest amount of thiamethoxam leachate late in each growing season (Fig. 2) when compared with other types of insecticide delivery.

Early season rainfall was not exceptionally heavy in either year of this experiment (Fig. 3). The accumulation of leachate detections in lysimeters likely is reflected by the steady application of irrigation water and rainfall. One clear exception to this pattern occurred in 2012 at 155–156 days after planting when 89 mm of

Table 2. Neonicotinoid concentration from irrigation water, 2011 and 2012.

| Date | Days after planting | Insecticide concentration (µg/L)[a] | |
		clothianidin	thiamethoxam
28 June 2011	39	-	0.310
1 September 2011	114	-	0.327
10 July 2012	60	-	0.533
15 August 2012	96	0.225	0.580

[a]Samples obtained from irrigation pivots while under operation in potato fields containing lysimeter experiments.

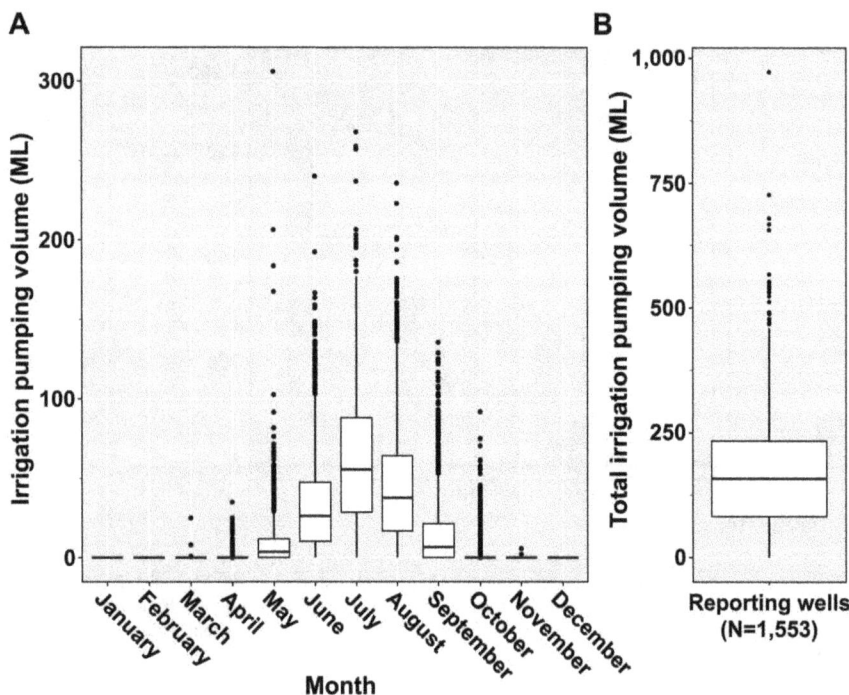

Figure 4. Reported irrigation inputs in the Central Wisconsin River Water Management Unit. Average reported agricultural pumping (megaliters, ML) in the Central Wisconsin River Water Management Unit for 2012. Monthly pumping records were reported by growers to the Wisconsin Department of Natural Resources Bureau of Drinking Water and Groundwater. Upper and lower whiskers extend to the values that are within 1.5*Inter-quartile range beyond the first (25%) and third (75%) percentiles. Data beyond the end of whiskers indicate outlier values and have been plotted as points.

rain fell within a 24-hour period. Peak detections of thiamethoxam in 2012 began to trend upward following this rain event, however the timing of similar detections across treatments in 2011 occurred at about the same time. One additional explanation may be that increased levels of pesticide losses are associated with plant death or senescence. In each year of this study, the largest proportion of pesticide detections in leachate occurred after vine killing with herbicide in the potato crop. Vine killing in commercial potato production is a common practice designed to aid the tubers in developing a periderm. Perhaps the rapid loss in root function following plant death permits excess pesticide to be solubilized and washed through the soil profile more quickly in root channels. In both seasons of this study, however, large episodic rain events did not occur early in the growing season. These results do appear, however, to document low to moderate levels of leaching losses that occur throughout the season even when the crop is managed at nominal evapo-transpirative need.

Untreated control plots also yielded low-level detections of thiamethoxam throughout both seasons. To better understand these insecticide detections in control plots, we sampled water directly from the center pivot irrigation system providing irrigation directly to the potato crop. Samples were taken while the systems were operational from lateral spigots mounted on the well casings. In both years, samples revealed low concentrations of thiamethoxam present in the groundwater at two time points in each sample season (Table 2) from which irrigation water was being drawn. Clothianidin was also present at a single time point in 2012 (Table 2). These positive detections of low-dose thiamethoxam were obviously being unintentionally applied directly to the crop through irrigation and this information is new to the producers in the Central Sands of Wisconsin. Although systemic neonicotinoids have recently been detected from surface water runoff and catch

basins associated with irrigated orchards [10], [39], to our knowledge no other study has documented the occurrence of neonicotinoids in subsurface groundwater being recycled through operating irrigation wells. Currently, the known exposure pathways for insecticide residues are most often associated with direct application or systemic movement of insecticides in floral structure and guttation water [8],[9],[40].

The implications for non-target effects resulting from these groundwater contaminants is currently unknown, but could be important considering the scale of irrigation ongoing in the Central Sands potato agroecosystem in Wisconsin (Fig. S2). Using a combination of aerial photography and NASS Cropland Data Layers, we identified 2,530 different irrigated field units distributed within the Central Wisconsin River Water Management Unit (Fig. S2). In all, 71,864 hectares of irrigated cropland were identified within the extent of the water management unit. Average irrigated field unit size was 28.4 ± 17.7 hectares (min. 1, max 138). Irrigation use patterns demonstrated clear increases in the summer months of the 2012 growing season (Fig. 4). Average annual pumping volume reported to the Wisconsin Department of Natural Resources in 2012 was 170.6 ± 115.6 megaliters (ML) of irrigation water (min. 0.00001, max 972.1) distributed over 1,553 reporting wells. Peak pumping volumes occurred in the month of July, averaging 61 ± 43.3 ML (min. 0, max 286.4). The timing of peak pumping correspond with crop demands for and reproductive phases of common open and closed pollination crops grown in the region.

While considerable attention has been focused on the positive attributes of the neonicotinoids [1–3], an increasing body of research suggests substantial negative impacts not only in terms of pest resistance development (e.g., Colorado potato beetle), but also impacts on non-target organisms and surrounding ecosystems

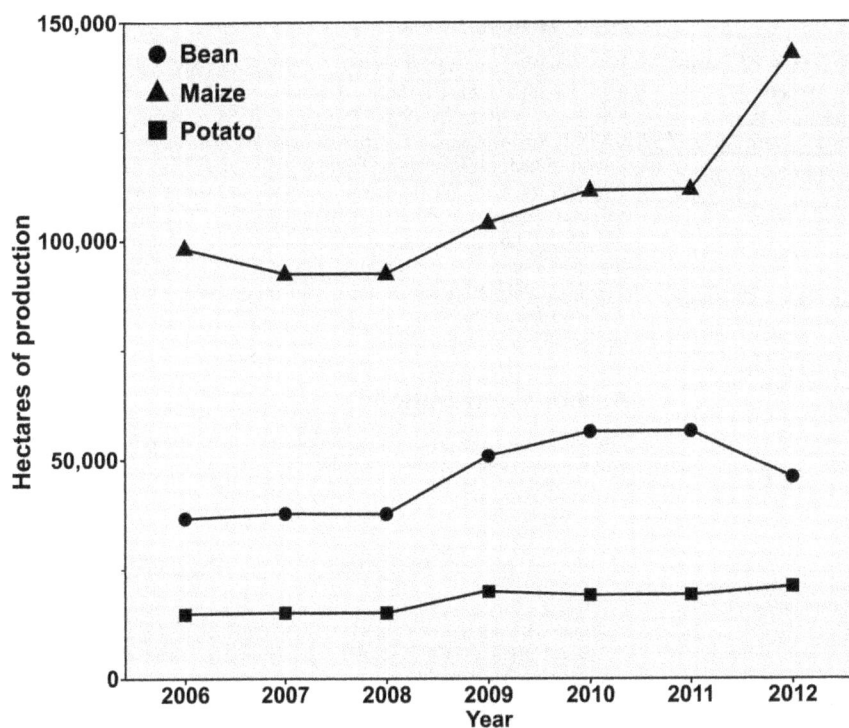

Figure 5. Crop area grown in the Central Wisconsin River Water Management Unit. Cropping trends in the Central Wisconsin River Water Management Unit from 2006–2012. Crop groups are often planted with a soil-applied neonicotinoid insecticide for insect pest management. Crop totals within the water management unit were tabulated from annual USDA-NASS Cropland Data Layers [26].

[8],[10],[41–44]. Recent studies have documented the negative influence of neonicotinoids on pollinator population health (both native and managed) which, in turn, created substantial concern about the long-term sustainability of these pesticides in agriculture [7],[11],[43],[45–49]. Exposures to pollinators reportedly occur through chronic, sub-lethal contact with low concentrations of neonicotinoid residues in pollen, nectar, waxes, and guttation drops of common crop plants [50–53]. Gill et al. [43] and Whitehorn et al. [54] found that low concentrations ($\leq 10~\mu g~L^{-1}$) of imidacloprid significantly reduced colony-level health in bumblebee (*Bombus terrestris* L.). Imidacloprid residues measured by those authors are consistent with insecticide concentrations found in nectar and pollen of flowering crops, further supporting the direct crop-pollinator toxicological pathway hypothesis [47],[52],[54],[55]. Though they have received much less attention, many closed pollination crops also provide resources for pollinators (e.g., pollen, water)[56],[57]. These crops also rely on neonicotinoids and may have currently undescribed risks for non-target organisms through indirect contaminant pathways in the agroecosystem [51],[58].

Possible exposure related to a high frequency of irrigation could drive the exposure of non-target arthropods to low concentrations of neonicotinoid insecticides in irrigation water. Although such impacts have yet to be documented directly, new comprehensive reviews of neonicotinoid environmental impacts have demonstrated numerous unanticipated impacts occurring at the ecosystem scale [9],[58]. In the Wisconsin agroecosystem, neonicotinoids are used on a large proportion of crops grown with irrigation [28],[29]. Trends in production show increased maize production over the past six years in the Central Wisconsin River Water Management Unit (Fig. 5). As a result of common neonicotinoid seed treatment on maize, accelerating production may partially explain the increased frequency of neonicotinoid detection in

groundwater. Unfortunately, little crop-specific pesticide information exists for individual neonicotinoids at the watershed scale [26]. Although measurement of specific contributions of crops to measured insecticide contamination is currently not available, this study demonstrates a research approach to better understand leaching from different application methods. Improved understanding of crops and insecticide delivery that results in greater risk of insecticide leaching will inform targets to reduce aquifer contamination and recirculation of soil-applied insecticides. Area-wide application of neonicotinoid insecticides through irrigation water applications may have considerable unanticipated or undocumented environmental impacts for non-target organisms through chronic low-dose exposure to insecticides.

Conclusions

To gain a better understanding of the seasonal cycle of neonicotinoids moving from the potato system, this study used an experimental approach to document the leaching potential of common neonicotinoid application methods. Results presented here benefit both potato producers and regulators by identifying trends in leachate losses for these commonly used, water-soluble insecticides. Lysimeter experiments documented loss of thiamethoxam following the application of vine desiccants at the conclusion of the potato production season. Leachate losses did vary among the different delivery methods over time indicating some variability in the patterns of pesticide leachate throughout the season. Quantification of crops commonly using neonicotinoid soil applications in the Central Wisconsin Water Management Unit highlights the need to research leaching potential from soil-applied neonicotinoids in other commodities. Documentation of several neonicotinoids in irrigation water suggests a new candidate pathway for non-target environmental impacts of insecticides.

Supporting Information

Figure S1 Chemical structures and properties of common neonicotinoid insecticides. Chemical structures were drawn using ChemDraw (version 13, Perkin Elmer Inc., Waltham, MA). Properties of each active ingredient were accessed from the National Center for Biotechnology Information PubChem online interface. Available: https://pubchem.ncbi.nlm.nih.gov/. Accessed 2014 Mar 20.

Figure S2 Irrigated field locations in the Central Wisconsin River Water Management Unit. Distribution of fields irrigated with high capacity wells (n = 2530) in the Central Wisconsin River Water Management Unit [27]. Points indicate locations of individual irrigation units identified from aerial photography using ArcGIS.

References

1. Jeschke P, Nauen R, Schindler M, Elbert A (2010) Overview of the status and global strategy for neonicotinoids. J Agric Food Chem 59: 2897–2908.
2. Elbert A, Haas M, Springer B, Thielert W, Nauen R (2008) Applied aspects of neonicotinoid uses in crop protection. Pest Manag Sci 64: 1099–1105.
3. Jeschke P, Nauen R (2008) Neonicotinoids–from zero to hero in insecticide chemistry. Pest Manag Sci 64: 1084–1098.
4. United States Environmental Protection Agency (2003) Imidacloprid; pesticide tolerances. Fed Regist 68: 35303–35315.
5. United States Environmental Protection Agency (2012) What is the conventional reduced risk pesticide program? Available: http://www.epa.gov/opprd001/workplan/reducedrisk.html. Accessed 2012 Oct 17.
6. Szendrei Z, Grafius E, Byrne A, Ziegler A (2012) Resistance to neonicotinoid insecticides in field populations of the Colorado potato beetle (Coleoptera: Chrysomelidae). Pest Manag Sci 68: 941–946.
7. Cresswell JE, Desneux N, vanEngelsdorp D (2012) Dietary traces of neonicotinoid pesticides as a cause of population declines in honey bees: An evaluation by Hill's epidemiological criteria. Pest Manag Sci 68: 819–827.
8. Blacquiere T, Smagghe G, Van Gestel CA, Mommaerts V (2012) Neonicotinoids in bees: A review on concentrations, side-effects and risk assessment. Ecotoxicology 21: 973–992.
9. Goulson D (2013) An overview of the environmental risks posed by neonicotinoid insecticides. J Appl Ecol 50: 977–987.
10. Starner K, Goh KS (2012) Detections of the neonicotinoid insecticide imidacloprid in surface waters of three agricultural regions of California, USA, 2010–2011. Bull Environ Contam Toxicol 88: 316–321.
11. Miranda GR, Raetano CG, Silva E, Daam MA, Cerejeira MJ (2011) Environmental fate of neonicotinoids and classification of their potential risks to hypogean, epygean, and surface water ecosystems in Brazil. Human and Ecological Risk Assessment: An International Journal 17: 981–995.
12. Gupta S, Gajbhiye V, Agnihotri N (2002) Leaching behavior of imidacloprid formulations in soil. Bull Environ Contam Toxicol 68: 502–508.
13. Papiernik SK, Koskinen WC, Cox L, Rice PJ, Clay SA, et al. (2006) Sorption-desorption of imidacloprid and its metabolites in soil and vadose zone materials. J Agric Food Chem 54: 8163–8170.
14. Chiovarou ED, Siewicki TC (2008) Comparison of storm intensity and application timing on modeled transport and fate of six contaminants. Sci Total Environ 389: 87–100.
15. Gupta S, Gajbhiye V, Gupta R (2008) Soil dissipation and leaching behavior of a neonicotinoid insecticide thiamethoxam. Bull Environ Contam Toxicol 80: 431–437.
16. Juraske R, Castells F, Vijay A, Muñoz P, Antón A (2009) Uptake and persistence of pesticides in plants: measurements and model estimates for imidacloprid after foliar and soil application. J Hazard Mater 165: 683–689.
17. Wisconsin Department of Agriculture, Trade and Consumer Protection (2010) Fifteen years of the DATCP exceedence well survey. WI-DATCP, Madison, WI.
18. Wisconsin Department of Agriculture, Trade and Consumer Protection (2011) Agrichemical Management Bureau annual report – 2011. Available: http://datcp.wi.gov/Environment/Water_Quality/ACM_Annual_Report/. Accessed 2012 Jul 10.
19. United States Department of Agriculture - Natural Resources Conservation Soil Service (2013) Web Soil Survey. USDA-NRCS, Washington, DC. Available: http://websoilsurvey.sc.egov.usda.gov. Accessed 2014 Jan 8.
20. Cooley ET, Lowery B, Kelling KA, Speth PE, Madison FW, et al. (2009) Surfactant use to improve soil water distribution and reduce nitrate leaching in potatoes. Soil Sci 174: 321–329.
21. Wisconsin Department of Natural Resources (2013) DNR drinking water system: high capacity wells. WI DNR, Madison, WI. Available: http://dnr.wi.gov/topic/wells/highcapacity.html. Accessed 2013 Aug 22.
22. Wickham H (2007) Reshaping data with the reshape package. Journal of Statistical Software 21: 1–20. Available: http://www.jstatsoft.org/v21/i12/. Accessed 2011 Jan 15.
23. Bussan A, Colquhoun J, Cullen E, Davis V, Gevens A, et al. (2012) Commercial vegetable production in Wisconsin. Publication A3422. University of Wisconsin-Extension, Madison WI.
24. Seccia S, Fidente P, Barbini DA, Morrica P (2005) Multiresidue determination of nicotinoid insecticide residues in drinking water by liquid chromatography with electrospray ionization mass spectrometry. Anal Chim Acta 553: 21–26.
25. United States Department of Agriculture - National Agricultural Imagery Program (2010) Wisconsin NAIP. USDA-NAIP, Washington, DC. Available: http://datagateway.nrcs.usda.gov/. Accessed 2011 Jan 15.
26. United States Department of Agriculture - National Agricultural Statistics Service Cropland Data Layer (2012) Wisconsin Cropland data layer. USDA-NASS, Washington, DC. Available: http://nassgeodata.gmu.edu/CropScape/. Accessed 2013 May 10.
27. Wisconsin Department of Natural Resources (2002) Wisconsin DNR 2003 watersheds. Wisconsin DNR, Madison, WI. Available: http://dnr.wi.gov/maps/gis/documents/dnr_watersheds.pdf. Accessed 2013 Jun 12.
28. Thelin GP, Stone WW (2013) Estimation of annual agricultural pesticide use for counties of the conterminous United States, 1992–2009. US Department of the Interior, US Geological Survey.
29. Stone WW (2013) Estimated annual agricultural pesticide use for counties of conterminous United States, 1992–2009. U.S. Geological Survey Data Series 752, 1-p. pamphlet, 14 tables.
30. Cullen EM, Davis VM, Jensen B, Nice GRW, Renz M (2013) Pest management in Wisconsin field crops. Publication A3646.University of Wisconsin-Extension, Madison WI. Available: http://learningstore.uwex.edu/pdf/A3646.PDF as of 08/18/2013. Accessed 2013 Aug 23.
31. Team R Core (2011) R: A language and environment for statistical computing (Version 2.15.2). Vienna, Austria: R foundation for statistical computing; 2012. Available: http://cran.r-project.org. Accessed 2012 Jun 15.
32. Pinheiro J, Bates D, DebRoy S, Sarkar D (2007) Linear and nonlinear mixed effects models. R package version 3.1–108.
33. Pinheiro J, Bates D (2000) Mixed-effects models in S and S-PLUS. New York: Springer.
34. Mossbarger Jr W, Yost R (1989) Effects of irrigated agriculture on groundwater quality in corn belt and lake states. J Irrig Drain Eng 115: 773–790.
35. Kraft GJ, Stites W, Mechenich D (1999) Impacts of irrigated vegetable agriculture on a humid North-Central US sand plain aquifer. Ground Water 37: 572–580.
36. Saad DA (2008) Agriculture-related trends in groundwater quality of the glacial deposits aquifer, central Wisconsin. J Environ Qual 37: 209–225.
37. Postle JK, Rheineck BD, Allen PE, Baldock JO, Cook CJ, et al. (2004) Chloroacetanilide herbicide metabolites in Wisconsin groundwater: 2001 survey results. Environ Sci Technol 38: 5339–5343.
38. Agrian Inc. (2013) Advanced product search. Available: http://www.agrian.com/labelcenter/results.cfm. Accessed 2013 Mar 21.
39. Hladik ML, Calhoun DL (2012) Analysis of the herbicide diuron, three diuron degradates, and six neonicotinoid insecticides in water–Method details and application to two Georgia streams: U.S. Geological Survey Scientific Investigations Report 2012–5206.
40. Hopwood J, Vaughan M, Shepherd M, Biddinger D, Mader E, et al. (2012) Are neonicotinoids killing bees? A review of research into the effects of neonicotinoid insecticides on bees, with recommendations for action. Xerces Society for Invertebrate Conservation, USA.
41. Casida JE (2012) The greening of pesticide–environment interactions: Some personal observations. Environ Health Perspect 120: 487–493.

Acknowledgments

We thank the cooperating growers for generously allowing us to conduct lysimeter studies on their farm. We thank Amy DeBaker, Rick Graham, Jeff Postle, Wendy Sax, Stan Senger, and Steve Sobek of Wisconsin DATCP for their support of this project. We thank Dave Johnson and Robert Smail of Wisconsin DNR-Water Bureau for providing 2012 irrigation use data. We thank Birl Lowery and Mack Naber for input on lysimeter design and installation. We thank Scott Chapman, Ken Frost, and David Lowenstein for their help installing lysimeters. We thank Claudio Gratton, George Kennedy, Jessica Petersen, and Wesley Stone for their insightful comments on earlier versions of this manuscript. We thank the Wisconsin Potato and Vegetable Growers Association for continued support of our research efforts.

Author Contributions

Conceived and designed the experiments: ASH RLG. Performed the experiments: ASH. Analyzed the data: ASH. Wrote the paper: ASH RLG

42. Krupke CH, Hunt GJ, Eitzer BD, Andino G, Given K (2012) Multiple routes of pesticide exposure for honey bees living near agricultural fields. PLoS ONE 7: e29268.

43. Gill RJ, Ramos-Rodriguez O, Raine NE (2012) Combined pesticide exposure severely affects individual-and colony-level traits in bees. Nature 491: 105–108.

44. Seagraves MP, Lundgren JG (2012) Effects of neonicotinoid seed treatments on soybean aphid and its natural enemies. J Pest Sci 85: 125–132.

45. Cresswell JE, Page CJ, Uygun MB, Holmbergh M, Li Y, et al. (2012) Differential sensitivity of honey bees and bumble bees to a dietary insecticide (imidacloprid). Zoology 115: 365–371.

46. Henry M, Beguin M, Requier F, Rollin O, Odoux J, et al. (2012) A common pesticide decreases foraging success and survival in honey bees. Science 336: 348–350.

47. Stoner KA, Eitzer BD (2012) Movement of soil-applied imidacloprid and thiamethoxam into nectar and pollen of squash (*Cucurbita pepo*). PloS ONE 7: e39114.

48. Tapparo A, Marton D, Giorio C, Zanella A, Soldà L, et al. (2012) Assessment of the environmental exposure of honeybees to particulate matter containing neonicotinoid insecticides coming from corn coated seeds. Environ Sci Technol 46: 2592–2599.

49. Tomé HVV, Martins GF, Lima MAP, Campos LAO, Guedes RNC (2012) Imidacloprid-induced impairment of mushroom bodies and behavior of the native stingless bee *Melipona quadrifasciata anthidioides*. PloS ONE 7: e38406.

50. Chauzat M, Faucon J, Martel A, Lachaize J, Cougoule N, et al. (2006) A survey of pesticide residues in pollen loads collected by honey bees in France. J Econ Entomol 99: 253–262.

51. Girolami V, Mazzon L, Squartini A, Mori N, Marzaro M, et al. (2009) Translocation of neonicotinoid insecticides from coated seeds to seedling guttation drops: A novel way of intoxication for bees. J Econ Entomol 102: 1808–1815.

52. Laurent FM, Rathahao E (2003) Distribution of (14C) imidacloprid in sunflowers (*Helianthus annuus* L.) following seed treatment. J Agric Food Chem 51: 8005–8010.

53. Mullin CA, Frazier M, Frazier JL, Ashcraft S, Simonds R, et al. (2010) High levels of miticides and agrochemicals in North American apiaries: Implications for honey bee health. PLoS ONE 5: e9754.

54. Whitehorn PR, O'Conner S, Wackers FL, Goulson D (2012) Neonicotinoid pesticide reduces bumble bee colony growth and queen production. Science 336: 351–352.

55. Dively GP, Kamel A (2012) Insecticide residues in pollen and nectar of a cucurbit crop and their potential exposure to pollinators. J Agric Food Chem 60: 4449–4456.

56. Free JB (1993) Insect pollination of crops. London: Academic Press.

57. Klein A, Vaissière BE, Cane JH, Steffan-Dewenter I, Cunningham SA, et al. (2007) Importance of pollinators in changing landscapes for world crops. Proc R Soc B 274: 303–313.

58. Sánchez-Bayo F, Tennekes HA, Goka K (2013) Impact of systemic insecticides on organisms and ecosystems. *In* Stanislav T, editor, Insecticides - Development of Safer and More Effective Technologies. Rijeka: InTech. 367–416.

20

Distribution and Frequency of *kdr* Mutations within *Anopheles gambiae* s.l. Populations and First Report of the *Ace.1*G119S Mutation in *Anopheles arabiensis* from Burkina Faso (West Africa)

Roch K. Dabiré[1]*, Moussa Namountougou[1], Abdoulaye Diabaté[1], Dieudonné D. Soma[1], Joseph Bado[1], Hyacinthe K. Toé[1,2], Chris Bass[3], Patrice Combary[4]

1 IRSS (Institut de Recherche en Sciences de la Santé), Centre Muraz, Bobo-Dioulasso, Burkina Faso, **2** Department of Vector Biology, Liverpool School of Tropical Medicine, Liverpool, United Kingdom, **3** Biological Chemistry and Crop Protection Rothamsted Research, Harpenden, United Kingdom, **4** National Malaria Control Programme, Ministry of Health, Ouagadougou, Burkina Faso

Abstract

An entomological survey was carried out at 15 sites dispersed throughout the three eco-climatic regions of Burkina Faso (West Africa) in order to assess the current distribution and frequency of mutations that confer resistance to insecticides in *An. gambiae* s.l. populations in the country. Both knockdown (*kdr*) resistance mutation variants (L1014F and L1014S), that confer resistance to pyrethroid insecticides, were identified concomitant with the *ace*-1 G119S mutation confirming the presence of multiple resistance mechanisms in the *An. gambiae* complex in Burkina Faso. Compared to the last survey, the frequency of the L1014F *kdr* mutation appears to have remained largely stable and relatively high in all species. In contrast, the distribution and frequency of the L1014S mutation has increased significantly in *An. gambiae* s.l. across much of the country. Furthermore we report, for the first time, the identification of the *ace*.1 G116S mutation in *An. arabiensis* populations collected at 8 sites. This mutation, which confers resistance to organophosphate and carbamate insecticides, has been reported previously only in the *An. gambiae* S and M molecular forms. This finding is significant as organophosphates and carbamates are used in indoor residual sprays (IRS) to control malaria vectors as complementary strategies to the use of pyrethroid impregnated bednets. The occurrence of the three target-site resistance mutations in both *An. gambiae* molecular forms and now *An. arabiensis* has significant implications for the control of malaria vector populations in Burkina Faso and for resistance management strategies based on the rotation of insecticides with different modes of action.

Editor: Basil Brooke, National Institute for Communicable Diseases/NHLS, South Africa

Funding: This work was supported by the National Malaria Control Program (NMCP) of Burkina Faso. The funders had no role in study design, data collection and analysis, decision to publish, or preparation of the manuscript.

Competing Interests: The authors have declared that no competing interests exist.

* Email: dabire_roch@hotmail.com

Introduction

The pyrethroid class of insecticides have become a mainstay for vector control since the ban of DDT due to off-target toxicity and the development of resistance. They have been most widely used to treat bed nets (ITNs) dedicated to personal and community protection [1,2,3]. Unfortunately, knock down resistance (*kdr*) to pyrethroids, which also confers cross-resistance to DDT, was first reported in *Anopheles gambiae* populations from Côte d'Ivoire [4]. Resistance likely resulted from the earlier intensive use of DDT and selection from pyrethroid use in crop protection particularly in cotton areas [5,6]. *kdr* was initially shown to result from a point mutation (L1014F) in the pyrethroid target protein the voltage-gated sodium channel [7]. Based on a simple PCR diagnostic developed in the first report of the *kdr* mutation [7] several studies have been carried out on the distribution and the frequency of this

mechanism throughout Africa. Initial studies showed that L1014F *kdr* was most widely distributed in West African *An. gambiae* s.l. populations [6,8,9]. This mutation was observed initially in the S molecular form of *An. gambiae* s.s. reaching high frequency but was not found either in sympatric mosquitoes of the M molecular form or *An. arabiensis* populations [5]. This provided further evidence of reproductive barrier between the M and S molecular forms [10,11] and the two molecular forms of *An. gambiae* s.s. were recently confirmed as two distinct species termed *Anopheles coluzzii* for the M form and *Anopheles gambiae* for the S form [12]. However, a few years after the initial finding of the *kdr* mutation in the S molecular form, this mutation was also reported in the M form from the littoral of Benin and Côte d'Ivoire [13]. In-depth investigations carried out later in these geographic regions confirmed that this phenomenon was frequently observed in littoral but was rare inland [11]. DNA sequencing of these mosquitoes suggested that the mutation emerged in the M form by genetic introgression from the S form [14,15]. In contrast, the emergence of the Leu-Phe *kdr* mutation within *Anopheles arabiensis* resulted from a *de novo* mutation event [15]. An extensive monitoring program in Burkina Faso has revealed that the L1014F *kdr* mutation initially detected in low frequency in the *An. gambiae* M molecular form and *An. arabiensis* [11,15] has spread throughout the country and is observed in mosquito populations at relatively high frequency [16,17]. Recently the L1014S *kdr*, which initially predominated in East Africa [18,19], was reported in West Africa, first in Benin and then Burkina Faso within *An. arabiensis* populations [20,21]. More recently this mutation was reported in a small number of individuals of the M and S forms of *An. gambiae* in Burkina Faso [22]. Taken together these results provide fundamental insight into the evolutionary processes underlying resistance in *Anopheles gambiae s.l.* Furthermore from an applied perspective, the emergence of resistance has significant implications for vector control programmes, especially those focused on the use of ITNs/Long-Lasting Insecticidal Nets (LLINs) or indoor residual sprayings (IRS). Although LLINs had shown good control of certain pyrethroid resistant populations [23] reduced efficacy of treated nets against *An. gambiae* populations with *kdr* resistance has since been reported [24].

Other insecticides belonging to the organophosphate (OP) and carbamate (CM) classes have been investigated to be used in mosaic, or in combination, with pyrethroids for bednet impregnation [25]. In addition to the use of LLINs, bendiocarb was recently used in IRS applications in West Africa through the President's Malaria Initiative (PMI) roadmap [26]. Initially described in *Culex* populations from Côte-d'Ivoire [27] reduced susceptibility to OPs and CMs was observed in *An. gambiae* populations in the North of Côte d'Ivoire and related to the domestic use of insecticide [28]. *An. gambiae* populations from Benin with resistance to the CM bendiocarb were reported after just three year of IRS use [29]. A common mechanism of resistance to OP and CM insecticides results from a single point mutation (termed *ace-1^R*)in the target protein the acetylcholinesterase enzyme [30]. This mutation results in a glycine to serine replacement at amino acid position 119 and can be detected by a simple PCR-Restriction Fragment Length Polymorphism (RFLP) diagnostic [31]. This approach has been used to examine the frequency and distribution of this mutation in Burkina Faso where it was found predominately in the *An. gambiae* S form and in low frequency in the M form [9,16,32]. A recent study suggested that the mutation had introgressed from one form to the other but the precise origin of the introgression could not be determined due to the small sample size [33]. Since then, extensive country-wide surveys were performed in Burkina Faso from 2008 to 2010 and no case of *An. arabiensis* carrying this mutation was reported, although sample sizes for this species were sometimes small [16,17].

However insecticide resistance may also occur by other physiological mechanisms such as metabolic detoxification through increased enzyme activities (monooxygenases, esterases, or glutathione S- transferases) [34,35].

Burkina Faso is composed of three agro-climatic areas which exhibit different patterns of insecticide use especially in relation to crop protection. The present study provides an update on the distribution and the prevalence of the *kdr* L1014 and L1014S and *ace-1^R* mutations in *An. gambiae s.l.* populations throughout the 13 health regions dispersed across these different agro-climatic areas. We report here, for the first time, the occurrence of the *ace-1^R* mutation at remarkably high frequencies in *An. arabiensis*.

Materials and Methods

Study sites

Burkina Faso covers three ecological zones, the Sudan savannah zone in the south and west where rainfall is relatively heaviest (5–6 months), the arid savannah zone (Sudan-sahelian) which extends throughout much of the central part of the country and the aridland (Sahel) in the north. The northern part of the country has a dry season of 6–8 months. The varied ecological conditions are reflected in the different agricultural systems practiced throughout the country, from arable to pastoral lands. The western region constitutes the main cotton belt extending to the south where some new cotton areas have been cultivated since 1996. All ecological zones support the existence of *Anopheles* species that vector malaria and the disease is widespread throughout the country. Larvae were sampled from 15 sites dispersed throughout the three ecological zones (Table 1). The GPS coordinates were incorporated in Table 1.

Mosquito sampling

Larvae of *An. gambiae* s.l. were collected from at least 10 breeding sites dispersed throughout each sampling site mainly comprising pools of standing water and other small water collections. Larvae were pooled to constitute a colony, which was reared in the insectary to adulthood. A sample of 100 adult females were randomly sorted, killed and kept on silica gel in 1.5-ml tubes and stored at −20°C prior to PCR analysis. Anopheline species were identified morphologically using the standard identification keys of Gillies and Coetzee [36].

PCR analyses

An average of 30 mosquitoes was sampled per site by PCR analysis. Genomic DNA was extracted from single specimens and used as template for PCR to determine the species within the *An. gambiae* complex using the protocol SINE 200 of Santalomazza *et al.* [37] that allows the concomitant identification of *An. gambiae* M and S (respectively known as *Anopheles coluzzii* and *Anopheles gambiae*) and *An. arabiensis*. The same individuals were then tested for both the L1014F and L1014S *kdr* mutations using the protocols of Martinez-Torres *et al.*[7] (using specific primers Agd1, Agd2, Agd3 and Agd4) and Ranson *et al.* [18] (using Agd1, Agd2, Agd4 and Agd5) respectively:

- Agd1: 5′-ATAGATTCCCCGACCATG-3′;
- Agd2: 5′-AGACAAGGATGATGAACC-3′;
- Agd3: 5′-AATTTGCATTACTTACGACA-3′;
- Agd4: 5′-CTGTAGTGATAGGAAATTTA-5′;
- Agd5: 5′-TTTGCATTACTTACGACTG-3′.

The *ace-1^R* mutation was detected from the same samples by PCR according to the protocol of Weill *et al.* [31] using specific primers *Ex3AGdir* (GATCGTGGACACCGTGTTCG) and *Ex3-AGrev* (AGGATGGCCCGCTGGAACAG). Then the PCR products were digested using *Alu 1* enzyme at 37°C for 3 hours.

Statistical analysis

Data were compared between ecological zones and pooled for each species to compare the genotypes frequency between *An. gambiae* species by Chi^2 tests. The genotypic frequencies of L1014F and L1014S and *ace-1^R* in mosquito populations were compared to Hardy-Weinberg expectations using the exact test procedures implemented in GenePOP (ver.3.4) software [38].

Table 1. Distribution of *Anopheles gambiae* s.l. from 15 sites in Burkina Faso.

Study sites	Geographic references	Social environment	Climatic areas	Agricultural practices	Date of collection	*An. gambiae* s.l. N	*An. gambiae* n1	*An. gambiae* %	*An. coluzzii* n2	*An. coluzzii* %	*An. arabiensis* n3	*An. arabiensis* %
Gaoua	10°40'N; 3°15'W	sub-urban	Sudanian	cereals, cotton, old area	30/10/2012	43	39	90,69	1	2,33	3	6,98
Banfora	10°40'N; 3°15'W	sub-urban	Sudanian	cereals, cotton, old area	09/07/2012	30	24	80,00	6	20,00	0	0
Sindou	10°40'N; 3°15'W	rural	Sudanian	cotton, old area	01/10/2012	35	24	68,57	6	17,14	5	14,29
Orodara	10°40'N; 3°15'W	sub-urban	Sudanian	fruits, cotton, old area	23/19/2012	28	23	82,14	4	14,29	1	3,57
Dioulassoba	10°40'N; 3°15'W	traditional-urban	Sudanian	swamp	23/11/2012	29	4	13,79	5	17,24	20	68,97
Soumousso	10°40'N; 3°15'W	rural	Sudanian	cotton, old area	30/12/2012	30	20	66,67	3	10,00	7	23,33
Boromo	10°40'N; 3°15'W	sub-urban	Sudan-sahelian	cotton, old area	08/10/2012	33	16	48,48	0	0	17	51,52
Dédougou	10°40'N; 3°15'W	sub-urban	Sudan-sahelian	cotton, old area	06/10/2012	30	12	40,00	2	6,67	16	53,33
Koudougou	10°40'N; 3°15'W	urban	Sudan-sahelian	cotton, since 1996	07/11/2012	37	19	51,35	5	13,51	13	35,14
Nanoro	10°40'N; 3°15'W	rural	Sudan-sahelian	cereals	09/07/2012	32	4	12,50	24	75,00	4	12,50
Koupela	10°40'N; 3°15'W	sub-urban	Sudan-sahelian	cotton since 1996	06/10/2012	30	14	46,67	8	26,67	8	26,67
Fada	10°40'N; 3°15'W	sub-urban	Sudan-sahelian	cotton since 1996	25/08/2012	60	19	31,67	27	45,00	14	23,33
Kaya	10°40'N; 3°15'W	sub-urban	Sahelian	cereals, vegetables	03/10/2012	32	15	46,88	5	15,63	12	37,50
Ouahigouya	10°40'N; 3°15'W	sub-urban	Sahelian	cereals, vegetables	08/10/2012	31	20	64,52	10	32,26	1	3,23
Dori	10°40'N; 3°15'W	sub-urban	Sahelian	cereals, vegetables	01/10/2012	33	12	36,36	5	15,15	16	48,48

N: number total of mosquitoes.
n1: number of *An. gambiae*.
n2: number of *An. coluzzii*.
n3: number of *An. arabiensis*.

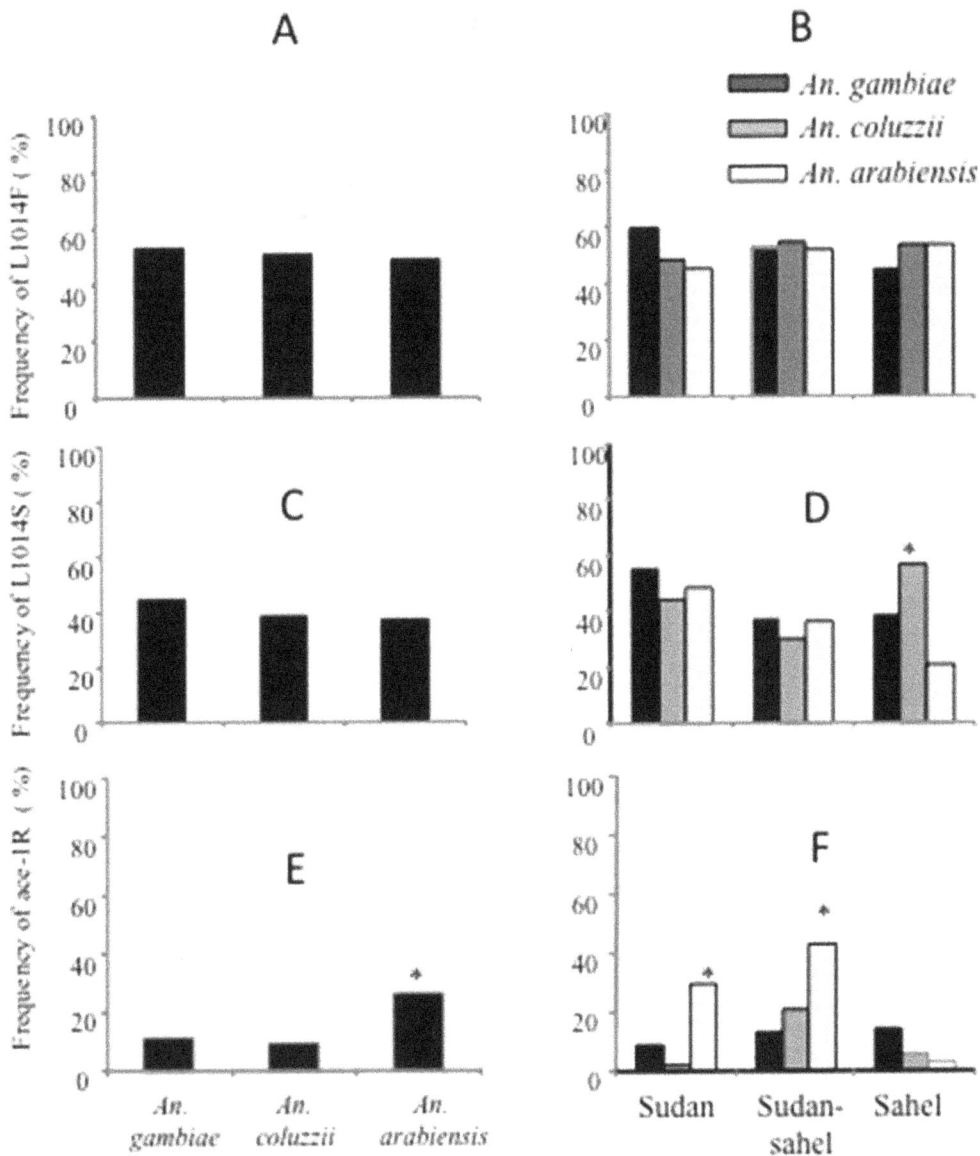

Figure 1. Comparison of allele frequencies of 1014F, 1014S and *ace-1^R* mutations within *Anopheles gambiae*, *An. coluzzii* and *An. arabiensis* populations from 15 sites dispersed across the 3 agro-ecological regions of Burkina Faso.

Ethical issues

Ethical approval was not required in this study.

This study was not carried out on private land. For each, no permission was required our study does not degrade the environment. No permission was required for these locations/activities as the field activities did not involve damaged of protected species. We did not use any vertebrate during this study.

Results

Out of 516 mosquitoes analysed in PCR, 513 successfully scored (less than 5% failure rate). Overall species composition of the collected mosquitoes comprised a higher proportion of *An. gambiae* (51.7%) than *An. coluzzii* (21.6%) and *An. arabiensis* (26.7%) (Table 1). The species repartition across the three ecological regions revealed that *An. gambiae* was the predominant species in all regions including, in the Sahel where it comprised more than 49% of the *An. gambiae s.l.* population. *Anopheles*

arabiensis was the second most predominant vector found in samples collected from the three regions. Somewhat *An. coluzzii* was found at a relatively low proportion of less than 15%. The central areas were characterised by an overlapped repartition of the three species 38.4%, 27.81% and 33.75% for *An. gambiae*, *An. coluzzii* and *An. arabiensis* respectively and proportions did not differ significantly ($\chi^2 = 1.95$, df = 1, $P > 0.05$). In the Sahel region, *An. gambiae* also predominated (49.75%) and the proportions of the two other species did not differ significantly at 21.01% and 29.74% for *An. coluzzii* and *An. arabiensis* respectively ($\chi^2 = 4.88$, df = 1, $P > 0.05$).

The overall frequency of the L1014F mutation averaged 50% and did not significantly differ between species (Figure 1A) whatever the ecological zone (Figure 1B) ($\chi^2 = 0.14$, df = 1, $P > 0.05$) even though the highest values were observed in the sudan zone (Figure 2). However some deviation from Hardy-Weinberg expectations was observed within the *An. arabiensis* populations in Dedougou and Dori and within *An. coluzzii* populations in Fada,

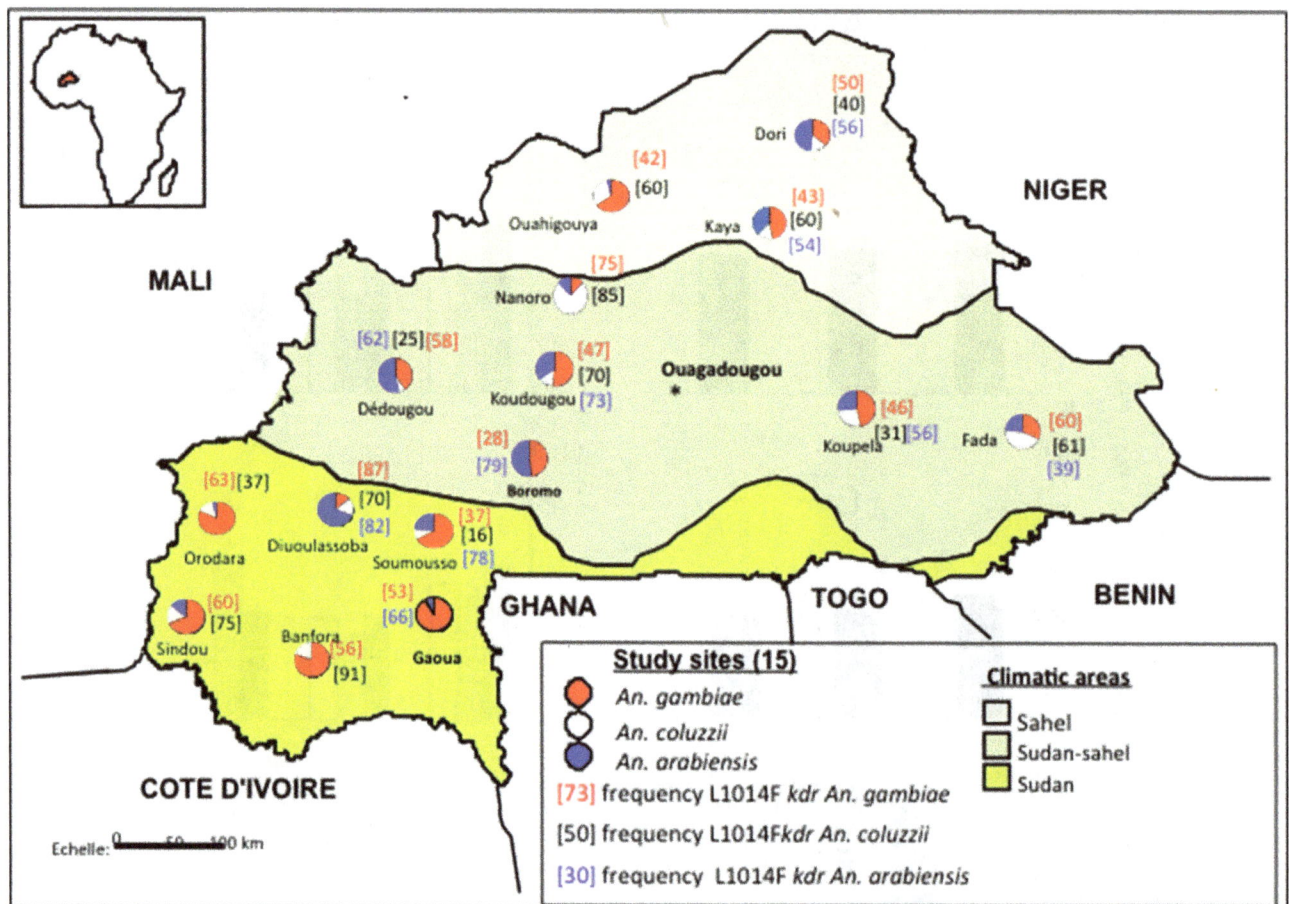

Figure 2. Distribution the 1014F *kdr* allele frequency from 15 sites dispersed across Burkina Faso.

Kaya, Ouahigouya and Dori with an excess of resistant homozygous alleles (Table 2). The same patterns were found in seven sites for *An. gambiae* (Gaoua, Banfora, Sindou in the West, Dedougou, Koudougou and Koupela in the central region and Ouahigouya in the Sahel) ($P < 0.05$).

The overall allele frequency of the L1014S *kdr* mutation (Figure 3) was relatively higher in *An. gambiae* (48%) followed by *An. coluzzii* (38%) and *An. arabiensis* populations (37%) with no significant difference between the last two ($x^2 = 3.24$, df = 1, $P > 0.05$) (Figure 1C). Comparing between ecological regions, L1014S *kdr* frequency did not differ significantly between species, except in the Sahel where it was significantly higher in *An. coluzzii* than *An. arabiensis* ($x^2 = 10.21$, df = 1, $P < 0.001$) and *An. gambiae* ($P < 0.04$) (Figure 1D). The observed genotypic frequencies were not significantly different from Hardy-Weinberg expectations at the 95% confidence level (Table 2) in populations from any site except in the *An. gambiae* populations from Orodara, Soumousso, Koupela, Fada, and in the *An. arabiensis* populations from Dioulassoba and Kaya where a heterozygous deficit was observed ($P = 0.005$) and *An. gambiae* populations in two sites (Dedougou and Kaya) where an excess of heterozygotes was observed ($P < 0.05$).

The *ace-1^R* mutation (Figure 4) was recorded in all the 15 sites under study with a wider distribution within the *An. gambiae* populations (Table 3). The overall allele frequency of *ace-1^R* was significantly higher in *An. arabiensis* (0.26) than in *An. gambiae* (0.11) ($x^2 = 14.4$; df = 1, $P = 0.001$) and *An. coluzzii* (0.09)

($x^2 = 11.77$, df = 1, $P = 0.006$) (Figure 1E) with no significant difference between the last two ($x^2 = 0.37$, df = 1, $P = 0.54$). Compared between zones, the *ace-1^R* allele frequency in *An. arabiensis* was higher than that of *An. coluzzii* ($x^2 = 8.15$, df = 1, $P = 0.004$) and *An. gambiae* ($x^2 = 9.79$, df = 1, $P < 0.001$) in the Sudan and Sudan-sahelian savannah (with respectively $x^2 = 6.89$, df = 1, $P < 0.008$ and $x^2 = 17.34$, df = 1, $P < 0.0003$) (Fig. 1F). In the Sahel no significant difference was observed between the three species ($x^2 = 0.89–0.021$, df = 1, $P > 0.05$). The observed genotypic frequencies were significantly different from Hardy-Weinberg expectations at the 95% confidence level (Table 3) in *An. gambiae* population from Orodara, Soumousso, Koudougou, Fada, Ouahigouya, Dori and Dioulassoba, Koudougou and Kaya for *An. arabiensis* where a heterozygote deficit was observed ($P = 0.005$). Furthermore, the percentage of homozygous resistant individuals was significantly higher in *An. arabiensis* (25%) than in *An. gambiae* (6.25%). No homozygous resistant individual was recorded in *An. coluzzii* from any site.

Discussion

This study provides current information on the distribution of three members of the *Anopheles gambiae* complex across Benin and the frequency and distribution of three important target-site resistance mechanisms in these populations. In regards to the distribution of *An. gambiae* species throughout the country, the most significant finding is that *An. arabiensis* appears to be spreading in the Sudan whereas in the past it comprised only

Table 2. Allelic and genotypic frequencies at the *kdr* 1014F and 1014S locus in *An. gambiae s.l* populations.

Species	Sites	N	Genotypes						Genotypes				
			1014L/1014L	1014L/1014F	1014F/1014F	f(L1014F)	[95%CI]	p(HW)	1014L/1014L	1014L/1014F	f(L1014F)	[95%CI]	p(HW)
An. arabiensis	Gaoua	5	1	0	2	0.66	[8.5–9.82]	-	0	2	0.66	[8.5–9.82]	0.2000
	Banfora	0	0	0	0	-	-	-	0	0	0.9	-	-
	Sindou	10	5	0	0	0	-	-	1	4	0	[7.38–9.18]	-
	Orodara	1	1	0	0	0	-	0.4678	0	0	0.45	-	-
	Dioulassoba	30	1	5	14	0.82	[3.13–4.71]	0.2308	2	8	0.42	[2.34–3.38]	0.0003
	Soumousso	11	1	1	5	0.78	[5.74–7.3]	0.0956	2	2	0.37	[4.37–5.21]	0.2914
	Boromo	17	2	3	12	0.79	[3.42–5.00]	0.000	6	3	0.28	[2.31–3.35]	0.3405
	Dédougou	23	6	0	10	0.62	[3.23–4.47]	0.1652	5	2	0	-	0.3213
	Koudougou	13	2	3	8	0.73	[3.9–5.36]	-	0	0	0.5	[6.41–7.41]	-
	Nanoro	6	4	0	0	0	-	0.4406	0	2	0.5	[4.39–5.39]	0.0857
	Koupela	13	2	3	3	0.56	[4.61–5.73]	0.2970	2	3	0.53	[4.39–5.39]	0.1795
	Fada	25	6	5	3	0.39	[2.87–3.65]	0.0933	8	3	0.57	[3.42–4.48]	0.9035
	Kaya	17	4	3	5	0.54	[3.61–4.69]	-	1	4	0.37	[3.07–3.81]	0.0061
	Ouahigouya	1	0	1	0	0.5	[3.32–4.32]	0.0031	0	0	0	[18.5–20.5]	-
	Dori	22	6	2	8	0.56	[3.1–4.22]	-	4	2	0.26	[2.32–2.84]	0.2260
An. coluzzii	Gaoua	1	1	0	0	0	[18.5–20.5]	-	0	0	0	[18.5–20.5]	-
	Banfora	7	0	5	0	0.91	[6.69–8.51]	-	0	1	0.16	[3.04–3.36]	0.0909
	Sindou	12	1	1	4	0.75	[6.15–/.65]	0.2727	1	5	0.91	[6.69–8.51]	-
	Orodara	5	2	1	1	0.37	5.58–6.32	0.4286	1	0	0.12	[3.27–3.51]	-
	Dioulassoba	9	1	1	3	0.7	[6.61–8.01]	0.3333	1	3	0.7	[6.61–8.01]	0.3333
	Soumousso	4	2	1	0	0.16	[4.36–4.68]	-	0	1	0.33	[6.16–6.82]	0.2000
	Boromo	0	0	0	0	-	-	-	0	0	-	-	-
	Dédougou	3	1	1	0	0.25	[6.67–7.17]	-	0	1	0.5	[9.28–10.28]	0.6190
	Koudougou	7	0	3	2	0.7	[6.6–8.01]	1	2	0	0.2	[3.72–4.12]	-
	Nanoro	39	1	5	18	0.85	[2.82–4.52]	0.3983	0	3	0.37	[2.06–2.8]	0.3333
	Koupela	9	3	5	0	0.31	[3.54–4.16]	1	1	0	0.06	[1.64–1.76]	0.7446
	Fada	46	7	7	13	0.61	[2.33–3.55]	0.0186	17	2	0.38	[1.94–2.7]	0.0817
	Kaya	8	2	0	3	0.6	[6.17–7.37]	0.0476	2	1	0.4	[5.13–5.93]	0.3333
	Ouahigouya	17	4	0	6	0.6	[4.19–5.39]	0.0017	2	5	0.6	[4.19–5.39]	1
	Dori	9	3	0	2	0.4	[5.13–5.93]	0.0476	1	3	0.7	[6.61–8.01]	-
An. gambiae	Gaoua	74	14	8	17	0.53	[3.75–2.81]	0.0002	0	35	0.92	[2.12–3.96]	1
	Banfora	29	7	7	10	0.56	2.43–3.55	0.0434	3	2	0.14	[1.36–1.64]	0.1518
	Sindou	46	8	3	13	0.6	[2.49–3.69]	0.0003	5	17	0.81	[2.78–4.4]	0.0611

Table 2. Cont.

Species	Sites	N	Genotypes			f(L1014F)	[95%CI]	p(HW)	Genotypes		f(L1014F)	[95%CI]	p(HW)
			1014L 1014L	1014L 1014F	1014F 1014F				1014L 1014L	1014L 1014F			
	Orodara	33	5	7	11	0.63	[2.6–3.86]	0.0904	1	9	0.41	[2.2–3.02]	0.0420
	Dioulassoba	8	0	1	3	0.87	[8.239.97]	-	2	2	0.75	[7.71–9.21]	0.3257
	Soumousso	29	8	9	3	0.37	[2.29–3.63]	0.5690	5	4	0.32	[2.16–2.8]	0.0000
	Boromo	25	8	7	1	0.28	[2.31–2.87]	0.7912	4	5	0.43	[2.78–3.64]	0.1201
	Dédougou	19	5	0	7	0.58	[3.72–4.88]	0.0004	7	0	0.29	[2.75–3.33]	0.0150
	Koudougou	26	9	2	8	0.47	[2.61–3.55]	0.0005	4	3	0.26	[2.03–2.55]	1
	Nanoro	5	1	0	3	0.75	[7.71–9.21]	0.1429	0	1	0.25	[4.64–5.14]	0.1429
	Koupela	24	7	1	6	0.46	[3.08–4.00]	0.0013	4	6	0.57	[3.37–4.51]	0.0003
	Fada	30	3	9	7	0.6	[2.87–4.07]	0.6254	5	6	0.44	[2.54–3.42]	0.0473
	Kaya	19	5	7	3	0.43	[2.88–3.74]	0.5785	3	1	0.16	[1.86–2.18]	0.0000
	Ouahigouya	30	10	3	7	0.42	[2.4–3.25]	0.0020	2	8	0.45	[2.48–3.38]	0.0632
	Dori	18	4	4	4	0.5	[3.49–4.49]	0.2300	1	5	0.55	[4.03–5.13]	0.0520

N: number of mosquitoes.

f(1014F): frequency of the kdr W resistant allele.

f(1014S): frequency of the kdr E resistant allele.

p(HW): probability of the exact test for goodness of fit to Hardy Weinberg equilibrium.

-: not determined.

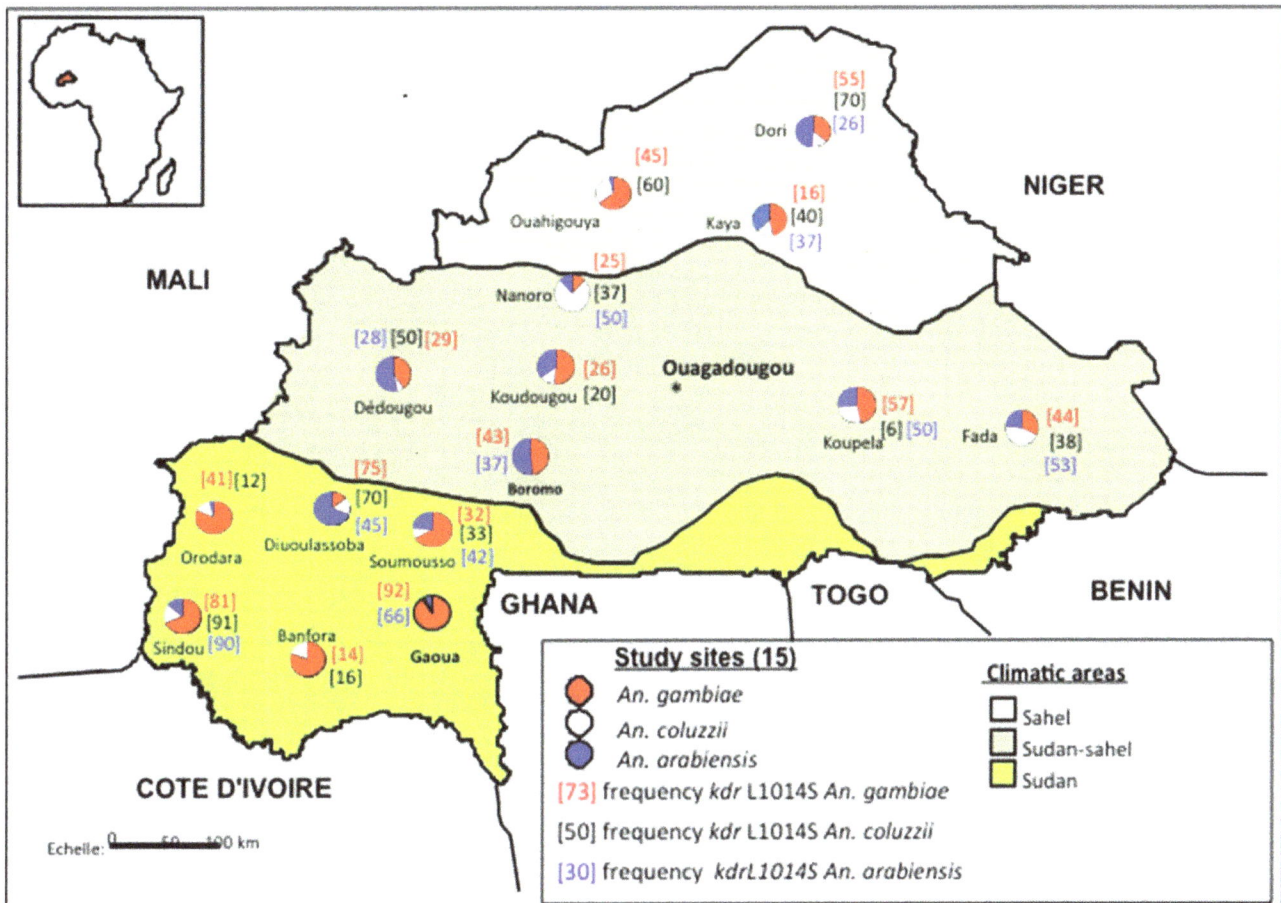

Figure 3. Distribution the 1014S *kdr* allele frequency from 15 sites dispersed across Burkina Faso.

around 5% of the *An. gambiae* complex species [6]. Furthermore, this species is now present in Sindou at 14.29% (nearest the frontier of Cote-d'Ivoire) where it was absent a decade ago [9]. The reason for this is not clear but could be related to climatic changes, such as irregularities in rainfall observed in the boundaries of the Sudan region that may make the landscape more favourable to the establishment of this species.

Across sampling covering 15 sites we identified the L1014F and L1014S *kdr* mutations concomitant with the *ace*-1 G119S mutation confirming the presence of multiple resistance mechanisms in the *An. gambiae* complex in Burkina Faso [16,17]. The distribution and the prevalence of the L1014F *kdr* mutation in *An. gambiae* species including *An. gambiae, An. coluzzii* and *An. arabiensis,* has been well documented in Burkina Faso for over a decade [9,16]. Many studies reported this mutation at high frequency within *An. gambiae* and *An. coluzzii* populations especially in *An. gambiae* populations from the Sudan area where mutation frequency was approaching fixation [9,15,16]. Over recent years the frequency of this mutation has increased within both *An. coluzzii* and *An. arabiensis*. In this study although the L1014F mutation remains widespread in all three ecological regions and is present at relatively high frequency within the three species (averaging 50%), the frequencies reported in this current study were lower in the Sudan ecological regions (West and South West covering the old cotton belt) than those from previous studies [9,16,22]. For the other climatic zones i.e. central and northern regions the allele frequencies of L1014F varied within the three

species with particularly high frequencies in *An. arabiensis*. The reason(s) for the reduction of L1014F frequency in *An. gambiae* populations in the Sudan area is not known, however, a similar trend was recently observed in the Western region of Burkina Faso where transgenic and biological control practices have been implemented for crop protection of cotton over the last four years (a long side conventional crop protection approaches) (Namoun-tougou, unpublished). These alternative cotton-growing practices would be expected to reduce the quantity and frequency of insecticide use in agriculture and this may in turn reduce the selection pressure experienced by local mosquito populations. The analysis of observed genotypic frequencies revealed a heterozygote deficit for the L1014F mutation in the three species of *An. gambiae* s.l. from many sites especially in the Sahel for *An. coluzzii* and *An. arabiensis* and in the Sudan and Sudan-Sahel for *An. gambiae* which deviated significantly from Hardy-Weinberg expectations. This finding is not surprising as the same patterns were observed in the West (Orodara and Soumousso) four years ago [9] in combination with a novel mutation, N1575Y, in the voltage-gated sodium channel, recently reported in *An. gambiae* s.l. populations in Soumousso [39].

The L1014S *kdr* mutation was recently recorded at highest frequency in *An. arabiensis* populations in the centre on the country [21] and in Bobo-Dioulasso at frequencies averaging 38% [40]. Previous studies have recorded only a few individuals of *An. gambiae* and *An. coluzzii* from the Centre-East part of the country [17] carrying this mutation in the heterozygous form. The present

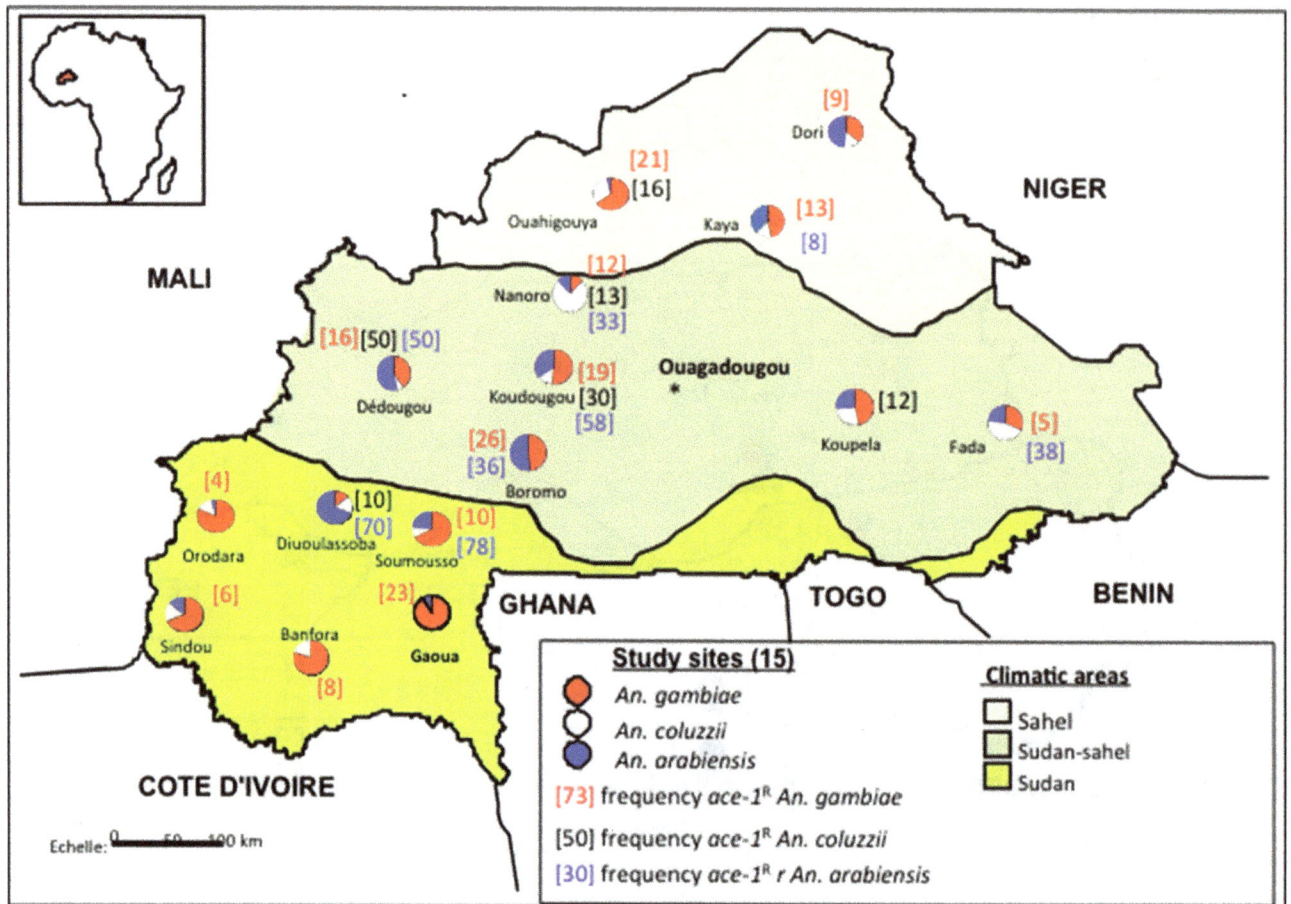

Figure 4. Distribution the *ace-1^R* allele frequency from 15 sites dispersed across Burkina Faso.

study reveals that this mutation has since spread across the whole country and is now observed at relatively high and similar frequencies (40%) between the three species. The comparison of the observed genotypic frequencies of this mutation with that expected for Hardy-Weinberg equilibrium indicated, depending on the site, a deficit or excess of heterozygotes, mainly for *An. gambiae* populations. The occurrence of the L1014F *kdr* mutation in *An. coluzzii* had been suggested to have occurred by introgression from *An. gambiae* and via a *de novo* mutation event in *An. arabiensis* [15], however, the origin of the L1014S mutation in *An. gambiae*, *An. coluzzii* and *An. arabiensis* species in West Africa is not so clearly understood. The proximity of Burkina Faso from the Benin frontier where the L1014S mutation was first reported in *An. arabiensis* populations [20] suggests that it arrived in Burkina Faso via migration of *An. arabiensis* carrying the mutation from Benin, however, the origin of this mutation in *An. gambiae* and *An. coluzzii* populations in Burkina Faso remains to be elucidated.

In this study we report, for the first time, the presence of the *ace.1* G119S mutation in *An. arabiensis* populations from eight sites: Dioulassoba, Soumousso in the West, Boromo, Dédougou, Koudougou, Nanoro and Fada in the Centre-North and East and Kaya in the North. In these sites *An. arabiensis* was observed as the second major vector after *An. gambiae* except at Fada and Nanoro where the proportion of *An. arabiensis* was lower than that of *An. coluzzii*. To confirm this finding, we repeated the PCR amplification of *ace.1^R* for our *An. arabiensis* specimens and used,

as a control, 30 specimens of *An. Arabiensis* which we had confirmed in a previous study do not have this mutation. No false positives were observed in these samples suggesting our data is robust. The *ace.1^R* allele was observed in this study in *An. arabiensis* at varying frequency reaching a maximum value of 78% in populations from Dioulassoba and the lowest value in Kaya at 8%. Except for samples from Soumousso and Nanoro where the sample size was not sufficient (n<10) to compare genotype frequencies, deviations from Hardy-Weinberg equilibrium were observed at three sites (Dioulassoba, Koudougou and Kaya) as a result of a high heterozygote deficit. The same pattern was observed in *An. gambiae* from Orodara, Soumousso, Koudougou, Fada, Ouahigouya and Dori. The deficit of heterozygous genotypes observed in Orodara and Soumousso is not new as Dabiré *et al.* [41] reported similar results from the these areas from which the duplicated allele (*ace.1^D*) was reported by Djogbenou *et al.* [33]. It is possible that this duplicated allele *ace.1^D* is also present within *An. arabiensis* especially in Dioulassoba where the proportion of homozygous mutants was atypically high (60%). The high frequency of this mutation in Dioulassoba populations is intriguing as recent studies failed to find any L1014F *kdr* or *ace-1^R* in *An. arabiensis* population from this site [40,42]. As for the L1014S mutation, additional sequence analysis of the region flanking the *ace.1* locus are necessary to confirm whether the *ace.1* mutation in *An. arabiensis* has evolved along the same pathway as *kdr* e.g. as a *de novo* mutation or introgression from *An. gambiae* or *An. coluzzii*. Unfortunately our PCR data is not backed up by

Table 3. Allelic and genotypic frequencies at the ace-1 locus in *An. gambiae s.l* populations from 15 sites in Burkina Faso.

Species	Sites	N	Genotypes			f(119S)	[95%CI]	p(HW)
			119G 119G	119G 119S	119S 119S			
An. arabiensis	Gaoua	3	3	0	0	0	-	-
	Banfora	0	0	0	0	-	-	-
	Sindou	5	5	5	0	0	-	-
	Orodara	1	1	1	0	0	-	-
	Dioulassoba	20	4	4	12	0.7	[2.95–7.13]	0.0264
	Soumousso	7	1	1	5	0.78	[5.74–7.57]	0.2308
	Boromo	15	5	9	1	0.36	[2.67–5.42]	0.9488
	Dédougou	14	4	6	4	0.5	[3.19–7.25]	0.0444
	Koudougou	12	5	0	7	0.58	[3.72–9.1]	0.0004
	Nanoro	3	2	0	1	0.33	[6.16–17.45]	0.2000
	Koupela	8	8	0	0	0	-	-
	Fada	13	4	8	1	0.38	[2.96–6.26]	0.9449
	Kaya	12	11	0	1	0.08	[1.52–2.27]	0.0435
	Ouahigouya	1	1	0	0	0	-	-
	Dori	14	14	0	0	0	-	-
An. coluzzii	Gaoua	1	1	0	0	0	-	-
	Banfora	6	6	0	0	0	-	-
	Sindou	6	6	0	0	0	-	-
	Orodara	4	4	0	0	0	-	-
	Dioulassoba	5	4	1	0	0.1	[2.67–4.71]	-
	Soumousso	3	3	0	0	0	-	-
	Boromo	0	0	0	0	-	-	-
	Dédougou	2	0	0	2	0.5	[9.28–34.65]	1
	Koudougou	5	2	3	0	0.3	[4.49–10.78]	1
	Nanoro	23	17	6	0	0.13	[1.34–2.04]	1
	Koupela	8	6	2	0	012	[2.28–3.9]	1
	Fada	27	27	0	0	0	-	-
	Kaya	5	5	0	0	0	-	-
	Ouahigouya	9	6	3	0	0.16	[2.64–4.39]	1
	Dori	5	5	0	0	0	-	-
An. gambiae	Gaoua	36	22	11	3	0.23	[1.33–2.2]	0.2811
	Banfora	24	20	4	0	0.08	[1.05–1.46]	1
	Sindou	24	21	3	0	0.06	[0.92–1.23]	1
	Orodara	23	22	0	1	0.04	[0.74–0.99]	0.0222

Table 3. Cont.

Species	Sites	N	Genotypes			f(119S)	[95%CI]	p(HW)
			119G 119G	119G 119S	119S 119S			
	Dioulassoba	4	4	0	0	0	-	-
	Soumousso	20	18	0	2	0.1	[1.29-1.88]	0.0021
	Boromo	15	9	4	2	0.26	[2.32-4.31]	0.2260
	Dédougou	12	8	4	0	0.16	[2.1-3.59]	1
	Koudougou	18	14	1	3	0.19	[1.82-3.07]	0.0029-
	Nanoro	4	3	1	0	0.12	[3.27-6.29]	-
	Koupela	12	12	0	0	0	-	-
	Fada	19	18	0	1	0.05	[0.96-1.27]	0.0270
	Kaya	15	11	4	0	0.13	[1.69-2.62]	1
	Ouahigouya	19	14	2	3	0.21	[1.85-3.16]	0.0096
	Dori	11	10	0	1	0.09	[1.68-2.59]	0.0476

N: number of mosquitoes.

f(119S): frequency of the 119S resistant ace.1 allele.

p(HW): probability of the exact test for goodness of fit to Hardy Weinberg equilibrium.

-: not determined.

insecticide susceptibility bioassays and so we cannot assess the correlations between *kdr* and *ace*-1 mutations and the phenotypic expression of resistance.

The emergence of the *ace-1^R* mutation in *An. gambiae* s.l. population from the cotton-growing areas may be linked to the agricultural use of OP and CM insecticides used for crop protection. Other sources of selection pressure outside the cotton belt include insecticide use for vegetable growing and domestic use of insecticide in public health. Bioassays performed in 2012 on *An. gambiae* populations from sites located in the cotton belt of the West of Burkina Faso revealed the development of resistance to CMs and OPs especially to benidocarb (Dabiré, unpublished) correlating with the prevalence and frequency of genetic resistance revealed in the present study. However, further bioassays on a wider scale are now required in order to understand the implications of the current status of the *ace-1^R* mutation for the efficacy of OP and CM insecticides in vector control in Burkina Faso. The information provided by such studies combined with the genetic data presented here is a prerequisite for the informed use of CM and OP based-combinations for bednet impregnation and/or indoor residual spraying.

Acknowledgments

This work was supported by the National Malaria Control Program (NMCP) of Burkina Faso.

Author Contributions

Conceived and designed the experiments: RKD AD PC. Performed the experiments: DDS JB HKT. Analyzed the data: RKD MN. Wrote the paper: RKD CB. Supervised field work: MN. Revised the manuscript: MN AD CB. Performed PCR analyses: DDS JB HKT. Assured the financial support of the study through the Ministry of Health: CB. Read and approved the final version of the manuscript: RKD MN AD DDS JB HKT CB PC.

References

1. Carnevale P, Robert V, Boudin C, Halna JM, Pazart L, et al. (1998) La lute contre le paludisme par des moustiquaires imprégnées de pyrthrinoides au Burkina Faso. Bull Soc Pathol Exot 81: 832–846.
2. D'Alessandro U, Olaleye BO, McGuire W, Langerock P, Bennett S, et al. (1995) Mortality and morbidity from malaria in Gambian children after introduction of an impregnated bednet programme. Lancet 345: 479–483.
3. Binka FN, Kubaje A, Adjuik M, Williams LA, Lengeler C, et al. (1996) Impact of permethrin impregnated bednets on child mortality in Kassena-Nankana district, Ghana: a randomized controlled trial. Trop Med Int Health 1: 147–154.
4. Elissa N, Mouchet J, Riviere F, Meunier JY, Yao K (1993) Resistance of Anopheles gambiae s.s. to pyrethroids in Cote d'Ivoire. Ann Soc Belg Med Trop 73: 291–294.
5. Chandre F, Darrier F, Manga L, Akogbeto M, Faye O, et al. (1999) Status of pyrethroid resistance in Anopheles gambiae sensu lato. Bull World Health Organ 77: 230–234.
6. Diabate A, Baldet T, Chandre F, Akogbeto M, Guiguemde TR, et al. (2002) The role of agricultural use of insecticides in resistance to pyrethroids in Anopheles gambiae s.l. in Burkina Faso. Am J Trop Med Hyg 67: 617–622.
7. Martinez-Torres D, Chandre F, Williamson MS, Darriet F, Berge JB, et al. (1998) Molecular characterization of pyrethroid knockdown resistance (*kdr*) in the major malaria vector Anopheles gambiae s.s. Insect Mol Biol 7: 179–184.
8. Awolola TS, Oyewole IO, Amajoh CN, Idowu ET, Ajayi MB, et al. (2005) Distribution of the molecular forms of Anopheles gambiae and pyrethroid knock down resistance gene in Nigeria. Acta Tropica 95: 204–209.
9. Dabire KR, Diabate A, Namountougou M, Toe KH, Ouari A, et al. (2009a) Distribution of pyrethroid and DDT resistance and the L1014F *kdr* mutation in Anopheles gambiae s.l. from Burkina Faso (West Africa). Trans R Soc Trop Med Hyg 103: 1113–1120.
10. Favia G, Lanfrancotti A, Spanos L, Siden Kiamos I, Louis C (2001) Molecular characterization of ribosomal DNA polymorphisms discriminating among chromosomal forms of Anopheles gambiae s.s. Insect Mol Biol 10: 19–23.
11. Diabate A, Baldet T, Chandre C, Dabire KR, Kengne P, et al. (2003) KDR mutation, a genetic marker to assess events of introgression between the molecular M and S forms of Anopheles gambiae (Diptera: Culicidae) in the tropical savannah area of West Africa. J Med Entomol 40: 195–198.
12. Coetzee M, Hunt R, Wilkerson R, Della Torre A, Coulibaly BM, et al. (2013) Anopheles coluzzii and Anopheles amharicus, new members of the Anopheles gambiae complex. Zootaxa 3619: 246–274.
13. Fanello C, Akogbeto M, della Torre A (2000) Distribution of the knock down resistance gene (kdr) in Anopheles gambiae s.l. from Benin. Trans R Soc Trop Med Hyg 94: 132.
14. Weill M, Chandre F, Brengues C, Manguin S, Akogbeto M, et al. (2000) The kdr mutation occurs in the Mopti form of Anopheles gambiae s.s. through introgression. Insect Molecular Biology 9: 451–455.
15. Diabate A, Brengues C, Baldet T, Dabire KR, Hougard JM, et al. (2004) The spread of the Leu-Phe kdr mutation through Anopheles gambiae complex in Burkina Faso: genetic introgression and de novo phenomena. Trop Med Int Health 9: 1267–1273.
16. Dabiré KR, Diabaté A, Namountougou M, Djogbenou L, Wondji C, et al. (2012a) Trends in Insecticide Resistance in Natural Populations of Malaria Vectors in Burkina Faso, West Africa: 10 Years' Surveys. In: Perveen F, editors. Insecticides - Pest Engineering. ISBN. InTech: 479–502.
17. Namountougou M, Diabate A, Etang J, Bass C, Sawadogo SP, et al. (2013) First report of the L1014S kdr mutation in wild populations of Anopheles gambiae M and S molecular forms in Burkina Faso (West Africa). Acta Tropica 125: 123–127.
18. Ranson H, Jensen B, Vulule JM, Wang X, Hemingway J, et al. (2000) Identification of a point mutation in the voltage-gated sodium channel gene of Kenyan Anopheles gambiae associated with resistance to DDT and pyrethroids. Insect Mol Biol 9: 491–497.
19. Verhaeghen K, Van Bortel W, Roelants P, Backeljau T, Coosemans M (2006) Detection of the East and West African kdr mutation in Anopheles gambiae and Anopheles arabiensis from Uganda using a new assay based on FRET/Melt Curve analysis. Malaria J 5: 16.
20. Djegbe I, Boussari O, Sidick A, Martin T, Ranson H, et al. (2011) Dynamics of insecticide resistance in malaria vectors in Benin: first evidence of the presence of L1014S kdr mutation in Anopheles gambiae from West Africa. Malaria J 10: 261.
21. Badolo A, Traore A, Jones CM, Sanou A, Flood L, et al. (2012) Three years of insecticide resistance monitoring in Anopheles gambiae in Burkina Faso: resistance on the rise? Malaria J 11: 232.
22. Namountougou M, Simard F, Baldet T, Diabate A, Ouedraogo JB, et al. (2012) Multiple insecticide resistance in Anopheles gambiae s.l. populations from Burkina Faso, West Africa. PLoS One 7: e48412.
23. Henry MC, Assi SB, Rogier C, Dossou-Yovo J, Chandre F, et al. (2005) Protective efficacy of lambda-cyhalothrin treated nets in Anopheles gambiae pyrethroid resistance areas of Cote d'Ivoire. Am J Trop Med Hyg 73: 859–864.
24. N'Guessan R, Corbel V, Akogbeto M, Rowland M (2007) Reduced efficacy of insecticide-treated nets and indoor residual spraying for malaria control in pyrethroid resistance area, Benin. Emerg Infect Dis 13: 199–206.
25. Guillet P, N'Guessan R, Darriet F, Traore-Lamizana M, Chandre F, et al. (2001) Combined pyrethroid and carbamate 'two-in-one' treated mosquito nets: field efficacy against pyrethroid-resistant Anopheles gambiae and Culex quinquefasciatus. Med Vet Entomol 15: 105–112.
26. Ossè R, Aikpon R, Padonou GG, Oussou O, Yadouleton A, et al. (2012) Evaluation of the efficacy of bendiocarb in indoor residual spraying against pyrethroid resistant malaria vectors in Benin: results of the third campaign. Parasit Vectors 5: 163.
27. Chandre F, Darriet F, Doannio JM, Riviere F, Pasteur N, et al. (1997) Distribution of organophosphate and carbamate resistance in Culex pipiens quinquefasciatus (Diptera: Culicidae) in West Africa. J Med Entomol 34: 664–671.
28. N'Guessan R, Darriet F, Guillet P, Carnevale P, Traore-Lamizana M, et al. (2003) Resistance to carbosulfan in Anopheles gambiae from Ivory Coast, based on reduced sensitivity of acetylcholinesterase. Med Vet Entomol 17: 19–25.
29. Aikpon R, Agossa F, Osse R, Oussou O, Aizoun N, et al. (2013) Bendiocarb resistance in Anopheles gambiae s.l. populations from Atacora department in Benin, West Africa: a threat for malaria vector control. Parasit vectors 6: 192.
30. Weill M, Lutfalla G, Mogensen K, Chandre F, Berthomieu A, et al. (2003) Comparative genomics: Insecticide resistance in mosquito vectors. Nature 423: 136–137.
31. Weill M, Malcolm C, Chandre F, Mogensen K, Berthomieu A, et al. (2004) The unique mutation in ace-1 giving high insecticide resistance is easily detectable in mosquito vectors. Insect Mol Biol 13: 1–7.
32. Djogbenou L, Dabire R, Diabate A, Kengne P, Akogbeto M, et al. (2008) Identification and geographic distribution of the ACE-1R mutation in the malaria vector Anopheles gambiae in south-western Burkina Faso, West Africa. Am J Trop Med Hyg 78: 298–302.
33. Djogbenou L, Chandre F, Berthomieu A, Dabire R, Koffi A, et al. (2008) Evidence of introgression of the ace-1(R) mutation and of the ace-1 duplication in West African Anopheles gambiae s.s. PLoS ONE 3: e2172.
34. Scott JG (1996) Cytochrome P450 monooxygenase-mediated resistance to insecticides. J Pest Sci 21: 241–245.

35. Hemingway J, Karunaratne SH (1998) Mosquito carboxylesterases: a review of the molecular biology and biochemistry of a major insecticide resistance mechanism. Med Vet Entomol 12: 1–12.

36. Gillies MT, Coetzee M (1987) A supplement to the Anophelinae of Africa south of the Sahara. Pub. South Afr. Inst Med Res 55: 143.

37. Santolamazza F, Calzetta M, Etang J, Barrese E, Dia I, et al. (2008) Distribution of knock-down resistance mutations in *Anopheles gambiae* molecular forms in west and west-central Africa. Malar J 7: 74.

38. Raymond M, Rousset F (1995) GENEPOP Version 1.2 A population genetics software for exact tests and ecumenicism. J Hered: 248–249.

39. Jones CM, Toe HK, Sanou A, Namountougou M, Hughes A, et al. (2012a) Additional selection for insecticide resistance in urban malaria vectors: DDT resistance in *Anopheles arabiensis* from Bobo-Dioulasso, Burkina Faso. PLoS One 7: e45995.

40. Jones CM, Liyanapathirana M, Agossa FR, Weetman D, Ranson H, et al. (2012) Footprints of positive selection associated with a mutation (N1575Y) in the voltage-gated sodium channel of *Anopheles gambiae*. Proc Nati Acad Sci U S A 109: 6614–6619.

41. Dabire KR, Diabate A, Namountougou M, Djogbenou L, Kengne P, et al. (2009b) Distribution of insensitive acetylcholinesterase (*ace*-1R) in *Anopheles gambiae s.l.* populations from Burkina Faso (West Africa). Trop Med Int Health 14: 396–403.

42. Dabire RK, Namountougou M, Sawadogo SP, Yaro LB, Toe HK, et al. (2012) Population dynamics of *Anopheles gambiae s.l.* in Bobo-Dioulasso city: bionomics, infection rate and susceptibility to insecticides. Parasit vectors 5: 127.

21

Controlling Malaria Using Livestock-Based Interventions: A One Health Approach

Ana O. Franco[1,2*], **M. Gabriela M. Gomes**[1], **Mark Rowland**[2], **Paul G. Coleman**[2], **Clive R. Davies**[†2]

1 Instituto Gulbenkian de Ciência, Oeiras, Portugal, **2** Faculty of Infectious and Tropical Diseases, London School of Hygiene and Tropical Medicine, London, United Kingdom

Abstract

Where malaria is transmitted by zoophilic vectors, two types of malaria control strategies have been proposed based on animals: using livestock to divert vector biting from people (zooprophylaxis) or as baits to attract vectors to insecticide sources (insecticide-treated livestock). Opposing findings have been obtained on malaria zooprophylaxis, and despite the success of an insecticide-treated livestock trial in Pakistan, where malaria vectors are highly zoophilic, its effectiveness is yet to be formally tested in Africa where vectors are more anthropophilic. This study aims to clarify the different effects of livestock on malaria and to understand under what circumstances livestock-based interventions could play a role in malaria control programmes. This was explored by developing a mathematical model and combining it with data from Pakistan and Ethiopia. Consistent with previous work, a zooprophylactic effect of untreated livestock is predicted in two situations: if vector population density does not increase with livestock introduction, or if livestock numbers and availability to vectors are sufficiently high such that the increase in vector density is counteracted by the diversion of bites from humans to animals. Although, as expected, insecticide-treatment of livestock is predicted to be more beneficial in settings with highly zoophilic vectors, like South Asia, we find that the intervention could also considerably decrease malaria transmission in regions with more anthropophilic vectors, like *Anopheles arabiensis* in Africa, under specific circumstances: high treatment coverage of the livestock population, using a product with stronger or longer lasting insecticidal effect than in the Pakistan trial, and with small (ideally null) repellency effect, or if increasing the attractiveness of treated livestock to malaria vectors. The results suggest these are the most appropriate conditions for field testing insecticide-treated livestock in an Africa region with moderately zoophilic vectors, where this intervention could contribute to the integrated control of malaria and livestock diseases.

Editor: Thomas A. Smith, Swiss Tropical & Public Health Institute, Switzerland

Funding: Ana O. Franco was funded by the Portuguese Fundação para a Ciência e Tecnologia (FCT - SFRH/BD/9605/2002), co-financed by the Programa Operacional Ciência e Inovação 2010 (POCI 2010) and Fundo Social Europeu (FSE), and by EPIWORK - European Commission (Grant Agreement 231807). The publication fees were paid by the London School of Hygiene and Tropical Medicine. The funders had no role in study design, data collection and analysis, decision to publish, or preparation of the manuscript.

Competing Interests: The authors have declared that no competing interests exist.

* Email: afranco@igc.gulbenkian.pt

† Deceased

Introduction

In the last few decades there has been increasing recognition of the need for an integrated public health and veterinary approach, accounting for the surrounding social-ecological system, to face many of the most challenging disease threats: the so-called 'One Health' approach [1]. Broadly speaking, animals play an important role in the epidemiology of several of the most important diseases of man, where they can act as a reservoir source for infectious pathogens, and/or a source of blood-meal to arthropod vectors of human disease. The recognition of this relationship has led to the implementation of human disease control strategies targeted at animal populations. These control opportunities have been investigated both empirically and theoretically. Yet, our knowledge on what determines the public health benefits of many of these veterinary interventions remains limited.

A case study of the 'One Health' concept is human malaria in regions where its mosquito vectors (*Anopheles* spp.) also feed on animals, since the presence of livestock close to the household can affect the rate of vector-human contacts and consequently the risk of disease transmission among people. As the *Plasmodium* malaria parasites that infect humans are not infective to livestock, it has since long been proposed that animals could be used to divert the malaria vector biting from humans, a control intervention known as zooprophylaxis [2,3]. However, despite the large number of studies performed worldwide for over a century to try to assess the value of this strategy in the fight against malaria (reviewed in [4,5,6,7,8,9]), the available evidence is still contradictory and no consensus exists on the prophylactic effect of animals. Indeed, although in several situations the presence of livestock has been referred to as a protective factor for malaria vector-human contact and/or disease, such as in Papua New Guinea [10,11] and Sri Lanka [12], the opposite has been reported in various other studies, where livestock were shown to be a risk factor, such as Pakistan [13,14], Philippines [15,16], and Ethiopia [17,18] (throughout this work the term livestock is used to refer to cattle and other domestic large and small ruminants - buffalos, sheep, goats -, as well as donkeys, horses, and swine).

The apparently contradictory outcomes of the numerous studies conducted result from a combination of several possible effects of livestock on malaria. On one hand, livestock may divert the blood-seeking mosquito vectors from humans, thereby decreasing the biting on people [10,11,19] and, as a result, decreasing the transmission of the malaria parasite [20] and preventing its amplification in people (i.e. the basis for the zooprophylaxis concept). But on the other hand, livestock can provide additional blood-sources and/or larval breeding sites [21,22,23,24,25], which can increase vector survival and/or density [26], consequently increasing the probability of the vector surviving the parasite extrinsic incubation period and becoming infectious, as well as increasing biting on people [2,6,27]. Additionally, livestock may attract more mosquitoes, which, once in the vicinity of the human dwellings, may end up biting humans rather than animals [14,15,16,18]. The resulting net impact of livestock on malaria risk therefore depends on the relative contribution of each of those effects.

In areas where the presence of livestock near people increases malaria transmission, an apparently simple solution could be to change livestock management in order to deploy the animals away from people's houses, between village and vector breeding site [7]. However, in Pakistan as well as in some Ethiopian regions, for instance, this is not likely to be a feasible strategy, given that livestock are such an important source of household income that people prefer to keep the animals near their houses to prevent them from being stolen [13,28,29,30] and to facilitate husbandry practices, such as milking the lactating animals. An alternative solution has therefore been proposed: target the non-human host of the zoophilic mosquito, by treating livestock with insecticides/acaricides [13] (hereafter referred globally as 'insecticides' for simplicity). This strategy has since long been effectively used to control ectoparasites and the diseases they transmit to animals (and often also to humans), as well as to reduce the direct economic losses they cause due to decrease in productivity (e.g. lower efficiency of feed conversion, weight gain and milk production) [31]. Namely, insecticide treatment of livestock has been applied against tsetse flies transmitted animal and human trypanosomiasis in sub-Saharan Africa [32,33,34,35], tick-borne diseases worldwide (such as anaplasmosis, babesiosis, theileriosis) [33,36], and a variety of other biting and/or nuisance flies [37,38], mosquitoes [38,39], biting midges [40], mites, and lice.

The effectiveness of insecticide-treated livestock (ITL) against malaria was successfully tested by a community-randomised trial in Pakistan [41], where the main vectors, An. stephensi and An. culicifacies, are highly zoophilic [42]. Notably, following the treatment of virtually all domestic animals (93% of the population of cattle, sheep and goats) with a solution of the pyrethroid deltamethrin applied by sponging, Plasmodium falciparum malaria incidence decreased by 56% (95% CI 14%–78%), and prevalence decreased by 54% (95% CI 30–69%). Moreover, efficacy was comparable to that of traditional indoor insecticide spraying but with 80% less costs. Livestock previously infested with ectoparasites also improved in weight and milk yield productivity, enhancing community uptake of the programme [41]. Additional studies have followed to explore whether this strategy could also be applied in sub-Saharan Africa, for integrated control of malaria and animal trypanosomiasis and tick-borne diseases. Notably, bioassays of deltamethrin applied by spot-on and by spray have been conducted in Ethiopia [43], and in Tanzania [44], respectively, to assess the effects of ITL on the mortality and behaviour of malaria vectors. However, despite the encouraging results from these bioassays, the impact of ITL on malaria transmission at the community level is yet to be formally assessed in Africa, where the disease burden is the greatest, but the dynamics and determinants of infection differ from Asia.

A possible concern with ITL is repellency of mosquitoes, which may increase vector feeding on untreated livestock or unprotected humans, and make the intervention detrimental. It is known that certain insecticides exert not only (1) a toxic or direct insecticidal effect, killing mosquitoes that contact with an insecticide-impregnated surface, but also (2) behavioural avoidance responses. These sub-lethal behavioural effects include a) contact-mediated irritancy, inhibiting mosquitoes from remaining on the treated surface, thereby stimulating them to exit prematurely (common with pyrethroid insecticides), and b) non-contact or spatial repellency, which acts from a distance of the treated surface inhibiting mosquitoes from entering treated areas [45,46]. Hereafter, the latter two responses will be referred together as repellency, since any of them could cause mosquitoes diversion to another host, in analogy with the shift in host feeding from humans to domestic animals that has occasionally been associated with the use of pyrethroid-treated nets [47,48,49,50,51]. Additionally, a case-control study in the Pokot territory of Kenya and Uganda [52] found that people with ITL had a higher risk of Visceral Leishmaniasis, suggesting that the insecticide might have repelled sandflies attempting to feed on animals and diverted them to feed on humans. Although, to the best of our knowledge, such behavioural shift has not been reported for ITL and anopheline mosquitoes, the possibility of it occurring should not be disregarded and is therefore important to investigate, particularly because the most promising insecticides tested on livestock to target malaria vectors have been pyrethroids [19,39,43,44,53,54,55]. The popularity of pyrethroids is due to their high insecticidal action associated with low mammalian toxicity [56,57] which makes them safe for both the treated animals and for the consumers of animal products.

An additional concern with using ITL against malaria in Africa is that, even in areas where the moderately zoophilic An. arabiensis vector (which can easily feed on humans or livestock, depending on host abundance and accessibility) predominates over more anthropophilic vectors such as An. gambiae s.s., the ITL intervention is still likely to achieve a smaller reduction in malaria transmission than in Pakistan (and other areas of South Asia), where the vectors are highly zoophilic, taking most of their bloodmeals upon livestock. A possible way to overcome this problem could be to artificially increase the attractiveness of insecticide-treated animals to the malaria vector. Although such has not been tested in the field yet, the use of synthetic attractants to lure anopheline vectors towards baits or traps and away from humans is an area of increasing research [58,59].

This work aims to clarify the different effects of livestock on malaria and to understand under what circumstances livestock-based interventions could play a role in malaria control programmes. This was achieved by, firstly, developing a mathematical model that predicts the apparently contradictory outcomes that have been associated with the presence of untreated livestock in different ecological settings, and secondly, by expanding the model to incorporate insecticide treatment of livestock and fitting it to data from Pakistan (where the ITL trial was performed [41]) and from Ethiopia (where a field study was conducted [9]) to investigate the potential and limitations of ITL. We focus on livestock-based interventions, without comparing their effect with other malaria control interventions, such as insecticide-treated bednets and indoor spraying with residual insecticides. The model characterizes situations where livestock by itself can lead to a decrease, increase, or no net impact on malaria transmission to humans, and it further indicates that treating livestock with insecticide can be a useful complementary tool to control malaria, not only in Asia, but also in sub-Saharan Africa.

Materials and Methods

Malaria model

A mathematical model for the transmission dynamics of human malaria was developed based on the Ross and Macdonald models [60,61], where humans are compartmentalized into either susceptible (uninfected and not immune), or infected/infectious (SIS model), and mosquito vectors are divided into susceptible (uninfected and not immune), exposed/latent (have been infected but are not yet infectious) or infectious (SEI model). Here, the Ross-Macdonald model is extended by discriminating the feeding behaviour of the vector on its alternative hosts: livestock and human populations, and by incorporating the treatment of livestock with insecticide as a potential *novel* method to control human malaria. The new model explicitly incorporates the effects of untreated and insecticide treated livestock on the vector population feeding behaviour, mortality and population density, allowing exploration of the impact of livestock-based interventions on malaria transmission dynamics. A diagrammatic flow chart of the model is presented in Figure 1. Throughout the article, the human, vector and livestock populations will be referred to with the subscripts h, v and l, respectively.

The model is formally represented by a system of ordinary differential equations as follows. For the dynamics of infection in the human population, we have

$$\frac{dS_h}{dt} = -\left(aqb\frac{I_v}{N_h}\right)S_h + rI_h, \tag{1}$$

$$\frac{dI_h}{dt} = \left(aqb\frac{I_v}{N_h}\right)S_h - rI_h,$$

where $N_h = S_h + I_h$ (total human population). Transmission of infection from vectors to humans depends on the number of infected vectors per human, I_v/N_h, the vector blood feeding rate on any host, a, (the interval between bloodmeals on any host is $1/a$), the proportion q of feeds taken on humans (so-called human blood index - HBI), the probability b that a human will become infected following the bite of an infectious vector, and the number of susceptible hosts, S_h. Once susceptible humans are infected the parasite undergoes a period of latency before infective gametocytes appear, but as this period is short compared to the duration of infection, it is not represented explicitly in the model [62]. Infected individuals, I_h, recover from infection at a rate r, eventually becoming fully susceptible to re-infection (the average duration of infection is $1/r$). It is therefore assumed that there is no boosting immunity due to repeated infections, as done for simplification in earlier zooprophylaxis models [27,63,64,65,66]. Human natural mortality and reproductive rates are omitted from the model because humans have a long life expectancy relative to other time periods used in the model (such as the latent period, infectious period and vector life span). We also assume no disease-induced death and therefore, the human population size remains constant.

The disease dynamics in the vector population is represented by

$$\frac{dS_v}{dt} = \rho N_v - \left(aqc\frac{I_h}{N_h} + \mu\right)S_v,$$

$$\frac{dL_v}{dt} = \left(aqc\frac{I_h}{N_h}\right)S_v - (\omega + \mu)L_v, \tag{2}$$

$$\frac{dI_v}{dt} = \omega L_v - \mu I_v,$$

where $N_v = S_v + L_v + I_v$ (total vector population). The vector population comprises only adult female anopheline mosquitoes, since males do not blood feed. Transmission of infection from

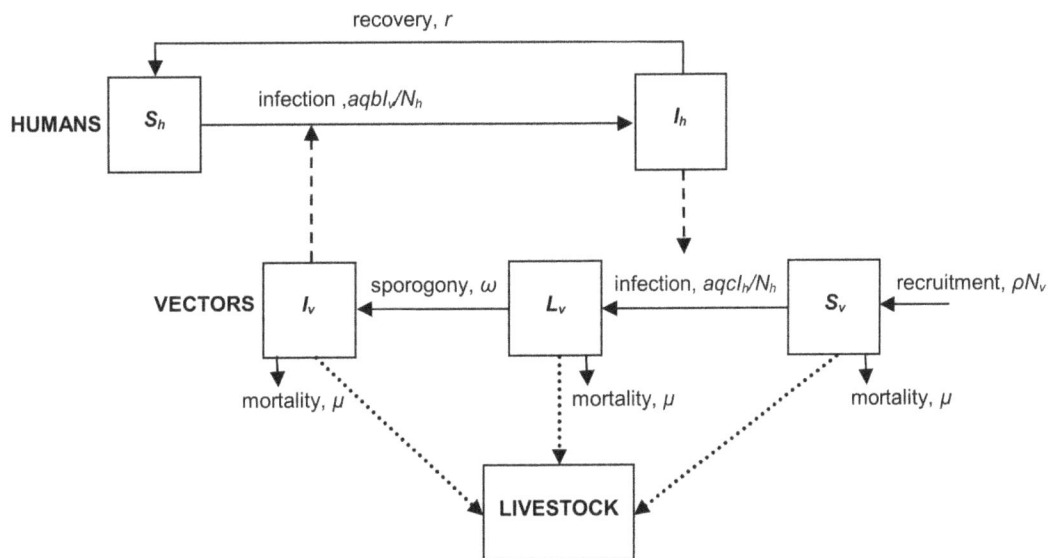

Figure 1. Schematic representation of the malaria model. Horizontal solid lines denote transitions between epidemiological states, and dashed lines represent transmission of infection between human hosts and mosquito vectors. Dotted lines denote vectors feeding on livestock. The vector population consists of adult female anopheline mosquitoes.

humans to vectors depends on the proportion of infectious humans, I_h/N_h, the vector feeding rate on humans, aq, and the probability c that a vector will become infected after feeding upon an infectious human. Infected latent mosquitoes, L_v, become infectious after a sporozoite maturation period (latent period $= 1/\omega$). Anopheline vectors are assumed to remain infectious throughout their life, as usually observed. Infection is assumed to have no impact on vector feeding behaviour, reproduction, nor mortality, as in most malaria models. Although some effects from infection have been described [67], they are not considered in the approximation adopted here.

The vector life expectancy is often about the same order of magnitude as the latent period in the vector. Consequently, only a minority of the infected vector population survives to become infectious, and therefore the model must incorporate the class of latent vectors as well as vector mortality and recruitment. The mortality rate of adult vectors, μ, is assumed to be age independent, such that the average vector life-span is $1/\mu$. We consider two implementations for the recruitment rate. One where the vector population is kept constant by assuming that recruitment and mortality rates are equal $\rho = \mu$.

Another where the density-dependent regulation of the adult vector population due to competition within the larval stages, which depends on the abundance and extent of breeding sites, is explicitly modelled. Following Lord et al. [68] and Kawaguchi et al. [66], the recruitment rate of newly emerged female adults entering the susceptible class is given by

$$\rho = \rho_0 - \rho_s N_v \text{ and } \rho_s = \frac{\rho_0}{K},$$

$$\Leftrightarrow \rho = \rho_0 \left(1 - \frac{N_v}{K}\right),$$

where ρ and ρ_s are the vector recruitment rate in the absence of density-dependence constraints and the strength of the density-dependence in recruitment, respectively, and K is the vector-carrying capacity of the ecosystem. It is assumed that the number and capacity of the breeding sites (and therefore, the vector-carrying capacity) remain the same independently of the hosts' abundance and availability. For instance, the potential increase in breeding sites due to livestock hoof prints is not considered here. While density dependence is essential for the systematic investigation of the zooprophylactic effects of livestock populations with different sizes and characteristics, this is no longer a focus on the investigation of insecticide treatment later on. As such, variable vector population is primarily used in the first part of the Results (Untreated Livestock) and the constant vector population is the implementation of choice throughout the second part (Insecticide Treated Livestock).

As done for simplification in previous malaria models, we assume that vectors take one bloodmeal per gonotrophic cycle, and therefore, the interval between bloodmeals corresponds to the length of the gonotrophic cycle. Similarly, female mosquitoes are assumed to feed homogenously with a fixed preference for humans and/or animals.

Formulation of livestock effects

In the absence of insecticide treatment, the effects of livestock on the human blood index, q, follow what has been proposed by Sota and Mogi [63] and are defined by:

$$q = \frac{N_h A_h}{N_h A_h + N_l A_l},$$

which can be simplified to:

$$q = \frac{1}{1 + \dfrac{N_l}{N_h}\dfrac{A_l}{A_h}},$$

where A_h and A_l are the proportional availabilities of the human and livestock hosts, respectively, and can take any value between 0 and 1, inclusive. The term availability encompasses all the factors that can influence the likelihood of the vector feeding on a given type of host, when two types of alternative hosts are present in equal numbers. Namely, these factors include the accessibility of each host to the vector (which can vary with distance between vector breeding sites and location of humans/livestock at night, whether located indoors or outdoors, under a bednet or not, or livestock enclosed inside a shed or not), and on the intrinsic propensity to feed upon humans *versus* animals (anthropophily *versus* zoophily), and to feed in the location where the host resides (endophagy *versus* exophagy), which can be modified by vector genetics and learning. In the presence of insecticide treatment, the expression for the human blood index is generalized as

$$q = \frac{1}{1 + \dfrac{N_l}{N_h}\dfrac{A_l(1-\varepsilon\alpha)}{A_h}}, \tag{3}$$

where ε is the proportion of livestock population treated with insecticide, hereafter referred as treatment coverage, and α is the diversion probability, defined as the probability that a host-seeking mosquito will be diverted away from ($\alpha > 0$, repellency) or towards ($\alpha < 0$, attractancy) an insecticide-treated animal. Therefore, the insecticide treatment of livestock only affects the human blood index if the intervention has some diversion effect upon the vectors, either repellency or attractancy (see **Text S1.1** for more details).

The baseline mortality rate of the vector is decomposed as being the sum of the minimum mortality rate (μ_m) due to causes other than searching for a bloodmeal host (i.e. mortality due to hazards during the act of feeding on a host, the gestation period, the search for oviposition sites, and the underlying aging process), and the mortality due to searching for a bloodmeal host (μ_s). The search-related mortality is assumed to be proportional to the length of the searching period, which is inversely related to the abundance and availability of potential blood meal hosts. These assumptions follow previous models by Saul [27] and Killeen and Smith [65].

When no livestock are treated with insecticide, the expression for the vector mortality rate therefore becomes

$$\mu = \mu_m + \left(\frac{1}{(N_h A_h + N_l A_l)j}\right)a,$$

where the last term is the search-related mortality, μ_s. Parameter j is a factor to scale the proportional availabilities (A_h, A_l) of hosts to the mosquito vectors into absolute availability values. As in most previous malaria models, it is assumed that the feeding success of malaria vectors is independent of the density of vectors per available host [69].

When livestock are treated with insecticide the mortality rate is generalized as

$$\mu = \mu_m + \mu_s + \mu_k$$

$$= \mu_m + \left(\frac{1}{(N_h A_h + (1 - \epsilon\alpha)N_l A_l)j} \right)a + \left(\frac{\epsilon(1 - \alpha)N_l A_l}{N_h A_h + N_l A_l}k \right)a. \quad (4)$$

In the μ_s term, as seen for the human blood index, if there is repellency $(\alpha > 0)$ it is as if the availability of livestock became reduced by the proportion $\epsilon\alpha$, which corresponds to the proportion of bites attempted on a given animal that will be diverted to another animal or human host. This will cause an increase in the time it takes for the vector to find a bloodmeal host, with consequent increase in the search-related vector mortality. Conversely, if there is attractancy $(\alpha < 0)$, it is as if the availability of livestock became increased by the proportion $\epsilon\alpha$, which corresponds to the proportion of bites attempted on a given animal that were diverted from another animal or human host. This will decrease the time for the vector to find a bloodmeal host, thereby decreasing the search-related vector mortality. The μ_k term accounts for the direct lethal effect of insecticide applied on livestock, and is a function of the vector biting rate on livestock, the treatment coverage, ϵ, the diversion probability, α, and the insecticidal probability, k. The daily biting rate, a, needs to be included in the expressions for μ_s and μ_k, since the additional mortalities, either due to searching for a bloodmeal host or due to attempting to feed on insecticide-treated livestock, are only suffered by the vector when it attempts to blood feed (see **Text S1.2** for more details). Our model assumes that the insecticide effects (diversion and insecticidal probabilities) are constant, therefore reflecting average values of what would be observed throughout the year.

Simulations

The system of equations (1)–(2) was analysed symbolically for the derivation of endemic equilibrium solutions (see **Text S2**) and numerically for the simulation of dynamical trajectories over time. Numerical integration was performed using BERKELEY MA-DONNA v. 8.3.9, with the built-in method *fourth order Runge-Kutta*. The equilibrium solutions were further explored with MATLAB v. R2011a.

We first investigate the effects of untreated livestock in malaria transmission and then move to explore the impact of treating livestock with an insecticide that has lethal and possible diversionary effects (repellency or attractancy) upon malaria vectors. For this purpose, a range of simulations was performed with system (1)–(2), focusing on scenarios of endemic *Plasmodium falciparum* malaria.

Threshold derivation

We also determined the threshold conditions required for persistence of malaria, by analyzing the equilibria of the model represented by system (1)–(2). The average number of secondary cases generated by a single infectious individual introduced in a population of fully susceptible individuals, is known as the basic reproduction number, denoted by R_0 [61,70]. This threshold quantity expresses the transmission potential of an infectious disease and must exceed unity for the infection to be maintained in the population. The expression for R_0 was derived by linearization around the disease-free equilibrium (DFE), based on the next-generation operator approach [71,72]. We then explored the impact of ITL on R_0 for different intervention scenarios. Namely,

by setting $R_0 = 1$, we obtained the critical proportion of the livestock population that must be treated with insecticide, and assessed how this critical coverage would be affected by the insecticide diversionary properties.

Parameterization

Parameters values for the untreated livestock model were obtained directly or derived from the literature and are provided in Table 1. The effects of insecticide treatment were explored using parameter values that were either extracted or derived from empirical data from the index studies in the North-West Frontier Province of Pakistan (ITL trial conducted by Rowland et al. [41]) and in the Konso district of South-West Ethiopia (field study by Franco [9]), or from previous studies within or near the area of the index studies, as listed in Table 2. See **Text S3** for details on parameterization.

Results

Untreated livestock

Here we explore the effects that varying the abundance and/or availability untreated livestock could have on different outcome measures of malaria transmission. All the simulations used parameter values as listed in Table 1, unless otherwise specified.

Figure 2 shows the vector population density and prevalence of human infection over time as livestock are introduced in a setting where previously only humans and no livestock were present. Simulations were performed assuming a fix human density $(N_h = 100)$ and 1 head of livestock per person $(\theta_{Nl} = 0.25)$. The proportional availability of livestock to vectors was the same as that of humans for all plots in this figure $(A_l = 0.5)$, illustrating the case of a moderately zoophilic vector, like *An. arabiensis* in Ethiopia. Additional scenarios of host density and availability were also explored.

Firstly, we simulate a modified model with the best case scenario where the vector population is kept constant $(N_v(t) = 1000)$ by assuming that recruitment and mortality rates are the same $(\rho = \mu)$ (black line), and secondly, the carrying capacity was set to a higher level $(K = 5,000$ to $100,000)$ and the vector population density increased from its initial equilibrium $(N_v(0) = 1000)$ towards carrying capacity (coloured lines). As we would expect, in the case of constant vector density, the introduction of livestock leads to consistent reductions in the prevalence of human cases by diverting vector feeds to livestock (black line). Overall, the higher the numbers and/or availability of the introduced livestock, the stronger is the predicted zooprophylactic effect on malaria transmission. When the vector population density is allowed to increase, however, the prevalence of human cases might increase (coloured lines). For a given density and availability of livestock, the higher the carrying capacity is in relation to the initial vector population density, the higher the vector density and consequently malaria transmission levels in the new endemic equilibrium. In all simulated scenarios in **Figure 2** the system reaches a new equilibrium in less than 3 years after the introduction of livestock.

Figure 3 examines various outcome measures of malaria transmission that characterize the new endemic equilibrium that is reached under a range of relative livestock to human density $(\theta_{Nl}$ varying from 0 to 1), when the proportional availability of livestock to vectors is either the same $(A_l = 0.5)$ or nine times higher $(A_l = 0.9)$ as that of humans, the latter resembling a scenario of a highly zoophilic vector, $\sim$ *An. culicifacies* in Pakistan. The outcome measures investigated include the human blood index (HBI, designated as q in our model), daily overall vector mortality (μ), vector density (N_v), daily entomological inoculation rate (EIR),

Table 1. Parameter values for modelling the effects of untreated livestock on malaria.

Symbol	Definition	Value	[Reference]
a	Vector daily biting rate on any host	0.5	[89]
b	Probability that humans become infected from the bite of an infectious vector	0.04	[90]
c	Probability that vectors become infected after biting on an infectious human	0.3	[90]
r	Human daily recovery rate from infection (1/average duration of infection)	0.05	[29,30,90]
ω	Daily rate at which infected mosquitoes become infectious (1/latent period)	0.07	[91]
μ	Overall average vector daily mortality rate ($\mu_m + \mu_s$)	Varied	Derived
μ_h	Vector daily mortality rate in absence of available livestock	0.1	[89]
μ_m	Vector daily minimum mortality rate when there are no hazards due to search for a bloodmeal host	0.05**	[89]
μ_s	Vector daily mortality rate due to searching for a bloodmeal host	Varied**	Derived
ρ	Overall average vector daily recruitment rate	Varied	Derived*
ρ_0	Vector daily recruitment rate in the absence of density-dependence constraints	Varied	Derived*
ρ_s	Strength of the density-dependence in recruitment (/day)	Varied	Derived*
K	Carrying capacity of the vector population (/ha)	10^3 to 10^5	-
$N_v(0)$	Initial vector density, prior to change in livestock abundance and/or availability (/ha)	10^3	-
N_h	Human density (/ha)	100	-
θ_{Nl}	Relative density of livestock:humans (N_l/N_h)	0 to 20	-
A_l	Proportional availability of livestock to vectors	0 to 1	-
A_h	Proportional availability of humans to vectors ($= 1-A_l$)	0 to 1	-
q	Proportion of vector bloodmeals on humans (Human Blood Index)	0 to 1	Derived
j	Scaling factor to transform proportional availabilities into absolute availabilities	Varied	Derived

*For simulations with constant vector population density: $\rho = \mu$; for variable vector density: $\rho = \rho_0(1 - N_v/K)$ and $\rho_0 = \mu_h K/(K - N_v(0))$.
**The relative magnitudes of μ_s and μ_m were varied in a sensitivity analysis.

and prevalence of infection in humans (I_h). The EIR is the number of infective mosquito bites received by a human per unit time, estimated multiplying the daily human-biting rate (HBR) by the proportion of mosquitoes with sporozoites in their salivary glands (I_v/N_v). The HBR is the total number of mosquito bites received by a human, per day, and is calculated as the product of the number of vectors per human and the number of daily bites on humans per vector (HBR $= (N_v/N_h)a$HBI). The figure illustrates the effects of livestock on decreasing the human blood index while decreasing vector mortality (**Figure 3A,B**) and increasing vector population density (**Figure 3C,D**). The combination of these effects may lead to situations where the presence of livestock increases, decreases, or has no significant impact on malaria transmission (all other panels in **Figure 3**). The introduction of livestock is predicted to have a zooprophylactic effect, i.e. decrease malaria transmission, in two situations. One is that the vector population density does not increase as a result of livestock introduction. The other is that although the vector population density increases as a result of livestock introduction, the livestock numbers and availability to vectors are sufficiently high, such that the increase in vector density is counteracted by the diversion of bites from humans to animals (**Figure 3**). Otherwise, the introduction of livestock is predicted to increase malaria transmission.

Impact of vector search-related mortality on the effects of untreated livestock. For the purpose of illustrating the model behaviour, the simulations for untreated livestock assume that the vector-search mortality when no livestock are available (i.e. when $N_l = 0$ or $A_l = 0$), has the same value as the vector minimum mortality rate ($\mu_s = \mu_m = \mu_h 0.5 = 0.05$/day). A sensitivity analysis was done to explore the impact of different relative magnitudes of

the vector search-related mortality (**Figure S1** in **Text S4.1**). If the vector search-related mortality is already negligible before livestock are introduced, then introducing livestock will have no impact on the vector mortality, and will simply decrease HBI, consequently decreasing malaria transmission. Conversely, if the vector search-related mortality is considerable, introducing livestock can considerably decrease vector mortality, which can increase the proportion of vectors surviving the extrinsic incubation period to become infectious, and thereby counteracting the decrease of the HBI due to diversion of mosquito bites from humans to livestock, consequently increasing malaria transmission. After a certain threshold of livestock density, further increasing their abundance produces negligible reduction on vector mortality.

Insecticide treated livestock

To explore the effects of ITL on malaria the model was fitted to *P. falciparum* malaria transmitted by the highly zoophilic *An. culicifacies* in Pakistan and the more anthropophilic *An. arabiensis* in Ethiopia. Parameter values are listed in Table 2. The main differences in the malaria transmission parameters between the Asian and African settings are as follows. In Ethiopia, livestock were 8.1 times more abundant, although with an estimated 56.8 times lower availability to the main malaria vector, than in Pakistan, resulting in a predicted HBI over 4 times higher in the African than in the Asian setting. Additionally, the estimated duration of the latent period in vectors was slightly shorter, while the vector life expectancy was 75% higher in Ethiopia than in Pakistan. The initial density of vectors per human and the probability of infection in vectors were set to be, respectively, 3.3 and 13.6 times higher in Pakistan than in Ethiopia.

Table 2. Parameter values for modelling the effects of insecticide-treated livestock on malaria.

Symbol	Definition	Value		[Reference]	
		Pakistan	Ethiopia	Pakistan	Ethiopia
a	Vector daily biting rate on any host (1/gonotrophic cycle)	0.4	0.4	[92]	[93,94]
b	Probability that humans become infected from the bite of an infectious vector	0.5	0.5	[95,96]	[95,96]
c	Probability that vectors become infected after biting on an infectious human	0.95	0.07	**	**
r	Human daily recovery rate from infection (1/average duration of infection)	0.05	0.05	***	***
ω	Daily rate at which infected mosquitoes become infectious (1/latent period)	0.057	0.064	Derived from [9,91]	Derived from [9,91]
μ	Overall average vector daily mortality rate ($\mu_m + \mu_s + \mu_k$)	Varied	Varied	Derived	Derived
μ_0	Vector daily natural mortality rate in the absence of ITL (1/natural life expectancy)	0.22	0.12	Derived from [41,92]	Derived from [93,94,97]
μ_m	Vector daily minimum mortality rate when there are no hazards due to search for a bloodmeal host (1/vector maximum life expectancy)	0.11****	0.06****	-	-
μ_s	Vector daily mortality due to searching for a bloodmeal host*	0.11****	0.06****	Derived	Derived
μ_k	Vector daily mortality due to the direct lethal effect of insecticide applied on livestock	Varied	Varied	Derived	Derived
ρ	Overall average vector daily recruitment rate	$= \mu$	$= \mu$	-	-
N_v	Vector density (/ha)	5000	1500	**	**
N_h	Human density (/ha)	100	100	-	-
θ_{Nl}	Relative density of livestock:humans (N_l/N_h)	0.14	1.13	[41]	[9]
θ_{Al}	Relative availability of livestock:humans (A_l/A_h)	53.24	0.938	Derived from [98]	Derived from [99]
A_l	Proportional availability of livestock to vectors ($\theta_{Al}/(1+\theta_{Al})$)	0.982	0.484	Derived	Derived
A_h	Proportional availability of humans to vectors ($1-A_l$)	0.018	0.516	Derived	Derived
q	Proportion of vector bloodmeals on humans*	0.118	0.485	Derived	Derived
j	Scaling factor to transform proportional availabilities into absolute availabilities	Varied	Varied	Derived	Derived
ε	Treatment coverage: proportion of livestock population that is treated with insecticide	0 to 1	0 to 1	-	-
k	Insecticidal probability	0.1 (0 to 0.9)	0.1 (0 to 0.9)	Derived from [54]	-
α	Diversion probability ($\alpha > 0$, repellency; $\alpha < 0$, attractancy)	0 to 1	−1 to 1	-	-

Malaria vectors: *An. culicifacies* in Pakistan, *An. arabiensis* in Ethiopia.
*Parameter values pre-intervention that will be affected if livestock are treated with an insecticide with diversion properties.
**Values chosen to produce malaria prevalence similar to the observed in the index study areas.
***M. Rowland unpublished data.
****The relative magnitudes of μ_s and μ_m were varied in a sensitivity analysis.

It was assumed that, prior to the ITL intervention, an endemic equilibrium of malaria transmission had been reached and, as in most previous zooprophylaxis models [27,65,66], vector population density was at its equilibrium level, and remained constant throughout the intervention (i.e. vector recruitment and mortality rates are the same). We therefore consider the scenario where the insecticide has no impact on the overall vector population density. Thus, the beneficial impact of an ITL intervention with a non-diversionary insecticide is assumed to be due only to the decrease on vector survival caused by the toxic insecticidal effect, and consequent reduction in the proportion of vectors that become infectious. When the insecticide additionally has some repellent properties, there is some beneficial effect from increasing the vector search-related mortality, which partially counteracts the increase in vector bloodmeals on humans. Conversely, when there

is attractancy, there is the greater benefit of decreasing the bloodmeals in humans, which counteracts the decrease in vector search mortality.

Impact on malaria prevalence. We started by exploring the predicted impact of ITL on the prevalence of human infection. This is represented in terms of the prevalence ratio (PR), which is defined as the ratio between the prevalence under a given coverage of insecticide-treated livestock (ε) and the prevalence pre-intervention. The proportional reduction on the pre-intervention prevalence is given by 1-(prevalence ratio).

Simulations were initially performed to estimate the coverage of treated livestock (ε) and insecticidal probability (k) required to obtain the 54% reduction in *P. falciparum* prevalence observed in the Pakistan ITL trial, for the Pakistan and the Ethiopian simulated scenarios, assuming the use of an insecticide with no diversion properties ($\alpha = 0$, **Figure 4**). For any given intervention effort, ITL is predicted to cause a stronger reduction in malaria prevalence in Pakistan than in Ethiopia. Nevertheless, the same reduction in prevalence could be achieved in Ethiopia, if using higher treatment coverage and/or a product with stronger or longer lasting insecticidal properties. For instance, for the scenario of $k = 0.1$ (estimated value from Pakistan data, as detailed in **Text S3.2**), the predicted treatment coverage required to obtain the observed reduction in prevalence (PR $= 0.46$) is $\varepsilon = 15\%$ in Pakistan and 25% in Ethiopia (**Figure 4**).

We also investigated whether by increasing the attractiveness of insecticide treated livestock to vectors it would be possible to obtain in Ethiopia the same reduction in prevalence as observed in the Pakistan trial, with the same intervention effort used in the Asian setting (**Figure 5**). To achieve in Ethiopia the same PR $= 0.46$ with similar coverage as predicted for Pakistan ($\varepsilon = 15\%$, for $k = 0.1$, assuming no repellency) would require an attractancy of 20%, while with attractancy of 10% or 30%, the coverage would be approximately 19% or 13%, respectively.

We then explored how the coverage would be affected if the insecticide had a repellency effect upon vectors, and what might be the repellency probability above which the intervention could become deleterious, by causing prevalence to increase above the pre-intervention level. Not surprisingly, the intervention benefits considerably decrease if the insecticide has repellency properties. Considering again the case of the estimated $k = 0.1$ (**Figure 5**), to achieve the observed reduction in prevalence (PR $= 0.46$) with the 93% coverage that was actually applied in the Pakistan trial, the model suggests that a repellency probability of $\sim 17\%$ would need to be acting in Pakistan. If that same repellency level was acting in Ethiopia, the required coverage was predicted to be 60%. For repellency above 17% in Pakistan or above 21% in Ethiopia, the achieved reduction in prevalence is expected to be always smaller than the observed (i.e. the prevalence ratio, PR, would always be >0.46), even if all livestock are treated ($\varepsilon = 1$). The intervention would become deleterious (PR >1) for repellency above 20% in Pakistan and above 28% in Ethiopia (**Figure 5**). The smaller the coverage (for a given k), or the greater the k (for a given coverage), the higher is the repellence threshold above which ITL will start becoming detrimental (PR >1) (**Figure S2** in **Text S4.2**).

Threshold phenomena. The derived basic reproduction number for the malaria model is given by:

$$R_0 = \frac{N_v}{N_h} \frac{(aq)^2 bc}{r\mu} \frac{\omega}{(\omega + \mu)} \quad (5)$$

where q is given by expression (3) and μ is given by expression (4).

By setting $R_0 = 1$ in (5) we see that the critical proportion of the livestock population that must be treated with insecticide in order to interrupt malaria transmission is

$$\varepsilon_c = \frac{\sqrt{r\omega\left(4\frac{N_v}{N_h}(aq')^2bc + r\omega\right)} - r\left(\omega + 2\left(\mu_m + \frac{a}{(N_hA_h + N_lA_l)j}\right)\right)}{2ra(1 - q')k}, (6)$$

where q' is the HBI in the absence of insecticide treatment:

$$q' = \frac{1}{1 + \frac{N_l}{N_h}\frac{A_l}{A_h}}.$$

This expression for ε_c is valid for the best-case scenario regarding repellence and the worst case scenario regarding attractancy, i.e. when using an insecticide without any diversionary effects upon vectors ($\alpha = 0$).

To explore how repellency ($\alpha > 0$) or attractancy ($\alpha < 0$) would impact ε_c numerical simulations were performed (**Figure 6**). The stronger the insecticidal probability (k), the smaller is the critical proportion of treated livestock (ε_c) required to potentially reduce R_0 below unity, for any given α in Ethiopia, and for $\alpha < 0.4$ in Pakistan. In the Asian scenario, for $\alpha \geq 0.4$, above a certain treatment coverage and insecticidal probability, there could be a shift from $R_0 < 1$ to $R_0 > 1$. For instance, for $\alpha = 0.4$ and $k = 0.4$, R_0 becomes less than 1 if coverage is above 36% and below 90%, while for coverage above 90% then R_0 increases to greater than 1. Similarly, for $\alpha = 0.5$ and $k = 0.6$, R_0 is reduced to less than 1 for coverage between 29% and 70%, but for coverage above 70% the R_0 becomes above 1. For any given k, the stronger the repellence, the higher is the critical coverage, while the stronger the attractancy, the lower is the critical coverage. Furthermore, for any given k, with or without repellency, the critical coverage is always higher for Ethiopia than for Pakistan (**Figure 6**).

Impact of vector search-related mortality on the effects of insecticide-treated livestock. The baseline simulations for ITL assume that the background vector search-related mortality (pre-livestock treatment) has the same value as the vector minimum mortality rate ($\mu_s = \mu_m = \mu_0.0.5$). Additional simulations were done to explore the sensitivity of the findings to alternative search-related vector mortality values (**Figure 5** and **Figure 6** can be contrasted with **Figure S3** and **Figure S4** in **Text S4.2.1**, respectively).

Although there is uncertainty about its exact value, the relative magnitude of the background vector search-related mortality will only affect the intervention impact if the insecticide has diversionary properties. Namely, decreases in the background search mortality will counteract the only benefit of repellence (which was an increase on the search-associated vector mortality), and consequently decrease the beneficial impact of an ITL intervention. In general, the smaller the background search-related mortality, the stronger is the detrimental effect of any given repellency probability ($\alpha > 0$) on malaria prevalence or R_0, and consequently, the greater is the coverage required to achieve a given reduction in prevalence or R_0, and the lower is the repellence threshold above which the intervention would become deleterious (and *vice-versa*). For instance, comparing the baseline scenario with the worst-case scenario of null background vector search-related mortality, the repellence threshold would decrease from 20% to 13% in Pakistan and from 28% to 19% in Ethiopia (**Figure S3** in **Text S4**). This relationship becomes however increasingly non-linear with increase in the insecticidal effect (k) of a treatment with repellency, namely in Pakistan (**Figure S4** in **Text S4**). For a given attractancy probability ($\alpha < 0$), the smaller the background vector-search related mortality, the stronger are the intervention benefits, and consequently, the smaller is the coverage required to achieve a given reduction in prevalence or R_0 (**Figure S3** and **Figure S4** in **Text S4**).

Discussion

By combining a mathematical model with field data we have explored the different effects that livestock can have on human malaria in areas where the disease is transmitted by zoophilic vectors, allowing us to understand under which circumstances livestock-based interventions could play a role in malaria control programmes.

Our model predicts that the presence of untreated livestock will have a zooprophylactic effect in two scenarios. One is when vector population density does not increase as a result of livestock introduction. The other is when although the vector population

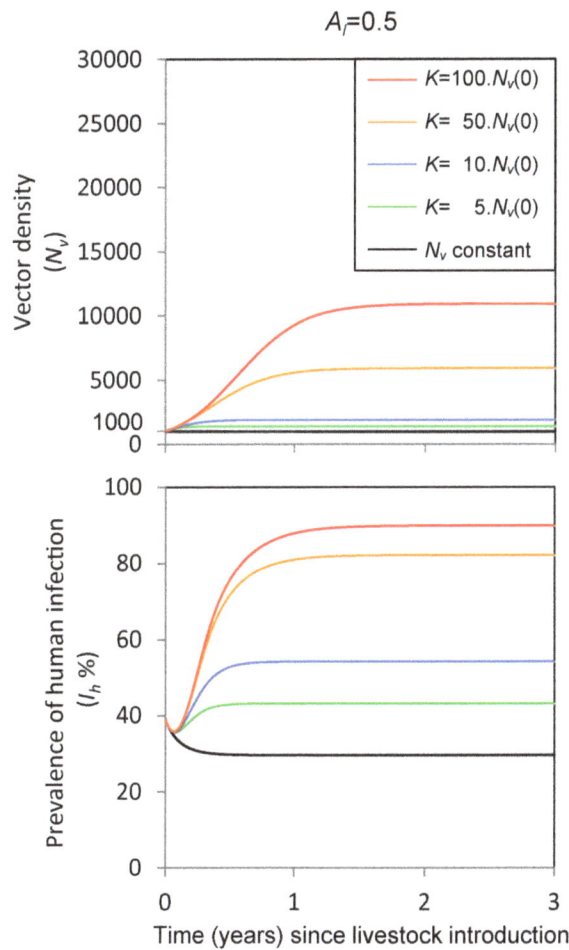

Figure 2. Temporal effect of introducing livestock in a setting with endemic malaria. Effect of introducing livestock in a setting where only humans were present, when: N_v remains constant (black line), and when N_v increases until reaching a maximum, which depends on the carrying capacity, K (increasing from $K = 5,000$ (green) to 100,000 (red)). $N_v(0) = 1000$, $N_h = 100$ and $A_l = 0.5$: the availability of livestock to vectors is the same as that of humans; $\theta_{Nl} = 0.25$ (1 head of livestock per 4 persons). To achieve the same initial equilibrium N_v (and I_h) for various K values, the vector recruitment rate in the absence of density-dependence constraints was set to vary accordingly: $\rho_0 = \mu_h K / (K - N_v(0))$. Other parameters are as in Table 1.

density increases, the numbers and availability of livestock to vectors are sufficiently high (such that the resulting diversion of bites from humans to livestock can counteract the increase in vector density), or the vector mortality related with host-search pre-livestock introduction was sufficient low (such that introducing livestock causes no significant decrease on the already small search mortality). Otherwise, the introduction of livestock is predicted to increase malaria transmission.

These results are in agreement with the insights from two previous zooprophylaxis models [27,63]. Namely, Sota & Mogi [63] also identified as key determinants of the beneficial *versus* detrimental effect of untreated livestock on malaria transmission whether the vector population had reached its maximum possible density prior to livestock introduction, and whether the density and/or availability of animal hosts were sufficiently high. It is worthwhile mentioning that these features are captured by both the present model and the Sota & Mogi [63] model although the

two works differ in the approach used to model the potential detrimental impact of livestock on malaria transmission. The present work explicitly models the effect of animal or human hosts' abundance and availability on vector mortality, with consequent impact on the dynamics and density of adult vectors. Instead, Sota & Mogi [63] assumed a constant vector mortality rate, and modelled the effect of hosts abundance and availability on the probability of successful blood feeding of the vector, with consequent impact on the number of eggs laid and density of adult vectors in the future generations. Aside from the work by Sota & Mogi [63], two other previous zooprophylaxis models that addressed the effect of untreated livestock on malaria transmission [27,65], have also explicitly modelled the effect of animal or human hosts abundance and availability on vector mortality. Saul [27] also highlighted that the effect of untreated livestock on malaria greatly depended on the magnitude of the search-related vector mortality: when this is significant, increase in livestock density could lead to increased malaria transmission, which is consistent with our results.

Regarding the insecticide-treatment of livestock, when using an insecticide without diversionary properties, any given intervention effort is predicted to achieve a stronger reduction in malaria transmission in a setting with highly zoophilic vectors (exemplified by Pakistan) than with the more anthropophilic *An. arabiensis* (illustrated by Ethiopia), as expected. Yet, the same reduction in malaria prevalence could be achieved in Ethiopia, if treating a high proportion of the livestock population with a product that has stronger and/or lost lasting insecticidal effect than what was used in Pakistan. The predicted intervention effort required to achieve a given reduction in prevalence with a non-repellent insecticide, is however, surprisingly low, and most likely unrealistic. In the Pakistan trial, a 54% reduction in prevalence was obtained, following treatment of 93% of the livestock population in the trial villages (cattle, goats, and sheep) [41]. Our results suggest that, to achieve the observed reduction in prevalence with such high treatment coverage, the insecticidal effect would need to be extremely small.

When accounting for a possible repellency effect of the insecticide, the expected benefits of the intervention decrease considerably in both settings, requiring more realistic parameter values to obtain the results observed in the Pakistan trial. Repellency threshold probabilities were identified above which the intervention could become detrimental, increasing the prevalence of human infection above the pre-intervention levels. For repellency probability below those thresholds any vector diversion to humans was predicted to be overcompensated by the insecticidal (direct lethal) effect and the increased search-related mortality of the mosquitos attempting to blood feed on insecticide-treated animals. Within that range of repellency probability for which ITL is likely to still reduce malaria prevalence, a greater benefit may be observed in Pakistan or in Ethiopia, depending on the repellency and coverage levels.

The results indicate that repellency has a stronger detrimental impact on malaria (prevalence or R_0) in Pakistan than in Ethiopia, and therefore, it would take a smaller level of repellency for ITL to start becoming deleterious in the Asian setting. Above the repellency threshold the intervention becomes always more detrimental in settings with higher availability of livestock to vectors, like in Pakistan and other settings with highly zoophilic vectors.

The repellency level of the insecticide applied to animals can thus have an important effect on the intervention outcome. For a given treatment coverage, the stronger and/or longer lasting the insecticidal effect, the higher is the repellency threshold above

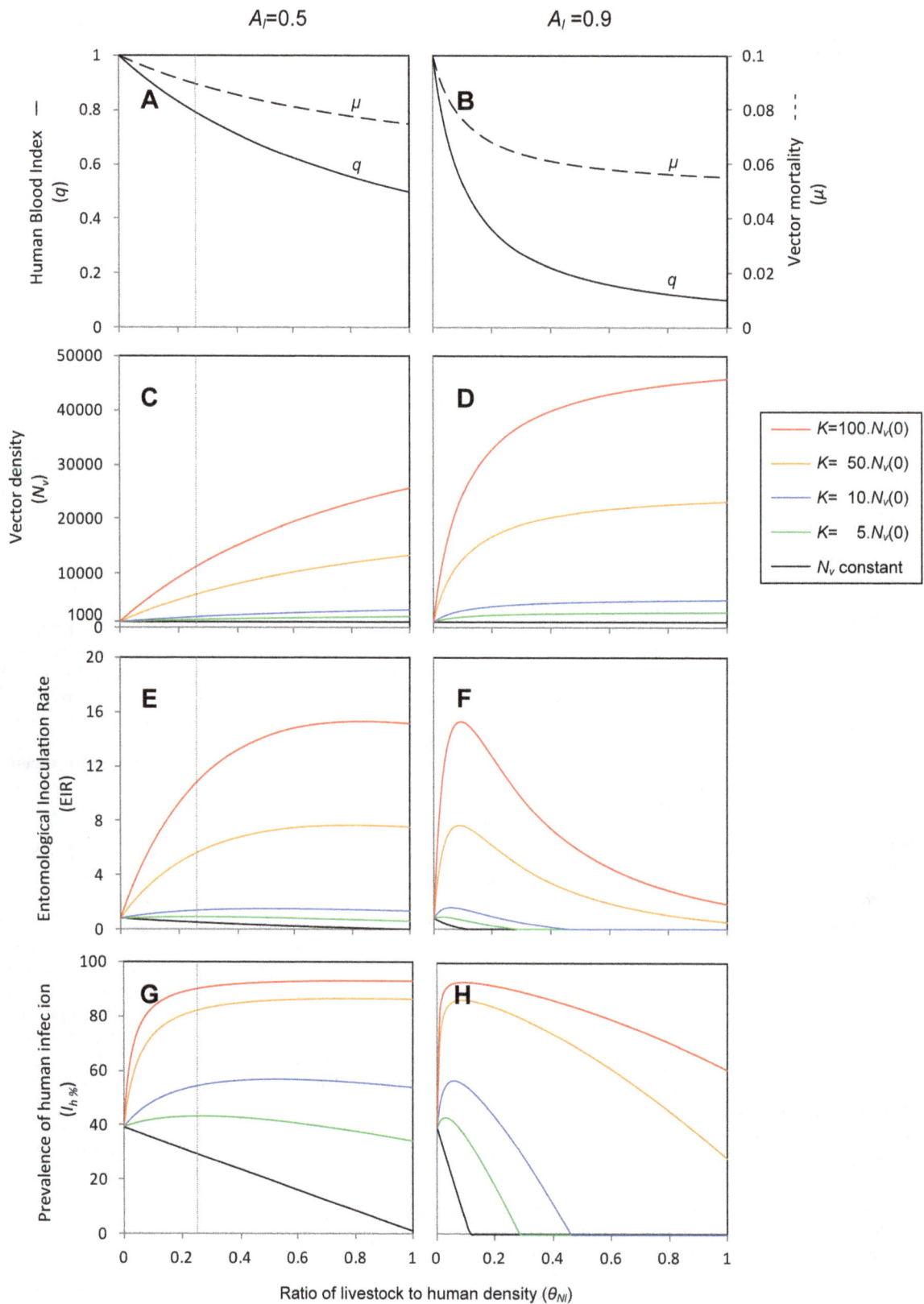

Figure 3. Effect of altering the relative livestock to human density, for different vector density scenarios, at the new endemic equilibrium. Comparing a scenario where the availability of livestock to vectors is the same as that of humans (left, $A_l = 0.5$) *versus* where it is 9 times higher than that of humans (right, $A_l = 0.9$). Along the x-axis, representing $\theta_{Nl} = N_l/N_h$, the livestock density N_l is varied relative to a fixed human density $N_h = 100$. $N_v(0) = 1000$. Effect of introducing livestock when: N_v remains constant (black line), and when N_v increases until reaching a maximum, which depends on the carrying capacity, K (coloured lines: K increasing from 5,000 (green line) to 100,000 (red line)). The effects of introducing livestock on the human blood index (HBI) and on the vector mortality rate (μ) are independent from the vector density scenarios (A, B).

The vertical line in the left panels highlights the new endemic equilibrium that is reached after the introduction of 1 head of livestock per 4 persons ($\theta_{Nl} = 0.25$), corresponding to the end of the timeline in Figure 2. Other parameters are as in Table 1.

which the intervention starts becoming detrimental. Additionally, when considering doing ITL interventions with high treatment coverage of the livestock population, researchers should be aware that the higher the intervention coverage, the greater the detrimental effect from a given repellency level, and the greater the benefits from reducing repellency. A small decrease in repellency could greatly improve the intervention benefits, with the effect being greater in scenarios with more zoophilic mosquitoes. Interestingly, this is the opposite from the case of insecticide-treated nets (ITNs), where repellency could be beneficial in some circumstances. Namely, the greater the proportion of the human population covered with ITNs the greater are the expected benefits from repellency, and, conversely, the smaller the coverage the greater the likelihood that malaria vectors might be diverted from ITN-protected people to those unprotected (if the density and/or availability of animal hosts to the mosquito vector are small) [65].

In general therefore, the smaller the repellency, the greater the benefits of an ITL intervention. The benefits in settings with moderately zoophilic vectors, such as *An. arabiensis* in sub-Saharan Africa, could be further improved by artificially increasing the attractiveness of livestock to the malaria vector.

If the insecticide has diversionary properties upon the malaria vectors, the magnitude of the vector mortality related with host searching was predicted to considerable affect the model results. Namely, the smaller the vector search-related mortality pre-intervention, the stronger are the insecticide diversionary effects upon malaria prevalence or R_0, be it the detrimental effect of a given repellency probability on transmission, or the reduction in transmission obtained with a given attractancy probability. Given the influential role of the vector search-related mortality upon the effects of untreated and insecticide-treated livestock on malaria

transmission, obtaining field estimates for this component of vector mortality is an important challenge that future research should address.

The repellency threshold above which the intervention might become detrimental could be as low as 13% in Pakistan and 19% in Ethiopia, if assuming all livestock population is treated with an average direct insecticidal effect of 10%, under the worst case scenario of null vector search-related mortality. The smaller the treatment coverage, and/or the stronger the insecticidal effect or the search-related mortality, then the higher the repellency level at which ITL can still be safely used.

To our knowledge, this is the first modelling approach that explicitly explores the potential effects of repellency and attractancy in the context of ITL and malaria transmission. The present work is an improvement in relation to previous malaria models of the impact of applying insecticide on animals [27], on animal sheds [66], or on bednets [65]. None of the former two models [27,66], explored a repellent or attractant effect of the insecticide, and although work by Killeen and Smith [65] has looked at repellency and livestock applied to an African setting, it did so in the context of insecticide-treated bednets and diversion of malaria vectors to humans and/or untreated cattle, without referring to insecticide-treatment of cattle.

Considerations on modelling repellency

The present work assumes that when a mosquito tries to bite on an insecticide-treated animal and is repelled, it will be diverted to bite on another host. Nonetheless, it could be that the mosquito is not able to find a successful bloodmeal and does not feed in that night, ending up either feeding only on the following night, or dying earlier. The impact of repellency on vector mortality is

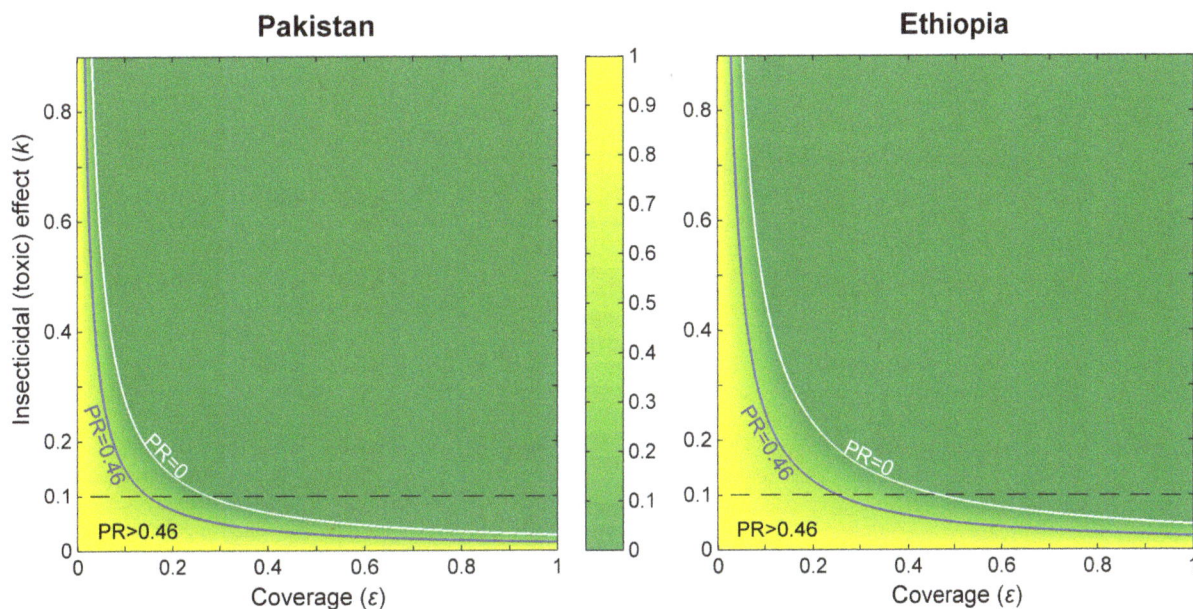

Figure 4. Predicted impact of Insecticide Treatment of Livestock on malaria prevalence, without diversion ($\alpha = 0$). This figure shows the combination of values of coverage and insecticidal probability required to achieve a given prevalence ratio (PR: prevalence with ITL / baseline prevalence). Blue line: PR = 0.46 (like the observed in the Pakistan trial); White line: PR = 0; Dashed line: $k = 0.1$, as estimated for the Pakistan trial. The colour bar shows the scale of PR values, from 0 to 1. Other parameters are as in Table 2.

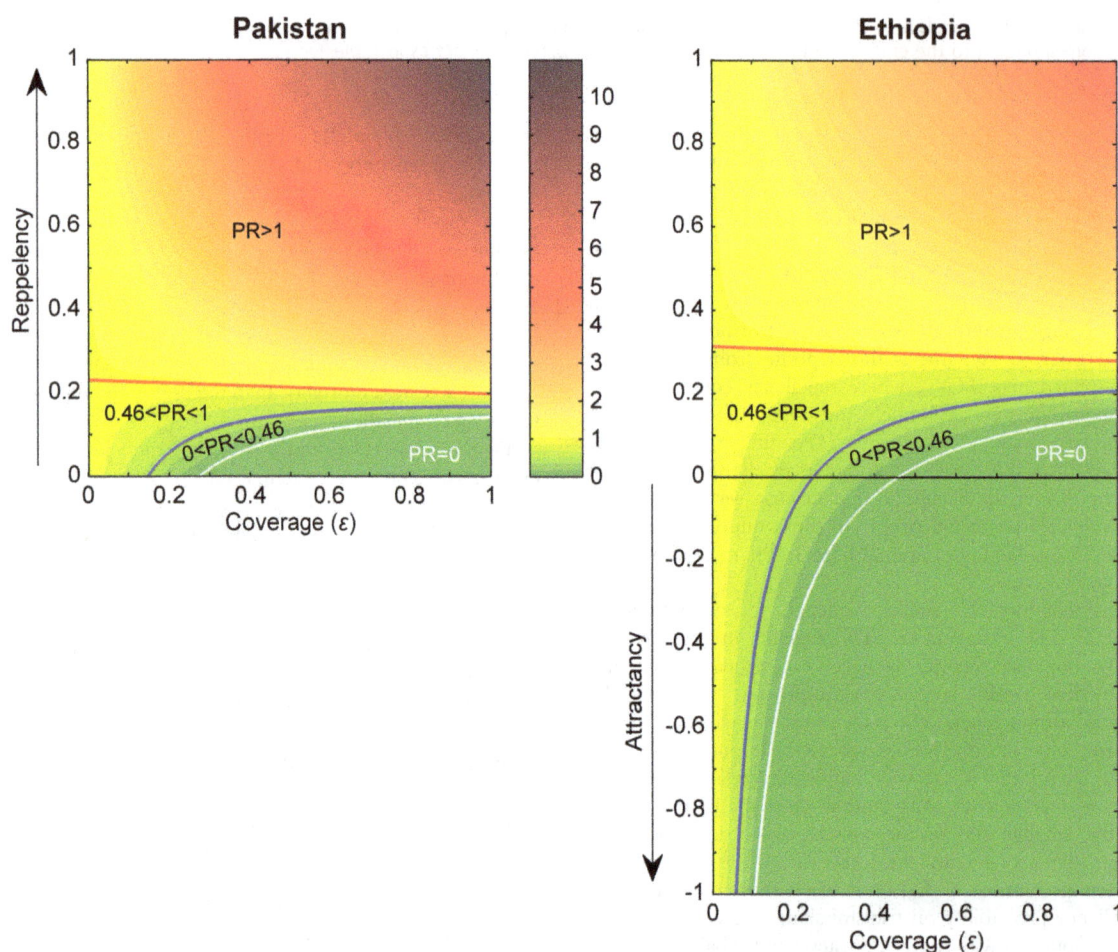

Figure 5. Predicted impact of Insecticide Treatment of Livestock on malaria prevalence – with repellency ($\alpha > 0$) or attractancy ($\alpha < 0$) for $k = 0.1$. This figure shows how the diversionary properties of the insecticide affect the coverage required to achieve a given prevalence ratio (PR: prevalence with ITL / baseline prevalence). Blue line: PR = 0.46 (like the observed in the Pakistan trial); White line: PR = 0; Red line: PR = 1 (above which treating livestock increases malaria prevalence). Along the y axis, α is varying from no diversion ($\alpha = 0$) to maximum repellency ($\alpha = 1$) or maximum attractancy ($\alpha = -1$). The colour bar shows the scale of PR values, from 0 to ≈ 11 in Pakistan and up to ≈ 5 in Ethiopia. Other parameters are as in Table 2.

captured by the model, since repellency reduces the availability of treated livestock, increasing the time required to find a bloodmeal host and consequently increasing the vector search-related mortality. The impact of repellency increasing the interval between bloodmeals is something that could be explored by extending the model to explicitly account for that possibility.

We also assume that the probability of vectors being repelled to humans after attempting to bite on livestock, depends only on the proportion of livestock population that is treated with insecticide (coverage, ε), on the repellency probability of the insecticide ($\alpha > 0$), and on the relative number and availability of livestock or human hosts. Additionally, the model assumes that repellency and coverage are independent. In reality, however, the occurrence of a repellent effect can depend on additional factors such as characteristics of the: a) insecticide (chemical compound, formulation and concentration); b) intervention (concentration of the insecticide on the animal's coat, which will eventually decrease with time after application); and c) mosquito vector [73]. Also, the insecticide concentration is likely to be heterogeneous throughout the animal's surface, and the place where mosquitoes land on the animals can therefore be determinant.

With regards to the mode of action of insecticides applied on livestock depending on the properties of the insecticide itself, some pyrethroids are more toxic to vectors than repellent (e.g. deltamethrin, used in the Pakistan ITL trial), other pyrethroids are more repellent than toxic (e.g. permethrin), and other classes of insecticides (e.g. organophosphates) are just toxic and non-repellent. Yet, even the typical toxic deltamethrin tends to be repellent at low dosages. Namely, as the applied dose of deltamethrin decays over time it goes from being toxic to non-toxic but repellent and then to just repellent.

Due to this, a big concern during the Pakistan ITL trial [41] was that mosquitoes would be repelled onto humans as the dosage of deltamethrin decayed, but it appears malaria was still controlled because the insecticide was reapplied regularly before there was too much decay. This explanation is consistent with the findings from the present work where, on one hand, when accounting for repellency the model results are more compatible with the observed Pakistan trial results, than when assuming that the insecticide had no diversion effect. On the other hand, the predictions suggest that the stronger and/or longer lasting the insecticidal effect, the highest is the repellency threshold above which the intervention is likely to become detrimental. Addition-

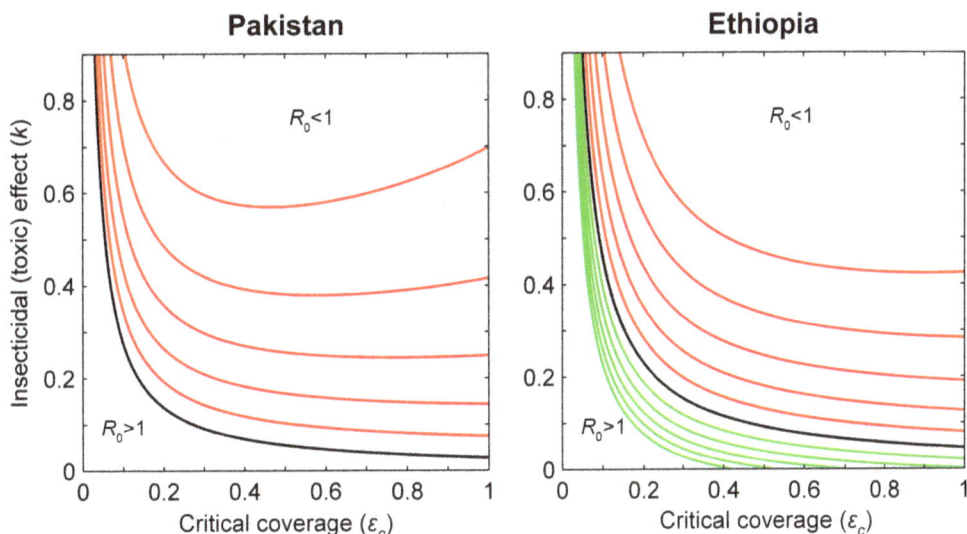

Figure 6. Critical proportion of ITL as a function of the insecticidal (k), and diversionnary effect (α). The lines show the combination of values of coverage and insecticidal probability required to achieve $R_0 = 1$, above which R_0 will be decreased below 1, for a given diversion probability (α). Black line: $\alpha = 0$, no repellency or attractancy (is the same as the white line in Figure 4); Red lines: $\alpha > 0$, repellency increasing from 0.1 to 0.5 (top), at intervals of 0.1; Green lines: $\alpha < 0$, attractancy increasing from -0.1 to -0.5 (bottom), at intervals of 0.1. Other parameters are as in baseline simulations (Table 2).

ally, at the high treatment coverage applied in the trial, there is only a small difference between the repellency level with which the observed reduction in prevalence would be achieved, and the repellency threshold. This supports the hypothesis that the intervention effort applied in the Pakistan trial was sufficiently high to make the repellency effects non evident.

Increasing livestock attractiveness to vectors

By increasing the attractiveness of insecticide treated animals to malaria vectors, it could be possible to further enhance the impact of ITL in malaria control in settings with more opportunistic vectors, such as *An. arabiensis* in Ethiopia, as shown in this work. This could eventually enable extending the geographic regions where ITL might reduce malaria burden, to include also areas with more anthropophilic vectors, such as *An. gambiae s.s.*, the most competent malaria vector in sub-Saharan Africa. Given the potential benefits this could bring, it would be worthwhile further exploring this hypothesis in future work.

In practice, however, insecticides tend to be non-attractant (i.e. neutral or repellent). Therefore, to artificially increase livestock attractiveness would require developing an insecticide that has also attractancy properties (in addition to its toxic insecticidal effect), or alternatively, treat livestock with an attractant substance on top of applying a standard insecticide. Although this may sound somewhat speculative, it is not much different from what has been successfully tested in other systems, where synthetic attractants have been applied to baits or traps to increase their attractiveness to tsetse flies [74,75,76], anopheline mosquitoes [58,59], and other insects of medical and veterinary importance [77].

Regarding possible detrimental implications of artificially attracting more mosquitoes into livestock, these are likely to be minimal. Attracting a mosquito to a cow does not necessarily mean the mosquito will succeed in biting/blood feeding as it may be killed or knocked down by exposure to the insecticide before taking up blood, and that is usually the case, namely with

pyrethroids. The expected reduction in mosquito survival due to increased exposure to the insecticide toxicity should actually lead to less biting. Therefore, it is unlikely there would be additional disease burden or economic costs, as long as the attractancy would be specific for malaria vectors and would not cause increased number of biting flies or other arthropods that are vectors of pathogens to livestock, and would also not cause a reduction in the animal's blood through excessive biting that could decrease milk or meat yield.

Optimizing insecticide-treated livestock interventions

It is important to highlight that, although we explored the impact that treating livestock with insecticides could have on malaria transmission, this intervention has been traditionally used with a veterinary purpose, to control tsetse flies, ticks and other ectoparasites, and the diseases they transmit to animals, improving livestock health and productivity, such as milk and meat yield. Therefore, when evaluating the cost-effectiveness of the intervention, both the animal health benefits and the public health benefits need to be captured ('One health'). Given the potential double side benefits of veterinary interventions like this, and given the central role of livestock in poor tropical settings, to control human disease and improve livestock health will have disproportionate economic impact that needs to be captured, as accounting for it could promote the wider implementation of the intervention. Namely, if the costs of ITL are allocated to the human health and the animal health sectors in proportion to the benefits, the intervention might be profitable and cost-effective for both sectors. Here lies a challenge to the Public Health community, which will require strengthened collaboration with the Animal Health community.

In addition to the animal health/productivity benefits, ITL uses much less insecticide than traditional malaria control methods, such as indoor spraying of houses with residual insecticide, making ITL very cheap from a human disease control perspective. In the Pakistan trial, sponging livestock with deltamethrin was shown to achieve a reduction in malaria burden similar to indoor residual spraying but with 80% less campaign costs. Furthermore, if accounting for the increase in milk production by the treated cattle, associated with clearance of tick infestations, the economic gain would be enough to cover all insecticide and labor costs [41].

The mathematical model developed here could be used to examine the economic aspects of the 'One Health' approach to disease control, encompassing both human and animal health benefits at a societal level. The model provides a framework for quantifying the benefits of ITL as a reduction in the human health burden (expressed as prevented DALYs, Disability Adjusted Life-Years), associated reduction in health care costs (expressed as $), as well as the improvements in animal health and productivity (expressed as $). The cost-effectiveness of ITL, accounting for both the human and veterinary benefits, could then be compared with other interventions that deliver only human health benefits, such as indoor residual spraying (IRS) and insecticide-treated bednets (ITNs), and the relative attractiveness of ITL across epidemiological settings and animal production systems examined.

The use of any animal-based intervention for malaria control will only be a component of the broader integrated malaria control approach, and will have to be deployed alongside case detection, treatment and prevention. The relative importance of animal-based interventions within the broader approach will vary between settings. Also, the adoption of any recommended intervention is intimately related to the socio-economics of the setting and it is therefore vital to understand the drivers for adoption by the target population.

In Pakistan very high treatment coverage was achieved with a free campaign and the animal owners were enthusiastic because they could see the benefits of tick elimination and improved milk and meat yield [41]. Previously to the campaign the insecticide treatment of livestock for ectoparasites was normally *ad hoc* done by householder according to perceived need, which would lead to only partial coverage at any one time. Therefore, a subsidized campaign approach is recommended, similarly to the externally funded campaigns of IRS.

Although the empirical evidence for Africa is lacking, some inferences can be made from tsetse control work. In particular, the experience from controlling human sleeping sickness in southeast Uganda by targeting the cattle reservoir of the human infective parasite shows that large scale campaigns can also reach very high (>80%) coverage levels with insecticide and trypanocidal treatment. Additionally, reducing the volume of insecticide, and so the price of ITL treatment, through restricted application protocols to target insecticide use to those areas of the cattle where tsetse or anopheline mosquitoes preferentially feed, have the potential to drive routine ITL adoption by small-holder farmers [35,43]. To drive the private uptake of ITL usage, farmers need to see a direct benefit to their animals. Experience from the sleeping sickness work shows that for effective control through ITL it is important that the insecticide products used work against both ticks and the human disease vectors (such as synthetic pyrethroids), as tick control is often the main motivation for farmers to use ITL [34].

ITL is particularly useful for malaria control where vectors, in addition to bloodfeeding on livestock, are (or have became) exophagic (feeding outdoors, therefore escaping to ITNs exposure) and/or exophilic (resting outdoors, and thereby evading IRS).

One cannot rule out that long term and intensive use of ITL may lead to selection for anthropophily, with a consequent shift in preference from animals to humans (assuming that host preference is determined by genetic polymorphisms [78,79,80]). Therefore, changes in the HBI (as a proxy for host preference) should be monitored in regions where repeated campaigns are undertaken [14,41]. Additionally, selection for anthropophily could be countered by combining ITL with indoor strategies to control anthropophilic and endophilic mosquitoes, like ITNs and IRS [19].

At the time of the field studies in the settings to which to which the ITL model was parameterized (Konso region of Ethiopia and NWFP in Pakistan) most people were not using bednets. Future work could expand the present model to investigate the use of livestock-based interventions alongside ITNs or IRS, to provide additional insights to the potential impact that combining these strategies might have on malaria transmission.

A concern inherent to any vector control intervention based on insecticides is the potential development of resistance. Namely, pyrethroid resistance is becoming increasingly wide spread across anopheline mosquitoes [81,82] and several other arthropods that feed on livestock, such as ticks [83,84]. It has been argued that the treatment of livestock with pyrethroids is not likely to induce stronger selection pressure for resistance in malaria vectors than insecticide-treated bednets or indoor residual spraying of houses and cattle sheds, but nevertheless, appropriate monitoring of the vector populations is required if wide scale and long term ITL interventions are implemented [14,41,54].

It has also been recommended that research efforts should target the identification of alternative non-pyrethroid insecticides for livestock treatment [54]. Possible candidates have recently been suggested from the avermectins class of insecticides, which have since long been used in veterinary and human medicine against several helminths and arthropod pests [85] and were latest shown to be also toxic to anopheline mosquitoes. Namely, feeding on bovine blood treated with ivermectin reduced survivorship and fecundity of *An. gambiae s.s.* and *An. arabiensis* [86,87] and may possibly also inhibit the sporogony of *P. falciparum* as it was recently shown in treated humans [88]. Another promising avermectin is the more recent eprinomectin which has similar antihelminthic and ectoparasiticidal action as ivermectin in cattle, but with much less mammary excretion, allowing its use in pregnant and lactating animals, on the contrary of ivermectin [87]. Any of these avermectines could overcome the problems of pyrethroid resistance as well as repellency upon malaria vectors, and could be administered as part of mass livestock vaccination campaigns, simultaneously benefiting animal and human populations. Additionally, while pyrethroids can only be administered topically, both ivermectin and eprinomectine are available topically (as pour-on) and also as injectable formulation (subcutaneous administration), which could surmount the difficulty faced with pyrethroids of achieving high enough concentrations of product throughout the animal's skin. Malaria vectors would however need to bite the animal and take a bloodmeal to be exposed to the insecticide, but every biting mosquito would be exposed and die more promptly, therefore requiring a smaller dose, compared to pyrethroids. Further studies are needed to assess the effects of livestock treated with the recommended dose of ivermectin or eprinomectine upon wild populations of malaria vectors.

Conclusions

A mathematical model was developed to predict the different effects of untreated and insecticide-treated livestock in malaria outcomes in different regions. Similarly to previous work, our model indicates that the zooprophylactic effect of untreated livestock depends on whether 1) the pre-existing malaria vector population had reached its maximum density, 2) livestock abundance and availability to the vector is sufficiently high, and 3) vector mortality related with host-search pre-livestock introduction was sufficiently low. We additional find that, as expected, the insecticide-treatment of livestock is likely to be more beneficial to humans in settings with highly zoophilic malaria vectors as in Pakistan and other areas of South Asia, than in settings with moderately zoophilic vectors, as *An. arabiensis* in sub-Saharan African. Nevertheless, the intervention could also substantially decrease malaria burden in the latter settings, under certain conditions, as illustrated here with the predictions for Ethiopia. Namely, in regions with moderately zoophilic vectors the benefits of the intervention will be maximized if 1) treating most of the livestock population with a product that has a stronger or longer lasting toxic insecticidal effect than what was used in the Pakistan trial, and that has little (ideally null) repellency effect (such as the non-pyrethroids ivermectin or eprinomectin), or 2) if the attractiveness of the treated animals to malaria vectors could be increased.

It is hoped that this work may lead to increasing awareness about the non-linear effects of livestock on malaria transmission, and to the implementation of a community-based trial of insecticide-treated livestock in an African region where *An. arabiensis* predominates, and where this strategy could potentially contribute to the integrated control of human malaria and livestock diseases.

Supporting Information

Text S1   Formulation of livestock effects. S1.1. Livestock effects on Human blood index. S1.2. Livestock effects on vector mortality.

Text S2   Malaria model endemic equilibrium solutions.

Text S3   Model parameterization. S3.1. Untreated livestock model parameterization. S3.2. Insecticide-treated livestock model parameterization for Pakistan and Ethiopia.

Text S4   Sensitivity analyses. S4.1. Sensitivity analysis of the effects of untreated livestock. *Figure S1. Sensitivity analysis of the effects of livestock availability/density vs. vector search-related mortality on several malaria outcomes, at the new endemic equilibrium.* S4.2. Sensitivity analysis of the effects of insecticide-treated livestock. *Figure S2. Sensitivity analysis of the effects of repellency/attractancy vs. insecticidal probability (k) on the prevalence ratio.* S4.2.1. Sensitivity analysis of the impact of vector search-related mortality upon the effects of insecticide-treated livestock. *Figure S3. Sensitivity analysis of the effects of repellency/attractancy vs. vector search-mortality on the prevalence ratio. Figure S4. Sensitivity analysis of the effects of repellency/attractancy vs. vector search-mortality on the critical proportion of insecticide-treated livestock.*

Acknowledgments

This work is dedicated to Professor Clive R. Davies, who supervised Ana O. Franco's PhD when this work started but sadly passed away in March 2009. We are grateful to Iñaki Tirados for help with the Ethiopian data, Simon Brooker and Steve Torr for useful discussions and comments on an earlier version of this work, and the Collective Dynamics group at the Instituto Gulbenkian de Ciência (Portugal) for helpful suggestions.

Author Contributions

Conceived and designed the experiments: AOF PGC CRD MGMG. Performed the experiments: AOF. Analyzed the data: AOF. Contributed reagents/materials/analysis tools: AOF MR. Wrote the paper: AOF MGMG PGC MR.

References

1. Zinsstag J, Schelling E, Waltner-Toews D, Tanner M (2011) From "one medicine" to "one health" and systemic approaches to health and well-being. Prev Vet Med 101: 148–156.
2. Escalar G (1933) Applicazione sperimentale della zooprofilassi in Ardea. Rivista di Malariologia 12: 373–380.
3. W.H.O. (1982) Manual on environmental management for mosquito control with special emphasis on mosquito vectors. Offset Publication No. 66. Geneva: World Health Organization. 283 p.
4. Hacket LW (1937) Malaria in Europe. Oxford, UK: Oxford University Press.
5. Brumpt E (1944) Revue critique: Zooprophylaxie du paludisme. Annales de Parasitologie Humaine et Comparée 20.
6. Service MW (1991) Agricultural development and arthropod-borne diseases: a review. Revista de Saúde Pública 25: 165–178.
7. W.H.O. (1991) Joint WHO/FAO/UNEP/UNCHS Panel of Experts on Environmental Management for Vector Control (PEEM): Report on the Ninth (1989) and Tenth (1990) Meetings. Geneva, Switzerland: World Health Organization. WHO–CWS/91.7 WHO–CWS/91.7. 28 p.
8. Bettini S, Romi R (1998) [Zooprophylaxis: old and new problems]. Parassitologia 40: 423–430.
9. Franco AIO (2010) Effects of livestock management and insecticide treatment on the transmission and control of human malaria. PhD thesis. London School of Hygiene and Tropical Medicine, University of London, United Kingdom.
10. Charlwood JD, Dagoro H, Paru R (1985) Blood-feeding and resting behaviour in the *Anopheles punctulatus* Dönitz complex (Diptera: Culicidae) from coastal Papua New Guinea. Bulletin of Entomological Research 75: 463–475.
11. Burkot TR, Dye C, Graves PM (1989) An analysis of some factors determining the sporozoite rates, human blood indexes, and biting rates of members of the *Anopheles punctulatus* complex in Papua New Guinea. The American Journal of Tropical Medicine and Hygiene 40: 229–234.
12. van der Hoek W, Konradsen F, Dijkstra DS, Amerasinghe PH, Amerasinghe FP (1998) Risk factors for malaria: a microepidemiological study in a village in Sri Lanka. Transactions of the Royal Society of Tropical Medicine and Hygiene 92: 265–269.
13. Bouma M, Rowland M (1995) Failure of passive zooprophylaxis: cattle ownership in Pakistan is associated with a higher prevalence of malaria. Transactions of the Royal Society of Tropical Medicine and Hygiene 89: 351–353.
14. Hewitt S, Kamal M, Muhammad N, Rowland M (1994) An entomological investigation of the likely impact of cattle ownership on malaria in an Afghan refugee camp in the North West Frontier Province of Pakistan. Medical and Veterinary Entomology 8: 160–164.
15. Russel PF (1934) Zooprophylaxis failure. An experiment in the Philippines. Rivista di Malariologia 13: 610–616.
16. Schultz GW (1989) Animal influence on man-biting rates at a malarious site in Palawan, Philippines. The Southeast Asian Journal of Tropical Medicine and Public Health 20: 49–53.
17. Ghebreyesus TA, Haile M, Witten KH, Getachew A, Yohannes M, et al. (2000) Household risk factors for malaria among children in the Ethiopian highlands. Transactions of the Royal Society of Tropical Medicine and Hygiene 94: 17–21.
18. Seyoum A, Balcha F, Balkew M, Ali A, Gebre-Michael T (2002) Impact of cattle keeping on human biting rate of anopheline mosquitoes and malaria transmission around Ziway, Ethiopia. East African Medical Journal 79: 485–490.
19. Habtewold T (2004) Interaction between *Anopheles*, cattle and human: exploration of the effects of various cattle management practices on the behaviour and control of *Anopheles arabiensis* in Ethiopia. PhD Thesis. Greenwich: University of Greenwich, U.K. 249 p.
20. Subramanian S, Manoharan A, Sahu S, Jambulingam P, Govardhini P, et al. (1991) Living conditions and occurrence of malaria in a rural community. Indian Journal of Malariology 28: 29–37.
21. Service MW (1993) Mosquito ecology. Field sampling methods. London: Chapman & Hall.
22. Gillies MT, De Meillon B (1968) The Anophelinae of Africa south of the Sahara. Publications of the South African Institute for Medical Research, No. 54. South African Institute for Medical Research, Johannesburg.
23. White BN, Magayuka SA, Boreham PFL (1972) Comparative studies on sibling species of the *Anopheles gambiae* Giles complex (Diptera Culicidae) bionomics and vectorial activity of species A and species B at Segera, Tanzania. Bulletin of Entomological Research 62: 295–317.
24. Charlwood JD, Edoh D (1996) Polymerase chain reaction used to describe larval habitat use by *Anopheles gambiae* complex (Diptera: Culicidae) in the environs of Ifakara, Tanzania. Journal of Medical Entomology 33: 202–204.
25. Minakawa N, Mutero CM, Githure JI, Beier JC, Yan G (1999) Spatial distribution and habitat characterization of anopheline mosquito larvae in

Western Kenya. The American Journal of Tropical Medicine and Hygiene 61: 1010–1016.

26. McLaughlin RE, Focks DA (1990) Effects of cattle density on New Jersey light trap mosquito captures in the rice/cattle agroecosystem of southwestern Louisiana. Journal of the American Mosquito Control Association 6: 283–286.

27. Saul A (2003) Zooprophylaxis or zoopotentiation: the outcome of introducing animals on vector transmission is highly dependent on the mosquito mortality while searching. Malaria Journal 2: art. no.32.

28. Habtewold T, Walker AR, Curtis CF, Osir EO, Thapa N (2001) The feeding behaviour and Plasmodium infection of Anopheles mosquitoes in southern Ethiopia in relation to use of insecticide-treated livestock for malaria control. Transactions of the Royal Society of Tropical Medicine and Hygiene 95: 584–586.

29. Gupta S, Swinton J, Anderson RM (1994) Theoretical studies of the effects of heterogeneity in the parasite population on the transmission dynamics of malaria. Proceedings of the Royal Society of London Series B-Biological Sciences 256: 231–238.

30. Collins WE, Jeffery GM (2003) A retrospective examination of mosquito infection on humans infected with Plasmodium falciparum. The American Journal of Tropical Medicine and Hygiene 68: 366–371.

31. USDA (1976)Control of insects affecting livestock. 99 p.

32. Thomson MC (1987) The effect on tsetse flies (Glossina spp.) of deltamethrin applied to cattle either as a spray or incorporated into ear-tags. Tropical Pest Management 33: 329–335.

33. Bekele J, Asmare K, Abebe G, Ayelet G, Gelaye E (2010) Evaluation of Deltamethrin applications in the control of tsetse and trypanosomosis in the southern rift valley areas of Ethiopia. Vet Parasitol 168: 177–184.

34. Bardosh K, Waiswa C, Welburn SC (2013) Conflict of interest: use of pyrethroids and amidines against tsetse and ticks in zoonotic sleeping sickness endemic areas of Uganda. Parasit Vectors 6: 204.

35. Torr SJ, Maudlin I, Vale GA (2007) Less is more: restricted application of insecticide to cattle to improve the cost and efficacy of tsetse control. Medical and Veterinary Entomology 21: 53–64.

36. George JE (2000) Present and future technologies for tick control. Ann N Y Acad Sci 916: 583–588.

37. Foil LD, Hogsette JA (1994) Biology and control of tabanids, stable flies and horn flies. Revue Scientifique et Technique (International Office of Epizootics) 13: 1125–1158.

38. Schmidtmann ET, Lloyd JE, Bobian RJ, Kumar R, Waggoner JW, et al. (2001) Suppression of mosquito (Diptera: Culicidae) and black fly (Diptera: Simuliidae) blood feeding from Hereford cattle and ponies treated with permethrin. Journal of Medical Entomology 38: 728–734.

39. Nasci RS, McLaughlin RE, Focks D, Billodeaux JS (1990) Effect of topically treating cattle with permethrin on blood feeding of Psorophora columbiae (Diptera: Culicidae) in a southwestern Louisiana rice-pasture ecosystem. Journal of Medical Entomology 27: 1031–1034.

40. Standfast HA, Muller MJ, Wilson DD (1984) Mortality of Culicoides brevitarsis (Diptera: Ceratopogonidae) fed on cattle treated with ivermectin. Journal of Economic Entomology 77: 419–421.

41. Rowland M, Durrani N, Kenward M, Mohammed N, Urahman H, et al. (2001) Control of malaria in Pakistan by applying deltamethrin insecticide to cattle: a community-randomised trial. Lancet 357: 1837–1841.

42. Reisen WK, Milby MM (1986) Population dynamics of some Pakistan mosquitoes: changes in adult relative abundance over time and space. Annals of Tropical Medicine and Parasitology 80: 53–68.

43. Habtewold T, Prior A, Torr SJ, Gibson G (2004) Could insecticide-treated cattle reduce Afrotropical malaria transmission? Effects of deltamethrin-treated Zebu on Anopheles arabiensis behaviour and survival in Ethiopia. Medical and Veterinary Entomology 18: 408–417.

44. Mahande AM, Mosha FW, Mahande JM, Kweka EJ (2007) Role of cattle treated with deltamethrin in areas with a high population of Anopheles arabiensis in Moshi, Northern Tanzania. Malaria Journal 6: 109.

45. Chareonviriyaphap T (2012) Behavioral Responses of Mosquitoes to Insecticides, Insecticides - Pest Engineering, Dr. Farzana Perveen (Ed.). InTech. Available from: http://www.intechopen.com/download/get/type/pdfs/id/28262 (Accessed 1 August 2013).

46. Lines JD, Myamba J, Curtis CF (1987) Experimental hut trials of permethrin-impregnated mosquito nets and eave curtains against malaria vectors in Tanzania. Medical and Veterinary Entomology 1: 37–51.

47. Takken W (2002) Do insecticide-treated bednets have an effect on malaria vectors? Tropical Medicine & International Health 7: 1022–1030.

48. Charlwood JD, Graves PM (1987) The effect of permethrin-impregnated bednets on a population of Anopheles farauti in coastal Papua New Guinea. Medical and Veterinary Entomology 1: 319–327.

49. Magesa SM, Wilkes TJ, Mnzava AE, Njunwa KJ, Myamba J, et al. (1991) Trial of pyrethroid impregnated bednets in an area of Tanzania holoendemic for malaria. Part 2. Effects on the malaria vector population. Acta Tropica 49: 97–108.

50. Githeko AK, Adungo NI, Karanja DM, Hawley WA, Vulule JM, et al. (1996) Some observations on the biting behavior of Anopheles gambiae s.s., Anopheles arabiensis, and Anopheles funestus and their implications for malaria control. Experimental Parasitology 82: 306–315.

51. Bøgh C, Pedersen EM, Mukoko DA, Ouma JH (1998) Permethrin-impregnated bednet effects on resting and feeding behaviour of lymphatic filariasis vector mosquitoes in Kenya. Medical and Veterinary Entomology 12: 52–59.

52. Kolaczinski JH, Reithinger R, Worku DT, Ocheng A, Kasimiro J, et al. (2008) Risk factors of visceral leishmaniasis in East Africa: a case-control study in Pokot territory of Kenya and Uganda. International Journal of Epidemiology 37: 344–352.

53. McLaughlin RE, Focks DA, Dame DA (1989) Residual activity of permethrin on cattle as determined by mosquito bioassays. Journal of the American Mosquito Control Association 5: 60–63.

54. Hewitt S, Rowland M (1999) Control of zoophilic malaria vectors by applying pyrethroid insecticides to cattle. Tropical Medicine & International Health 4: 481–486.

55. Vythilingam I, Ridhawati, Sani RA, Singh KI (1993) Residual activity of cyhalothrin 20% EC on cattle as determined by mosquito bioassays. Southeast Asian J Trop Med Public Health 24: 544–548.

56. Elliot M (1989) The pyrethroids: early discovery, recent advances and the future. Pesticide Science 27: 337–351.

57. W.H.O. (1990) Deltamethrin. Environmental Health Criteria 97. Geneva: World Health Organization.

58. Okumu FO, Killeen GF, Ogoma S, Biswaro L, Smallegange RC, et al. (2010) Development and field evaluation of a synthetic mosquito lure that is more attractive than humans. PLoS One 5: e8951.

59. Jawara M, Awolola TS, Pinder M, Jeffries D, Smallegange RC, et al. (2011) Field testing of different chemical combinations as odour baits for trapping wild mosquitoes in The Gambia. PLoS One 6: e19676.

60. Ross R (1911) The Prevention of Malaria. London: Murray.

61. Macdonald G (1952) The analysis of equilibrium in malaria. Tropical Diseases Bulletin 49: 813–829.

62. Nåsell I (1985) Hybrid Models of Tropical Infections. In: Levin S, editor.Lecture Notes in Biomathematics. New York: Springer-Verlag. pp. 46–50.

63. Sota T, Mogi M (1989) Effectiveness of zooprophylaxis in malaria control: a theoretical inquiry, with a model for mosquito populations with two bloodmeal hosts. Medical and Veterinary Entomology 3: 337–345.

64. Killeen GF, McKenzie FE, Foy BD, Bogh C, Beier JC (2001) The availability of potential hosts as a determinant of feeding behaviours and malaria transmission by African mosquito populations. Transactions of the Royal Society of Tropical Medicine and Hygiene 95: 469–476.

65. Killeen GF, Smith TA (2007) Exploring the contributions of bed nets, cattle, insecticides and excitorepellency to malaria control: a deterministic model of mosquito host-seeking behaviour and mortality. Transactions of the Royal Society of Tropical Medicine and Hygiene 101: 867–880.

66. Kawaguchi I, Sasaki A, Mogi M (2004) Combining zooprophylaxis and insecticide spraying: a malaria- control strategy limiting the development of insecticide resistance in vector mosquitoes. Proceedings of the Royal Society of London Series B-Biological Sciences 271: 301–309.

67. Hurd H (2003) Manipulation of medically important insect vectors by their parasites. Annual Review of Entomology 48: 141–161.

68. Lord CC, Woolhouse MEJ, Heesterbeek JAP (1996) Vector-borne diseases and the basic reproduction number: a case study of African horse sickness. Medical and Veterinary Entomology 10: 19–28.

69. Charlwood JD, Smith T, Kihonda J, Heiz B, Billingsley PF, et al. (1995) Density independent feeding success of malaria vectors (Diptera: Culicidae) in Tanzania. Bulletin of Entomological Research 85: 29–35.

70. Anderson RM, May RM (1991) Infectious Diseases of Humans: Dynamics and Control. Oxford: Oxford University Press.

71. Diekmann O, Heesterbeek JA, Metz JA (1990) On the definition and the computation of the basic reproduction ratio R0 in models for infectious diseases in heterogeneous populations. Journal of Mathematical Biology 28: 365–382.

72. van den Driessche P, Watmough J (2002) Reproduction numbers and sub-threshold endemic equilibria for compartmental models of disease transmission. Mathematical Biosciences 180: 29–48.

73. IVCC. Proceedings of the Innovative Vector Control Consortium (IVCC) - Insect Repellent Workshop; 2007 22–23 January; London, UK.

74. Rayaisse JB, Tirados I, Kaba D, Dewhirst SY, Logan JG, et al. (2010) Prospects for the development of odour baits to control the tsetse flies Glossina tachinoides and G. palpalis s.l. PLoS Negl Trop Dis 4: e632.

75. Vale GA, Lovemore DF, Flint S, Cockbill GF (1988) Odour-baited targets to control tsetse flies, Glossina spp. (Diptera: Glossinidae), in Zimbabwe. Bulletin of Entomological Research 78: 31–49.

76. Vale GA, Hall DR (1985) The role of 1-octen-3-ol, acetone and carbon dioxide in the attraction of tsetse flies, Glossina spp. (Diptera: Glossinidae), to ox odour. Bulletin of Entomological Research 75: 209–218.

77. Chaniotis BN (1983) Improved trapping of phlebotomine sand flies (Diptera: Psychodidae) in light traps supplemented with dry ice in a neotropical rain forest. J Med Entomol 20: 222–223.

78. Coluzzi M, Sabatini A, Petrarca V, Di Deco MA (1979) Chromosomal differentiation and adaptation to human environments in the Anopheles gambiae complex. Transactions of the Royal Society of Tropical Medicine and Hygiene 73: 483–497.

79. Donnelly MJ, Townson H (2000) Evidence for extensive genetic differentiation among populations of the malaria vector Anopheles arabiensis in Eastern Africa. Insect Molecular Biology 9: 357–367.

80. Petrarca V, Nugud AD, Ahmed MA, Haridi AM, Di Deco MA, et al. (2000) Cytogenetics of the *Anopheles gambiae* complex in Sudan, with special reference to *An. arabiensis*: relationships with East and West African populations. Medical and Veterinary Entomology 14: 149–164.

81. Curtis CF, Miller JE, Hodjati MH, Kolaczinski JH, Kasumba I (1998) Can anything be done to maintain the effectiveness of pyrethroid-impregnated bednets against malaria vectors? Philosophical Transactions of the Royal Society of London Series B, Biological Sciences 353: 1769–1775.

82. Ranson H, N'Guessan R, Lines J, Moiroux N, Nkuni Z, et al. (2011) Pyrethroid resistance in African anopheline mosquitoes: what are the implications for malaria control? Trends Parasitol 27: 91–98.

83. Beugnet F, Chardonnet L (1995) Tick resistance to pyrethroids in New Caledonia. Veterinary Parasitology 56: 325–338.

84. Rodriguez-Vivas RI, Alonso-Díaz MA, Rodríguez-Arevalo F, Fragoso-Sanchez H, Santamaria VM, et al. (2006) Prevalence and potential risk factors for organophosphate and pyrethroid resistance in *Boophilus microplus* ticks on cattle ranches from the State of Yucatan, Mexico. Veterinary Parasitology 136: 335–342.

85. Wilson ML (1993) Avermectins in arthropod vector management - prospects and pitfalls. Parasitology Today 9: 83–87.

86. Fritz ML, Siegert PY, Walker ED, Bayoh MN, Vulule JR, et al. (2009) Toxicity of bloodmeals from ivermectin-treated cattle to *Anopheles gambiae s.l.* Annals of Tropical Medicine and Parasitology 103: 539–547.

87. Fritz ML, Walker ED, Miller JR (2012) Lethal and sublethal effects of avermectin/milbemycin parasiticides on the African malaria vector, Anopheles arabiensis. J Med Entomol 49: 326–331.

88. Kobylinski KC, Foy BD, Richardson JH (2012) Ivermectin inhibits the sporogony of Plasmodium falciparum in Anopheles gambiae. Malar J 11: 381.

89. Warrel DA, Gilles HM (2002) Essential Malariology. London: Hodder Arnold. 348 p.

90. Nedelman J (1985) Some New Thoughts About Some Old Malaria Models - Introductory Review. Mathematical Biosciences 73: 159–182.

91. Molineaux L (1988) The epidemiology of humans malaria as an explanation of its distribution, including some implications for its control. In: Wernsdorfer WH, McGregor SI, editors. Malaria, Principles and Practice of Malariology. London: Churchill Livingstone. pp. 913–998.

92. Mahmood F, Reisen WK (1981) Duration of the gonotrophic cycles of *Anopheles culicifacies* Giles and *An. stephensi* Liston, with observations on reproductive activity and survivorship during winter. Mosquito News 41: 22–30.

93. Krafsur ES (1977) The bionomics and relative prevalence of *Anopheles* species with respect to the transmission of *Plasmodium* to man in western Ethiopia. Journal of Medical Entomology 14: 180–194.

94. Krafsur ES, Armstrong JC (1982) Epidemiology of *Plasmodium malariae* infection in Gambella, Ethiopia. Parassitologia 24: 105–120.

95. Verhage DF, Telgt DS, Bousema JT, Hermsen CC, van Gemert GJ, et al. (2005) Clinical outcome of experimental human malaria induced by *Plasmodium falciparum*-infected mosquitoes. The Netherlands Journal of Medicine 63: 52–58.

96. Rickman LS, Jones TR, Long GW, Paparello S, Schneider I, et al. (1990) *Plasmodium falciparum*-infected *Anopheles stephensi* inconsistently transmit malaria to humans. The American Journal of Tropical Medicine and Hygiene 43: 441–445.

97. Taye A, Hadis M, Adugna N, Tilahun D, Wirtz RA (2006) Biting behavior and *Plasmodium* infection rates of *Anopheles arabiensis* from Sille, Ethiopia. Acta Tropica 97: 50–54.

98. Reisen WK, Boreham PF (1982) Estimates of malaria vectorial capacity for *Anopheles culicifacies* and *Anopheles stephensi* in rural Punjab province Pakistan. Journal of Medical Entomology 19: 98–103.

99. Tirados I, Costantini C, Gibson G, Torr SJ (2006) Blood-feeding behaviour of the malarial mosquito *Anopheles arabiensis*: implications for vector control. Medical and Veterinary Entomology 20: 425–437.

Permissions

All chapters in this book were first published in PLOS ONE, by The Public Library of Science; hereby published with permission under the Creative Commons Attribution License or equivalent. Every chapter published in this book has been scrutinized by our experts. Their significance has been extensively debated. The topics covered herein carry significant findings which will fuel the growth of the discipline. They may even be implemented as practical applications or may be referred to as a beginning point for another development.

The contributors of this book come from diverse backgrounds, making this book a truly international effort. This book will bring forth new frontiers with its revolutionizing research information and detailed analysis of the nascent developments around the world.

We would like to thank all the contributing authors for lending their expertise to make the book truly unique. They have played a crucial role in the development of this book. Without their invaluable contributions this book wouldn't have been possible. They have made vital efforts to compile up to date information on the varied aspects of this subject to make this book a valuable addition to the collection of many professionals and students.

This book was conceptualized with the vision of imparting up-to-date information and advanced data in this field. To ensure the same, a matchless editorial board was set up. Every individual on the board went through rigorous rounds of assessment to prove their worth. After which they invested a large part of their time researching and compiling the most relevant data for our readers.

The editorial board has been involved in producing this book since its inception. They have spent rigorous hours researching and exploring the diverse topics which have resulted in the successful publishing of this book. They have passed on their knowledge of decades through this book. To expedite this challenging task, the publisher supported the team at every step. A small team of assistant editors was also appointed to further simplify the editing procedure and attain best results for the readers.

Apart from the editorial board, the designing team has also invested a significant amount of their time in understanding the subject and creating the most relevant covers. They scrutinized every image to scout for the most suitable representation of the subject and create an appropriate cover for the book.

The publishing team has been an ardent support to the editorial, designing and production team. Their endless efforts to recruit the best for this project, has resulted in the accomplishment of this book. They are a veteran in the field of academics and their pool of knowledge is as vast as their experience in printing. Their expertise and guidance has proved useful at every step. Their uncompromising quality standards have made this book an exceptional effort. Their encouragement from time to time has been an inspiration for everyone.

The publisher and the editorial board hope that this book will prove to be a valuable piece of knowledge for researchers, students, practitioners and scholars across the globe.

List of Contributors

Gao Hu, Fang Lu, Bao-Ping Zhai and Xiao-Xi Zhang
Key Laboratory of Integrated Management of Crop Diseases and Pests (Ministry of Education), College of Plant Protection, Nanjing Agricultural University, Nanjing, China

Ming-Hong Lu and Wan-Cai Liu
Division of Pest Forecasting, China National Agro-Tec Extension and Service Center, Beijing, China

Feng Zhu
Plant Protection Station of Jiangsu Province, Nanjing, China

Xiang-Wen Wu
Plant Protection Station of Shanghai City, Shanghai, China

Gui-Hua Chen
Plant Protection Station of Jinhua City, Jinhua, China

Kassahun Alemu
Department of Environmental and Occupational Health and Safety, Institute of Public Health, College of Medicine and Health Sciences, University of Gondar, Gondar, Ethiopia

Alemayehu Worku
Department of Epidemiology and Biostatistics, School of Public Health, College of Health Sciences, Addis Ababa University, Addis Ababa, Ethiopia

Yemane Berhane
Addis Continental Institute of Public Health, Addis Ababa, Ethiopia

Abera Kumie
Department of Health Management, Environmental Health, and Behavioral Sciences, School of Public Health, College of Health Sciences, Addis Ababa University, Addis Ababa, Ethiopia

Kate H. Macneale, Julann A. Spromberg, David H. Baldwin and Nathaniel L. Scholz
Northwest Fisheries Science Center, National Marine Fisheries Service, National Oceanic and Atmospheric Administration, Seattle, Washington, United States of America

Elliud Muli
The International Centre of Insect Physiology and Ecology (icipe), Nairobi, Kenya

Department of Biological Sciences, South Eastern Kenya University (SEKU), Kitui, Kenya

Harland Patch, Maryann Frazier, James Frazier, Tracey Baumgarten, James Tumlinson and Christina Grozinger
Department of Entomology, Center for Pollinator Research, Pennsylvania State University, University Park, Pennsylvania, United States of America

Baldwyn Torto, Joseph Kilonzo, James Ng'ang'a Kimani, Fiona Mumoki and Daniel Masiga
The International Centre of Insect Physiology and Ecology (icipe), Nairobi, Kenya

Minghui Zhao and Xin Ran
Beijing Institute of Microbiology and Epidemiology, State Key Laboratory of Pathogens and Biosecurity, Department of Vector Biology and Control, Beijing, China
Anhui Medical University, Hefei, China

Yande Dong, Zhiming Wu, Xiaoxia Guo, Yingmei Zhang, Dan Xing, Ting Yan, Gang Wang, Xiaojuan Zhu, Hengduan Zhang, Chunxiao Li and Tongyan Zhao
Beijing Institute of Microbiology and Epidemiology, State Key Laboratory of Pathogens and Biosecurity, Department of Vector Biology and Control, Beijing, China

Simone Dealtry, Guo-Chun Ding, Viola Weichelt and Kornelia Smalla
Julius Kühn-Institut – Federal Research Centre for Cultivated Plants (JKI), Institute for Epidemiology and Pathogen Diagnostics, Braunschweig, Germany

Vincent Dunon and Dirk Springael
Division of Soil and Water Management, KU Leuven, Heverlee, Belgium

Andreas Schlüter
Center for Biotechnology (CeBiTec), Institute for Genome Research and Systems Biology, Bielefeld University, Bielefeld, Germany

María Carla Martini, María Florencia Del Papa and Antonio Lagares
IBBM (Instituto de Biotecnología y Biología Molecular), CCT-CONICET-La Plata, Departamento de Ciencias Biológicas, Facultad de Ciencias Exactas, Universidad Nacional de La Plata, La Plata, Argentina

Gregory Charles Auton Amos, Elizabeth Margaret Helen Wellington and William Hugo Gaze
School of Life Sciences, University of Warwick, Warwick, United Kingdom

Detmer Sipkema
Laboratory of Microbiology, Wageningen University, Wageningen, The Netherlands

Sara Sjöling
Södertörns högskola (Sodertorn University), Inst. för Naturvetenskap, Miljö och medieteknik (School of Natural Sciences, Environmental Studies and media tech), Huddinge, Sweden

Jan Dirk van Elsas
University of Groningen, Groningen, The Netherlands

Christopher Thomas
School of Biosciences, University of Birmingham, Edgbaston, Birmingham, Warwick, United Kingdom

Michael T. White and John Marshall
MRC Centre for Outbreak Analysis and Modelling, Imperial College, London, United Kingdom

Dickson Lwetoijera
Ifakara Health Institute, Ifakara, Tanzania

Geoffrey Caron-Lormier
University of Nottingham, Sutton Bonington, Leicestershire, United Kingdom

David A. Bohan
INRA, UMR 1347 Agroécologie, Pôle ECOLDUR, Dijon, France

Ian Denholm
University of Hertfordshire, Hatfield, Hertfordshire, United Kingdom

Gregor J. Devine
QIMR Berghofer Medical Research Institute, Brisbane, Australia

Francisco Sanchez-Bayo
Faculty of Agriculture and Environment, The University of Sydney, Eveleigh, New South Wales, Australia

Koichi Goka
National Institute for Environmental Sciences, Tsukuba, Ibaraki, Japan

Nicolás Jaramillo-O, Idalyd Fonseca-González and Duverney Chaverra-Rodríguez
Instituto de Biología, Facultad de Ciencias Exactas y Naturales, Universidad de Antioquia UdeA, Medellín, Colombia

Neeraj Kumar, Subodh Gupta, Nitish Kumar Chandan, Md. Aklakur, Asim Kumar Pal and Sanjay Balkrishna Jadhao
Department of Fish Nutrition, Biochemistry and Physiology, Central Institute of Fisheries Education, Versova, Mumbai, Maharashtra, India

Martina G. Vijver
Institute of Environmental Sciences (CML), Leiden University, Leiden, The Netherlands

Paul J. van den Brink
Alterra, Wageningen University and Research centre, Wageningen, The Netherlands
Wageningen University, Wageningen University and Research centre, Wageningen, The Netherlands

Emiliane Taillebois, Abdelhamid Beloula, Sophie Quinchard, Steeve H. Thany and Hélène Tricoire-Leignel
Laboratoire Récepteurs et Canaux Ioniques Membranaires (RCIM), UPRES EA 2647 USC INRA 1330, SFR QUASAV 4207, Universitéd'Angers, Angers, France

Stéphanie Jaubert-Possamai and Denis Tagu
Institut de Génétique Environnement et Protection des Plantes (IGEPP), Institut National de la Recherche Agronomique (INRA), UMR 1349, Le Rheu, France

Antoine Daguin
Groupement Interprofessionnel de Recherche sur les Produits Agropharmaceutiques (GIRPA), Angers, France

Denis Servent
Institut de Biologie et Technologie (iBiTecS), Service d'Ingénierie Moléculaire des Protéines (SIMOPRO), Commisariat à l'Energie Atomique (CEA), Gif-sur-Yvette, France

Hyeong-Moo Shin and Deborah H. Bennett
Department of Public Health Sciences, University of California Davis, Davis, California, United States of America

Thomas E. McKone
Environmental Energy Technologies Division, Lawrence Berkeley National Laboratory, Berkeley, California, United States of America

School of Public Health, University of California, Berkeley, California, United States of America

Michael D. Sohn
Environmental Energy Technologies Division, Lawrence Berkeley National Laboratory, Berkeley, California, United States of America

Yizhou Chen, Daniel R. Bogema, Idris M. Barchia and Grant A. Herron
Elizabeth Macarthur Agricultural Institute, NSW Department of Primary Industries, Menangle, New South Wales, Australia

Eun Shil Cha and Won Jin Lee
Department of Preventive Medicine, College of Medicine, Korea University, Seoul, South Korea

Young-Ho Khang
Department of Preventive Medicine, University of Ulsan College of Medicine, Seoul, South Kore
Institute of Health Policy and Management, Seoul National University College of Medicine, Seoul, South Korea

Cordelia Forkpah, Luke R. Dixon and Olav Rueppell
Department of Biology, University of North Carolina, Greensboro, North Carolina, United States of America

Susan E. Fahrbach
Department of Biology, Wake Forest University, Winston- Salem, North Carolina, United States of America

Angelina Mageni Lutambi
Epidemiology and Public Health, Swiss Tropical and Public Health Institute, Basel, Switzerland University of Basel, Basel, Switzerland
Ifakara Health Institute, Dar es Salaam, Tanzania

Nakul Chitnis
Epidemiology and Public Health, Swiss Tropical and Public Health Institute, Basel, Switzerland, University of Basel, Basel, Switzerland
Fogarty International Center, National Institutes of Health, Bethesda, Maryland, United States of America

Olivier J. T. Briët, Thomas A. Smith and Melissa A. Penny
Epidemiology and Public Health, Swiss Tropical and Public Health Institute, Basel, Switzerland, University of Basel, Basel, Switzerland

You-Hui Gong, Xin-Rui Yu, Xue-yan Shi and Xi-Wu Gao
Department of Entomology, China Agricultural University, Beijing, China

Qing-Li Shang
College of Plant Science and Technology, Jilin University, Changchun, China

Anders S. Huseth
Department of Entomology, Cornell University, New York State Agricultural Experiment Station, Geneva, New York, United States of America

Russell L. Groves
Department ofEntomology, University of Wisconsin-Madison, Madison, Wisconsin, United States of America

Roch K. Dabiré , Moussa Namountougou1, Abdoulaye Diabaté, Dieudonné D. Soma and Joseph Bado
IRSS (Institut de Recherche en Sciences de la Santé), Centre Muraz, Bobo-Dioulasso, Burkina Faso

Hyacinthe K. Toé
IRSS (Institut de Recherche en Sciences de la Santé), Centre Muraz, Bobo-Dioulasso, Burkina Faso
Department of Vector Biology, Liverpool School of Tropical Medicine, Liverpool, United Kingdom

Chris Bas
Biological Chemistry and Crop Protection Rothamsted Research, Harpenden, United Kingdom

Patrice Combary
National Malaria Control Programme, Ministry of Health, Ouagadougou, Burkina Faso

Ana O. Franco
Instituto Gulbenkian de Ciência, Oeiras, Portugal
Faculty of Infectious and Tropical Diseases, London School of Hygiene and Tropical Medicine, London, United Kingdom

M. Gabriela M. Gomes
Instituto Gulbenkian de Ciência, Oeiras, Portugal

Mark Rowland, Paul G. Coleman and Clive R. Davies
Faculty of Infectious and Tropical Diseases, London School of Hygiene and Tropical Medicine, London, United Kingdom

Index

www.ingramcontent.com/pod-product-compliance
Lightning Source LLC
Chambersburg PA
CBHW080251230326

41458CB00097B/4268